Mountain Timberlines

ADVANCES IN GLOBAL CHANGE RESEARCH

VOLUME 36

For other titles published in this series, go to
www.springer.com/series/5588

Friedrich-Karl Holtmeier

Mountain Timberlines

Ecology, Patchiness, and Dynamics

Springer

Prof. Dr. Friedrich-Karl Holtmeier
Dionysiusstr. 6
48329 Havixbeck
Germany

ISBN 978-1-4020-9704-1 e-ISBN 978-1-4020-9705-8

Library of Congress Control Number: 2008942996

Printed on acid-free paper

9 8 7 6 5 4 3 2 1

springer.com

CONTENTS

PREFACE

For more than 40 years I have been engaged in timberline research. Thus, one could suppose that writing this book should not have been too difficult. It was harder, however, than expected, and in the end I felt that more questions had arisen than could be answered within its pages. Perhaps it would have been easier to write the book 30 years ago and then leave the subject to mature. Lastly it was the late Prof. Heinz Ellenberg who had convinced me to portray a much needed and complete picture of what we know of the timberline with special respect to its great physiognomic, structural and ecological variety. The first version of this book was published in the German language (Holtmeier, 2000). Nevertheless, I was very delighted when Prof. Martin Beniston encouraged me to prepare an English edition for the series 'Advances in Global Change Research', which guaranteed a wider circulation.

Timberline is a worldwide and very heterogeneous phenomenon, which can only be presented by way of examples. My own field experience is necessarily limited to certain timberline areas, such as the Alps, northern Scandinavia, northern Finland and many high mountain ranges in the western United States and Canada. However, my own observations and the results of my and my previous collaborators research were essential for developing the concept of the book and became integrated into the picture of timberline that is presented in the following chapters. Since the most thorough-going study of the literature cannot compensate for lack of one's own field experience and observation, the main discussion is focused on the timberline regions in the northern hemisphere where I have carried out extensive field research. Nevertheless, tropical timberlines and temperate timberlines in the southern hemisphere are also considered based on information from literature and on communication with colleagues.

This book is a conclusive synthesis of my own and my collaborators studies, and the evaluation of a wealth of literature. The intellectual conception of this volume has not principally changed compared to its first English edition. However, since the first English edition of this book was published much new material has accumulated. I have incorporated relevant new results of timberline research and other useful information into this edition. The text has been partly reorganized. Moreover, I added a few photos and graphs. Several graphs have been modified. Different from the first edition, examples of the influence of animals on timberline are given now in separate chapters. The ample reference list has grown again to now about 1.500 titles. However, in view of the worldwide existence of mountain timberline and of the many possible scientific approaches to timberline (botany, ecology, climatology, soil science, forestry, historical science, etc.) this list cannot be expected to be complete, and surely the reader will miss inclusion of one or another familiar publication. On the other hand, the references include many papers and books that were written in German. Anglophone researchers seldom or never refer to these papers. But these publications are reflecting a long tradition in timberline research by German speaking scientists and provide much useful information that I have tried to make available to those researchers not acquainted with the extensive 19th and 20th century publications written in German. I have also included older publications to show that many ideas on causes of altitudinal and northern timberlines are not as new as one might believe in view of the reference lists given in many recent papers citing only the most recent publications on the particular topic just being considered.

As in the previous edition, instructive illustration has been kept up. It is part of the conception of this book. Most of the photos were taken in my special research areas. In addition, friends and colleagues provided a few pictures from timberline regions that I did not visit by myself. All my photos were taken exclusively for documentation of timberline, which means that they do not show landscapes where also timberline was portrayed more or less accidentally. Still as ever there is no other comparable timberline-specific photo collection. The illustration is to give a detailed picture of the great physiognomic variety of timberline, which also mirrors its heterogeneity and ecological variety. This should be underlined as in a time of increasing modelling the visual element has been generally neglected. I hope this book will contribute to better understand the interplay of the many factors causing mountain timberlines and their great variety and dynamics.

Havixbeck, December 2008

Friedrich-Karl Holtmeier

ACKNOWLEDGEMENTS

This book could not have been written without revising by friends, colleagues and collaborators. Although being retired for almost 5 years I still feel deeply obliged to my academic teacher Prof. Dr. Dr. h. c. Carl Troll (1899–1975, Institute of Geography, University of Bonn) and also to Prof. Dr. Ulrich Schweinfurth (1925–2005, Institute of Geography, South Asia Institute, University of Heidelberg). Both introduced me into the wide field of landscape ecology and particularly into timberline problems already when I was a student. My thanks also go to Prof. Dr. Dr. h.c. Joachim Blüthgen (1912–1973, Institute of Geography, University of Münster), himself a pioneer in northern tree line research, who gave me the first chance to carry out field research in northern Finnish Lapland in the late 1960s. Moreover, I gratefully remember my friend Prof. Dr. Paavo Kallio (1914–1992, Kevo Subarctic Research Institute, University of Turku) who watched my tree line studies in Lapland with great interest and provided use of all facilities at the Kevo Subarctic Research Station in northern Finnish Lapland.

My special thanks go to those friends and colleagues who essentially contributed to developing the conception of this book or supported my fieldwork by their great hospitality and good company in the field. First of all I mention Prof. Dr. Gabriele Broll (Division of Geoecology, University of Vechta, Germany), Dr. James B. Benedict (Center for Mountain Archeology, Ward, Colorado), Mrs. Audrey DeLella-Benedict (Cloudy Ridge Naturalists, Ward, Colorado) and Dr. Wyman C. Schmidt (Research Scientist Emeritus, Intermountain Forestry Sciences Laboratory, Montana State University, Bozeman). I am also obliged to Dr. Maaike Y. Bader (Institute for Biology and Environmental Sciences, University of Oldenburg), Dr. Robert Brandes (Institute for Geography, University of Erlangen), Prof. Dr. Frank Klötzli (Institute of Geobotany, University of Zürich), Prof. Dr. Ernst Löffler (Physical Geography, University Saarbrücken), Prof. Dr. Hermann Mattes (Institute of Landscape Ecology, University of Münster), Prof. Dr. Richard Pott (Institute of Geobotany, University of Hannover), Dr. M. Daud Rafiqpoor (Institute of Geography, University of Bonn), Prof. Dr. Michael Richter (Institute of Geography, University of Erlangen), Prof. Dr. Udo Schickhoff (Institute of Geography, University of Hamburg), Dr. Hans-Uwe Schütz (Schöppingen), Dr. Andreas Vogel (Institute of Landscape Ecology, University of Münster) and Prof. Dr. Masatoshi Yoshino (Institute of Geography, Auchi University, Toyohashi-City, Japan). These colleagues provided me with pictures from timberlines that I did not visit myself. Not least I want to thank my former collaborators, in particular Dr. Kerstin Anschlag (Institute of Geography, University of Bonn) and Dr. Andreas Müterthies (EFTAS, Münster), who were a great help to me in search for literature and preparation of the many figures and graphs as I prepared the German version and the first English edition of this book. Finally my thanks go to my wife Waltraud Holtmeier and my son Jan-Ulrich Holtmeier for occasional assistance in the field and their patience with the 'fellow lost to timberline' having not recovered from his 'obsession' even after retirement. I am also very grateful to the Deutsche Forschungsgemeinschaft, who funded most of the fieldwork abroad and the European Community (Research Infrastructure Action under the FP6 'Structuring the European Research Area Programme, LAPBIAT') who repeatedly supported our field work in Finnish Lapland during the last years. Moreover, I thank Prof. Dr. Martin Beniston (University of Fribourg) having invited me to contribute to this book series and the Springer Verlag for publishing this second edition. I am also grateful to Prof. Dr. Robert M. M. Crawford (St. Andrews, Scotland) and to Mrs. Lillian Harris (Institute for Landscape Ecology, University of Münster) for occasional help in wording the text. My particular thanks are due to Mrs. Margaret Deignan (Springer-Verlag) for help in editing the book.

1 INTRODUCTION

This book is motivated by studies of global climate change and its impact on the environment, a topic which is increasingly being discussed among scientists as well as by the general public. One major point of concern is the possible shift of vegetation zones to higher altitudes and greater latitudes. In particular, attention is being given to northern and mountain timberlines, which are readily visible and for which past records are available. Undoubtedly, the upper timberline is the most conspicuous vegetation limit in high-mountain areas of all continents, with the sole exception for the Antarctic. Actually, impressive maps and graphs projecting the future positions of vegetation zones and altitudinal belts have already been published.

These predictions, however, are based on very simple assumptions that cannot encompass the ecological complexity and great heterogeneity of such boundaries. The expected shift of timberline is estimated by extrapolating the rough coincidence of the existing timberlines and present thermal conditions, usually described in terms of temperature sum, number of growing-degree days, or mean temperature of the warmest month (Section 4.3.1). However, timberline is not an organism that will individually respond to any change of the environment, nor is it a line that will advance or retreat parallel to an altitudinal shift of any isotherm considered to be essential to tree growth. Timberline is a biological boundary, a more or less wide ecotone that has to be understood as a space- and time-related phenomenon that will not respond linearly to changing temperatures or other environmental factors (Chapter 5).

The physiological response of tree vegetation to the environment ('plant functional types') is only one aspect of change among many others, such as the distribution pattern of dendroid and other vegetation, soils, plant communities, growth forms, number of tree species, regeneration (sexual, vegetative), successional stages, survival of seedlings and young growth, influence of animals, fire, site mosaic and, not least, human impact. Also, orographic influences, such as boulder fans, steep rock walls, and avalanches must be considered. They usually prevent forest from advancing to higher elevation. Moreover, the actual position and spatial pattern of timberline as well as the age structure of tree vegetation in the timberline ecotone reflect site history rather than present climatic conditions (Holtmeier, 1985a, b, 1994b, 1995a, 1999a, 2000). Extreme climatic events in the past (drought, snow-rich or snow-poor winters, late and early frosts, mass outbreaks of herbivorous and pathogenous insects, game, forest fires and other agents

F.-K. Holtmeier, *Mountain Timberlines: Ecology, Patchiness, and Dynamics*,
Advances in Global Change Research 36, 1–4.

have influenced timberline more or less. Human impact (burning, cattle and sheep grazing, lumbering, mining activities, etc.) plays an important role at almost all mountain and northern timberlines (Figure 1; see also Holtmeier, 1999a, c, 2000). As a consequence of these activities not only did the timberline

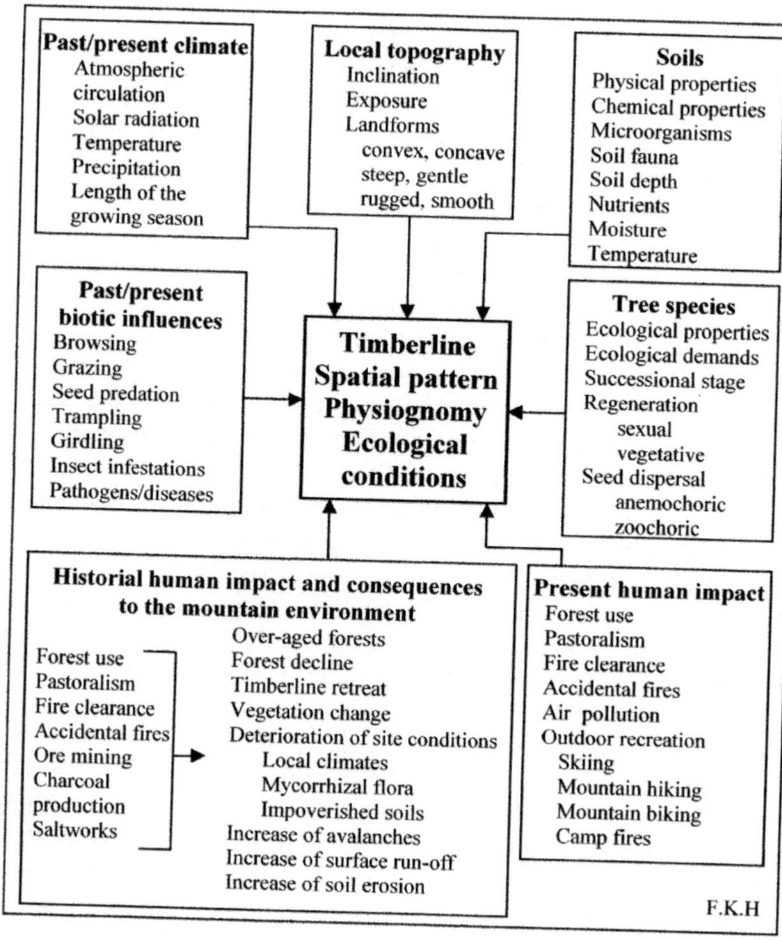

Figure 1. Factors influencing spatial pattern and physiognomy of timberline and ecological conditions in the timberline ecotone.

become lower, but also species composition and structure of the mountain forests changed considerably (Section 4.3.14.2). Although timberline is widely located below its climatic limit the advance of trees to higher formerly forested elevation is strongly hampered by adverse site conditions, even under a warmer climate (Holtmeier, 1965, 1974; Müterthies, 2002). However, it is difficult and frequently impossible to identify and assess these historical influences (occasionally visible in growth rings for example) on the present timberline physiognomy and ecology. In summary, the present timberline is away from being caused only by the present climate (e.g., mean air or soil temperature). The after-effects of the current situation will influence future changes of timberline position and spatial patterns (Figure 90).

Thus, a complex view is needed to understand the spatially varying heterogeneity and dynamics of timberline. A complex view does not only mean consideration of the functional interactions between the many timberline-relevant factors (Figure 1) in different environments but also switching between global and finer (regional, local) scales of consideration (Figure 2) to match the

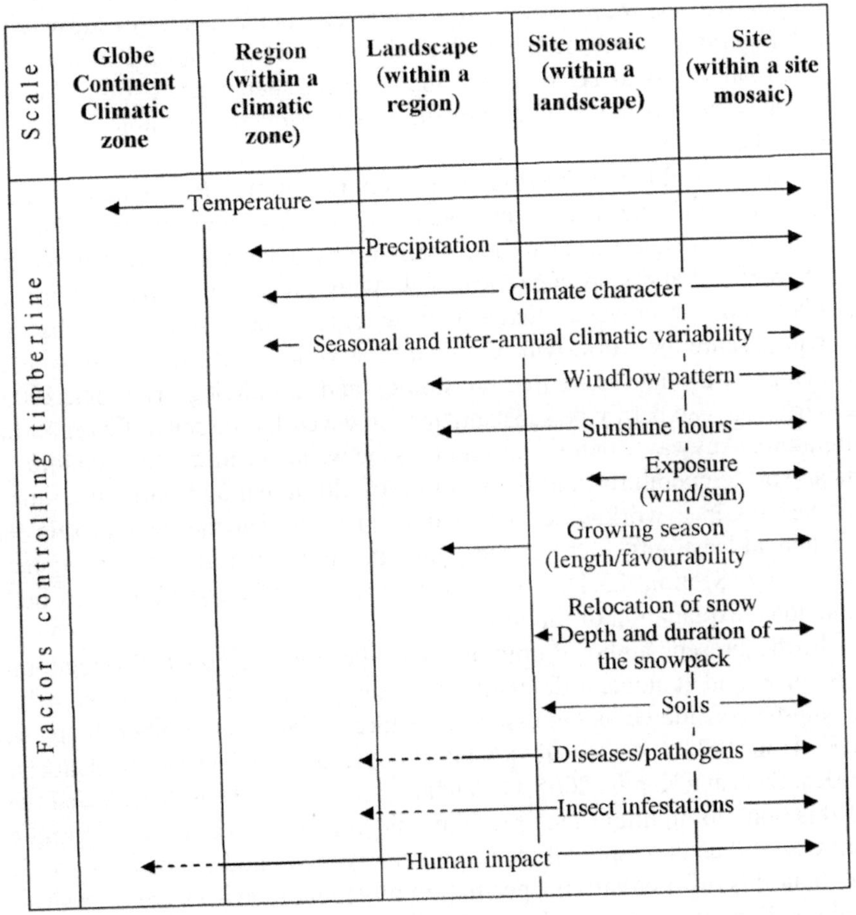

Figure 2. Timberline-controlling factors at different scales. Modified from Holtmeier and Broll (2005).

particular underlying factors and processes (see also Meentemeyer and Box, 1987). Operating at different scales (e.g., Curran et al., 1997) has been and probably still is the main problem for timberline researchers from different disciplines to better understand the complex nature of their common research object and also each others. Timberline heterogeneity increases from the global to the regional, landscape and local scales. Factors and processes at one scale may not be as important at another scale (Turner, 1989). Lack of soil moisture or waterlogging, for example, may control tree growth and timberline pattern at landscape and finer scales. In a global or zonal view, however, the effects of climatic variables such as temperature and precipitation may be more important. Conversely, coincidences of mean air or soil temperatures and the position of the altitudinal or northern timberline at a global or zonal scale reflect their general dependence from thermal-deficiency but do not provide any information on the possible role of the many other factors.

Deeper insight into the spatial and temporal timberline dynamics and a better understanding of the functional relationships between the timberline-relevant factors and trees can be expected only when considering timberline at the regional and landscape/local scale within different climatic regions. It has been argued, however, that a 'narrow regional perspective' has obscured or will obscure the world-wide dominant role of heat deficiency and its direct and indirect influences on tree physiology and morphology (Körner, 2003b, 2007b) what can hardly be substantiated, however, by the scientific timberline literature. Anyway, 'better' correlations between mean air temperatures or mean soil temperatures and the position of altitudinal and northern climatic timberlines in a worldwide view would just confirm the general rule that treeline, at least outside the tropics, is related to thermal-deficiency in one or other way (Section 4.3.1). This, however, will hardly contribute to a better functional explanation of timberline.

In the present author's opinion, it is the great regional physiognomic, biological and ecological diversity of the upper and northern treeline, which should be considered as the essential feature in the global timberline pattern (e.g., Troll, 1973; Wardle, 1974; Arno, 1984; Holtmeier, 1989; Callaghan et al., 2002a, Broll and Keplin, 2005; Callaghan et al., 2002b). Thus, it is the objective of this book to highlight the physiognomic and ecological variety of timberlines as well as their spatially varying heterogeneity and temporal dynamics. Without this differentiation, speculation on the more or less great sensitivity and possible response of timberline to the changing environment might result in confusion and too broad implications.

2 HISTORY AND PRESENT STATE OF TIMBERLINE RESEARCH

The upper timberline is the most conspicuous vegetation limit in high-mountain areas of all continents, except for the Antarctic. Timberline is also an important ecological boundary, marked by a change in site conditions and plant communities when crossing the forest limit. For example, above the closed forest topoclimatic conditions, soil distribution patterns, intensity of soil erosion, etc. are totally different from the forest belt. This also holds true for the habitat conditions of the forest-alpine tundra ecotone. It is characterized by a greater habitat variety compared to the closed mountain forest. The fact is that no other vegetation limit has a comparable effect on the high-mountain environment, making it all the more astounding that scientists have addressed timberline studies only relatively recently. In general, timberlines that have been disturbed by human impact are best investigated because access mostly is relatively easy.

2.1 Early timberline research

The earliest reports on timberline or treeline are usually based on more or less accidental observations, usually originating from general regional and local geographic studies. They mostly provide general remarks to the effect that mountain forests end at a certain altitude. The first reliable data on the altitudinal position of timberline are hardly older than 200 years (e.g., Hacquet, 1779; Zschokke, 1804, 1805, 1806; Kasthofer, 1818, 1822). Systematic timberline research began about 150 years ago. Early timberline research was reviewed by Imhof (1900), Marek (1910), Däniker (1923), and Holtmeier (1965, 1974).

When it became apparent from the increasing number of observations that climatic timberline and tree line are mainly caused by heat deficiency, researchers began to focus on thermal conditions (Sendtner, 1854; Kerner, 1864/1865). Some authors (Supan, 1994; Drude, 1890; Andersson, 1902; Köppen, 1919, 1920) emphasized the conspicuous coincidence of the polar timberline and the 10°C-isotherm of the warmest month (July). Kasthofer (1822) and the brothers Schlagintweit and Schlagintweit (1854) first mentioned the positive effect of mass-elevation (mountain-mass effect, Merriam effect) raising the altitudinal limits of vegetation, snow, and also agriculture and human settlements. Later, the mass-elevation effect on the

F.-K. Holtmeier, *Mountain Timberlines: Ecology, Patchiness, and Dynamics*,
Advances in Global Change Research 36, 5–10.

position of the upper timberline (Imhof, 1900) and other altitudinal limits such as snow line (Jegerlehner, 1903), and upper limit of human settlement (Flückiger, 1906) was systematically studied. Enquist (1929, 1933) calculated the duration and thresholds of certain temperatures that he believed to be the factors controlling the altitudinal limits of timberline forming tree species. Moreover, some coincidence was found between the altitudinal position of timberline or tree line and mean temperature of the 3 or 4 warmest months (tritherm or tetratherm), and the total sum of temperature of the growing season. Alternatively, the use of degree-days (base 10°C) has replaced the use of isotherms when close correspondence between degree–days and major vegetation zones was found (e.g., Hopkins, 1959). Attempts were also made to discover whether timberline is correlated to the amount of precipitation, an idea that Maurer (1915) refuted.

Altogether, those studies were necessarily based on the interpolation of data from meteorological stations far distant from each other. This holds particularly true for the northern timberline. Thus, all these attempts should be considered as approximations by data available in those days. The correspondence that had been found between the altitudinal position of timberline and mean temperatures of the growing season, air temperature sums, etc., clearly reflect the influence of heat deficiency, although mean temperature cannot be considered a causal factor (Section 4.3.1).

The monographs of Imhof (1900) and Marek (1910) improved the knowledge on timberline considerably. Imhof was concerned with the timberline in the Swiss Alps, where he systematically studied the mass-elevation effect on timberline, while Marek (1910) did the same for the Austrian Alps. Finally, Brockmann-Jerosch (1919) provided a synthesis of the knowledge of that time, and instead of the effects of single climatic factors he considered the influence of the climate character on the position of timberline in a worldwide view. Almost 40 years later, the geographer Hermes (1955) published a comprehensive monograph, based on a thorough review of literature, on the upper timberline and its distance from snow line in the mountains of the world. There are many other local and regional studies on high-altitude forests, on the requirements of the tree species forming the timberline, on forest history, forest use, and others. Although mainly descriptive, they all have contributed to a better understanding of timberline.

However, they could not provide deeper insight into the ecological causalities controlling position, spatial structure, and dynamics of timberline. In this respect, Kihlman's monograph on the northern tree line on the Kola Peninsula (Kihlman, 1890) was more progressive. Even today it is surprising how he assessed the ecological conditions and the effects of tree line-affecting factors such as winter desiccation, for example, by careful observation and consideration. For the most part his hypotheses should become evidenced by experimental research half a century later.

2.2 Modern timberline research

In the Alps it was Däniker (1923) who first studied timberline with special regard to ecological conditions. Modern experimental ecological timberline research began in the 1930s with the studies of Pisek and Cartellieri (1939), Michaelis (1934a, b, c, d), Steiner (1935), and Schmidt (1936). Timberline research in the Alps was stimulated by heavy avalanche catastrophes that occurred during the winter of 1951/1952 and 1953/1954. The high frequency and the high destructiveness of the avalanches and also of debris and mud flows were attributed for the most part to deforestation of the mountain slopes by humans (alpine pastures, mining, salt works, etc.) and to the bad condition of the over-used and over-aged high-altitude forests. Kasthofer (1822) and Landolt (1862) already emphasized the protective function of mountain forests and their reports encouraged forest restoration in some areas. In view of the many destructive avalanches that had occurred in the two consecutive snow-rich winters in the middle of the 1950s, extensive research programs were initiated in Austria and Switzerland to create scientific fundamentals for assessment and appropriate, site-adapted management (microclimate, soils, physiology, regeneration, etc.) of the mountain forests up to the potential timberline. Research stations were established close to timberline in Switzerland (1959, Stillberg in the Dischma Valley, near Davos) and Austria (1953, Obergurgl). Particularly in those areas many experimental field studies on topoclimate, plant communities, soils, snow fungi, mycorrhiza, and on the ecology and aptitude of tree species for high-altitude reforestation were carried out.

Also, many local and regional studies on timberline were published in regional monographs or in journals and periodicals of botany, geography, and forest sciences. Moreover, many publications are concerned with more special aspects such as altitudinal shifts of timberline over the course of time, tending or restoration of the protective functions of mountain forests. Further information is provided by investigations on the distribution and ecological requirements of the timberline-forming tree species and high-elevation forest management. The recently published books on 'Nordic mountain birch ecosystems' (edited by Wielgolaski, 2001) and 'Plant ecology, herbivory, and human impact in Nordic mountain birch forests' (edited by Wielgolaski et al., 2005), constitute a very valuable reference and summary of the present scientific knowledge on these birch forests, although timberline itself is more casually considered. Timberline in the Swedish Scandes has been studied in particular by Kullman (see reference list in this book). Brandes (2007) has published a many-faceted thesis on timberline in the high-mountains of Greece. This thesis is a valuable contribution to a better understanding of the Mediterranean timberlines in general as it shows timberline physiognomy, altitudinal position and dynamics being influenced more by natural factors than might be expected in view of historical human impact (pastoral use, fire, etc.) which affected timberline for thousands of years. Also, a great regional variety of timberline in the study area becomes apparent. Compared to the European Alps, no other mountain region has been covered by so many studies on the ecology of the timberline-forming tree species and on practical application of the results at high-altitude afforestation and restoration of mountain forests (Turner, 1985). However, hardly less numerous are the investigations of upper and northern timberline in North America and Scandinavia. Many of those studies are concerned with regional timberline dynamics as influenced by climatic fluctuations (references in Chapter 5).

In 1979 Tranquillini, a pioneer in experimental research on ecophysiology of timberline tree species, compiled the results of his own and others research in his book 'Ecological physiology of alpine timberline'. This book represents the state of the art at that time; it still is a good source for information and undoubtedly is most often quoted by English-speaking timberline researchers. Almost 30 years later, Wieser and Tausz (2007) edited a treatise on 'Trees at their upper limit' which continues the tradition of Tranqillini's timberline book. Unlike this book the new treatise is a cooperative work of nine experts. Although focussing mainly on the ecophysiological aspect of treeline in the European Alps it also includes altitudinal treelines on temperate mountains worldwide. The book gives a concise and thorough overview of the present state of knowledge on tree ecophysiology relevant to the altitudinal timberline in temperate mountains. However, the possibilities of transferring the results of experimental studies in laboratories or on field plots and of any other local investigations to other areas are limited, because of the great variety and heterogeneity of timberline. Even within a single mountain region, such as the Alps, we face problems in this respect due to the very locally varying site conditions. In particular, difficulties will increase in applying local results to timberline of distant mountain ranges, such as the Himalayas, for example (e.g., Miehe, 1982).

Altogether, the ecophysiological situation of trees growing at the upper timberline in the Alps, in other high mountains and in the Subarctic of North America and northern Europe can be considered the best investigated. Compared to these timberlines the knowledge on timberlines in the tropics is till fragmental (Miehe and Miehe, 1994, 1996). Recently, Schickhoff (2005) gave an extensive overview of the timberline in the Himalayas, Hindukush and Karakoram based on his own studies and a widely scattered literature on altitudinal position, physiognomy and floristics. Research on timberline ecological conditions in these mountain systems is still in its infancy. The same holds true for the timberlines at middle latitudes in the southern hemisphere, except for New Zealand. Most information refers to the East-African mountains (Hauman, 1933; Fries and Fries, 1948; Klötzli, 1958, 1975, 1977; Hedberg, 1964; Coe, 1967; Plesnik, 1980; Bussmann, 1994; Miehe and Miehe, 1994, 1996, 2000), to Mexico and the South-American Andes (Troll, 1959, 1973; Beaman, 1962; Lauer, 1973; Klink et al., 1973; Lauer and Klaus, 1975a; McQueen, 1976; Klink and Lauer, 1978; Hueck and Seibert, 1981; Bauman, 1988; Hildebrand-Vogel et al., 1990; Seibert and Meinhofer, 1991; Jordan, 1996; Vogel, 1996; Wardle, 1998), and to New Guinea (Van Steenis, 1953; Hope, 1976; Löffler, 1979; Smith, 1980). Recently, Bader (2007) published a detailed treatise (doctoral thesis) on tropical altitudinal treelines with emphasis on the ecological processes and factors controlling spatial timberline patterns, physiognomy and dynamics. The studies refer to the South American Andes and to Haleakala volcano (Hawaii). Schweinfurth (1966, 1980) and in particular Wardle (1985a, b, c, 1991, 2007) investigated the timberline in New Zealand and also reviewed older studies. The less advanced exploration of timberline in the 'southern' mountain regions may be partly explained by the fact that pressures to restore high-altitude forests have not been as strong so far as in the Alps, for example. Actually, human impact (lumbering, grazing, burning, bark-stripping, etc.) on mountain forests in subtropical and tropical regions is continuously increasing due to rapidly growing population (Haffner, 1982; Schickhoff, 1995a, b, 1996; cf. Section 4.3.14.2). In a not too distant future restoration of high elevation forests might be the only way to prevent humans from the negative effects of man-caused forest decline.

Besides the extensive studies of Brockmann-Jerosch (1919) and Hermes (1955) there are only a few recent and relatively concise contributions comparing timberlines on a world-wide scale (e.g., Ellenberg, 1963; Troll, 1973; Wardle, 1974, 1993; Holtmeier, 1985b, 1989; Plesnik, 1991; Körner, 1998a, b, 2007b), some given in connection with a presentation of the ecological situation of alpine vegetation in general (Crawford, 1989, 2008; Körner, 1999).

Arno (1984) has provided the most comprehensive modern compilation. Although mainly referring to timberlines in North America and written for the general public it also is very useful to any scientist interested in timberline. However, the author refers almost without exception to literature written in English. Ohsawa (1990), Tuhkanen (1993), Malyshev (1993) and Miehe and Miehe (1994, 1996) provide outlines of timberline of larger regions, mainly with respect to the influence of thermal conditions on the timberlines.

Global climate change has a stimulating effect on timberline research as is reflected in the rapidly growing number of publications. In a series of publications, Körner and co-authors, for example, have revived the age-old discussion on the 'ultimate cause' of altitudinal treeline. They consider low mean temperatures in the rooting zone during the growing season or all-year-round (at tropical treelines) to be the critical factor controlling directly worldwide treeline position (Körner, 1998a, b, 1999, 2003a, b for review; Paulsen et al., 2000; Hoch and Körner, 2003; Körner and Paulsen, 2004; Hoch and Körner, 2006; Körner, 2007b). This theory is considered applicable everywhere and to steer modern treeline research around to the supposed 'right' direction (e.g., Hoch and Körner, 2003).

Undoubtedly, a well-defined threshold soil temperature controlling treeline worldwide would put out timberline researchers from the misery of inscrutable timberline complexity and would also make modelling the future timberline position a lot easier (see also Körner, 2007b). However, many uncertainties and inconsistencies are left (Section 4.3.5) and the discussion continues. Modelling treeline response and possible consequences of treeline advance to greater altitude and more northern locations has become a modern instrument to approach the timberline phenomenon (Chapter 6).

3 DEFINITIONS, TERMINOLOGY

Undoubtedly, more attempts have been made to define timberline than other vegetation limits, in particular for correlating the location of this prominent vegetation- and landscape-limit to certain isotherms or other altitudinal limits, such as snow line, for example (e.g., Hermes, 1955). Most definitions refer to a certain minimum tree height or minimum forest cover. Aas (1964) and Mork (1968b), for example, consider a forest to be a closed forest if the average distance between the trees does not exceed 30 m. The critical minimum heights range from 2 to 8 m (Table 1), the minimum cover from 30% to 40% (Jenic and Locvenc, 1962; Ellenberg and Muller-Dombois, 1967; Ellenberg, 1978).

Table 1. Minimum height as criterion for identifying upper treeline

Author	Tree height (m)
Brockmann-Jerosch (1919)	5
Schröter (1926)	4–5
Leibundgut (1938)	5
Vincent (1938)	8
Rubner (1953)	6–8
Hermes (1955)	5
Plesnik (1959)	8
Jenic and Locvenc (1962)	5
Ellenberg (1963)	2
Wardle (1964, 1965a)	1
Holtmeier (1965)	>average depth of winter snowpack
Hustich (1966)	5
Müller-Dombois and Ellenberg (1974)	2
Wardle (1974, 1981a)	2
Bernadzki (1976)	2
Braathe (1977)	2.5 (at timberline 3)
Kullman (1979 onwards)	2
Little (1979)	4
Piussi and Schneider (1985)	2
Timoney et al. (1992)	3–4
Hofgaard (1997a)	2
Paulsen et al. (2000)	3

Minimum tree height as criterion is also differently used as far as different tree species are concerned. For example, Aas and Faarlund (1996) require a minimum height of 2.5 m to consider a birch (*Betula tortuosa*) growing at the upper timberline in northern Scandinavia a tree, while the

F.-K. Holtmeier, *Mountain Timberlines: Ecology, Patchiness, and Dynamics*,
Advances in Global Change Research 36, 11–28.

critical height of a pine (*Pinus sylvestris*) should be at least 5 m. Kullman (1987), on the other hand, included birch and spruce higher than 2 m when monitoring tree line in the southern Swedish Scandes, while the corresponding stem height of pine was only 1 m. In the temperate mountains a minimum height of 2 m appears to be adequate to the particular climatic and ecological situation, as a birch or any other tree species growing taller will be exposed to the harsh climatic influence above the winter snow cover, whereas smaller individuals are fairly well protected. However, particularly for this reason and with respect to the varying thickness of the snow cover in the forest-alpine tundra ecotone, the present author (Holtmeier, 1965, 1974) has objected to an absolute minimum height as criterion. Thus, a pine or birch should be considered a tree as soon as it projects beyond the average snow cover typical of the specific site (Table 1; see also Däniker, 1923). At tropical mountain timberline, however, this does not work, because a long-lasting seasonal snow cover is lacking. Yet, also a certain minimum tree height as criterion for defining tree line might make sense in so far as, for example, a 2 m high 'tree' would be more decoupled from the climate near the soil surface and thus would be growing in a different environment than the lower vegetation. Kessler (1995), on the other hand, holds the view that a tree may be smaller than 2 m provided that it shows typical tree habitus (one or several stems and a crown). However, where tall growing forests, such as 9 m high *Erica* stands on Moun. Kenya, grade into 'low' and 'dwarf forest' and finally into shrub, defining forest limit and tree line turns out to be rather difficult and remains arbitrary (Coe, 1967; Miehe and Miehe, 1994). For further differentiation of the upper forest belt Miehe and Miehe (1994, 1996), for example, make a difference between 'low forests' (>5 to 10 m) and 'dwarf forests' (<5 m). Salomons (1986), on the other hand, calls the 3 to 8 m high stands of *Gynoxis*, *Hesperomeles* and some attributed species occurring above the closed forest belt in the Páramo in central Columbia 'Andean dwarf forest'.

To be sure, defining the upper limit of mountain forests by a minimum tree height will result in a 'line' that can easily be correlated to any average temperature or temperature sums, growing degree days, etc. supposed to be controlling factors. However, studying the response of this 'line' will not provide any deeper insight into the ecological situation and spatial and temporal structures (Holtmeier, 1965, 1974; Stugren and Popovici, 1991).

This becomes particularly clear in view of the dynamics in the forest-alpine tundra ecotone. There has been a long discussion as to whether a natural climatic forest limit would be sharper than a transition zone (ecotone) at all. The existence of a transition zone is explained by unfavourable edaphic conditions and/or human impact (Scharfetter, 1938; Ellenberg, 1963, 1966, 1978; Schiechtl, 1967; Nägeli, 1969; Mayer, 1970; Köstler and Mayer, 1970; Kral, 1971). This concept has been taken over in textbooks on plant geography (Ellenberg, 1978; Klink and Mayer, 1983). From his own experience in mountain areas in and out of Europe and in the Subarctic the present author does not agree with this opinion, since the mountain and polar timberline are so heterogeneous that they should not be generalised to such an extent (Holtmeier, 1985b). In fact, in various high mountains that have not, or only randomly, been influenced by humans, the closed forest ends abruptly at its upper climatic limit. However, in many other high-mountain areas as well as in the Subarctic, the climatic timberline forms a more or less wide ecotone, extending from the closed forest to isolated stunted trees within the lower alpine belt or tundra. On gently sloping fjelds in central and northern Finnish Lapland, for example, broad Savannah-like mountain birch forests (*Betula tortuosa*) with sporadic several metres high trees (Photo 16) form the timberline ecotone. Thus, the position of the tree limit depends on how a tree is defined. There is much evidence that also the original upper birch forests were once open woodland (e.g., Oksanen et al., 1995; Kankaanpää, 1999; Holtmeier and Broll, 2006). Dead wood, for example, partly found in densely spaced peat hummocks (10–30 cm high) above the present timberline support this hypothesis (Holtmeier and Broll, 2006). The hummocks developed from organic matter accumulated around the stem base of the birches as can be observed below the present timberline.

The existence of these ecotones cannot be primarily and everywhere attributed to human interference and/or unfavourable pedological conditions, but must be explained as the result of the complex influences of the actual and previous climate, fire, biotic factors and site history on tree growth and ecological conditions. In many cases the existence of a timberline ecotone is the result of oscillations of the climate, persistence of tall (mature) trees and regeneration under changing conditions. A general warming may be followed by advance of the forest to greater altitude and northern latitude (Chapter 5), while cooling will cause decay and retreat of the forest in the long-term, followed by change of the site conditions in the former forested area.

Moreover, mountain timberlines are formed by tree species with different ecological properties and requirements. Spruce, for example, is more shade-tolerant than pine or larch and thus forms comparatively dense forests while light-demanding pine or larch forests are usually more open. Some species, such as fir and spruce, are able to reproduce and propagate by layering (formation of adventitious roots) under conditions that would prevent sexual regeneration completely. In this respect most timberline forming pines, for example, are at a clear disadvantage compared to spruce or fir (Holtmeier, 1985b, 1986a, 1993a, Section 4.3.10). Moreover, due to the local and regional history of climate, vegetation and, not least, human impact, tree stands at timberline may be different as to successional stage, age classes, composition and ecological dynamics.

In view of the great physiognomic variety and heterogeneity of mountain timberlines and with respect to the many possible scientific approaches to timberline, it is not surprising that a general precise and practicable definition, meeting all aspects, is hardly possible. Nevertheless, from a global view at least four different types of timberlines can be distinguished (Figure 3).

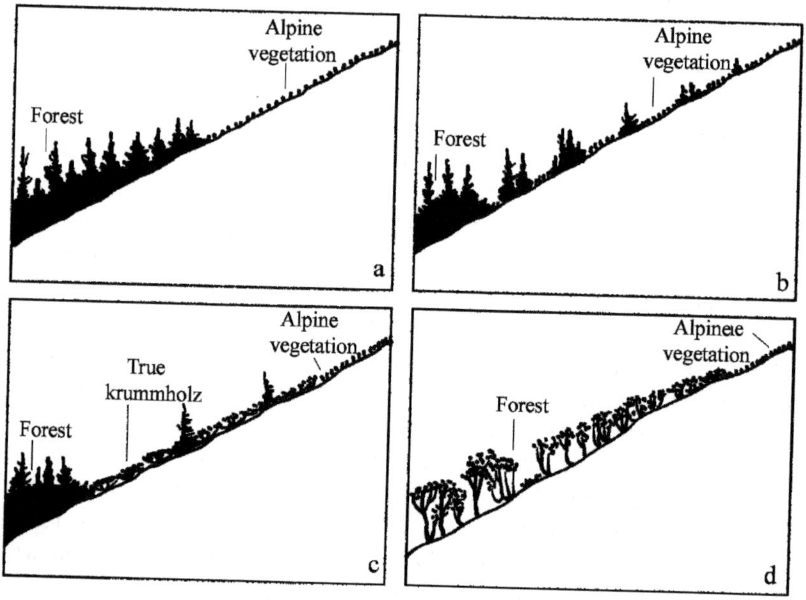

Figure 3. Main types of timberline. **a** – abrupt forest limit bordering alpine vegetation, **b** – transition zone (ecotone), **c** – true krummholz belt (e.g., *Pinus mugo*, *Pinus pumila*) above the upright growing forest, **d** – gradual transition from high-stemmed forest to crippled trees of the same species bordering alpine vegetation (e.g., *Nothofagus solandri* var. *cliffortioides*). Modified from Norton and Schönenberger (1984).

Timberline may occur as a line (e.g., Photos 1, 3, 4, 55, 111), or as a more or less wide ecotone (e.g., Photos 2, 32, 38, 51–53, 61, 67, 76, 88, 94, 95, 122; see Table 4 for synonyms). Different types may occur in close proximity to each other, even on a single mountainside (Photos 1 and 2). In other areas the forest gradually merges into alpine scrub formed by wooden species other than the tree species in the forest. Occasionally, high-stemmed forest stands border the alpine vegetation (Figure 3a, Photo 3) while in some areas tree height decreases approaching the upper limit of tree growth, and the most advanced individuals are more or less stunted (Figure 3a; Photo 4).

These climatically shaped individuals (e.g., Photos 24, 29, 38, 58, 59, 64, 66–77, 84, 87, 90, 91, 104) of the normally upright-growing tree species are usually called 'krummholz' in English. This popular term has been introduced from the German language. Originally it meant contorted, gnarled, twisted

Photo 1. Abrupt forest limit (*Picea engelmannii*, *Abies lasiocarpa*) on the WNW-exposed slope of Goliath Mountain (Mt. Evans area, Colorado) at about 3.500 m (view SW). F.-K. Holtmeier, 19 July 1994.

Photo 2. The same slope as above (view N). In this section, which provides higher soil moisture (being reflected in the distribution of willow shrubs), the timberline occurs as an ecotone located at about 3.500 to 3.540 m. F.-K. Holtmeier, 17 July 1997.

Photo 3. Abrupt limit of *Nothofagus solandri* var. *cliffortioides*-forest (evergreen) in the Craigieburn Range (New Zealand, South Island) at about 1.350 m. F.-K. Holtmeier, 24 November 1979.

Photo 4. Abrupt upper limit of *Nothofagus pumilio*-forest (deciduous) on an east-facing slope (550 to 580 m) on Isla Navarina (Tierra del Fuego). The uppermost trees exhibit dwarfed growth forms. A. Vogel, 3 March 1990.

and prostrate growing species such as *Pinus mugo*, *Pinus pumila*, *Alnus viridis*, *Alnus sinuata*, and *Alnus maximowiczii* (Figure 3c, Photos 5, 6; see also Photos 12, 18, 114, 115) the growth form of which is genetically predetermined. Thus, it should not be confused with climatically stunted 'krummholz'. Although krummholz in the proper sense (Holtmeier, 1973, 1981a) does not display tree habitus, Masuzawa (1985, see also Saiko and Masuzawa, 1987) calls the *Alnus maximoviczii*-belt above the larch forest on Mt. Fuji 'dwarf forest'.

Photo 5. Prostrate mountain pine belt (*Pinus mugo*) above Swiss stone pine forest (*Pinus cembra*) in the High Tatra near Strbske Pleso (Slovakia). F.-K. Holtmeier, August 1970.

In the following (see also Holtmeier, 1965, 1974), timberline is understood to be the transition zone between closed forest, the density of which depends on tree species represented and site conditions, and the most advanced individuals of the forest-forming tree species (see also Däniker, 1923; Pfister et al., 1977; Slatyer and Noble, 1992; Heikkinen et al., 1995; Smith et al., 2003). Ecotonal timberlines are characterized by a mosaic of tree clumps, scattered groves, isolated, more or less deformed tree individuals and treeless patches covered by low shrubs, herbs, and grasses. Here, it should be stressed again that, at close sight, abrupt timberlines usually reveal themselves as narrow ecotones. In the temperate and northern mountains, these outliers of tree growth are usually deformed, only a few decametres high and mostly but not everywhere protected by the snow cover from adverse climatic influences in the winter ('scrub line' in the sense of Arno, 1984, dwarf tree or cripple limit in the sense of Schröter, 1926).

Photo 6. Dwarf Siberian stone pine (*Pinus pumila*) overtopped by several *Abies nephrolepis* on Lisaja Shg (Sikote Aline, Russia) at about 1.200 m. H. Mattes, 25 August 1997.

Though these climatically stunted spruces, firs and pines will usually not meet the minimum height of a 'tree' the genetic disposition for becoming a tree is inherent, as is evidenced by prostrate-growing individuals having assumed or re-assumed tree-like features (single- or multi-stemmed, crown) or in the event of improved climatic conditions (Photos 72–75, Figure 58). This makes them different from shrubs, which thrive from the base (basitony) and show a sympodial ramification (Strasburger et al., 1991). This is one reason to distinguish genetically predetermined from climatically shaped 'krummholz' (Table 2; Holtmeier, 1973, 1974, 1981a). When the latter is considered in the following chapters it is put in quotation marks. Variation of growth form of tree individuals caused by the effects of oscillating climate is typical of the ecotone (Scott et al., 1997). Consequently, upright stems thriving from a mat-like growing 'tree' above 2 m height should not be lumped together with advance of timberline as it has been usual for several authors (e.g., Kullman, 1986a, 2000b, 2002; Lavoie and Payette, 1992; Lescop-Sinclair and Payette, 1995; see also Section 5.1).

As to the ecological situation, timberline ecotones are completely different from the closed forest and from the treeless alpine belt. While in the treeless alpine belt the microclimatic pattern is a function of the influence of microtopographical structure (convex, concave) on solar radiation and windflow near the soil surface, in the ecotone these climatic elements are influenced also by the scattered stands of trees, an aspect that has hardly

Table 2. Krummholz-terminology and its practical use

	True krummholz (genotype)	Environmentally induced krummholz (phenotype)
Growth form	Shrub, scrub	Environmentally shaped growth forms of the forest tree species
Example	*Pinus mugo* *Pinus pumila* *Alnus viridis* *Alnus sitchensis* *Podocarpus nivalis*	Flagged trees Table and flagged table trees, Mat-like growth etc., Identical or similar growth in different tree species at same external influences
Distribution	Usually more or less wide altitudinal belt above the forest, which is formed by other species also common in avalanche chutes	Controlled by the locally varying site conditions (e.g. wind-exposure, snow depth etc.) in the ecotone
English terms	Krummholz[1] Scrub[2] Subalpine scrub[3] Elfin wood[4] Dwarf forest[5]	Krummholz[6] Dwarfed krummholz[7] Wind-timber[8] Dwarf forest[9] Dwarf/matted trees[10] Crippled trees[11] Stunted trees[12] Brushwood[13]
German terms	Krummholz[14] Knieholz[15] Latschenbuschwald[16] Grünerlenbuschwald[17] Legföhrengebüsch[18] Alpenerlengebüsch[19] Krummholzwald[20]	Krüppelholz[21] Baumkrüppel[22] Krüppelbäume[23] Krummholz[24]

References: [1]Klikoff (1965); [2]Wardle (1973, 1977); [3]Wardle (1973, 1977); [4]Hansen-Bristow (1986); [5]Masuzawa (1985), Saiko and Masuzawa (1987); [6]Wardle (1968, 1973, 1974, 1993), Lamarche and Mooney (1972), Ives (1973b), Troll (1973b), Pfister et al. (1977), Komarkova (1976, 1979), Peet (1981), MacMahon and Andersen (1982), Arno (1984), Arno and Hoff (1989), Crawford (1989); [7]Habeck (1969), MacMahon and Andersen (1982); [8]Löve (1970); [9]Coe (1967), Salomons (1986), Miehe and Miehe (1994), Miehe and Miehe (1996); [10]Griggs (1938), Bliss (1963); [11]Arno (1984); [12]Arno (1984); [13]Cuevas (2002); [14]Schröter (1926), Hegi (1958), Braun-Blanquet (1964), Schmidt (1969), Ellenberg (1978), Franz (1979), Klink and Mayer (1983), Strasburger et al. (1991); [15]Hueck (1962), Klink (1973), Kuoch and Schweingruber (1975), Ellenberg (1978); [16]Ozenda (1988); [17]Ozenda (1988); [18]Rübel (1912); [19]Rübel (1912); [20]Ellenberg (1978); [21]Braun-Blanquet (1964); [22]Ellenberg (1978); [23]Geiger (1901), Ozenda (1988); [24]Marek (1910), Piussi and Schneider (1985).

been considered so far in timberline literature (Section 4.3.12). In the extremely wind-exposed forest-alpine tundra ecotone of the Rocky Mountains, for example, even the prostrate crippled trees cause by their effects on the wind-mediated relocation of snow a locally varying mosaic of snow-covered and snow-free patches which in turn influence site conditions (cf. Photo 87). The less broken the terrain, the more the wind, snow cover, radiation exchange and thus site conditions are influenced by the distribution pattern of stands of trees and openings. The influence of the mosaic of tree clumps and open areas on snow accumulation, for example, may result in a longer duration of snow cover in the ecotone (cf. Photo 32) compared to the forest (high interception) and the treeless alpine zone (deflation). The finely differentiated local site pattern that appears is partly cause and partly result of the way in which the tree stands are distributed. These ecological conditions are peculiar to the ecotone (Figure 4). In the closed forest, however, the influence of microtopography on solar radiation, wind velocity and direction is by far less important. Altogether, these effects of microtopography and trees on the patchiness of site conditions are by far more important for tree growth, regeneration and survival than altitudinal gradients of air temperature.

Consequently, a better understanding of the ecological situation and spatial and temporal dynamics in the ecotone requires extensive local and regional studies on microsite conditions and microsite patterns specific to the ecotone and cannot be achieved only by investigating physiological responses of mature tree growth to thermal conditions (Holtmeier, 1994b, 1999a). This also holds true for the tropical mountain timberlines. These appear to be as diverse and heterogeneous as the timberlines outside the tropics and obviously more different from each other than can be assumed in view of the tropical type of timberline, which Troll (1959, 1973) compared in a schematic sketch to the 'general' type of timberline in the temperate zones. This generalisation might have been useful for teaching differences in the effects of temperate and tropical climates on timberlines. Thus, it was not by chance that this sketch was adopted by many authors in their textbooks on geography or geobotany (e.g., Price, 1981; Klink and Mayer, 1983; Leser et al., 1991). However, from the landscape ecological view, this generalisation disguised the diversity that is typical of timberlines in the worlds' mountains, and it is this diversity that should be explored (Chapter 1).

Bader et al. (2007), for example, describe timberlines in the tropical Andes and on Haleakala volcano (Hawaii) as being mostly abrupt. However, their physiognomy varies locally. Most of these timberlines are fringed by tall shrubs, ferns or tall grasses (Photo 7). In other places, patchy timberlines occur with sharply-contoured dense tree stands alternating with Páramo vegetation (cf. Photo 113), thus forming a more or less wide ecotone. Moreover, 'gradual' timberlines occur where forest decreasing in height with altitude passes almost seamlessly into the Páramo, although this situation is rather rare. Richter et al. (2008) report considerable variation in the physiognomy of neo-tropical timberline in the Cordillera Real (Ecuador).

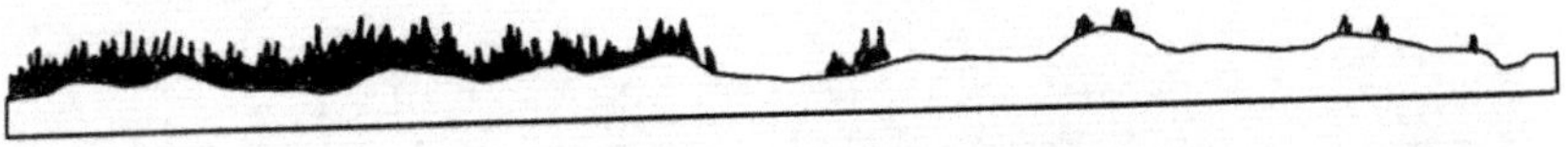

Forest	**Timberline ecotone**
Radiation transfer in the upper canopy surface	Radiation transfer at the surface of the ground vegetation, tree clumps, scattered trees and at the soil surface
Almost no influence of microtopography on solar radiation	Strong influence of microtopography, tree clumps and scattered trees on solar radiation
Relatively low temperatures	Locally varying temperature (High temperatures at sun-exposed and wind-protected sites)
Almost no effect of microtopography on wind velocities and directions	Wind velocities and directions modified by microtopography, tree clumps and scattered trees
Low wind velocities	High wind velocities
High interception of precipitation by the forest canopy	Low interception of precipitation by tree clumps, scattered trees and ground vegetation
Relatively even depth and length of the snow cover	Irregular depth and length of the snow cover due to the effects of microtopography and scattered trees on wind velocities and directions
Relatively high humidity	Locally varying humidity
Relatively even distribution of soil moisture	Locally varying soil moisture
'Forest soils' (in conifer forests mainly Podzols)	Mosaic of different soils related to microtopography and plant cover

Figure 4. Ecological conditions in the forest and in the forest-alpine tundra ecotone. Modified from Holtmeier (1979).

Thus, in tropical high mountains it may be hard to distinguish timberline and tree line in their proper sense, particularly if high-stemmed forests, such as the 9 m high *Erica* forest on Mt. Kenya for example, gradually merges into 'low forest', 'dwarf forest' and shrub without any change of tree species (Section 4.3.11). In this case a demarcation between tall and lower forest is arbitrary (Coe, 1967; Miehe and Miehe, 1994).

Photo 7. Abrupt timberline (at about 3.400 m) in the eastern cordillera of southern Ecuador. This timberline has not been influenced by fire for more than 10 years. M. Y. Bader, October 2002.

The great physiognomic and ecological variety and heterogeneity of mountain timberlines is reflected in timberline terminology (Table 3, see also Holtmeier, 1974, 2000). Some terms refer to the location of the timberline only (upper, lower timberline); others refer to the controlling factor or complex of factors (climatic, orographic, anthropogenic timberline) or to both location and causes (alpine, polar/subarctic/arctic, continental, maritime timberline). Thus, for example the alpine, the polar (subarctic, arctic, northern) and maritime timberlines are climatic timberlines. While the upper and northern timberlines are caused by heat deficiency, the maritime timberline is caused by strong winds and salt spray that adversely affect tree growth at the seashore (Brockmann-Jerosch, 1928). The maritime timberline also is a lower timberline (see also continental timberline). Consequently, the term 'maritime timberline' should not be confused with the comparatively low altitudinal timberline in mountains with a maritime climate, as did Pott (1993) for example. Also, the continental timberline is a lower timberline (Brockmann-Jerosch, 1919) that borders the steppe (Photo 8). Since the continental timberline is caused by insufficient moisture, it is also called dry timberline or drought-caused timberline. In arid and semiarid zones, mountains that are high enough to catch sufficient moisture from the air currents, an upper and a lower climatic timberline ('double timberlines' in the sense of Arno, 1984) occur: the upper timberline caused mainly by heat deficiency, the lower timberline by lack of moisture. The less moisture is available to the forest the higher is the dry timberline located (Schweinfurth, 1957; Troll, 1972; Arno, 1984; Jacobsen and Schickhoff, 1995).

Table 3. Terms related to causes and position of timberline

Terms (tl = timberline)	**Synonyms**	**German terms** (Wgr = Waldgrenze)	**Position**	**Causes**
Climatic tl	Alpine tl Alpine treeline ecotone Mountain tl	Obere Wgr Alpine Wgr	Altitudinal limit	Heat deficiency, short growing season (outside the tropics)
Polar tl	Subarctic tl Arctic tl Northern tl Northern cold tl Subantarctic tl Antarctic tl Southern tl Southern cold tl	Subarktische Wgr Arktische Wgr Nördliche Wgr Subantarkt. Wgr Antarktische Wgr Südliche Wgr	Horizontal tl, bordering the tundra	Heat deficiency, short growing season
Inverted tl	Valley tl Valley bottom tl Bottom tl	Inversions-Wgr Inverse Wgr	Lower tl in mountain valleys	Inversions with frequent early and late frost, waterlogged soil
Continental tl	Lower tl Drought-caused tl Dry tl	Kontinentale Wgr Trockengrenze Desertische Wgr	Lower tl in mountain, tl bordering the steppe	Moisture deficiency
Maritime tl	Coastal tl	Maritime Wgr	Lower/horizontal tl at the ocean coast	Influences adverse to tree growth at the coast (e.g. salt spray)
Historic tl	Max. postglac. tl Hypsithermal tl	Historische Wgr. Wärmezeitliche Wgr	Altitudinal tl	Heat deficiency, short growing season (outside the tropics)
Potential tl	Hypothetical tl	Potentielle Wgr	Altitudinal tl	Heat deficiency, short growing season (outside the tropics)
Orographic tl		Orographische Wgr	altitudinal tl, always below the climatic tl	Steep topography, rock walls, talus cones, boulder fans
Edaphic tl		Edaphische Wgr	Mostly altitudinal tl, always below climatic tl	Missing soil, waterlogged soils
Anthropogenic tl	Man-caused tl Artificial tl	Anthropogene Wgr	Mostly altitudinal tl, below climatic tl	Pastoral use, mining, salt works, firewood, incendiarism
Actual tl	Present tl	Akltuelle Wgr	Altitudinal tl, below climatic tl	Different causes

Photo 8. View from Montgomery Pass (Hwy 6) of the lower timberline (*Pinus monophylla* and *Juniperus osteosperma*) at the foot of the White-Inyo Range (California). The timberline is caused by lack of moisture. F.-K. Holtmeier, 25 July 1994.

Occasionally, in poorly ventilated mountain valleys with frequent stagnant cold air on the valley bottom, another type of lower timberline can be observed, the so-called inverted timberline (Photo 9). It is (mostly) caused by frost occurring within the cold air layers (Wardle, 1965b, 1971, 1973, 1974, 1980, 1985b, 1993, 2007; Moore and Williams, 1976; Costin, 1981; Paton, 1988; Slatyer, 1988; Banks and Paton, 1993). Also, the existence of inverted timberlines in tropical high mountain valleys is ascribed to frost (e.g., Fries and Fries, 1948; Hedberg, 1964; Wardle, 1971, 1974; Davidson and Reid, 1985) and/or to water-saturated soils (Smith, 1980; Davidson and Reid, 1987; Gilfedder, 1988; Young, 1993). Löffler (1979), however, referring to his studies in high mountains of New Guinea, rejects the frost hypothesis and considers water logging to be the controlling factor (Photo 10). An inverted timberline caused by water-saturated soils would be considered an edaphic rather than a climatic timberline. In addition, fire, mainly caused by humans for different purposes (Young, 1993; see Section 4.3.14), probably is an important agent preventing the valley bottoms in many tropical high mountains from being invaded by forest.

The northern (polar, subarctic, arctic) timberline caused by heat deficiency, borders the tundra or the subarctic dwarf shrub-lichen heath (Northern Europe). The ecotone is snaking its way for more than 13.000 km at varying width across northern Eurasia and northern North America. The farther north

Photo 9. Inverted timberline (*Picea engelmannii*, *Abies lasiocarpa*) in the Cache la Poudre Valley (Rocky Mountain National Park, Colorado) at about 3.170 m. Frequent stagnant cold air and frost very likely prevent the conifer forest from colonizing the valley bottom. F.-K. Holtmeier, September 1994.

Photo 10. Inverted timberline in the Mt. Wilhelm area (New Guinea). Grassland covers the valley bottom. Accumulation of cold air with frequent frost temperatures, waterlogging or human-caused fires are possible factors keeping the valley bottom treeless. E. Löffler.

the more the trees disappear from exposed interfluves and similar convex topography and are restricted to sheltered lowland and river valleys (Atkinson, 1981; Larsen, 1989). Thus, the northern timberline often occurs as an altitudinal boundary at comparatively low elevations. The varying width of the forest-tundra ecotone in Canada, for example, is partly ascribed to this effect of regional topography (Larsen, 1989; Timoney et al., 1992). Where high mountains are located precisely where the northern forest-tundra ecotone would otherwise exist, they eliminate the gradual transition zone and create an abrupt northern timberline as on the southern slope of the Brooks Range, for example (Larsen, 1989; Hobbie and Chapin III, 1998).

In contrast to the northern timberline, a southern (subantarctic, antarctic) timberline can be observed only on a few islands in high southern latitudes (Tuhkanen, 1993, 1999).

Besides climate many other factors such as steep rocky trough walls, talus cones, slope debris and avalanche chutes may prevent the forest from reaching its upper climatic limit. This is the orographic timberline (Photo 11, see also Figures 38–40 and Photo 43). Orographic timberlines, as is also true for the man-caused (anthropogenic) and edaphic timberlines, are always located more or less far below the elevation to which the forest would advance at the given climatic conditions (cf. Table 3).

Man has influenced mountain forests and timberline in many ways, such as forest pasture, clearing high-elevation forests to create alpine pastures, mining, burning, charcoal production, salt works, timber harvesting for construction wood, fuel and others (see Section 4.3.14.1). Anthropogenic timberlines may be abrupt, as can be particularly observed, for example, in many tropical mountains, where the high-elevation grasslands and forests are regularly burned (cf. Photo 113, see also Section 5.4). In other regions, the uppermost forests became open and over-aged due to over-utilization as in the European Alps, for example. However, a general characteristic that would be common to all anthropogenic timberlines does not exist.

By many authors the upper climatic timberline is also called alpine timberline (Jenic and Locvenc, 1962; Wardle, 1974;Tranquillini, 1979a; Leuschner and Schulte, 1991; Körner, 1998a, b; Bader, 2007; Bader et al., 2007; Wieser and Tausz, 2007). The present author, however, prefers the terms ‘upper timberline’ or ‘altitudinal timberline’ to ‘alpine timberline’, because the word ‘alpine’ does not meet the environmental conditions in the tropical high mountains. The same holds true for the treeless zone above the upper timberline in the tropics, which is often called ‘alpine zone’ (e.g., Klötzli, 1958; Young, 1993). Troll (1959) already explicitly referred to this problem. Thus, it seems less inappropriate to restrict the term alpine timberline, if inevitable, to the temperate mountains characterized by climates with cold and long winters, avalanches, strong influence of snow cover (distribution

Photo 11. Orographic timberline near Banff in the Canadian Rocky Mountains). F.-K. Holtmeier, 21 July 1994.

Table 4. Terms and synonyms concerning the transition zone between the closed forest and the upper limit of crippled trees

English terms	German terms
Timberline[1]	Kampfzone[14]
Timberline region[2]	Kampfgürtel[15]
Timberline zone[3]	Kampfwald[16]
Timberline ecotone[4]	Waldgrenzökoton[17]
Forest-tundra ecotone[5]	Subalpines Ökoton[18]
Forest-alpine tundra ecotone[6]	
Alpine timberline ecotone[7]	
Subalpine parkland[8]	
Subalpine forest[9]	
Subalpine belt[10]	
Patch forest[11]	
Meadow tree clump parkland[12]	
Treelimit region[13]	

References: [1]Wardle (1974), Price (1981), Heikkinen et al. (1995); [2]Franklin and Dyrness (1973); [3]Daly (1984); [4]Tranquillini (1979a), Holtmeier (1994b); [5]Clements (1936), Marr (1948); Marr and Marr (1973); [6]Wardle (1973); [7]Arno (1984); [8]Rochefort et al. (1994), Miller and Halpern (1998); [9]MacMahon and Andersen (1982); [10]Löve (1970); [11]Weisberg and Baker (1995); [12]MacMahon and Andersen (1982); [13]Marr (1967); [14]Schröter (1926), Holtmeier (1965, 1974), Mayer (1974), Tranquillini (1979a), Ozenda (1988), Piussi and Schneider (1985); [15]Geiger (1901), Schröter (1926), Scharfetter (1938), Schmidt (1969), Troll (1973), Tranquillini (1979a); [16]Strasburger et al. (1991); [17]Walter (1973), Holtmeier (1993a, 1995a); [18]Walter (1973).

pattern, thickness, duration) on site conditions (soil temperature, soil moisture, length of the growing season, etc.) and tree growth (Section 4.3.7). The upper climatic timberline in the tropics could be called 'tropical cold timberline'. If necessary, this term might be specified by using 'African cold timberline' (instead of 'Afroalpine timberline', Wesche, 2002) or 'Andean cold timberline', for example, instead of 'tropical alpine treeline' (Bader, 2007; Bader et al., 2007; Wesche et al., 2008).

The upper timberline at its maximum altitudinal position (postglacial optimum) is called historic timberline. It had advanced to considerably higher elevation and northern position than the present climatic timberline. Potential timberline means the altitudinal position forest could achieve at the present climate without being disturbed by human impact.

In general, the use of the terms timberline, tree line, forest limit etc. is rather ambiguous. Thus, in many cases it does not become clear whether the author refers to the upper limit of closed forest, to the entire ecotone or to the upper tree line. The last term, for example, may refer to the ecotone or to tree line in the proper sense as well (Table 4). Occasionally, however, clear differences are made. Habeck and Hartley (1968), for example, consider the upper limit of closed forest as 'forest line', the upper limit of upright growing trees (arborescent growth) as 'tree line' (see also Rochefort et al., 1994; Price, 1981) and the altitudinal limit of crippled tree individuals as 'scrub line' (see also Arno, 1984). By Rochefort et al. (1994) again 'scrub line' is called 'tree limit' (in contrast to 'tree line'), 'krummholz limit' (e.g., MacMahon and Andersen, 1982) or 'tree species limit' (e.g., Heikkinen, 1984). Griggs (1938) instead uses the term 'cripple line'. All these terms refer to climatically caused altitudinal limits of forest or trees. In case of controlling factors other than climate, adjectives such as 'anthropogenic' or 'orographic' are added.

4 PHYSIOGNOMIC AND ECOLOGICAL DIFFERENTIATION OF MOUNTAIN TIMBERLINE

As has been shown in the previous chapter mountain timberlines may be abrupt, gradual or occur as a transition zone bordering treeless vegetation such as dwarf shrubs and grasses (e.g., European Alps), dwarf shrub-lichen heath (fjell, Fennoscandia), mountain steppe (arid zone) or tussock grassland (e.g., in the tropics, New Zealand). On a global scale, the spectrum of trees species represented at timberline is very large and covers tree species of many genera and families. Locally, timberline is usually formed by one to three tree species but even more may occur (e.g., Wardle, 1971, 1974; Hope, 1976; Löffler, 1979; Corlett, 1984; Richter et al., 2008).

4.1 Tree species at timberline

Which and how many tree species occur at timberline depends on the climate zone and on the history of floral development. If timberline is formed by more than one species, the different ecological properties and requirements of the species (e.g. shade tolerant and intolerant, pioneer or climax species, animal or wind mediated seed dispersal, etc.) may play an important role as to the development of the tree stands, and also in respect of structure, physiognomy and climatically induced shifts of timberline. Competitive species such as beech (*Fagus sylvatica*), Norway or Engelmann spruce (*Picea abies*, *Picea engelmannii*), for example, often form dense stands and abrupt timberlines (cf. Photo 20), while in the case of less competetive species such as larch, most pines or juniper forests are comparatively open giving way gradually to grassland or other alpine vegetation (e.g., Walter, 1968; Armand, 1992; see also Photos 13 and 14). Some authors call such forests 'open forests' (e.g., Kessler, 1995; Miehe et al., 1998).

4.1.1 Influence of geological and floral history

Floristically, timberlines are more closely related to each other in the northern than in the southern hemisphere, which has to be ascribed to the geological history of the continents and the resulting geographic orientation of the mountain systems. While in the northern hemisphere, contiguous

F.-K. Holtmeier, *Mountain Timberlines: Ecology, Patchiness, and Dynamics*,
Advances in Global Change Research 36, 29–292.

approximately north–south or west–east oriented mountain ranges allowed floristic exchange over long distances, the situation in the southern hemisphere is somewhat different. After the break-up of the Gondwana continent, the flora of South America, Africa, Australia and New Zealand became isolated from each other. Consequently, also timberlines in these regions differ more from each other with respect to the tree species represented than is the case in the mountains of the northern hemisphere.

In South America and also in Southeast Asia and New Zealand, approximately north–south oriented mountain ranges allowed species from the temperate and subantarctic zone of the southern hemisphere to invade even the tropics where they form, together with tropical species, the upper timberline (e.g., *Podocarpus*, *Libocedrus*, *Metrosideros*, *Dracophyllum*, *Phyllocladus*, *Papuacedrus*; see also Troll, 1959, 1968; Bader, 1960; Walker and Flenley, 1979). As repeatedly demonstrated by Troll (1948a, 1959, 1968), the upper timberlines in the humid tropical high mountains exhibit certain climatic and ecological similarities to the strongly oceanic timberlines in the cool temperate zones of the southern hemisphere (Chile, Tierra del Fuego, New Zealand). Also, many genera and species are common to both these timberlines. Moreover, the physiognomy of the southern temperate timberline forests partly shows many correspondences to timberline forests in the humid tropical mountains (Section 4.3.11). In view of the existing differences, however, convergences should not be over-emphasized. For example, site conditions and growth forms of the trees at timberline in the southern Andes and in New Zealand are strongly influenced by the winter snow cover that does not exist at the tropical timberlines (Eskuche, 1973; Veblen et al., 1977; Wardle, 1998; see Section 4.3.7).

Not one of the tree species of the temperate and subantarctic zone has reached the upper timberline outside the northern hemisphere tropics, which may very likely be attributed to their low frost tolerance, compared to the boreal tree species (cf. Sakai et al., 1979; see also Tables 5–7). Boreal species, on the other hand, could invade mountains of the outer tropics. Comparatively open forests of *Pinus hartwegii* form the upper timberline on the high volcanoes in central Mexico between 4.000 to 4.200 m elevation (Troll, 1958, 1959; Beaman, 1962; Wardle, 1965b, 1993; Klink et al., 1973; Lauer, 1973; Lauer and Klaus, 1975a; Klink and Lauer, 1978; Perry et al., 1998; Velázquez et al., 2000; Biondi et al., 2005). In addition, small dense and wind-trimmed stands of *Juniperus monticola* occur above the upper pine limit on Iztaccihuatl and Popocatepetl (Beaman, 1962). In Southeast Asia *Pinus merkusii* crossed the equator. The southernmost occurrence is located at Mount Kerenji on Sumatra (2°S; Whitmore, 1975; Klötzli, 1984; Schweinfurth, 1988). Nowhere, however, do frost tolerant boreal tree species grow at timberline in the inner tropics, except for the genus *Juniperus* (cf. Troll, 1958; Bader, 1960; Von Wissmann, 1960, 1961, 1972; Bussmann, 1993; Miehe and Miehe, 1994). Very likely this can be ascribed to the lack of a warm and frost-free growing season (diurnal climate), which these species need to grow and reproduce.

It may be idle to ask how tropical timberline would have developed if boreal tree species had invaded. Undoubtedly, although extremely frost tolerant in winter, the boreal species would only have established themselves by successful gradual adaptation to the tropical high elevation climate with night frosts occurring throughout the year. Extreme frost hardiness, as is typical of the boreal tree species in winter, would not have been advantageous to them as they would not have tolerated frequent freezing during 'summer'. This hypothesis may be supported by experiments that were made with *Picea engelmannii* planted at timberline in New Zealand. This species is native to the continental Rocky Mountains. As mature trees, Engelmann spruce is highly frost tolerant in winter (cf. Table 5). During growing season, however, frost hardiness is comparatively low. Thus, many young planted spruces that did not adapt to the high oceanic timberline environment of New Zealand became heavily damaged or killed by light frost temperatures that frequently occur during and particularly in the end of the growing season. The same was observed in European larch (*Larix decidua*) (Wardle, 1968; Benecke and Havranek, 1980).

The native *Nothofagus* species (*Nothofagus solandri*, *Nothofagus menziesii*) that certainly were present on the Gondwana timberline (Wardle, 1993) could adapt to the environment while gradually advancing to greater elevation. In New Guinea, however, *Nothofagus* (16 different species, Van Steenis, 1953) was not able to advance beyond 3.000 m and thus does not occur at timberline, although frosts are not as strong as they are at the New Zealand timberline. However, freezing may occur at any time of the year and kill the leaves of *Nothofagus*, which are highly susceptible to frost (Wardle, 1973, 1984, 1993). That would mean that the situation of southern beech in New Guinea would be compared to a certain extent to that of Engelmann spruce or European larch at the New Zealand timberline.

On the other hand, introduced lodgepole pines (*Pinus contorta*) planted at and far above timberline in New Zealand do comparatively well. They may even exist 300 m above the closed *Nothofagus* forest even though they display crippled growth. Lodgepole pine seldom reaches the upper timberline in its native North America. In New Zealand it even regenerates up to an elevation of 1.570 m (Ledgard, 1980) and is spreading continuously. It is considered to become an increasing threat to the native high elevation grassland and to the alpine herb and shrub communities. In view of this process, Wardle (1986, 2007) concludes that the present alpine vegetation of New Zealand is not the best adapted alternative for climatic conditions in the New Zealand high mountains. If there had been a genetic exchange of the New Zealand mountain ranges with the mountains of the northern hemisphere, coniferous forest would presumably cover most of the present lower alpine belt. Smith

(1980) had a similar idea as to the altitudinal position of the timberline in New Guinea. He believes that timberline would be higher than present timberline because the young geological age of the mountains did not leave tree species enough time to adapt to tropical high-altitude climate.

4.1.2 Tree species at temperate and northern timberlines

In the northern hemisphere, upper and northern timberlines are mainly formed by different species of the genera *Pinus*, *Picea*, *Abies*, *Larix* and *Tsuga*, and locally also by arborescent juniper species. The prostrate pines *Pinus mugo* and also *Pinus pumila* deserve to be more closely considered. In some areas dense thickets (*Pinus mugo* in the Alps, the Dinaric Alps, Carpathian Mountains; *Pinus pumila* in Japan, in the Sikhote Alin Mountains, on Sachalin and Kamchatka) form a krummholz belt up to several hundred metres wide above the upper limit of the high-stemmed coniferous or birch forest (Photos 5 and 6; Kobayashi, 1971; Ishizuka, 1974; Gorchakovsky and Shiyatov, 1978; Hämet-Ahti, 1979; Miyawaki, 1979; Okitsu and Ito, 1984, 1989; Yanagimachi and Ohmori, 1991; Grishin, 1995; Grishin et al., 1996a; Takahashi, 2003). The prostrate growth of *Pinus mugo* and *Pinus pumila* is considered to be an adaptation allowing these pines to tolerate extremely low winter temperatures (e.g., Richardson and Rundel, 1998). This interpretation has to be seen in perspective, however (Section 4.3.7.2). In contrast to the arboreal conifers, both dwarf pine species may colonize avalanche tracks because of their high flexibility and resistance to breakage (e.g., Lämmermayr, 1932; Wilmanns et al., 1985) and their ability to regenerate by layering (Wardle, 1977; Okitsu and Ito, 1984; Wilmanns et al., 1985; Hafenscherer and Mayer, 1986), whereas in the other pines occurring at the upper timberline, layering is the exception (Section 4.3.10.2). Along avalanche tracks, dense dwarf pine thickets often extend from the krummholz belt above timberline down to the valley bottom (Photo 12; see also Fukarek, 1970; Plesnik, 1971, 1973). They may be considered a substitute formation of the high-stemmed forests in such places.

The northern timberline from western Siberia to the Kolyma River is formed mainly by larch: Siberian larch (*Larix sibirica* = *Larix russica*) ranging from western Siberia to the Pyasina River and Dahurian larch (*Larix dahurica*) occurring from there eastward to the Kolyma river. In the northeastern region of Asia larch is replaced by dwarf pine (*Pinus pumila*) (Kryuchkov, 1973).

In North America, the polar timberline is formed mainly by black and white spruce (*Picea mariana*, *Picea glauca*) (e.g., Tikhomirov, 1962; Hustich, 1966; Larsen, 1989). East of the Mackenzie River also *Larix laricina* occurs (Johnston, 1990; Packee, 1995; Schmidt, 1995). *Larix lyallii* is common to timberline forests in the northern Rocky Mountains, particularly west of the

Photo 12. Prostrate mountain pine (*Pinus mugo*) in an avalanche chute on the south-facing slope of the 'Mieminger Gruppe' (Tyrol). F.-K. Holtmeier, August 1970.

continental divide, and in the Cascades (Habeck, 1969; Arno and Habeck, 1972; Arno, 1984, 1990; Arno et al., 1995). Also, in some mountain regions of Eurasia larch is represented at the upper timberline, partly mixed with other conifers, as, for example in the central European Alps (*Larix decidua*), on the east slope of the Ural mountains (*Larix sibirica*) on the Putorana Plateau (Siberia, *Larix dahurica*) and northeast Yakutia (*Larix dahurica*). Larch also occurs at the upper timberline in the mountains of Transbaikalia and Subbaikalia (Lake Baikal region), in central Kamchatka (*Larix gmelinii*), in central Japan (*Larix leptolepis*), in Nepal, Bhutan and Tibet (*Larix griffithiana*), in South and West China (*Larix potaninii*; *Larix griffithiana*), in the eastern part of the Sajan mountains, in central Altai and in Changai mountains (*Larix sibirica*) (Egorov, 1967; Gorchakovsky and Shiyatov, 1978; Armand, 1992; Malyshev, 1993; Holtmeier, 1995b; Schmidt, 1995; Shimin and Shengxian, 1995; Takei, 1995; Lehmkuhl, 1997; Okitsu, 1997). *Larix olgensis* forms the upper timberline in the Changai Mountains, which are located at the border between China and Korea (Šrutek and Lepš, 1994). The genus *Tsuga* and *Chamaecyparis* are represented with only one species each (*Tsuga mertensina, Chamaecyparis nootkatensis*) at the upper timberline in north-western North America (Walter, 1971; Arno, 1984).

Arboreous juniper species (Photos 13 and 14) occur at the upper timberline of some semi-arid and arid mountain ranges, as for example in the High Atlas of Maroc (*Juniperus thurifera*, Rikli, 1946; Rauh, 1952; Messerli and Winiger, 1994), in the Himalayas, in the Karakoram, in the Hindukush and some other mountains of central Asia (*Juniperus excelsa, Juniperus semiglobosa, Juniperus seravschanica, Juniperus tibetica, Juniperus turkestanica, Juniperus recurva, Juniperus indica, Juniperus wallichiana*) (Troll, 1939, 1964; Schweinfurth, 1957; Freitag, 1971; Rathjens, 1972; Von Wissmann, 1972; Wardle, 1974; Succow, 1989; Schmidt-Vogt, 1990a, b; Miehe, 1982, 1991; Klötzli, 1991; Winkler, 1994, 1997; Jacobsen and Schickhoff, 1995; Schickhoff, 1995a, 1996, 2005; Miehe et al., 1996; Farjon et al., 2000; Richter, 2001). Nomenclature of *Juniperus* is very confusing. Thus, in an extended review on the upper timberline in the Himalayas, Hindukush and Karakoram Schickhoff (2005) subsumed all the taxa mentioned in literature under *Juniperus excelsa*. In eastern Tibet (uppermost Mekong area) juniper trees were found up to of 4.600 m (Von Wissmann, 1972). In south of southern Tibet (Xizang), the upper limit of arborescent juniper (*Juniperus tibetica*) is located at an altitude of 4.600 to 4.800 m (Miehe et al., 2000, 2003). Most of the relatively open juniper stands (4.750 m) are sacred forests and relics of former extended forests. Young juniper trees were found even at about 4.840 m (east of Nagarzê). At the drought and cold limit of *Juniperus tibetica* juniper trees give way to dwarf juniper (*Juniperus pingii* v. *wilsonii*). In western Xizang, dwarf juniper reaches its upper limit at about 5.000 m on sun-exposed slopes. In general, juniper is the most drought tolerant species among timberline forming tree species. *Juniperus polycarpos* (synonym *Juniperus seravchanina*) is widely distributed at the upper timberline in the mountains of Iran (Pourtahmasi et al., 2007). *Juniperus foetidissima* occurs at upper timberline on Kyllini (northern Peleponnesus; Photo 15). In the view of Klötzli (1991) the junipers at timberline in Eurasia fill a niche almost identical with that of bristlecone pine (*Pinus longaeva, Pinus aristata*) at upper timberline in North America.

Moreover, in many areas dicotyledonous tree species form the upper timberline or are at least represented there. Most of them are deciduous, such as the representatives of the genus *Betula*. In Fennoscandia and also on the Kola Peninsula, mountain birch (*Betula tortuosa*) forms the northern and upper timberline (Blüthgen, 1960; Hämet-Ahti, 1963, 1987; Gorchakovsky and Shiyatov, 1978; Aas and Faarlund, 1996). Typically, the birch shows multi-stemmed, contorted low growth forms (Photo 16). Because of morphological differences to *Betula tortuosa* Ledeb., described first from the Altai mountains, mountain birch in northern Europe is now considered to be a subspecies of *Betula pubescens*, which is called *Betula pubescens* spp. *czerepanovii* (Orlova) Hämet-Ahti (Hämet-Ahti, 1987). This subspecies results from introgressive hybridization between *Betula nana* and *Betula pubescens* (Kujala, 1929; Walters, 1964, Elkington, 1968; Kallio and Lehtonen, 1973; Vaarama and Valanne, 1973; Hämet-Ahti, 1987; Thórsson

Photo 13. Open *Juniperus semiglobosa*-forest on a south-facing slope in the Bagrot Valley (Karakoram) between 3.500 and 3.700 m. U. Schickhoff.

Photo 14. Open *Juniperus indica*-forest in the Hunza Karakoram at about 3.800 m. F. Klötzli, 15 August 1989.

Photo 15. Arboreous *Juniperus foetidissima* in the timberline ecotone on the southern slope of Kyllini (Peleponnesus) at about 1.750 m. The slope has been strongly influenced by grazing. R. Brandes.

et al., 2001). Since there is no general agreement, however, to this taxonomical differentiation and nomenclature (cf. Väre, 2001) the present author continues to call the mountain birch in Fennoscandia *Betula tortuosa*, as many other authors also do. South of the polar limit of conifers, mountain birch also occurs at the upper timberline in many areas, mixed with Norway spruce (*Picea abies*) and Scots pine (*Pinus sylvestris*).

East of Kola Peninsula, *Picea obovata* forms the timberline although birch occurs everywhere in the northern forests as far as middle Siberia. On the western slope of the northern Ural mountains, which is characterized by a milder and more humid climate compared to the eastern side, birch again forms the upper timberline while farther south it is mixed with *Picea obovata* and *Pinus sibirica* (Gorchakovsky and Shiyatov, 1978).

The *Betula ermanii*-forests along the east side of Eurasia can be considered the counterpart of the Fenno-Scandinavian mountain-birch forest

Photo 16. Multi-stemmed and crooked mountain birch (*Betula tortuosa*) on Staloskaidi (Finnish Lapland) at about 320 m. F.-K. Holtmeier, 3 September 2002.

(Hämet-Ahti and Ahti, 1969; Hämet-Ahti, 1987). They extend with some interruptions from Kamchatka down to Japan (Grishin, 1988, 1995; Ohsawa, 1990). Stands of birch (*Betula tortuosa*) also occur in southern Greenland. This birch is very likely genetically related to dwarf birch (*Betula glandulosa*). The upper limit of birch trees is located at an elevation of about 150 m. Stunted outliers have advanced to almost 250 m (Böcher, 1979). Birch is also quite common to the subalpine belt of some mountain regions in central Asia: for example in the Lake Baikal region (*Betula ermanii*), in Tien-Shan (*Betula saposhnikovii*), Alai mountains, Pamir (*Betula alaijica*), Altai (*Betula tortuosa*) and in the Himalayas (*Betula utilis*). Not least, birches occur in the western Caucasus (*Betula litwinowii*), in the High Tatras, in the European Alps (*Betula pubescens* ssp. *carpatica*), in the northwestern Cantabrian Mountains (*Betula verrucosa*) and in western Scotland (*Betula pubescens*) (e.g., Troll, 1939; Schweinfurth, 1957; Ern, 1966; Haffner, 1972; Kelletat, 1972; Zimina, 1973; Dolukhanov, 1978; Grebenchikov, 1978; Clapham et al., 1989; Miehe, 1991; Jacobsen and Schickhoff, 1995; Schickhoff, 1996).

In most of these regions birch is admixed as secondary tree species to the timberline forming conifers. In North America, birch occurs at timberline in a few places only, as for example on Mt. Washington (New Hampshire), where *Betula balsamifera* forms the upper timberline together with *Abies balsamea* and *Picea mariana* (Marchand and Chabot, 1978). Because of its high flexibility, birch is able to tolerate a heavy snow load and it may also sprout from its base after the main stems have been destroyed. Thus birch may colonize avalanche chutes in the same way as alders or willows for example (see below). Troll (1939, 1972, 1973), for example, has described such birch stands from the Nanga Parbat area (cf. Figure 35, see also Photo 40). The birches display decumbent growth with stems trailing downslope along the ground and rising at the tip. In the Canadian Rocky Mountains closed dense stands of birches (*Betula glandulosa*) mixed with willows (*Salix glauca*) locally occur in the upper part of avalanche endangered slopes (Johnson, 1987). Also, in the Alps small birch groves (*Betula pubescens*) can be found at such sites where they substitute for the high-stemmed coniferous forest, which is not able to survive there longer due to breakage (Photo 17, see also Holtmeier, 1965, 1967b, 1974). It is this competitive advantage that allows birch to develop persistent communities at such places (Hibsch-Jetter, 1994).

Shrub-like *Alnus* species such as green alder (*Alnus viridis*) in the central Alps or Sitka alder (*Alnus sitchensis* = *Alnus sinuata*) in the northern Rocky Mountains may form dense thickets above the forest limit or in avalanche chutes (Photos 18, 43 and 115; Figure 38). Both species are highly adapted to being repeatedly damaged by avalanches or sliding snow. Snow masses usually bend the flexible stems downslope, which will rise again to a more upright position after the snow has gone. Moreover, alders may ‘repair’ breakage by thriving basal sprouts and thus may survive at such sites where they replace the high-stemmed coniferous forest. Therefore, they can be considered a substitute formation at such sites, like prostrate mountain pine or birch. In Alaska, dense alder thickets locally form a scrub zone above the closed mountain forests (Hämet-Ahti, 1979; Arno, 1984). Also, varying mosaics of alder thickets and open patches covered by herbs and grasses locally occur above the forest limit (Mitchell, 1968). Alder thickets must be considered to be the only real krummholz (genetically predetermined scrub-like growth; Chapter 3) at the upper timberline in North American high mountains, unless also willow scrub (*Salix planifolia*, *Salix bebbiana*, *Salix brachycarpa*, *Salix glauca*) will be included. In Kamchatka *Alnus kamtschatica*

(*Alnus crispa*) forms a krummholz zone, partly together with *Pinus pumila*, above the upper limit of the birch forest that is located between 300 to 400 m elevation. The alder krummholz extends up to an elevation of 800 m (Hultèn, 1974; Grishin, 1995; Grishin et al., 1996a, b). The habitus of *Alnus*

Photo 17. Birches (*Betula pubescens*) in an avalanche chute on the east-facing slope of the Lüsenzer Valley (Tyrol, Austria). F.-K. Holtmeier, 24 April 1981.

Photo 18. Green alder (*Alnus viridis*) in an avalanche chute on Birgitzköpfl (Axamer Lizum, Tyrol) at about 1.920 m. F.-K. Holtmeier, 22 April 1981.

kamtchatika is the same as displayed by the other krummholz alders. The height of the alders decreases from about 3 to 4 m in the lower part of the krummholz zone to less than a few decametres at its upper limit where they grow mat-like close to the ground. In winter big snow masses pile up. They press the stems downhill and compress the alder thickets to less than 2 m height. Thus, they, and also the dwarf pines, get completely buried by snow (Hultèn, 1974). In the oceanic environment of the northern Kurils Siberian alder (*Alnus fruticosa*) and willow scrub (*Salix* spp.) form the timberline while *Pinus pumila* is found in the more continental regions (Crawford, 2008).

Quaking aspen (*Populus tremuloides*) is widespread in North America. In some areas it reaches even the northern tree line. At only a few places, this species can however be found at the upper timberline, as for instance on the Steens Mountains in Oregon (Faegri, 1966; Price, 1978) and some other isolated mountain ranges in the basin-range province of Nevada (Critchfield

Photo 19. Aspen (*Populus tremuloides*) in an avalanche chute on Berthoud Pass (Colorado Front Range). F.-K. Holtmeier, 7 September 1977.

Photo 20. Abrupt timberline formed by red beech (*Fagus sylvatica*) on the southern slope of Mt. Cusna (background) seen from Mt. Prado (Tusco-Emilian National Park, Northern Appenines, Italy). In this area, beech timberline is located between 1.650 and 1.770 m. G. Pezzi, 27 July 2004.

and Allenbaugh, 1969). However, in the timberline ecotone of the Rocky Mountains and some other mountain ranges of western North America, shrub-like growing aspen are represented, often occurring within or along avalanche chutes (Photo 19). They colonize such sites rapidly. It depends on the frequency of avalanches to which size the aspen will grow before being broken or thrown over by the snow masses. However, they usually survive by forming root suckers or thriving basal shoots. In the White Mountains (California), 1 to 2 m high aspen clumps reach the tree line. Elliott-Fisk and Peterson (1991) suppose the low growth forms ('dwarf aspen') to be genetically determined. In the Alaska Range, small stands of balsam poplar (*Populus balsamifera*) can be found 300 m above the upper spruce limit (Arno, 1966).

Poplars (*Populus suaveolens*) occur as pioneer trees on slopes covered by volcanic ash in Kamchatka (Grishin and Del Moral, 1996) and also on Mt. Fuji (Japan, Masuzawa, 1985). They usually form dense tree clumps that mainly originated from root suckers. Oaks are represented at some places at the timberline in the Caucasus (*Quercus macranthera*), in the Sierra Nevada and some other mountains ranges in southern Spain (*Quercus pyrenaica*; Ern, 1966; Troll, 1973). Arborescent *Rhododendron*-species are typical of the upper timberline in the eastern Himalayas (Schweinfurth, 1957, 1968; Haffner, 1972; Singh and Singh, 1987; Schmidt-Vogt, 1990a; Miehe, 1991).

Fagus sylvatica here and there form the upper timberline in the outer ranges of the southern and north-western European Alps, in the western Vosges (cf. Photo 99; Carbiener, 1969; Wilmanns, 1973; Ellenberg, 1978; Pott, 1992, 1993), in the Riesengebirge (Fanta, 1981), in the western Pyrenees (Plesnik, 1972), in Corsica (Pott, 1993) and Sicily (Rikli, 1946), in the outer ranges of the Dinarides (Koch, 1909; Fukarek, 1970; Lakusic, 1970; Puncer and Zupančič, 1970; Plesnik, 1972, Horvat et al., 1974), in the eastern Cantabrian Mountains, in the Apennines (Photo 20; Pezzi et al., 2007), in the Pontian Mountains and in the high mountains of Macedonia (Em, 1970; Nikolovski, 1970; Schreiber, 1998). In the western Caucasian mountains, *Fagus sylvatica* is replaced by *Fagus orientalis* (Plesnik, 1972; Grebenchikov, 1978; Armand, 1992).

4.1.3 Tree species at timberlines in the southern hemisphere and in the tropics

The situation in the southern hemisphere is totally different. Africa, Australia and New Zealand are more than 3.000 km away from the contiguous high-mountain ranges of the world. This isolated position resulted in a comparatively great variety of tree species at timberline that cannot be described in detail here.

Timberlines are mainly formed by evergreen broad-leaved trees and also by conifers native to the southern hemisphere. In New Guinea, for example, 13 tree species and tall shrub species occur at the upper timberline. They belong to at least five genera, among them *Podocarpus*, *Dacrycarpus*, *Quintinia*, *Rhododendron*, *Vaccinium*, *Rapanea*, *Olearia* and others (Wardle, 1971, 1974; Hope, 1976; Löffler, 1979; Corlett, 1984). On the other hand, tropical timberline may also by formed by one species only, such as *Podocarpus compactus* and *Libocedrus papuana* in New Guinea, *Erica arborea*, *Hagenia abyssinca* and *Hagenia leucoptychoides* in East Africa or representatives of the *Rosacea*-genus *Polylepis* in the South American Andes.

In Australia, the upper timberline is formed by *Eucalyptus pauciflora* (Costin, 1981), in Tasmania by *Eucalyptus coccifera*, deciduous *Nothofagus gunnii*, by *Athrotaxis*, and by the shrub-like conifers *Microcachrys*, *Phaerosphera* and *Diselma* (Wardle, 1973, 1974). In New Zealand, evergreen *Nothofagus solandri* (mountain beech) and *Nothofagus menziesii* (silver beech) are the main timberline forming tree species. *Nothofagus solandri* is typical of the more 'continental' regions of the south island, while *Nothofagus menziesii* occurs on the west side of the north island. Above the upper forest limit, dense subalpine scrub occurs formed by *Podocarpus*, *Dacrydium*, *Hoheria*, *Senecio*, *Olearia* and some other species. From the physiognomic and ecological point of view this subalpine scrub that can be found up to 100 m and even 300 m above the closed forest (Wardle, 1965b, 1971, 1974, 1980; Sakai et al., 1979) may be considered the equivalent to the mountain pine belt (true krummholz) of the Alps and the Carpathian Mountains, for example, or to the willow thickets of many North American high mountains.

In Central and East Africa, *Hypericum* (*Hypericum leucoptychoides*, *Hypericum revolutum* and *Hypericum lanceolatum*, *Philippia keniensis* and *Hagenia abyssinica*) form the upper timberline at elevations between 3.400 and 3.500 m. Oliver (1988), by the way, attributes *Philippia* to the genus *Erica*. Several studies (Hedberg, 1951; Troll, 1973; Plesnik, 1980; Miehe and Miehe, 1994, 1996) provide some evidence that the upper limit of the *Hagenia*-forests are caused by fire and grazing rather than by climate and should not considered a natural altitudinal timberline. However, Lange et al. (1997), referring to their studies on Mt. Kenya, consider burning to be of local importance only and take the upper limit of *Hagenia* forest for natural (climatically caused; see also Fries and Fries, 1948; Bussmann, 1994). Fires, usually starting in the dry heath and in the moorland above the forest belt, are reported to spread downhill. That, however, would be in contrast to the normal behaviour of fires that are usually driven upslope by upslope winds (Section 4.3.14; see also Laegaard, 1992). Locally, stands of arboreous *Erica* (*Erica arborea*, *Erica trimera*, *Erica keniensis*) advance beyond the *Hagenia*-belt to above 4.000 m elevation (Hauman, 1933; Ross, 1955; Miehe and Miehe, 1994). Towards the altitudinal limit of *Erica*, tree height gradually decreases, and finally the plants display shrub-like growth. Some authors include the ericaceous belt in the alpine zone (Fries and Fries, 1948; Klötzli, 1958; Coe, 1967; Walter and Breckle, 1984). In the Bale Mountains in southern Ethiopia, Miehe and Miehe (1996) divide the ericaceous belt in an '*Erica*-dwarf forest', with additional *Hypericum* and *Rapanea*, that merges into an '*Erica*-dwarf forest' with only one 'tree'-layer and into shrub-like *Erica* individuals that form the upper boundary of the ericaceous belt in this area.

Giant groundsels (Giant leaf rosettes; *Senecio*, *Dendrosenecio*; cf. Photo 25) up to 9 m tall occur even at 4,000 to 4,700 m altitude where they form forest-like stands (Fries and Fries, 1948, 1952; Hedberg, 1951, 1964; Klötzli, 1958; Coe, 1967; Weinert, 1985; Rehder et al., 1988; Beck, 1990; Miehe and Miehe, 1994, 1996; Fischer, 1996). They are the old-world counterparts to *Espeletia* in the tropical Andes (cf. Photo 26). However, it is a question of definition whether these trunk-forming composites will be called 'trees' (Hedberg, 1951, 1964; Hermes, 1955; Klötzli, 1958; Troll, 1959; Schmithüsen, 1968; Müller-Dombois and Ellenberg, 1974; Rauh, 1988; Wardle, 1993; Richter, 2001). The high-stemmed growth form, which makes a striking difference to the 'afro-alpine' vegetation, seems to indicate that they should be included. In other respects, however, one should refrain from considering these giant leaf rosettes being 'trees' (Richter, 2001), in particular with respect to comparing to tree growth forms at timberline and tree line in higher latitudes of the northern hemisphere (Hedberg, 1964). If these trunk-forming senecios were considered to be trees, an altitudinal shift of timberline and tree line in East Africa for about 800 or 400 m beyond the present upper limit of *Erica*-stands respectively would be the result.

In the Andes, which extend from the tropics to the cool-temperate southern latitudes, upper timberlines are formed by different species. Extremely species-rich timberlines occur at comparatively low elevation (about 3.400 m) in Costa Rica and western Panama. Another species-rich 'low neotropical timberline' (sensu Richter et al., 2008) is located between 3°S and 7°S in the Cordillera Real (Ecuador) within the Andean depression, where the *Polylepis* belt is missing except for a few stands in the southern part of this area (Baumann, 1988). In total 39 tree species are found (Richter et al., 2008). In the central cordillera of Columbia, dwarf forests (Salomons, 1986) consisting of *Gynoxis*, *Hesperomeles*, *Miconia* and some other species are to be found. The uppermost forest patches are located between 3.900 m (eastern slope) and 3.950 m elevation (western slope), solitary *Escallonia myrtilloides* may occur even a little higher. Four to five metres high groves of *Polylepis sericea* can be found at sheltered sites up to 4.300 m altitude. In the eastern cordillera of Quito, groves of *Polylepis pauta* (east side) and *Polylepis incana* (west slope) advance up to 4.100 to 4.200 m elevation (communication M.D. Rafiqpoor). Between Mexico and Tucuman (northern Argentina) deciduous *Alnus acuminata* (=*Alnus jorullensis*) or *Polylepis* are the timberline species (Hueck, 1953, 1966; Leonardis, 1976). *Polylepis* is a genus endemic to South America. It includes at least 15 species (Simpson, 1979). Kessler (1995) mentions even 20 species comprising eight subspecies (Kessler, 1995). Anyway, now as ever the taxonomy is controversial (Seibert and Menhofer, 1991). *Polylepis* ranges, with some gaps, from the Venezuelan Andes to Cordoba in Argentina (Baumann, 1988). Some species form small stands above the closed mountain forest. Seibert and Menhofer (1991) speak even of a particular forest belt surrounded by andine grass and shrub vegetation. In Bolivia, groves of *Polylepis tarapacana* (=*Polylepis tomentella*; Photo 21) are found at an altitude above 5.000 m (5.200 m; Simpson, 1979; Salgado-Labouriau, 1984; Baumann, 1988; Seibert and Menhofer, 1991; Goldstein et al., 1994; communication H. Ellenberg). Jordan (1996) reports *Polylepis* growing at 5.400 m. Troll considers *Polylepis* to be the old-world equivalent to *Hagenia abyssinica*. According to Hueck (1966) *Polylepis* shrub resembles European krummholz (*Pinus mugo* and *Alnus viridis*) in many respects. Some other wooden species are admixed to the *Polylepis* groves, for example representatives of the genera *Gynoxis*, *Hesperomeles* and *Escallonia*. *Gynoxis* and *Polylepis* are considered the highest climbing trunk-forming wooden plants at all (Baumann, 1988).

Photo 21. Open stands of *Polylepis tarapacana* (about 2 m high) on Sajama volcano (Bolivia). A field layer is missing, possibly due to lack of moisture. M. Y. Bader, 23 August 2003.

Some authors suppose the scattered occurrences of *Polylepis* to be dependent on comparatively favourable soil-climatic conditions (Troll, 1959; Koepke, 1961; Walter and Medina, 1969; Lauer and Klaus, 1975a; Lauer, 1979a, 1986; Walter and Breckle, 1984; Rauh, 1988). Ellenberg (1958, 1959, 1966), however, objected to this hypothesis since he had found such isolated groves on completely different substrates ranging from loamy to coarse material such as debris and boulders (Ellenberg, 1975). Also, he could not confirm that, as a rule, *Polylepis* groves occur mainly on special exposures, convex topography and mountain slopes or on wide valley bottoms and in gorges respectively (in contrast to Troll, 1959, 1973). Thus, he explained the *Polylepis* stands high above the present upper limit of closed forest to be relics of a former closely forested belt. Also Laegaard (1992), who studied the grass Páramo of Ecuador, could not find any dependence of the distribution pattern of *Polylepis* stands on specific microclimatic or soil conditions that could be taken as a rule (Photo 22).

In a recent study of the mosaic-like occurrence of shrubby relics of *Polylepis* in the grass Páramo of Papallacta (3.700 to 4.100 m; Ecuador) also Lauer (2000) mentions that soil conditions are not that different in this area to explain this distribution pattern. This is contrary to the earlier assumption of the same author (see also Lauer and Erlenbach, 1987; Lauer and Rafiqpoor,

Photo 22. Polylepis pauta-stands in the timberline ecotone on the east slope of the eastern cordillera of Quito (Páramo de Papallacta) between 4.100 and 4.200 m. The distribution of *Polylepis* stands does not seem to follow any rule, except that most topographic convexities are treeless. M. D. Rafiqpoor, April 1997.

2000), who now considers the isolated (solitary) stands of *Polylepis* to be the remains of forest belt that had existed during the postglacial climate optimum (7000 to 3000 BP). In Venezuela, on the other hand, palynological studies of lake sediments did not give any evidence of the existence of such a continuous forest belt dominated by *Polylepis* (Salgado-Labouriau, 1984). Local changes that became obvious from pollen analyses are ascribed to climatic change. Also, Di Pasquale et al. (2007) did not find evidence of former forests above the present timberline.

Ellenberg (1958, 1959) was the first to suppose that the high altitude forests disappeared mainly because of burning and grazing. The potential upper limit of continuous *Polylepis* forest could be identical with the present location of the relic stands: in the eastern ranges at about 4.100 m, in the western area at 4.350 m (Laegaard, 1992). That would mean that wide areas of the present Páramo and Puna are ancient forest land (e.g., Laegaard, 1992; Lauer and Rafiqpoor, 2000; Lauer et al., 2001; Bendix and Rafiqpoor, 2001; see also Figure 5). This hypothesis has not remained uncontradicted. Lack of evidence for this supposition and the climatically sheltered positions of the 'relic' *Polylepis* stands have led to an opposite interpretation as well as to a more differentiated view of Páramo history (e.g. Salgado-Labouriau, 1984; Di Pasquale et al., 2007; Cierjacks et al., 2008; see also Section 4.3.14).

Salgado-Labouriau (1984), for example, explicitly excludes any human impact. Bader et al. (2007, 2008b) suggest abrupt treelines in the tropical Andes to be caused at least partly by contrasting microclimates of forest and Páramo. This, however, is difficult to substantiate by direct evidence as natural tropical treelines that with certainty remained unaffected by humans are rare, if they really exist at all.

From 35°30′ to 55°S (Tierra del Fuego) mainly *Nothofagus pumilio* ('lenga') forms the upper timberline ecotone decreasing in altitude from approximately 2.000 m in the north (Alto de Vilches) to about 600 m in the south (Isla Navarino). Close to its upper limit 'lenga' exhibits stunted growth (Kalela, 1941a; McQueen, 1976, 1977; Hueck and Seibert, 1981; Hildebrand-Vogel et al., 1990; Vogel, 1996; Wardle, 1998, 2007; Barrera et al., 2000; Pollmann and Hildebrand, 2005). *Nothofagus antarctica* ('nirre') displaying also 'krummholz' growth forms is admixed to these stands. Both *Nothofagus* species shed their leaves in winter. 'Krummholz'-like growth forms seem to be more common at the timberline in the southern Andes than at *Nothofagus* timberline in New Zealand (Wardle, 1973, 1981b, 1989). This may be attributed to the more severe climate in the southern Andes being cooler throughout the year compared to the New Zealand Alps. Moreover, the southern Andes are much richer in snow, and winds are stronger and colder (Wardle, 1981b, 1998). On Sierra Nevada (38°37′) *Araucaria araucana* is widely distributed at the upper timberline but does not extent far beyond 40°S (Wardle, 1998). *Araucaria* retaining upright growth and umbrella-shaped crowns projects above 'nirre' and 'lenga' scrub (Photo 23). In the highly maritime mountains south of 41°30′S, which are deeply dissected by fjords and channels, thickets of stunted and crooked *Nothofagus antarctica* extend above the evergreen *Nothofagus dombeyi* forest that grows at lower elevations (Oberdorfer, 1960). Arborescent *Nothofagus pumilio* form the upper timberline here and there on the leeward side of the western mountains (Wardle, 1998). The *Nothofagus pumilio* timberline ecotone in the southern Andes is still less influenced by human impact than most timberlines in northern hemisphere (Pollmann and Hildebrand, 2005).

From the climatic-ecological aspect the timberline regions of southern South America and of the South Island of New Zealand may be compared to a certain extent and with great reservation to the climatic upper timberlines in the northern hemisphere. Some authors consider the deciduous Chilean *Nothofagus* forests to be the equivalent of the subarctic, subalpine and sub-maritime birch forests of the northern hemisphere (Skottsberg, 1910, 1916, 1960; Kalela, 1941a, b; Godley, 1960; Oberdorfer, 1960; Hämet-Ahti, 1963). In any case, structure and physiognomy of the *Nothofagus pumilio* timberline seem to be more similar to the northern hemisphere *Betula* and *Fagus* timberlines than to the *Nothofagus timberlines* in New Zealand (Pollmann and Hildebrand (2005).

Photo 23. Araucaria araucana stand at timberline (near Villarica, Chile) at 1.450 m. M. Richter.

The Falkland Islands and most of the other subantarctic islands are treeless, except for Ile Amsterdam (*Phylicia nitida*), the Auckland Islands (*Metrosideros lucida, Dracophyllum longifolium, Panax simplex*) and New Zealand (Brockmann-Jerosch, 1928; Tuhkanen, 1992, 1993). On Gough Island (Wace, 1961) and locally on Tristan da Cunha (Wace and Holdgate, 1958) *Phylicia nitida* forms dense thickets of 2 to 3 m height ('bush'). Solitary trees occur on wind-exposed ridges and slopes while no trees can be found on elevations above 450 m.

Little attention has been paid to the timberlines on the smaller oceanic islands in the tropical, subtropical and warm-temperate zones, such as the Azores, the Canary Islands, Madeira, Réunion, Grande Comore, Maui and Hawaii, Cape Verde Islands, Juan Fernandez Islands and others (Leuschner and Schulte, 1991; Leuschner, 1996, 1998). Compared to continental mountain ranges located in the same geographical latitude, these islands are characterized by high oceanic climates and isolated position, far off the continents. On most of these islands timberline is formed by one species only, as for example, on the Canary Islands by the endemic *Pinus canariensis*, on the Azores by *Erica azorica*, on Grand Comore by *Philippia comorensis*, on Yukushima (South Japan) by *Cryptomeria japonica*, on the great Hawaiian shield volcanoes by *Metrosideros polymorpha* or *Sophora chrysophylla* and on Tristan da Cunha (South Atlantic Ocean) by *Phylicia arborea*. In contrast, timberline on Réunion, Juan Fernandez Islands and on Madeira is formed by several tree species although one species usually prevails, as *Philippia montana* for instance on Réunion (Leuschner, 1996).

4.2 Relationship of timberline elevation to macroclimate, climate character, and the mass-elevation effect

In general, the upper timberline is caused by heat deficiency. It is highest in the subtropical mountains and drops from there to the equator and to the high latitudes as has been demonstrated by latitudinal transects (Figures 5 and 7).

In the high latitudes, the upper timberline is located close to sea level and is identical with the latitudinal polar timberline in some areas. In Eurasia, the upper timberline reaches its highest position in eastern and southern Tibet. In eastern Tibet, closed spruce forests occur on shaded mountain slopes up to an elevation of 4.600 m. Solitary trees may be found even above 4.700 m altitude (Schäfer, 1938; Von Wissmann, 1961; Li, 1993). In southern Tibet (Xizang), the uppermost juniper trees (*Juniperus tibetica*) occur even at an altitude of 4.800 m (Miehe et al., 2000, 2003). In Bolivia, however, *Polylepis* climbs even higher to 5.200 m and even to 5.400 m, making the highest tree line in the world. The climatic upper timberline in the tropics would be higher if humans had not removed the original forests to a large extent from elevations that then were colonized by Páramo and Puna vegetation. Thus, the latitudinal transect of vegetation zones presented by Troll (1948b, 1961) and adapted by many other authors (Walter, 1964; Schmithüsen, 1968; Schmidt, 1969; Price, 1981; Rathjens, 1982; Bramer, 1985) has been corrected according to Ellenberg (1996; cf. Figure 5). Also, the climatic timberline in East

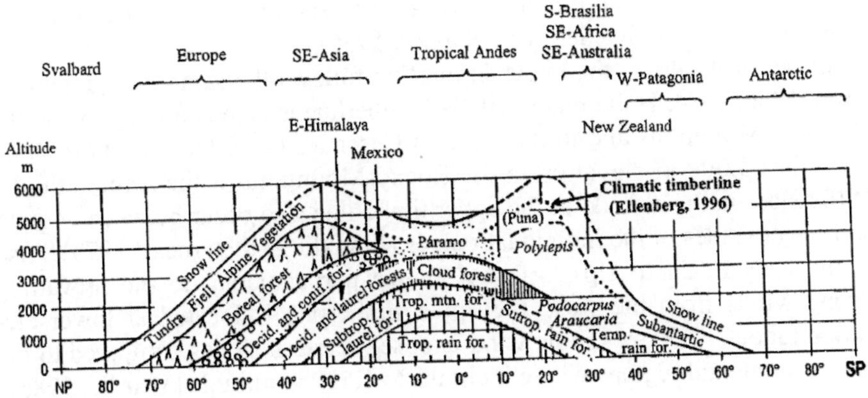

Figure 5. Meridional transect showing the altitudinal vegetation belts and snow line (after Troll, 1948). The altitudinal position of the climatic timberline is corrected by the author (after Ellenberg, 1996).

Africa is supposed to be located about 500 m above the present authropogenic timberline (Miehe and Miehe, 2000). Altogether, the location of the upper tropical timberline appears to be affected by human activities at least as much, if not to a greater extent, as the upper timberline in the temperate and boreal mountains (Section 4.3.14).

Some authors related the altitude of the climatic timberline to geographical latitude and calculated 'timberline gradients'. According to Daubenmire (1954), for example, timberline drops north for about 110 m per degree of latitude. However, such empirical gradients may be quite different because of the geographical location of the longitudinal (north–south oriented) transects they are related to. For example, the gradient between the northern border of Oregon and northern California is 184 m (after data from Arno, 1984), in the Appalachians 83 m (Cogbill and White, 1991), in the Ural Mountains 71 m, in Middle Siberia 89 m and in East Siberia 76 m per degree northern latitude (Malyshev, 1993). In the southern Andes between 35°S and 55°S, timberline declines south for about 80 m per degree of latitude (Crawford, 1989).

Timberline is usually higher in continental climates than in maritime regions. In Europe, upper timberline rises from 100 m near the west coast of Ireland to 700 m elevation in Wales up to 1.400 m in the Riesengebirge and 1.700 m in the High Tatras. In northern Norway timberline occurs at about 200 m altitude in the coastal region of Troms (almost 70°N), at about 700 m in inner Troms, and at 500 to 700 m elevation in more continental Swedish Lapland farther east. On the so-called 'high fjelds' in western Finnish Lapland (cf. Holtmeier, 1974), timberline is located at about 400 m to almost 600 m elevation. From there timberline gradually drops northward to less than 400 m elevation in northernmost Finnish Lapland. In southern Norway, timberline rises from 300 to 500 m in the fjordland to about 1.200 to 1.300 m altitude in the high fjelds of Jotunheimen (Figure 6; see also Aas, 1969; Aas and Faarlund, 2001; for Sweden see Kilander, 1965).

In North America, the upper timberline follows the same rule (Figure 7; Arno, 1966, 1984; Höllermann, 1980). Timberline is comparatively low in the Olympic Mountains and in the western Cascades (1.500 to 1.800 m) while it reaches 3.000 m elevation in the Rocky Mountains at the same latitude. Differences in the altitude of the upper timberline also occur between western and eastern sides of the individual ranges (Franklin and Dyrness, 1973). Also in the New Zealand Alps, oriented almost perpendicular to the prevailing strong winds from the west, the upper timberline is located at lower elevation (about 1.200 m) on the wind-exposed western slopes compared to the more continental regions where it climbs to 1.500 m altitude (Wardle, 1986).

The same principle can be observed in the southern Andes (Wardle, 1998). In Kamchatka, strong maritime influences from the south with big snow masses in winter causes a depression of the upper forest and krummholz limit (*Alnus kamtschatica*) in the southern part of the peninsula (Hultèn, 1974; Grishin, 1995; Grishin et al., 1996b).

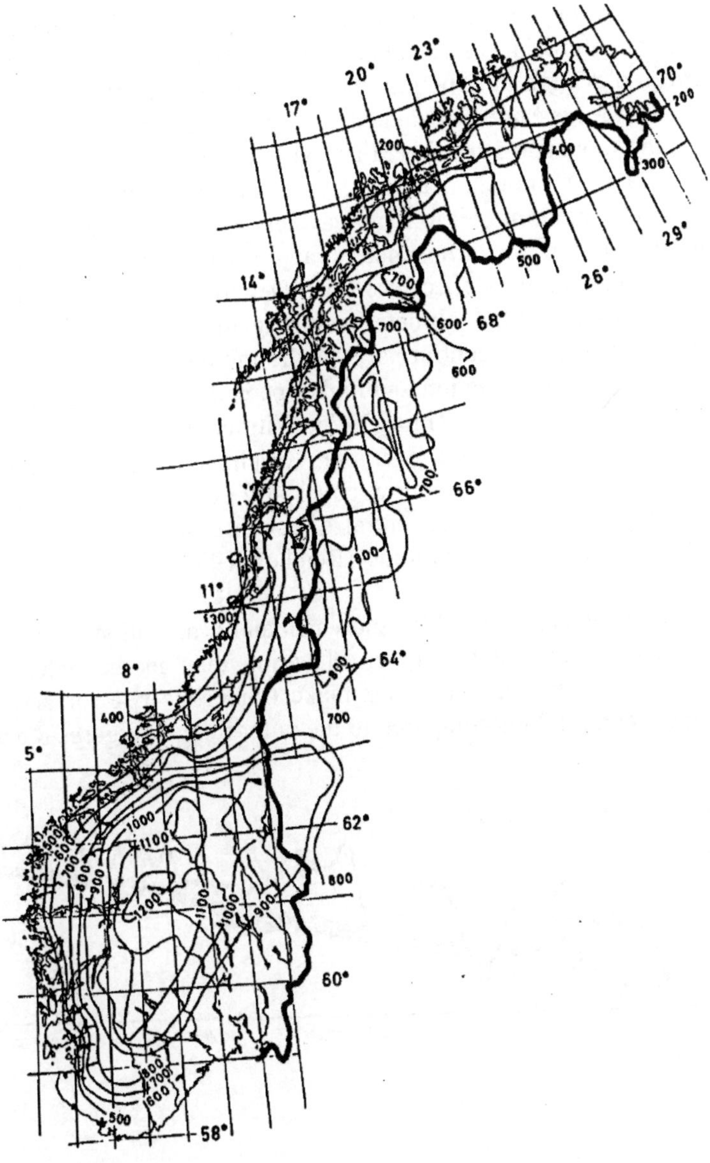

Figure 6. Altitudinal position oft he upper birch-forest limit (*Betula tortuosa*) in Norway. The maximum altitudinal timberline position is reached in the high fjelds of Jotunheimen (in the south) due to the combined timberline-raising effect of relatively southern latitude and great mass-elevation. After Aas (1964) in Aas and Faarlund (1996).

Additionally, large mountain massifs have a positive effect on the altitudinal position of the timberline (e.g., Tollner, 1949). The term 'mass-elevation' refers to the mean elevation of a mountain massif. For explanation: mean elevation of a mountain mass can be calculated by transforming it into a plateau without changing the mountain's basal area and volume. Terrain elevation is spatially averaged. Larger mountain massifs serve as a heating surface absorbing solar radiation and transforming it to long-wave energy. Consequently, temperature is higher than in the free atmosphere at any given elevation ('mass-elevation effect' or 'mountain-mass effect'; see also De Quervain, 1904). However, this effect is combined with a climate becoming increasingly continental from the mountain rim to the central parts of the mountain area, which are characterized by lower precipitation and higher percentages of sunshine compared to the outer mountain ranges (Fliri, 1975; Witmer et al., 1986).

In the northern Alps, for example, which are strongly influenced by moisture carrying air masses, snow cover at 2.000 m level lasts usually for about 280 days, whereas in the central Alps winter snow does not stay longer than 200 days at the same altitude (Reichel, 1931; Turner, 1961; Witmer et al., 1986).

Because less energy is used for snowmelt and evaporation, growing season is longer and warmer at any given elevation than in the outer mountain ranges. These favourable conditions make timberline rise for about 400 m higher in the central Alps compared to the outer ranges (Figures 8 and 9).

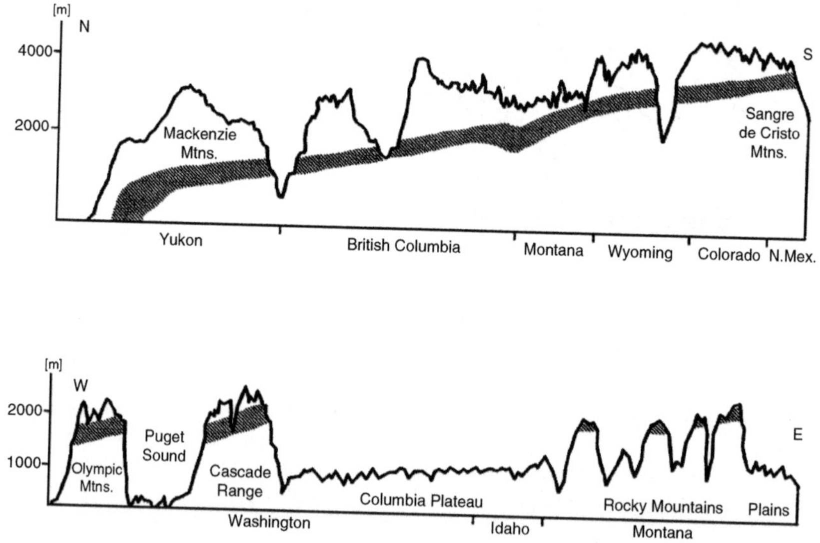

Figure 7. North–south (3.700 km) and west–east (965 km) transect showing the altitudinal position of the upper timberline in North America. While the rise of timberline from the Yukon area to the southern Rocky Mountains is due to latitude, the rise eastward results from increasingly continental climate. Modified from Arno (1984).

In mountains such as the Alps, which are exposed to air currents from all directions, it seems hard to separate the timberline-rising effects of continentality and mass-elevation because they both increase from the mountain rim to the interior part. However, the relatively continental climate (Gams, 1931) is of primary importance since it is the comparatively low cloudiness, frequent sunshine and little loss of energy through evaporation and snow melt that cause the heating effect of the elevated mountain surface (Turner, 1961). Gams (1931) has introduced the term of 'hygric continentality'. He showed that the distribution of the most important forest types in the Alps is related to the ratio of the area's altitude (m) to its annual precipitation (mm). The degree of 'hygric continentality' may be expressed as $\tan^{-1}$ which is 45° when altitude/precipitation = 1. When the angle is >45° the climate is considered hygric continental, whereas it is hygric maritime when the angle is <45° (Figure 10). Thus, for example, the distribution areas of larch-stone pine forests in the central Alps are related to high hygric continentality (65°, interior Ötztal; Austria). The red beech forests, on the other hand, are common to hygric maritime regions (about 31°). In a global view, the hygric continentality may reach 75° in the southern Rocky Mountains (Colorado Front Range), while it is only 15° to 20° in the high maritime regions of Scotland and Japan (Barry, 1992).

In the Pyrenees, the timberline-rising effects of continentality and great mass elevation, which both increase from the mountain rim to the interior parts of the mountain system overlap in a similar way, as was demonstrated above for the Alps and the Scandinavian mountains. Timberline gradually

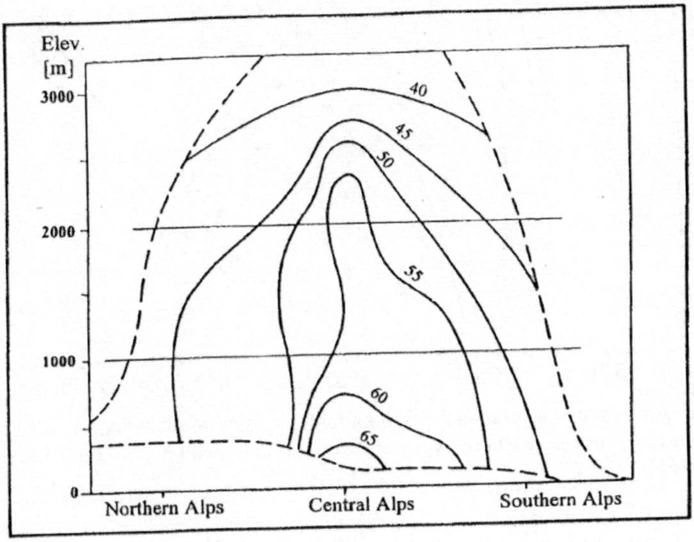

Figure 8. Mean annual duration of sunshine along a cross section through the Alps of Tyrol (Austria). Modified from Fliri (1975).

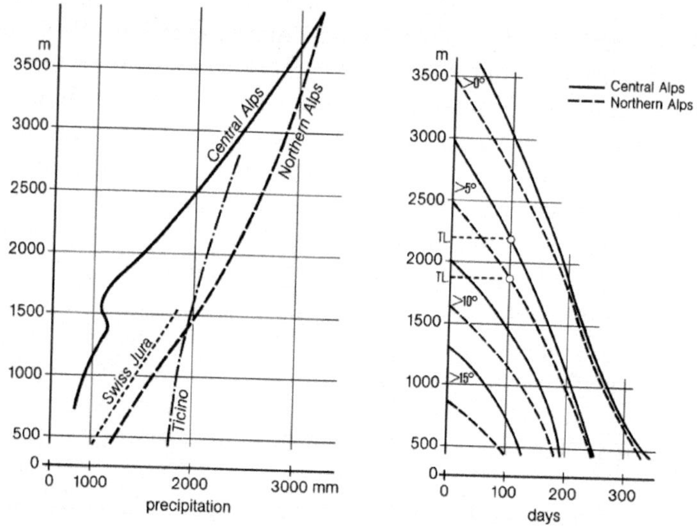

Figure 9. Annual precipitation (*left*) and number of days with mean temperatures above 0°C, 5°C, 10°C and 15°C (*right*) in the Central Alps and in the outer ranges. Modified from Ellenberg (1978).

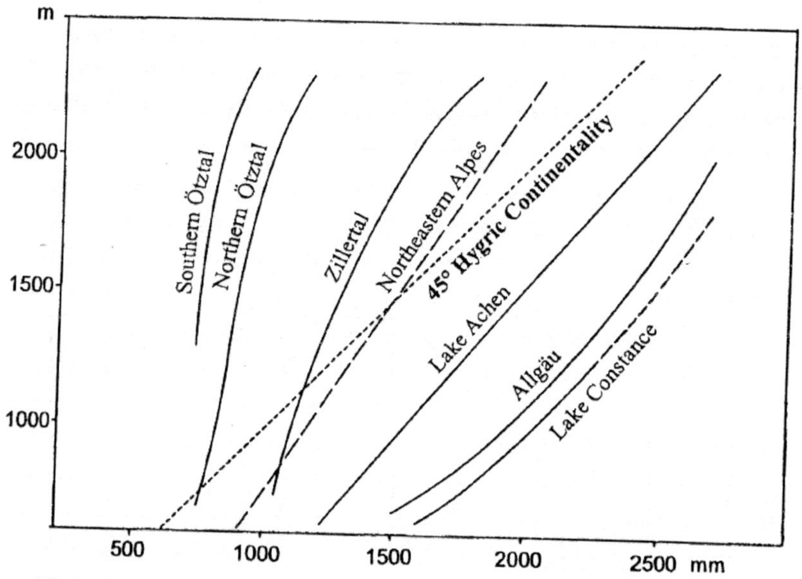

Figure 10. Amount of mean annual precipitation as related to elevation in some areas of the eastern Alps and in the northern Alps, and 'hygric continentality'. After Hader (1954), in Turner (1961).

ascends from the maritime west to the more continental east and also from the outer mountains towards the central parts. It reaches its highest altitude in eastern Andorra (almost 2.400 m) and in the Néouvielle-Group (2.300 m) which is fairly protected from advective influences (Figure 11; Höllermann, 1972). The same effect can be observed also in other climates. In the comparatively high and dry western cordillera of Bolivia, for example, the upper limit of the occurrences of *Polylepis tarapacana* (*P. tomentella*) is about 1.000 m higher than the upper limit of *Polylepis tarapacana* in the humid and lower eastern cordillera (Kessler, 1995).

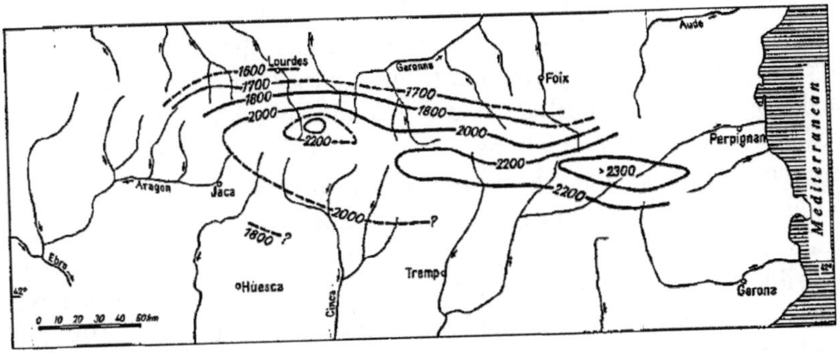

Figure 11. Altitude of timberline in the Pyrenees. Timberline reaches its highest position in the relatively continental central parts. Modified from Höllermann (1972).

In the tropics, the same relationship of the upper timberline and mass-elevation effect exists (Eyre, 1968; Hastenrath, 1968; Grubb, 1971; Van Steenis, 1972). For example, timberline rises from 3.800 to 3.900 m altitude in western Guatemala to over 4.000 m in the Mexican highland, counter to the general latitudinal gradient of air temperature. On Mt. Wilhelm (4.694 m, Central New Guinea), the upper timberline is higher (3.800 m) than on the smaller mountains in eastern New Guinea (McVean, 1973). Also in East Africa, the upper vegetation limits are located at higher altitude on the large volcanic mountain massifs compared to the lower ones (Hedberg, 1951; Schmitt, 1991). Though precipitation increases from south to north in the western Bolivian cordillera, the upper limit of the *Polylepis*-stands climbs higher in the north because of the greater mass-elevation of the northern mountain section (Kessler, 1995; Lauer and Rafiqpoor, 2000).

The situation on the remote mountainous oceanic islands is more difficult to assess. Remarkably, the upper timberline on some remote oceanic sub-tropical islands such as the Canary Islands, Hawaii, Tenerife, Madeira, New Caledonia and some others is located about 200 to 1.000 m lower compared to continental mountains at similar geographical latitude (Henning, 1974; Wardle, 1974; Höllermann, 1978; Leuschner and Schulte, 1991; Leuschner, 1996). This difference may be attributed to the oceanic climate and to aridity above the trade wind inversion. In respect to active volcanism on most of these islands, the relatively low position of the upper timberline might be supposed to be caused by volcanic eruptions and lava flows. However, when comparing these islands to each other and also to continental volcanic areas this hypothesis fails, as it becomes obvious that timberline is not necessarily lower on volcanoes than on other mountains (Leuschner, 1996). It is just on some inactive continental volcanoes, as in Mexico and Bolivia for example, that the upper timberline climbs to great elevation (Troll, 1959; Beaman, 1962; Klink et al., 1973; Lauer, 1973; Klink and Lauer, 1978).

Wardle (1971, 1973), on the other hand, believes that the low position of the timberline on these remote oceanic islands can be explained by the lack of tree species sufficiently adapted to high-elevation climate, as is the case with the species at timberline in continental mountains. His assumption is based on successful plantations of North American timberline forming conifer species above the upper limit of *Sophora chrysophylla* on Hawaii at 3.000 m elevation. Also, on some other islands, as for example on the Juan Fernandez Island and Tristan da Cunha, inadequate adaptation might be responsible for the low position of the timberline. On many other islands, however, this assumption can hardly be supported (Leuschner, 1996). Thus, the individual case requires a careful consideration of many factors that might be responsible for the altitudinal position of timberline. However, it will be hard to assess which factors are most important, because our knowledge of the local and regional conditions is still deficient. Recent studies on *Metrosideros polymorpha* on Mauna Loa (Cordell et al., 2000; cf. Section 4.3.3.2), for example, give some evidence that the limited frost resistance of this species may explain the comparatively low position of the timberline and would support the hypothesis of Wardle (1971, 1973).

At first sight, the varying altitudinal position of the upper timberline on the oceanic islands seems to be related to the mountain-mass, as, for example, timberline is considerably higher on the large Hawaiian shield volcanoes than on any other oceanic islands. However, when 15 remote islands were compared to each other no significant correlation between mountain-mass and timberline could be found. Also, the altitudinal differences in timberline of about 450 m between inactive Mauna Kea and neighboured active volcanoes Mauna Loa and Haleakala cannot be explained by climatic effects of different mountain-masses but must rather be ascribed to the influence of soil conditions. Pedogenesis and in particular substrate-depending soil moisture conditions appear to be factors controlling the upper timberline (Eggler, 1971; Henning, 1974; Leuschner, 1996; see also Section 4.3.8). The upper timberline (*Pinus canariensis*) on Pico de Teide (3.718 m) at an altitude of 2.000 and 2.300 m is considered to be very low compared to the latitudinal position of Tenerife (Höllermann, 1978). This may be attributed to volcanic events and also to human impact (over-utilization, cattle grazing, burning) (e.g., Höllermann, 1995). Moreover, some evidence is given that open stands of juniper (*Juniperus oxycedrus* spp. *grandifolia* = *Juniperus cedrus*) occurred above the present forest limit not that long ago. Such juniper groves can still be found close to timberline on La Palma (Canary Islands), Madeira and on the Azores (Höllermann, 1978).

The mass-elevation effect (=heating effect) on the regional climate is caused by large mountain masses only. However, also in the high latitudes, as on the isolated comparatively low fjelds of northern Lapland for example, timberline rises to greater elevations on the higher mountains than on the lower mountains (Hustich, 1937, 1942; Holtmeier, 1974). Because of the high latitudinal position north of the arctic circle, this cannot be attributed to a more or less pronounced heating effect, but is instead explained by the varying microtopographical conditions (convex topography alternating with concave topography), which provide better wind protection and more favourable thermal and moisture conditions (waterlogged sites excepted) for the trees at a given altitude on the high fjelds compared to the lower fjelds (Figure 12; Hustich, 1937; Holtmeier, 1974). In other words: the so-called 'Gipfelphänomen' (comparatively lower altitudinal vegetation limits on isolated summits; Scharfetter, 1938) is less pronounced on the higher fjelds.

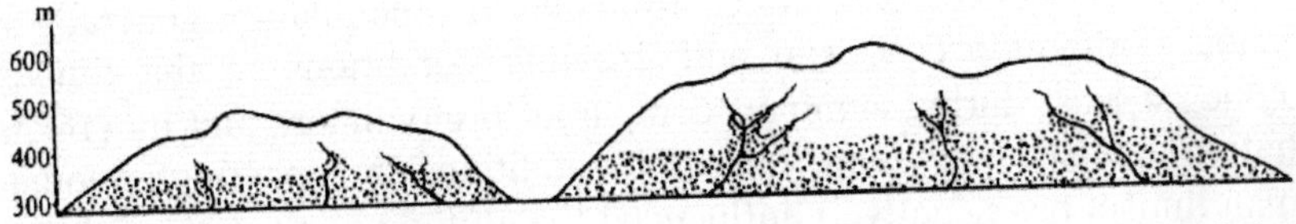

Figure 12. The effect of mass-elevation (schematic) on the altitudinal position of timberline on fjelds in northern Finnish Lapland.

Many big mountain massifs are characterized by a comparatively smooth topography exactly at the elevation where climate still allows tree growth. Thus, the ecotone may extend to greater altitudes compared to small and usually rugged mountain ranges where orographic influences (erosion, avalanches, land slides, etc.) prevent the forest from reaching its upper climatic limit. In the Colorado Front Range, for example, and also in other mountain ranges of the Rocky Mountains, broad timberline ecotones occur on old gently sculptured land surfaces up to 3.500 m elevation and locally even higher. In the central Alps, gentle topography is common to many mountain slopes at elevations between 2.000 to 2.400 m. Usually this gentle topography is identical with the trough shoulders of the glacially moulded valleys. If humans had not removed the forest for pastoral (Section 4.3.14) use the forest limit would reach its maximum altitudinal position in these sections.

4.3 Ecological conditions and processes at the timberlines

The timberline environment is characterized by site conditions often adverse to tree growth. It is generally accepted that heat deficiency is a keystone factor (e.g., Wilmanns, 1998). In mountains with long and more or less strong winters, the resistance of trees, in particular of seedlings and young growth, must be considered to be of primary importance to survival. However, even in a global view timberline cannot be attributed to thermal conditions only. In semiarid and arid climates, for example, low precipitation and also limited soil moisture supply may become a critical factor. Tree-rings from the timberline in high mountains of Greece, for example, provide evidence that tree growth in areas protected from moisture-carrying air masses is controlled more by limited moisture than by low temperatures during the growing season (Brandes, 2007; Brandes and Ise, 2007). On Tenerife, summer is often too dry for seedling survival (*Pinus canariensis*) at timberline. In the mild winter, on the other hand, seedlings and saplings suffer from frost because a protective snow cover is missing (Höllermann, 1978). On a microscale, however, lack of plant-available soil moisture may impede germination, seedling establishment and survival at timberline even at high latitudes more often than is usually expected at the given humid macroclimate.

Except for arid regions, direct influence of precipitation generally is of secondary importance to site and growing conditions in the timberline ecotone, at least during summer, compared to cloudiness and its effects on radiation balance and temperature. In addition, in most high mountains human impact has negatively influenced the altitudinal position of the upper timberline for hundreds or even thousands of years (Section 4.3.14).

4.3.1 Heat deficiency

Heat deficiency is a very complex factor and affects tree metabolism, growth, regeneration and survival in different ways: directly through low temperatures during the growing season, through shortness of the growing season, and particularly by extreme weather events such as frost in the growing season (e.g., Tranquillini, 1957, 1979a; Tranquillini et al., 1980; Wieser and Tausz, 2007). Heat deficiency can also be influenced by other site factors such as wind and duration of the winter snow cover and its effects on soil temperature, for example. Long-lasting snow cover and high soil moisture due to melt water keep the soil temperature low until early summer or even longer (e.g., Aulitzky, 1961; Turner et al., 1975; Neuwinger, 1980; Wolfsegger and Posch, 1980; Holtmeier, 1982, 1987b; Schreiber, 1998; Holtmeier and Broll, 1992; Müller, 1994; Schulte, 1994). On wind-exposed topography, air temperature, needle temperature, and soil temperature are usually low compared to better-protected sites and may impede regeneration and growth (Wardle, 1965b, 1968, 1993; Holtmeier, 1973, 1980, 1981b; Meurk, 1978; Tranquillini, 1979a; Dahms, 1992).

Tropical timberlines have long been supposed to be caused by permanently too low soil temperatures throughout the year (Walter and Medina, 1969; Walter, 1973; see Section 4.3.5). Recent observations, however, suggest that this hypothesis needs to be reconsidered (Miehe and Miehe, 1994; Bader et al., 2007). In general, we do not know much about physiological response of tree growth to high-altitude climate in the tropics. Instead of winter-hardening of newly formed tissues, physiological adaptation to permanent stress caused to the trees by the diurnal climate appears to be the controlling factor. Insufficient tolerance to excessive solar radiation, particularly when combined with low temperatures as they usually occur in the morning after cold nights, seems to be among the factors impeding seedling establishment in un-shaded open terrain above the closed forest (Bader et al., 2008b, 2008; Bader and Ruijten, 2008; see also Section 5.4).

The correspondence that has been found between the altitudinal position of timberline and mean temperatures of the growing season, air temperature sums, climate character, mass-elevation etc. reflect the controlling influence of heat deficiency (Holtmeier, 1974; Tuhkanen, 1980, 1993). However, mean temperatures do not exist in nature, and thus should be considered an indicator but not a causal factor. The same is true for mean annual soil temperatures. The variability of climatic conditions (late and/or early frost, drought, snow-rich or snow-poor winters, etc.) partly combined with burning or insect outbreaks are the decisive agents.

In a global view, the upper climatic timberline has been supposed to roughly coincide with the altitudinal position of the isotherm representing a mean temperature of the growing season ranging from 5°C to 7.5°C (Körner, 1998a, b, 2007b). However, if comparing mean temperatures at timberline, differences will become apparent.

At timberline in tropical mountains, for example, mean air temperature ranges between 5°C and 6°C throughout the year (Ellenberg, 1975; Lauer and Klaus, 1975a; Smith, 1975, 1980; Miehe and Miehe, 1994). The mean temperature of the warmest month at the timberline on subantarctic islands (Tuhkanen, 1992, 1993) and in the southern Andes (Wardle, 1968) is slightly higher (6°C to 7°C). Mean temperature at timberline near Ushuaia (southern Tierra del Fuego) was estimated at 5.7°C only (Tuhkanen, 1992). At the upper timberline in the southwestern part of New Zealand (Westland and Fjordland, Campbell Island) the mean temperature of the warmest month is said to exceed 8°C, while it is even about 10°C in other mountain regions of New Zealand (Zotov, 1938; Wardle, 1998).

In the continental high-mountains of the northern hemisphere the mean temperature of the warmest month ranges around 10°C (e.g., Brockmann-Jerosch, 1919; Daubenmire, 1954; Aulitzky, 1961), while it is up to 13°C at timberline in highly maritime mountains outside the tropics. Treeline on the Appalachian Mountains, for example, is correlated with a mean July temperature of 13°C (Cogbill and White, 1991). However, there are exceptions as on Mt. Fuji (Japan), for example. On the southern slope that had not been affected by the last eruption in 1707, forest has advanced to its upper climatic limit (about 2.800 m), which coincides with the 10°C-isotherm of the warmest month (Masuzawa, 1985).

Altogether, these mean temperatures differ too much to be considered appropriate indicators of the thermal conditions at upper timberlines. Temperature sums or the number of days with a minimum temperature of at least 5°C or the mean temperature of the 3 or 4 warmest months (tritherm, tetratherm) are more adequate. Helland (1912) already found that Scots pine no longer occurs in areas where the mean temperature for the period from June to September is below 8.4°C, nor does birch exist where this tetratherm falls below 7.5°C. In southern Norway, the upper birch-forest limit correlates best with the mean of the tritherm, which is 9.6°C on the west slope of the Scandes and 8.2°C on the eastern side (Aas, 1964; Aas and Faarlund, 1996). Odland (1996) again found that for the most part of Norway upper birch-forest limit correlates best to the mean maximum temperature of the warmest month (July, 15.8°C) or to the mean maximum temperature of the 3 warmest months (June through September, 13.2°C). Studies in northern Norway (Mook and Vorren, 1996) and in central southern Norway (Mork, 1968a) came to the same result (13.2°C and 13.3°C, respectively). The tetratherm (December through March) at the upper timberline in south-western New Zealand ranges from 9°C to 9.6°C and thus is 3°C to 4°C higher than at the upper timberline in Tierra del Fuego (Wardle, 1998).

Tuhkanen (1993) pointed out that the northern timberline in Canada does not better correlate with the mean maximum temperature of the warmest months than with the 10°C-isotherm of July. Above all, the position of this isotherm varies very much from year to year (Tuhkanen, 1980). Not least, it should be mentioned that the mean temperatures at timberline were usually not calculated from temperature recordings but were extrapolated from the nearest meteorological station (usually located in a valley or at the sea shore) to timberline elevation by using the saturated adiabatic laps rate.

Dahl and Mork (1959) found a close linear relationship between the growth of spruce (*Picea abies*) and the temperatures controlling dark respiration (respiration equivalent sums) of this species. This is supported by Skre (1972) who stated that the northern and upper distribution limit of spruce in Scandinavia better corresponds to the isolines of respiration sums than of any other factor. Referring to these observations Dahl (1986) concluded that too low temperatures negatively affect the production of ATP (essential to tree growth and development) and thus become a growth-limiting factor. This opened new prospects to the discussion on the effects of heat deficiency as the timberline-controlling factor.

Further attempts to correlate the altitudinal position of the upper timberline to any mean temperatures supposed to be crucial to tree growth will be disregarded here, because mean temperatures do not exist in nature and thus should not be considered a causal factor affecting climate or ecological conditions. Surprisingly, however, many authors still explain the position of the polar and/or upper climatic timberline to be caused by the position of the 10°C-isotherm of July (e.g., Costin, 1967; LaMarche, 1973; Plesnik, 1973; Ives and Hansen-Bristow, 1983; Pears, 1968; Grace, 1989; Thompson, 1990; Yanagimachi and Ohmori, 1991). If necessary, mean air temperatures may be used as indicators roughly describing the thermal conditions at the timberlines (Holtmeier, 1974; Tuhkanen, 1980, 1993; Ohsawa, 1990). The same is true for mean annual soil temperatures (Section 4.3.5).

4.3.2 Carbon balance, carbon limitation

It has been argued that the upper limit of tree growth might be caused by zero production of organic matter, which is impeded by the adverse climatic conditions at the timberline level. Additionally, the ratio of photosynthetically active leaf or needle mass and unproductive tissue is less favourable in trees than in dwarf shrub and other low vegetation at the same altitude and may negatively affect dry matter production (Boysen-Jensen, 1932, 1949; Ellenberg, 1975; Stevens and Fox, 1991; Slatyer and Noble, 1992; Körner, 1994; Cairns, 1998; Cairns and Malanson, 1998). In fact, carbon balance in bristlecone pines (*Pinus longaeva*) at timberline on the White Mountains (California) was occasionally negative during the winter (Schulze et al., 1967). Short-term studies (e.g. Schulze et al., 1967), however, do not allow predicting the long-term situation. Thus, Schulze et al. (1967) themselves pointed out that carbon loss during winter might be compensated in summer within 2 or 3 weeks at optimal photosynthesis. Dry matter loss in Swiss stone pine seedlings (*Pinus cembra*) in winter was estimated to be about one-eighth of the total weight of the plants. To compensate for this loss the seedlings needed to assimilate for 20 days after they had become snow-free (Tranquillini, 1979a). Wieser (1997), on the other hand, found that the carbon gain of a single day during the growing season can compensate for the total carbon lost by respiration during the 3 coldest winter months. The total carbon loss by respiration of a mature Swiss stone pine at timberline

(1.950 m) on Patscherkofel (south of Innsbruck, Austria) was calculated to be only 9% of the annual carbon gain (Wieser et al., 2005).

Occasionally, also at the northern timberline negative carbon balances were measured during the summer at overcast weather conditions and comparatively high night temperatures (Ungerson and Scherdin, 1968). But, neither there nor at the upper climatic timberlines were long-term negative or zero carbon balances measured that would determine the position of timberline (Tranquillini, 1959). Growth increment is obvious even in the extremely stunted outliers of tree growth in the upper timberline ecotone. In addition, abundant cone production may occur in these individuals from time to time (Photo 51), even though the seeds will for the most part not germinate (see also Figure 48). Also, the growth increment measured in *Nothofagus solandri* seedlings about 300 m beyond the mountain forest limit was far above the minimum required for seedling survival and does not correspond to the carbon hypothesis (Wardle, 1971). Young Swiss stone pines at timberline in the central Alps lost only about 33% carbon by respiration, mature larch 57% (Tranquillini, 2001). In general, the growth increment of the timberline forming tree species vary so much that an insufficient carbon balance and too low dry matter production are not very likely to be the factors causing timberline in general (Wardle, 1971).

On the other hand, from studies at the upper timberlines in the Olympic Mountains (Oregon) it became obvious that in seedlings of subalpine fir (*Abies lasiocarpa*) older than 3 years, the needle mass declines more at increasing dryness of the growing site than the water conductive capacity of the stems decrease. Thus, while the seedlings avoid drought stress by reduced transpiration, the carbon gain may be impeded by lower needle mass to such an extent that the young plants will not survive (Kuuluvainen et al., 1996). Also, in seedlings of subalpine fir and Engelmann spruce at the upper timberline in the Medicine Bow Mountains (Wyoming), negative carbon balances were occasionally observed during the day (Germino and Smith, 1995; see also Johnson et al., 2004). At sun-exposed sites, strong solar radiation impeded photosynthesis. A photosynthesis model predicted greater carbon acquisition during cloudy days (Johnson et al., 2004). However, carbon balance of shaded seedlings was even negative for a great part of the day. This had to be ascribed to insufficient water supply resulting from root competition and interception of light by the surrounding grasses and herbaceous vegetation.

Microsite facilitation (e.g., reduced sky and wind exposure; see also Photos 28, 68, and 89; Figures 55 and 63) was found to substantially increase photosynthetic carbon gain during the growing season resulting in enhanced root growth, amelioration of drought stress and seedling survival at the Rocky Mountain timberline (Wyoming). In contrast, seedlings unprotected by microsite facilitation usually died during the first or second year of growth. Seedling death was partly attributed to limited carbon uptake (Smith et al., 2003). Photooxidative stress and photoinhibition that occur at low temperature and excessive solar radiation loads in open patches are among the factors impeding carbon gain and thus carbon balance at both the temperate and the tropical timberlines (e.g., Wardle, 1965b; Ball et al., 1991; Germino and Smith, 1999; Johnson et al., 2004; Bader et al., 2008b; Tausz, 2007). Particularly after cold nights, situation may become critical to tree seedlings. Susceptibility to photooxidative stress and photoinhibition considerably differs between tree species (Germino and Smith, 1999; Tausz, 2007, there further literature).

From their investigations at the upper birch limit in the Torneträsk area (northern Sweden), Sveinbjörnsson et al. (1996) concluded that even a marginal carbon balance does not limit tree growth, though dry matter production may be drastically reduced from time to time by high needle loss and also through seed production. Also, the existence of the several thousand years old bristlecone pines (*Pinus longaeva*, *Pinus aristata*; see also Photos 69, 119) at the upper timberline in the White-Inyo Mountains (California), in the Snake Range (Nevada) and in the southern Colorado Front Range as well as the clonal groups (also called 'tree islands' in the following) of Engelmann spruce (*Picea engelmannii*) and subalpine fir (*Abies lasiocarpa*; Section 4.3.10.2) growing at timberline in many high-mountain ranges in North America stand in contrast to the hypothesis that insufficient carbon balance would determine the timberline. These trees and clones survived climatic periods with conditions far less favourable for a positive carbon balance than is the present climate (Ives, 1978; Arno, 1984; Holtmeier, 1985b, 1986a, 1995a, 1996). Also the fact that the CO_2 partial pressure (94%) at the very low maritime timberline in the Subarctic and Subantartic is twice as much as at the upper timberline in subtropical high mountains is not consistent with the carbon-balance hypothesis. Not least, in view of the large stands of trunk-forming giant groundsels (Section 4.1.3) some hundred metres above the real mountain forest limit in Ethiopia or on Mt. Kenya (Coe, 1967; Rehder et al., 1988 and others), for example, one can hardly imagine that the upper tree line is caused by an insufficient dry matter production. Miehe and Miehe (1996) did not even find stunted low growth forms of senecios or lobelias at their upper distribution limit in the Ethiopian mountains. Such growth forms, however, are common on Mt. Kenya (Rehder et al., 1988), and are typical of 'true trees' at their upper distribution limit in general. However, studies on the gas exchange and cold resistance of *Podocarpus oleifolius* and *Espeletia neriifolia* at timberline in the Venezuelan Andes showed that the altitudinal limit of *Podocarpus oleifolius* is caused by insufficient cold tolerance whereas the upper limit of *Espeletia neriifolia* is likely to result from insufficient carbon balance (Cavieres et al., 2000).

Not long ago, it has been hypothesized that a zero balance between production and all the losses of organic matter, including respiration, dead needles, die back of branches and roots, sets an altitudinal limit to tree growth (Paulsen et al., 2000; see also Körner, 2003a). Carbon allocation rather than carbon gain or balance would be the critical factor. Thus, trees at timberline responded to enriched CO_2 by increased carbon investment into their root system at the costs of needle mass and leaf area while they did not produce higher biomass (Hättenschwiler and Körner, 1998).

Relative higher amounts of root biomass compared to above ground biomass of timberline trees may be considered to be an adaptation to unfavourable climatic conditions at high elevation (Wieser and Tausz, 2007). Mountain beech (*Nothofagus solandri* var. *cliffortioides*), for example, produces only half as much above-ground biomass at high elevation than at low altitude, while production of root biomass is the same (Benecke and Nordmeyer, 1982). However, also other factors such as limited soil moisture and nutrient availability may have a similar effect (see also Hertel et al., 2008).

This we observed on rapidly draining deflation sites (cf. Photo 36 and Figure 31) in our timberline study areas in northernmost Finland, for example, where mountain birch seedlings (*Betula tortuosa*) respond to moisture stress and limited nutrient supply by developing comparatively large root systems (Holtmeier et al., 2003; Anschlag, 2006). Trees possibly respond to shortage of nutrient by investing more carbon in their root systems and by allocating more carbon to their above-ground tissue when nutrients are abundant (e.g., Bloom et al., 1985; Tilman, 1988; Gleeson and Tilman, 1992). This can also be concluded from studies on altitudinal gradients in South Ecuador. In the mountain forests a fivefold increase of the root/shoot ratio from 1.050 to 3.060 m has been observed. At the treeline (3.400 m), aboveground and belowground biomasses are expected to reach equal size (Moser et al., 2008). This probably reflects limitation of nutrient acquisition by the trees due to unfavourable site conditions (e.g., low temperature, low litter quality and slow litter decomposition at high elevation (Maraun et al., 2008). Thus, the trees seem to compensate limited nutrient availability by altitudinally increasing carbon and nutrient allocation to roots.

From the altitudinal increase of non-structural carbohydrates and lipids within the tree biomass it might be concluded that limitation of carbon use due to unfavourable thermal conditions during growing season prevents arborescent growth (Körner, 1999, 2003a; Hoch et al., 2002; Paulsen et al., 2000; Hoch and Körner, 2003; Shi et al., 2006). This hypothesis, however, has not remained uncontradicted (Smith et al., 2003; Johnson et al., 2004). Sveinbjörnsson (2000), for example, considers high carbohydrate concentrations at timberline an adaptation to frequent tissue loss. Clear evidence, however, is still lacking.

It had become obvious from field experience long ago (Tranquillini, 1967, 1979a; Holtmeier, 1971a, 1974; Wardle, 1971), that arborescent growth is restricted by direct, climatically caused damage before reaching an altitudinal limit set by carbon deficiency. Climatic injury, increasing with altitude and exposure (solar radiation, wind), might also be more important than limitation of carbon use due to low soil temperatures. This is clearly reflected in wind-shorn 'krummholz' growth forms gradually increasing in height from their wind-exposed side to their leeward edge where vertical leaders could develop into erect stems (see also Sections 4.3.4 and 4.3.11; Figures 53, 56 and 58).

At the treeline (2.200 m) in central Chile where *Kageneckia angustifolia*, a small tree species that forms open woodlands at the timberline, Piper et al. (2006) found higher concentrations of non-structural carbohydrates at the treeline in this tree species than at lower elevation (2.000 m). Thus carbon sink limitations may delimit the altitudinal distribution of this tree species. On the other hand, frost injury of the seedlings mainly at the end of the winter appears to be involved in treeline formation in this area.

Although there is much evidence that carbon reserves at the treeline will usually not be depleted (e.g., Körner, 2003a; Shi et al., 2006) it is still being discussed whether rising CO_2 concentration in the atmosphere will promote tree growth at timberline and timberline advance (cf. Section 5.3). Altogether, it seems unlikely that limited carbon supply causes the present altitudinal limit of tree life in the temperate zone (e.g., Körner, 2003a, 2007b; Hoch and Körner, 2003, 2005; Wieser, 2007).

4.3.3 Frost tolerance and damage

Frost regularly affects trees at the climatic timberline on temperate as well as on tropical mountains. In general, resistance to summer frost increases with elevation. Tree species with a higher frost resistance (e.g., *Pinus cembra*, *Larix decidua*, *Sorbus aucuparia*) usually have a higher altitudinal distribution limit than less frost-tolerant species (e.g., *Picea abies*) (Taschler and Neuner, 2004). As to the effects of frost on plants and plant response basic differences between tropical mountains and temperate mountains exist because of the diurnal (tropics) or seasonal (temperate zone) climates.

4.3.3.1 Temperate and northern timberlines

At timberline in temperate high mountains and at the northern timberline the resistance of trees to climatic influences during the winter and particularly in late winter is of paramount important to tree growth and survival. Frost and winter desiccation (Section 4.3.4) cause damage mainly to needles and shoots. For the most part, intensity and magnitude of these damages depend on the length and favourability of the preceding growing season. The opening of buds, flowering, and onset of shoot growth are delayed compared to lower elevation, while winter dormancy begins earlier. If only the growing season was long enough these organs would fully mature and acquire sufficient resistance to climatically caused injuries. In addition, the growing season must be long and favourable enough to guarantee carbon gain sufficient to compensate for respiration loss and loss of tissue through mechanical stresses (abrasion, breakage), biotic injuries (needle loss due to snow fungi) and damage by climatic influences.

In unfavourable years the growing season may be curtailed to an extent that newly formed plant tissue cannot completely develop and thus will be highly vulnerable to frost. Sudden drops of temperature during the growing season, often accompanied by frost, or drought (e.g., Hornstedt and Venn, 1980) may exacerbate the situation (Larcher, 1980b; Taschler et al., 2004; Taschler and Neuner, 2004). From the results of frost ring analysis in larches (*Larix decidua*) and Swiss stone pines (*Pinus cembra*) in the timberline ecotone (Upper Engadine, Switzerland; Müterthies, 2002, 2003) it became evident that frost damage increases with elevation. Above 2.300 m all larches and above 2.350 m all Swiss stone pines, taller than 2 m and thus projecting beyond the winter snowpack, were damaged by frost.

The tree species represented at timberline are highly frost tolerant during the winter, although frost tolerance is different in these species. In general, frost tolerance increases with altitude and declines at increasing oceanity of the climate (Sakai, 1983). Most northern hemisphere conifers (*Pinus*, *Picea*, *Larix* and some *Abies* species) are extremely frost tolerant (Table 5), while the trees at the upper timberline in the southern hemisphere are not nearly that resistant (Tables 6 and 7; cf. Slatyer, 1976; Sakai et al., 1979; Sakai, 1983; Alberdi et al., 1985) what might be ascribed to the different evolution of vegetation in both hemispheres (Sakai, 1983; Wardle, 1985c).

Generally, frost tolerance of the timberline forming tree species is so high that frost damage is rather unlikely at the winter temperatures occurring at timberline. Plants covered by snow in winter are usually less frost tolerant than plants that project above the snowpack (Pisek and Schiessl, 1946; Holzer, 1958; Tranquillini, 1958; Lenz, 1967; Schwarz, 1970; Neuner, 2007). Frost hardiness varies seasonally. It increases rapidly in autumn under the influence of decreasing temperatures and day length. Early frost may cause severe injuries to the trees. Particularly, new annual shoots and needles that could not sufficiently harden during the growing season may be killed.

In late winter and spring frost tolerance decreases again. From flushing until early summer the young needles, leaves and shoots are not yet fully developed (Ulmer, 1937; Pisek and Schiessl, 1946; Pisek, 1952; Tranquillini, 1958, 1979a, b; Wardle, 1981a, 1985c, 1993; Alberdi et al., 1985; Davidson

Table 5. Potential frost tolerance [°C] of some conifers and broad-leaved trees at the upper timberline in the northern temperate zone during the winter (modified from Larcher, 1985)

Species	Needles	Buds	Stem
Europe			
Pinus cembra	-50/-70	-70	-70
Picea abies	-40	?	-40
Larix decidua	deciduous	-35/-50	-40
Fagus sylvatica	deciduous	-27 (shoots -33)[2)]	?
North America			
Abies concolor	-70	-35	-70
Abies lasiocarpa	-50 (-80)[1)]	-45 (-40)[1)]	-40
Picea engelmannii	-45 (-70)[1)]	-40 (-60)[1)]	-45
Pinus aristata	-70	-70	-70
Pinus flexilis	?	?	-45
East Asia			
Abies veitchii	-35	-70	-70
Picea jezoensis	-40	-70	-70
Pinus pumila	-70	-70	-70
Tsuga diversifolia	-35	-70	-70
Larix leptolepis	deciduous	-30	-70
Betula ermanii	deciduous	-47 (shoots)[3)]	?

[1)] after Sakai and Weiser (1973)
[2)] after Tranquillini and Plank (1989)
[3)] after Gansert et al. (1999)

Table 6. Frost tolerance (°C) of some tree species at the upper timberline in New Zealand and Australia (Modified from Larcher, 1985; Sakai and Larcher, 1987)

Tree species	Leaves needles	Buds	Xylem	Region
Libocedrus bidwillii	–13	–13	–13	Otira, 900 m, NZ
Podocarpus nivalis[1]	–23	–22	–23	Arthur's Pass, 910 m, NZ
Phyllocladus alpinus[1]	–23	–20	–23	Arthur's Pass, 910 m, NZ
Podocarpus lawrencii[1]	–22	?	–22	Mt. Ginni, 1.962 m, Austr.
Nothofagus solandri	–12	?	?	New Zealand, 1.200 m
Eucalyptus pauciflora	–13	–13	–13	Australia, 1.330 m

[1] Shrubs.

Table 7. Frost-tolerance [°C] of some *Nothofagus* species in Chile (modified from Alberdi et al., 1985)

Species	Location	Autumn	Winter	Spring	Summer
Nothofagus antarctica	Paillacca (1.200 m)	decid.	-20,0*	-3,5*	-2,5*
	Entrelagos (220 m)	decid.	-17,5*	-2,5	-4,0 +
	Mallines (700 m)	decid.	-22,0	-4,5	-3,0
	Antillanca (1.080 m)	decid.	-22,0*	-4,0	-4,0
Nothofagus pumilio	Antillanca (1.040 m)	decid.	-18,0*	-2,5	-3,0
Nothofagus betuloides	Antillanca (1.000 m)	-8,6	-10,5 (-13,0*)	-4,0	-5,5
Nothofagus dombeyi	Antillanca (1.020 m)	-8,0	-10,5 (-13,5*)	-2,0	-2,0
	Agua Caliente (500 m)	-4,5	-8,5 (-10,0*)	-3,0	-2,0

decid. = deciduous * = leaf or leaf-bud temperature + = lethal damage

and Reid, 1985, 1987; Gansert et al., 1999). They may be seriously damaged already at temperatures only a few degrees below zero (≤−3°C) (Glerum, 1973; Tranquillini, 1979a, 1979b; Havranek, 1993). In this connection it has to be considered that the temperature of shoots and needles not heated by solar radiation usually ranges several degrees below ambient temperature (Tranquillini and Turner, 1961; Gross, 1989). Gansert (2002) assumes that the altitudinal position (2.800 m) of the birch treeline (*Betula ermanii*) on the southern slope of Mt. Fuji (Honshu) depends on freezing temperatures (−7°C) in spring and on cool periods in summer that may negatively affect maturing of overwintering tissue.

At timberline in the middle and high latitudes, mild frost can be expected at any time during the growing season due to radiation cooling at clear skies or advective introduction of cool air, respectively. For example, after the comparatively warm summer of 1995, a sudden drop in temperature caused severe frost damage in the trees on the eastern flank of the Colorado Front Range. Temperatures fell far below the freezing point. At the meteorological station of the Center for Mountain Archeology (2.565 m) near Ward, a temperature of −16°C was recorded on September 22. This was the absolute minimum temperature measured since the station had been established (1986). Similar frost temperatures were recorded at the meteorological station D-1 (3.743 m) on Niwot Ridge and also at the neighbouring station 'Saddle'

(3.536 m; Figure 13), which is located not far above the tree line. Trees and shrubs at all elevations were affected. The most severe injuries, however, were observed in *Picea engelmannii* and in *Abies lasiocarpa* in the upper part of timberline ecotone (3.530 m). Extreme needle loss and dieback of the apical shoots were observed especially in Engelmann spruce (communication J. B. Benedict; Photo 24). Although frost damage will not be a real threat to the survival of these trees it will cause distorted growth, however.

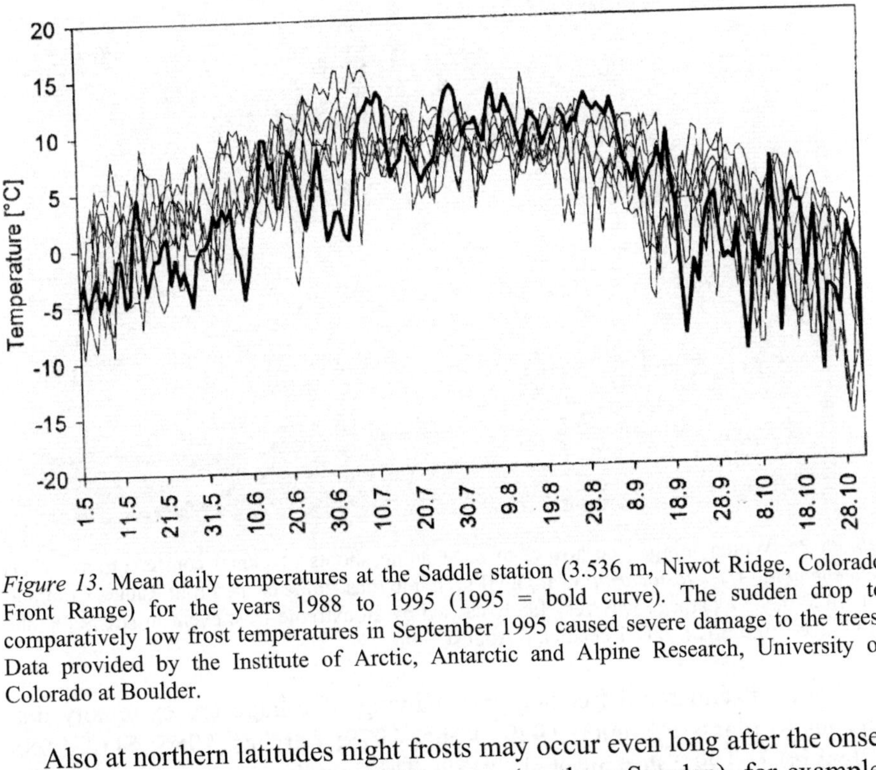

Figure 13. Mean daily temperatures at the Saddle station (3.536 m, Niwot Ridge, Colorado Front Range) for the years 1988 to 1995 (1995 = bold curve). The sudden drop to comparatively low frost temperatures in September 1995 caused severe damage to the trees. Data provided by the Institute of Arctic, Antarctic and Alpine Research, University of Colorado at Boulder.

Also at northern latitudes night frosts may occur even long after the onset of the growing season. In the Abisko area (northern Sweden), for example, dieback caused by frost during growing season seems to be a significant factor controlling birch growth at the birch tree line (Kauhanen, 1987).

Radiation frost affects mainly young growth and seedlings still growing close to the soil surface. In the Subalpine (3.230 m) of the Medicine Bow Mountains (Wyoming), night frost (8 cm above the surface) occurred on 26 of 67 days during the growing season 1993. Leaf temperature of a few centimetres high *Erigeron peregrinus* dropped below zero even at 41 days. The temperature of a young subalpine fir (27 cm above the surface) was below the freezing point at 25 days and was equal to the ambient temperature (Jordan and Smith, 1995).

Photo 24. Winter injury (see arrow) to the leading shoots of clonal conifer groups (*Picea engelmannii, Abies lasiocarpa*) near Devil's Thumb (east slope of the Front Range, Colorado) at 3.420 m. The damage was very likely caused by an extreme frost event in late September 1995 (see Figure 12). J. B. Benedict, 6 August 1996.

Frequent and rapid freezing and thawing of foliage are extremely dangerous to plants (Langlet, 1929; Venn, 1979; Larcher, 1985; Skre, 1988; Gross et al., 1991; Perkins et al., 1991). This may happen even on clear days with frost temperatures. The needles are heated by radiation far above air temperature and cool down very rapidly after sunset. During daytime, temporary cloudiness may cause the same effect (Tranquillini and Holzer, 1958; Weiser, 1970; Christersson, 1978; Christersson and Sandstedt, 1978; Gross et al., 1991). At the upper timberline in the Alps needle temperatures were recorded from February through May. Maximum temperatures in Norway spruce (Gross, 1989) were 11°C to 25°C higher than ambient temperature, and maximum temperatures in Swiss stone pine exceeded air temperature by 4.3°C to 11°C (Tranquillini, 1957, 1958; Havranek and Tranquillini, 1995). At tree line (2,400 m) in the Hida Mountains (central Honshu, Japan) the temperature of sun-exposed shoot tips of *Abies mariesii* at 1 m above the snow surface were 11.5°C warmer at clear sky than the air temperature

(Takahashi, 1944). Mean temperatures 5°C to 12°C (max. 20°C) higher than ambient temperature were recorded in leeward needles of mat-like growing Engelmann spruces in the timberline ecotone of the Rocky Mountains. At night, these needles cooled down 4°C below air temperature and fell even below zero in August (Hadley and Smith, 1986).

Frequently, frost damages can be observed after short periods of warm weather in winter. In some areas of the northern Swedish Scandes, for example, a sudden drop of temperature in April 1991 that had been preceded by warm weather in the first half of the month caused almost total loss of buds in the mountain birch (Tenow et al., 1995). The treetops were most heavily affected. Also Scots pines (*Pinus sylvestris*) were damaged and showed red needles at the northern side of the crown. Such discoloration of needles due to frost injury has been reported as a 'red belt' phenomenon from other areas in northern Europe and North America (e.g., Venn, 1970, 1979, 1993; Schmid et al., 1991; see Jalkanen, 1996, for further references). However, this kind of damage is not specific to timberline forest but may widely occur also within cold air layers (cold air pockets) at lower elevation mainly due to rapid and recurrent change of freezing and thawing temperatures. Studies on winter embolism (gas bubbles blocking the water transport system) in prostrate mountain pine (*Pinus mugo*) growing in the 'krummholz' belt revealed drought to be the main reason of damage. Freeze-thaw events seem to be of secondary importance (Mayr et al., 2003). Embolism occurred only in twigs projecting above the snowpack whereas those buried in the snow were not-affected.

Frost damage occurs mainly in seedlings and young growth (see also Holtmeier, 1971a; Sakai and Weiser, 1973; Slatyer, 1976; Höllermann, 1978; Wardle, 1986, 1993; Gilfedder, 1988; Sutinen et al., 2001; Treter et al., 2002; Block and Treter, 2003). The higher the altitude, the shorter the growing season and the lower the protection by the winter snow cover, the higher is the risk of the plants to be damaged. Extreme frost events may persistently impede natural regeneration in the timberline ecotone. Treter et al. (2002; see also Block and Treter, 2003), for example, suppose late frosts to be impeding advance of larch (*Larix sibiria*) to higher altitude in the mountains of northwestern Mongolia. Frost probably prevents seedling establishment beyond the upper edge of the *Nothofagus* forest in New Zealand thus causing an abrupt timberline (Wardle, 1986; Photo 3).

Total losses due to frost damage are rather rare in older trees, although they may be partially injured. Such damages will be reflected in more or less strong crown deformation or in generally stunted growth forms (see also Tranquillini and Plank, 1989). The situation at the upper timberline on oceanic islands is somewhat different. On Tenerife for instance, the seedlings of *Pinus canariensis* remain fully turgescent through the mild and humid winter. Because of little or no snow cover, however, the seedlings are highly sensitive to night frosts (about –6°C) frequently occurring at the soil surface (Höllermann, 1978).

Skre (1988) reviewed the results of frost-damage research (see also Bigras et al., 2001). Two main types of frost damage became apparent. Rapid and deep fall of temperature below zero may cause intracellular freezing and subsequent irreversible mechanical damage to the membrane systems. At slow cooling, on the other hand, cells are subjected to dehydration. The water withdrawn from the cells by osmotic forces may freeze in the intercellular space (Sakai, 1983). The winter hardening of the needles is triggered by the decrease of day length and first light frosts in autumn, and then gradually increases. Thus, needles and annual shoots that could fully develop during the growing season will be highly frost resistant. In late winter, increasing day length and low temperatures above zero break winter dormancy (Lavender and Silim, 1987). Frost tolerance, however, persists as long as temperatures are low (Schwarz, 1970; Havranek and Tranquillini, 1995).

Besides temperature and day length, also nutrient supply influences frost resistance. Nitrogen, for example, increases frost hardiness if not too much nitrogen is added. In the latter case, frost hardiness of needles and shoots declines because of earlier bud break in spring and prolonged annual shoot elongation in fall. Because of the resulting longer growing season the risk of late and/or early frost damage will increase. It also changes the allocation pattern of carbon and nutrients in the trees (Hinrichsen, 1986). More carbon will be allocated in the root system and more nutrients might be taken up. Thus, it depends on the local decomposition rate whether the additional nitrogen becomes a factor detrimental to tree growth at timberline. Excessive input of organic nitrogen and accelerated mineralisation in case of a general warming of the climate, for instance, might have persistent consequences to the timberline environment and tree growth.

Also, increased phosphorus supply may prolong the growing season, delay winter hardening and thereby increase frost damage. Potassium, which is an important agent in the water regulating mechanisms, may increase resistance of the plants to dehydration in late spring. Altogether, the ‘right’ balance between nutrients seems to be more important for frost tolerance than single nutrients (cf. Repo et al., 2001). Consequently, the role of the ‘minimum factor’ should be seen in perspective.

Knowledge of frost tolerance of tree roots is still very scarce. Generally, it is less than in the above ground parts of the plants (Parker, 1959; Lyr et al., 1967; Larcher, 1980a; Smit-Spinks et al., 1985; Coleman et al., 1992; Sutinen et al., 1997; Ryyppö et al., 1998; Repo et al., 2001). While fully hardened needles of *Picea abies* and *Pinus sylvestris*, for example, tolerate temperatures of −40°C and even below (cf. Table 5) the tree roots (60 to 80 years old trees) were damaged at temperatures between −13°C (−15°C) and

27°C (−22°C) (Korotaev, 1994; Sutinen et al., 1997; Table 8). While frost hardiness (50% of the roots Ø < 2 mm killed) in roots of seedlings of *Abies lasiocarpa* and *Abies amabilis*) was below −11°C, seedling roots of *Tsuga mertensiana* and *Pinus contorta* were already killed at −7.5°C (Coleman et al., 1991). This is surprising as the distribution of *Abies amabilis* is restricted to maritime areas with very high winter snow cover. The snow cover protects the seedlings from too low temperatures, and thus the seedlings roots actually do not depend on that comparatively high frost tolerance. In the rooting zone of the seedlings of the more widely distributed *Pinus contorta* and *Tsuga mertensiana*, however, deep frost temperatures regularly occur and may become critical to seedling survival at their natural growing sites.

Table 8. Frost-tolerance [°C] of the roots of some tree species (modified from Korotaev, 1994)

Tree species	Frost tolerance
Pinus sylvestris	-13 to -23
Picea abies	-15 to -27
Larix sibirica	-18 to -29
Betula pendula	-18 to -22

Frost tolerance of conifer roots does not increase as rapidly fast as in the needles (cf. Coleman et al., 1991; Sutinen et al., 1997). If snow comes late in winter or is blown off by strong winds, temperatures in the rooting zone of tree seedlings may drop far below frost hardiness of the roots. Moreover, compaction of the snow by grazing reindeer, for example, reduce the insulating effect of the snow and thus increase the risk of seedling roots growing close to the surface of being killed by too low soil temperatures. Fine roots are less frost resistant than the large more lignified roots (Smit-Spinks et al., 1985; Ryyppö et al., 1998).

4.3.3.2 Tropical timberlines

Trees at the tropical climatic timberline have to tolerate night frost all the year round and thus are exposed to changes from 'summer-like' to 'winter-like' conditions in less than 24 h. However, temperatures fall below the freezing point for a few hours only. Contrary to the trees at timberline in temperate mountains, trees at the tropical timberline do not have the possibility to exploit seasonal differences for the development of frost resistance and, in a sense, cannot 'escape' the most severe part of the annual environmental regime (Crawford, 1989). They are exposed to permanent diurnal environmental stress (Larcher, 1980b; Smith, 1974, 1980; Beck et al., 1984; Rada et al., 1985; Crawford, 1989, 1997; Goldstein et al., 1994;

Rundel, 1994). The strongest freezing temperatures occur during the dry season. Then temperatures of −10°C are not unusual. On Pico de Orizaba (Mexico) the mean minimum daily temperature ranges below freezing point all the year round (Lauer and Klaus, 1975a).

High altitude populations of *Metrosideros polymorpha*, which grows from sea level to timberline (2.500) on the east-flank of Mauna Loa (Hawaii), avoid freezing by permanent supercooling. Leaves did not show any damage until temperature dropped to below −8.5°C, and ice formed in the tissue. Consequently, leaf tissue damage did usually not occur up to 2.500 m elevation. This is exactly the present location of treeline. Probably, this limited freezing resistance is the cause of the relatively low position of treeline compared to continental regions where treeline can be up to 1.500 m higher at the same latitude (Cordell et al., 2000).

Information on adaptation of the trees to this diurnal stress and particularly on frost resistance of tree species at the upper timberline in the tropics is still scarce. Comparatively, we know much more about the genus *Polylepis*. It was mentioned already, that favourable local site conditions could not explain the scattered occurrences of *Polylepis*-groves high above the closed mountain forest as was supposed by many authors (Section 4.1.3). Instead, *Polylepis*-trees seem to be highly adapted to the severe high-altitude diurnal climate. In contrast to the giant groundsels *Senecio* (Africa) and *Espeletia* (Andes), which are highly frost tolerant because of morphological adaptation (e.g., Larcher, 1980b; Hinckley et al., 1985; Meinzer and Goldstein, 1986), *Polylepis* does not show any morphological features that could explain its high frost resistance. However, studies on *Polylepis sericea* (Rada, 1983; Rada et al., 1985, 1996; Squeo et al., 1991; Goldstein et al., 1994) suggest physiological adaptation to be the decisive factor. High photosynthetic capacity and a high dark respiration rate provide sufficient chemical energy for repairing cellular damages, opening of the stomata in the early morning, and production of substances that increase frost resistance. The leaves seem to be adapted to nocturnal freezing temperatures by physiological mechanisms. Thus, daily changing cell-sap concentration allows deep nocturnal undercooling (−6°C to −8°C) of the leaves of *Polylepis* and lower the freezing point of the cell sap. The supercooling capacity is highest during the warm and wet season. Thus, frost tolerance is at its maximum during the period of greatest metabolic activity (Rada et al., 2001). The Andean alder (*Alnus acuminata* = *Alnus jorullensis*), for comparison, does not enjoy such physiological mechanisms and thus is not able to advance to as high an altitude as *Polylepis* (Azocar, unpublished in Rada et al., 1985). However, also *Polylepis* does not tolerate freezing of its tissue (Rada et al., 1985). Seasonal changes in osmotic potential may maintain water uptake and turgor of *Polylepis* and thus a positive carbon balance during the 4 months dry season (Rada et al.,

1996). These mechanisms allow *Polylepis* a certain homeostasis in the harsh timberline environment, which is controlled by wide daily temperature ranges, diurnal fluctuation in water availability, and seasonal shortage of plant available soil moisture (Goldstein et al., 1994).

The giant groundsels (*Senecio*, *Espeletia*) are able to buffer recurrent nocturnal freezing temperatures effectively. The trunk with its wide frost-intolerant pith is well insulated by a thick cork layer and an outer mantel of dry leaves (Photos 25 and 26). Temperatures measured inside and outside the trunk of *Senecio keniodendron*, for example, at about 4.200 m on Mt. Kenya, clearly evidence the good insulation capacity of the leaf mantle and cork layer. Temperature in the marginal part of the pith was about +3°C while the ambient air temperature outside the leaf mantle was −5°C (Hedberg, 1964; Coe, 1967; Hedberg and Hedberg, 1979). Thus, the fluids in the stem do not freeze. Also, inside the dense rosette, formed by thick and hairy leaves that close at night (nyctinastic), the growing point is effectively protected from freezing temperatures (Hedberg, 1964). Although about 80% of the leaf water of *Senecio keniodendron* freezes at temperatures of −5°C, leaves will not be killed (Beck, 1990). Leaves of *Espeletia semiglobulata* may tolerate undercooling for hours and will not freeze before leaf temperature drops as low as −9°C or −10°C (Larcher, 1975).

4.3.4 Winter desiccation and abrasion

Winter desiccation (frost drought) is another effect of frost causing damages to trees in winter (Michaelis, 1934a, b, c, d; Schmidt, 1936; Müller-Stoll, 1954; Larcher, 1957, 1963, 1972, 1985; Tranquillini, 1965, 1974, 1976, 1979a; Holtmeier, 1971a, 1974; Lindsay, 1971; Baig et al., 1974; Baig and Tranquillini, 1980; Aulitzky et al., 1982; Larsen, 1993; Havranek and Tranquillini, 1995; Cairns, 2001). Winter desiccation has long been considered the main cause of the upper limit of tree growth in the high mountains outside the tropics (e.g., Turner, 1968; Tranquillini, 1976, 1979a, 1982; Klink and Mayer, 1983; Schwarz, 1983; Otto, 1994). Lauer and Rafiqpoor (2002) consider the northern timberline to be caused by winter desiccation. Even the decline of Swiss stone pine between 8500 and 7200 BP, which had been reconstructed from pollen diagrams, was explained to be caused mainly by winter desiccation (Burga, 1990), though no evidence could be given.

In contrast to frost damage, which is caused by sudden dehydration, winter desiccation results from gradual water loss through transpiration. The transpiration loss cannot be compensated for because of the frozen ground and partly frozen tissue (Sakai, 1970; Tranquillini, 1982; Larcher, 1963, 1985; Sakai and Larcher, 1987; Herrick and Friedland, 1991). Trees, in particular

Photo 25. Giant groundsel (*Senecio* sp.) on Kilimanjaro. H. Kleinn.

seedlings and young growth, at upper timberline are rather susceptible to this kind of damage since cuticle resistance to transpiration declines as cuticle thickness declines by altitude (Baig et al., 1974; Baig and Tranquillini, 1980; Tranquillini, 1979a, b; Delucia and Berlyn, 1984). At occasional warmer weather conditions in winter, however, desiccation may be delayed because some water is available to the plants (Hygen, 1965; Havis, 1970; Kincaid and Lyons, 1981). Also, water stored in the aboveground organs may mitigate water stress to a certain extent (e.g., Peschl, 1983; Grace, 1990). Occasionally, water is also taken up through bark (Katz et al., 1989) and cuticle (Stålfelt, 1944). Water uptake through the cuticle seems to occur at the same order of magnitude as water is lost via transpiration and thus might be an important factor controlling water balance (Härtel and Eisenzopf, 1953).

Photo 26. Espeletia grandiflora in the Páramo de Otún (Cordillera Central, Columbia) at 3.850 m. M. Richter.

Desiccation usually occurs in late winter when needles and shoots warm up far above air temperature under the influence of incoming radiation. At the upper timberline, needle temperatures reached 18.4°C in March and up to almost 30°C in April (Tranquillini and Turner, 1961). Higher temperatures did not even occur at midsummer. The resulting steep vapour gradient between foliage and the cold ambient air causes water loss by cuticular transpiration. Persistent sunny weather conditions in winter and particularly in late winter increase the risk of winter desiccation in the trees. The effects of desiccation, yellow–red or bright brown–red needles and shoots, do not become visible before late winter or in spring. Usually, the discoloured needles fall off in the beginning of the growing season. Damages caused by freezing or by winter desiccation are difficult to distinguish visually. Clear evidence can only be provided by permanent control of the plants through the winter (e.g., Wardle, 1981a). Desiccation only affects the foliage projecting beyond the winter snow cover, while seedlings and young growth are fairly well protected from any injurious climatic influences if sufficiently covered by snow. Needles and shoots that could not completely develop due to a short and/or unfavourable growing season are more susceptible to winter desiccation compared to fully matured needles and shoots (Wardle, 1965b, 1968, 1971; Holtmeier, 1971a, 1979c; Lindsay, 1971; Baig et al., 1974; Tranquillini, 1974, 1976, 1979a; Platter, 1976; Baig and Tranquillini, 1980). Shoots that are likely to be impaired or killed by winter desiccation can be predicted from their pale colour in the previous late summer or autumn as Wardle (1968) has demonstrated. Such shoots are usually less hardened than bright green shoots. Water loss will also increase if the stomata are not completely developed (Holtmeier, 1971a, 1974; Tranquillini, 1982). Fully developed needles and shoots, however, exhibit great cuticle resistance and usually survive, if not damaged by other agents (Marchand and Chabot, 1978; Marchand, 1980).

In the Swedish Scandes (Jämtland), exceptionally severe needle damage occurred in Scots pine and Norway spruce forests during the winter 1986/1987 which was extremely cold and almost devoid of snow. Deep soil freezing over a long period of time accompanied by clear weather and great amplitudes of temperature in late winter had very likely fostered these damages, which were ascribed to winter desiccation (Kullman, 1989a, b, 1993; Kullman and Högberg, 1989). Probably, the vitality of the trees had been reduced during the period from 1981 to 1987, which was characterized by persistently cool growing seasons (Lindgren et al., 1989). Thus, the trees were predisposed to heavy needle loss in the extreme winter 1986/1987 (Kullman, 1989a, 1991). Young growth, however, not taller than 20 cm, was not affected because the snow cover protected it. Dieback of more than 90% of the established pines (*Pinus sylvestris*) at timberline in the Handölan Valley (southern Swedish Scandes) during a 32-year monitoring period has been attributed to winter desiccation as the main factor (Kullman, 2006). Kullman (2006) suggests increased winter temperatures to have reduced winter desiccation as a 'chronic stressor and cause of mortality' in Scots pine at timberline in the Handölan Valley (see also Kullman, 2007b).

Damage caused by winter desiccation is common at sun-exposed topography, in the sun-facing part of the tree crowns and at sites with little snow cover (Sakai and Larcher, 1987; see also Cairns, 2001). However, they also occur at wind-exposed sites, though the cooling effect of the wind reduces the vapour gradient from the needle surface to the ambient air (Figure 14). Strong winds may reduce or even remove the boundary layer and replace the vapour-enriched air by relatively dry air (Tranquillini, 1982). Also in summer persistent strong winds from more or less one direction impede needle development at the wind-exposed side of the trees (Bernbeck, 1907; Grace, 1977; Wade and Hewson, 1979; Holtmeier, 1980, 1981b; Dahms, 1992; cf. Figures 29 and 54).

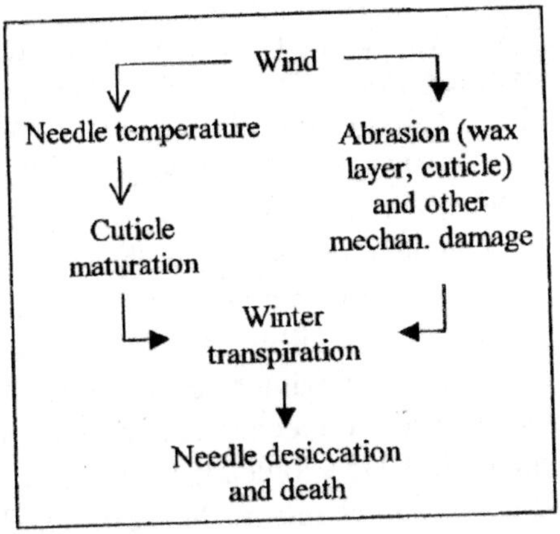

Figure 14. The influence of strong wind on conifer needles. Open arrow heads stand for negative effects. Full arrow heads mean positive effects. Modified from Hadley and Smith (1980, 1983, 1986, 1989, 1990).

Strong wind not only lowers temperature but also impairs CO_2 uptake (Tranquillini, 1979a) and photosynthesis (cf. Holtmeier, 1978, 1980, 1985a; Hadley and Smith, 1986, 1987, 1989; Dahms, 1992). It has become obvious from studies at the upper timberline in the Dischma Valley (near Davos, Switzerland) that the effect of wind velocity on young growth of European larch (*Larix decidua*) and Swiss stone pine (*Pinus cembra*) depends on the specific site conditions. At sites characterized by low global radiation a mean wind velocity of 2.5 m s^{-1} increased growth and survival rate of young trees, while at high radiation growth was reduced if the wind velocity exceeded 1.5 m s^{-1}. At a wind velocity of 3 m s^{-1}, height growth of young larches decreased to less than 50% (Turner, 1971). Also, long-term recordings (7 years) in the Sellrain Valley (Austria) revealed a negative correlation between height growth of Swiss stone pine during the growing season and increasing wind velocity (Kronfuss and Havranek, 1999).

At extremely wind-influenced sites, abrasion of the cuticle wax layer and other mechanical damages may increase water loss to a critical extent (cf. Figure 14). Ice particle abrasion of the wax layer reduces cuticle resistance. Even fully developed cuticles may be abraded (cf. Figure 14 and Photo 31; Hadley and Smith, 1983, 1986, 1990; Dahms, 1992). Not least, strong winds cause breakage, in particular to frozen needles, shoots and branches, thus exposing open-ended vascular tissue. Thereby, water loss is accelerated and may result in tissue death (Marchand and Chabot, 1978; Marchand, 1980; Van Gardeningen et al., 1991). Havranek (1993), however, considers removal of the protective snow cover and exposure of seedlings and young growth to be the most important wind effect.

While incomplete morphological development of needles and shoots have long been supposed to be the main causes of winter desiccation (e.g., Michaelis, 1934a, b; Müller-Stoll, 1954; Larcher, 1957; Wardle, 1968, 1971, 1974; Holtmeier, 1971a; Tranquillini, 1974, 1976, 1979a, 1980; Baig and Tranquillini, 1976; Platter, 1976; Schulze, 1980; Sowell et al., 1982; Barclay and Crawford, 1984; Delucia and Berlyn, 1984) the influence of ice particle abrasion and other mechanical damages on the needles and shoots was almost disregarded or considered unimportant (Holzer, 1959; Tranquillini, 1967, 1979a, Turner, 1968). Tranquillini (1967; see also Wieser and Tausz, 2007), for instance, even supposed winter desiccation to be a precondition to subsequent needle loss caused by mechanical wind effects (see in contrast Havranek and Tranquillini, 1995). Indeed, he referred without exception to experimental studies at timberline in the Alps on Patscherkofel (near Innsbruck) and near Obergurgl (Gurglertal, Tyrol), where winds are usually not that strong and persistently blowing from more or less one direction, as typical of timberline in many ranges of the Rocky Mountains, for example (Holtmeier, 1978, 1996).

Locally, however, extremely windy conditions also occur at the timberline in the Alps. Pru dal Vent (it means 'wind meadow'), for instance, a hillside (2.210 m) located on the southern side of Bernina Pass (Grison, Switzerland) is such a place. It is fully exposed to frequent and strong winds from the north and northwest (Figure 15; Holtmeier, 1971b).

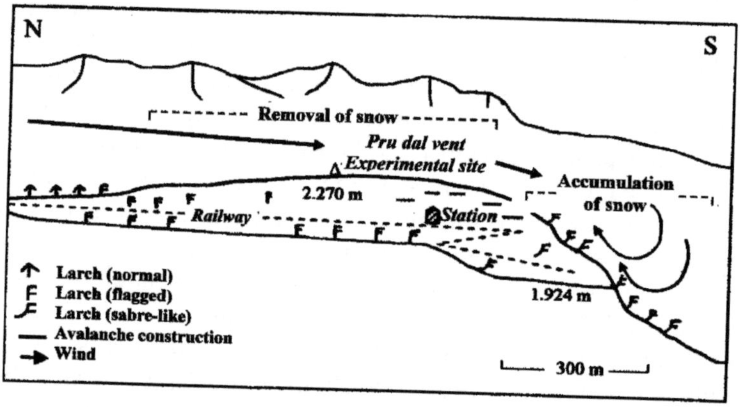

Figure 15. N-S-transect through Pru dal Vent (Bernina Pass, Grison). The strong wind controls snow relocation and the growth forms of the trees. Modified from Holtmeier (1971b).

In 1968, the present author found wind-exposed European larch (*Larix decidua*) and Swiss stone pine (*Pinus cembra*) severely damaged by ice and sand particle abrasion. Also, the wind-facing side of fence posts was strongly abraded by 35 mm within 48 years (about 0.73 mm·year^{-1}), except for the part in the soil (Photo 27). Following the suggestion of the present author,

H. Turner (Swiss Federal Institute of Forestry Research) established a long-term field experiment on top of Pru dal Vent. Young Swiss stone pines (2 years old) and European larches (6 years old) that had been planted there on the different experimental plots already showed first signs of abrasion after the first winter (1970/1971), other damages disregarded, that were, as usual, attributed to winter desiccation (communication H. Turner; see also Holtmeier, 1974).

Photo 27. The wind-facing side of this fence post on the wind-swept top of Pru dal was abraded for 35 mm within 48 years, except for the part that was in the ground. F.-K. Holtmeier, 29 September 1968.

In a project report, published almost 40 years later (Streule and Häsler, 2006), abrasion has been confirmed to be the main factor causing lethal damage to the young trees on Pru dal Vent. Swiss stone pine turned out to be more vulnerable to abrasion than larch. Young growth has survived only when protected by artificial windbreaks (wooden shields of different shapes, stones; Photo 28). Moreover, fatal winter desiccation occurred in young trees at sites lacking winter snow cover.

In the Cairngorm Mountains (Scotland), Grace (1990) found water loss from Scots pine (*Pinus sylvestris*) increasing with elevation. This is in accordance with the winter desiccation hypothesis. However, water loss is considered to be an effect of stomatal dysfunction caused by mechanic damage (Grace, 1990; Van Gardeningen et al., 1991) and direct damage to the cuticle by mutual abrasion of plant parts and wind-blown soil or ice particles rather than of a thin or less developed cuticle.

Photo 28. This young Swiss stone pine (*Pinus cembra*) has survived in the lee of the blocks which provide shelter from the permanent strong winds on top of Pru dal Vent. F.-K. Holtmeier, 7 June 2001.

Observations and studies at timberline in the Rocky Mountains (e.g., Lindsay, 1971; Holtmeier, 1980; Hadley and Smith, 1983, 1986, 1989, 1990; Dahms, 1992), on Mt. Washington (New Hampshire, Marchand and Chabot, 1978; Marchand, 1980, 1987), in the Sierra Nevada (Klikoff, 1965), at the northern tree line in Canada (e.g., Scott et al., 1987a, Scott et al., 1993) and at the upper birch limit (350 m) in southern Greenland (Quindudalen) (Kauhanen, 1987) provide evidence that at least in these areas ice particle abrasion and other mechanical damages caused to trees by wind occur usually prior to winter desiccation (Perkins et al., 1991).

Abrasion of foliage and stems projecting above the winter snow cover and fully exposed to the prevailing winds is caused mainly by wind-driven ice particles (Photos 29 and 30). However, wind-mediated soil particles also cause abrasive damages to foliage (cf. Figure 14). At treeline (2.500 m) of Mt. Fuji (Japan), for example, bark abrasion by wind-blown fine volcanic particles turned out to be the primary factor causing winter desiccation and formation of 'krumm-holz' growth forms in larch (*Larix leptolepis*) (Maruta, 1998). Dahms (1992) took pictures with a scanning electronic microscope of the surface of young needles that had been sampled from the windward and leeward shoots of mat-like and wedge-like subalpine firs and Engelmann spruces growing in the forest-alpine tundra ecotone on the Colorado Front Range. Distinct differences between the wind-exposed and wind-protected needles became apparent from these pictures. Less than 3 months after the needles had flushed, the fibrillated wax structures on the surface of the wind-exposed

Photo 29. The stems of this clonal group (*Picea engelmannii*) on Trail Ridge (Rocky Mountain National Park, Colorado) at about 3.400 m are heavily abraded by wind-driven ice particles from the west (right). F.-K. Holtmeier, 22 July 1987.

Photo 30. Detail of Photo 29. This photo of an abraded stem (cf. Photo 29) was taken from the opposite side. The varying resistance of the wood is reflected in the more or less intense abrasion. F.-K. Holtmeier, 1 July 1979.

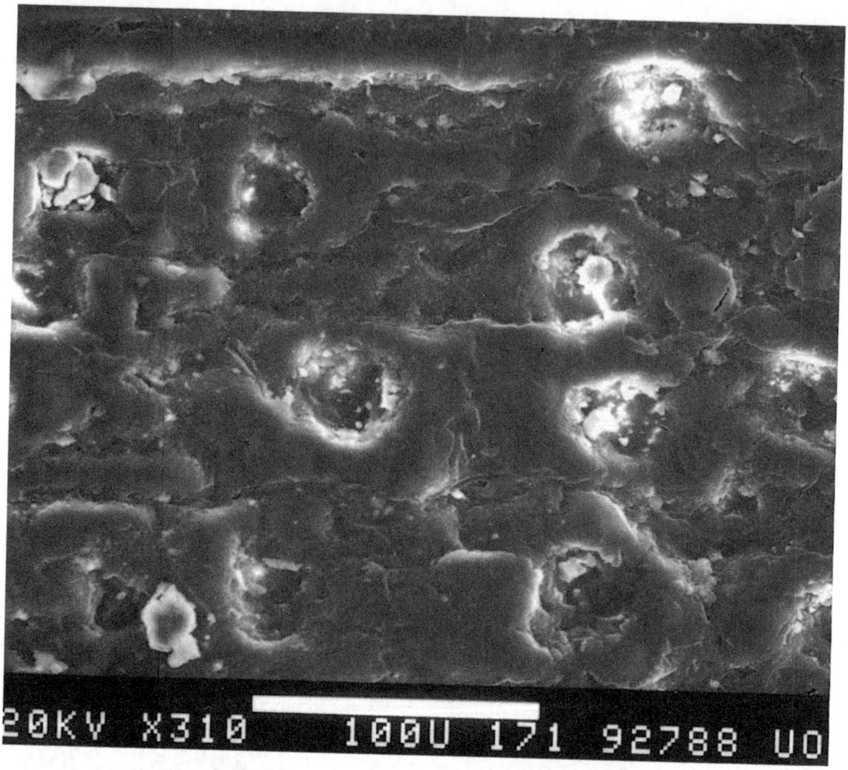

Photo 31. Picture taken by scanning electron microscope (310x) of the surface of a needle collected from the windward side of a wedge-like *Picea engelmannii* tree island in the forest-alpine tundra ecotone on Niwot Ridge (Front Range, Colorado) at 3.465 m. Abrasion, amorphic wax in the stomata, and also silt in the outer stomata are clearly to be seen. From Dahms (1992).

needles (30 to 40 cm above the soil surface) were partly removed by abrasion. The needles from the leeward side of the trees were not affected, however. After the second summer, abrasion of the epicuticular waxes had increased. Obviously, fine windblown soil was the abrasive agent. Mineral particles (7 to 17 μm) were found in the outer stomata (Photo 31). Very likely, the particles originated from loose material that had been brought to the soil surface by borrowing pocket gophers (*Thomomys talpoides*, Holtmeier, 1987b, 1999c; Schütz, 1998, 2005).

Winter desiccation occurs at timberlines outside the tropics only. It affects mainly evergreen conifers and dwarf-shrub vegetation (Pisek and Cartellieri, 1933; Larcher, 1957, 1985; Slatyer, 1976; Körner, 1994) but may also happen to leafless shoots of deciduous trees (Tranquillini, 1982). Peridermal transpiration may cause high water losses that, related to the surface area, may even exceed water loss from shoots with needles. Platter (1976), for example, found peridermal transpiration (related to dry weight) of larch shoots equal to water loss from *Picea abies* shoots at the same conditions, while transpiration of *Pinus cembra* was lower. At experimental plots (1.560 m, near Davos, Switzerland), rowan (*Sorbus aucuparia*) and green alder (*Alnus viridis*) suffered seriously from desiccation after the snow cover had been artificially removed (Frey, 1983). A marked increase of water stress during the coldest part of the winter was monitored in the buds of *Betula tortuosa* and *Sorbus aucuparia* at their upper limit in western Scotland (Barclay and Crawford, 1982), even though no fatal damages occurred. The buds of rowan in particular were highly resistant to desiccation despite the low water content of the tissue. Also in beech (*Fagus sylvatica*), winter desiccation is not very likely, because of low transpiration loss from the shoots. In contrast, damage caused by freezing of the tissue is common in winter and particularly in late winter, because the shoots of beech are by far less frost tolerant than conifer needles and shoots. This may be the main cause of dwarfed and crippled growth of beech at its altitudinal limit (Tranquillini and Plank, 1989).

Also, at the upper limit (2.040 m) of snow gum (*Eucalyptus pauciflora*) in the Snowy Mountains (Australia) winter desiccation is not a critical factor, though cuticular resistance is reduced as altitude increases, probably because of incomplete morphological development of the exposed shoots (Slatyer and Noble, 1992). Even several years' studies did not reveal any critical water stress that might have caused winter damage to the trees (Cochrane and Slatyer, 1988). However, after the shoots had lost their winter hardiness, the foliage fell victim to late frost, despite the winter being mild (Slatyer, 1976). Wardle (e.g., 1978, 1985b, 1991, 2007), in contrast, reports winter desiccation in evergreen *Nothofagus solandri* var. *cliffortioides* at wind-exposed topography at the upper timberline in New Zealand (see also Schönenberger, 1984), though he also considers frequent freeze–thaw events and damages caused to the leaves by ice particle abrasion to have been involved (Wardle, 1985c). In 'red belts' in the White Mountains (northern Appalachians) Herrick and Friedland (1991) observed water loss in *Picea rubens* needles during the winter. The authors consider the needles to have been predisposed to desiccation by other factors that were not recorded, however.

Altogether, in view of the many possible interacting agents causing winter injury at timberline every year, the role of winter desiccation has likely been a little overestimated. Winter desiccation may be fatal to seedlings unprotected by snow when resulting in complete needle loss. In most cases, however, desiccation is restricted to tissue projecting beyond the winter snow cover where it is one factor among others causing needle loss and shoot die-back and thus distorted growth (Section 4.3.11). In general, winter desiccation is not likely to be limiting tree growth at the timberline.

At the tropical upper timberline, winter desiccation cannot be expected because of the diurnal climate. As was already demonstrated, daily night frost causes severe physiological stress to the plants. Soil at timberline, however, may freeze at night for a few hours only and then rapidly warms up during the day. Thus, soil frost does not impede water uptake continuously, as is the case in the temperate mountains during winter. Thus, it seems hard to agree with Schwarz (1983) who by simulation predicted a higher climatic timberline for the Andes which would be determined by desiccation at frost temperatures. The author speculated that the lower position of the present upper forest limit might be ascribed to the lack of tree species adapted to the climate above the present forest limit. Nevertheless, Lauer (1986), who supposes *Polylepis* to be the real timberline forming tree species at undisturbed conditions, referred to this simulated timberline to substantiate his hypothesis, because the present uppermost groves of *Polylepis* coincide with the simulated frost-drought timberline. Braun (1988) speculated likewise.

4.3.5 Soil temperature

In the previous chapters air temperature and its effects on the trees at timberline and on the altitudinal position of timberline have been considered. Equally important is soil temperature. In contrast to data on air temperature, however, soil temperature data are not only rare but also often hardly comparable, because they were usually recorded at different conditions (depths, plant cover, moisture, etc.) and by different instrumentation. Moreover, soil temperature data available were often measured occasionally rather than continuously recorded. Existing recordings seldom cover several years. In a few cases, soil temperatures were measured along local topographical gradients at timberline (rib-groove structure, convex, concave topography, etc.; e.g., Aulitzky, 1961, 1963a; Turner, 1971; Kronfuss, 1972; Turner et al., 1975; Neuwinger, 1980). Occasionally, soil temperatures were measured along altitudinal transects to discover altitudinal soil-temperature gradients (e.g., Walter, 1973; Fröhlich and Willer, 1977; Winiger, 1979, 1980, 1981; Mook and Vorren, 1996). Recently, systematic world-wide measurements of soil temperature (measured at 10 cm depth and under compact canopies of 3 m high trees) at almost 50 timberline sites from different climatic zones were carried out to prove the hypothesis that low mean soil temperatures cause the position of the altitudinal timberline (Körner and Paulsen, 2004). The upper limit of tree growth and position of timberline in humid mountains was found to be worldwide best correlated with a mean soil temperature of 6.7°C ± 0.8°C (or 6.1°C ± 0.7°C; Hoch and Körner, 2003) during the growing season or all year round in the tropics (see also Walter and Medina, 1969;

Ellenberg, 1975; Lauer and Klaus, 1975a; Smith, 1980; Miehe and Miehe, 1994).

Soil temperatures are influenced by numerous factors such as soil texture, porosity, bulk density, humus content, soil moisture, heat conductivity, heat storage capacity, kind and structure of the plant cover, phenology, exposure to radiation and wind, soil water movement, length and duration of the winter snow cover, etc. The complex factor 'snow cover', however, does not play any role at tropical timberlines. These many factors are more or less interrelated and are inversely influenced by soil temperature. Although it seems difficult to set up a rule on soil temperatures at timberline, some aspects will be discussed in the following.

Mean soil temperature decreases as altitude increases. However, decrease in soil temperature is less than in air temperature. Thus, the thermal gradient between the soil surface and ambient air increases. This explains why low growth is advantageous in the upper timberline ecotone (Section 4.3.11). Mean soil temperature decreases more or less rapidly from the soil surface to greater depth. At the upper birch-forest limit (567 m) in Skibotndalen (northern Norway), for example, the mean daily maximum soil temperature (June to September) was 14.9°C at 1 cm depth and only 7.2°C at 20 cm (Mook and Vorren, 1996; Figure 16). Generally, at a few centimetres depth soil temperature is lower than aboveground air temperature (10 cm). The thermal gradient in the topsoil increases by elevation. In northern Norwegian mountains, for example, differences in the maximum temperature (June to September) at 1 and 4 cm depth were 5.5°C at an elevation of 300 m, and 7.2°C at 1.000 m (Figure 16; Mook and Vorren, 1996).

Temperature within the rooting zone is a factor decisive to plant growth and survival. Usually, tree seedlings root in the upper 10 to 20 cm of the soil. Low soil temperatures shorten the growing season, impede photosynthesis, root respiration, root growth, decomposition, nutrient uptake, germination of seeds, seedling growth, maturing of annual shoots, storage of reserves, etc. (e.g., Aulitzky, 1961; Lyr and Hoffmann, 1967; Spomer and Salisbury, 1968; Havranek, 1972; Kaufmann, 1975; Higgins and Spomer, 1976; Tranquillini, 1979a; Neuwinger, 1980; Turner and Streule, 1983; Davis et al., 1991; Schulze et al., 1994; Karlsson and Nordell, 1996; Weih, 1998; Bednorz et al., 1999; Häsler et al., 1999; Weih and Karlsson, 1999; Karlsson and Weih, 2001).

As biochemical processes become increasingly impeded at temperatures below +5°C low soil temperatures at timberline impair root growth and limit dry matter production (Tranquillini, 1979a; Lyr and Hoffmann, 1967; Turner and Streule, 1983; Tyron and Chapin III, 1983). Root growth of Norway spruce (*Picea abies*), for instance, starts already at soil temperatures between 2°C and 4°C. It remains low, however, until temperatures will rise above

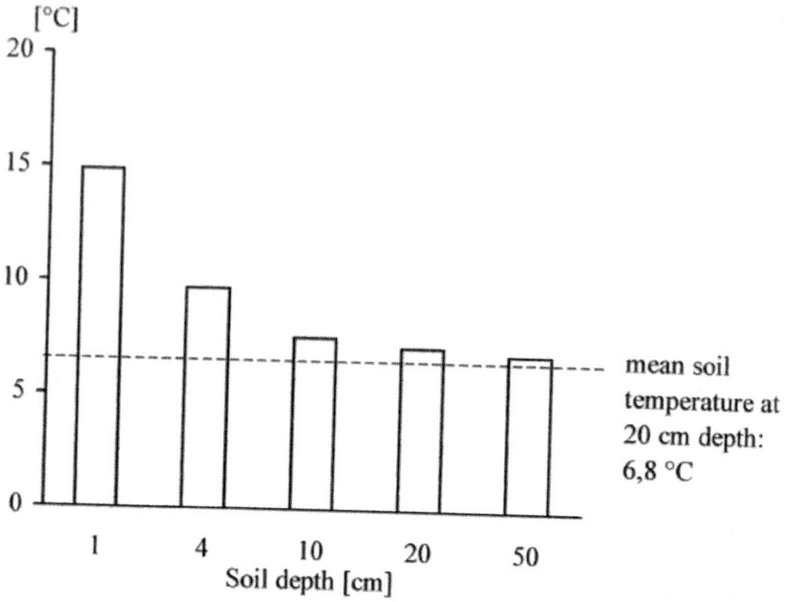

Figure 16. Mean daily maximum temperature (June–September, 1989–1992) at the upper birch-forest limit (ca. 570 m, southwest-exposed slope of Ádjit, Skibotn Valley, northern Norway). After data from Mook and Vorren (1996).

8°C. Good growths cannot be observed before temperatures exceed 14°C (Ott et al., 1997). Roots of 6 years old Swiss stone pine (*Pinus cembra*) near the timberline on Stillberg (Dischma Valley, Swiss central Alps) did not grow at soil temperatures (10 cm depth) lower than 3–5°C (Turner and Streule, 1983). Aulitzky (1961) found the photosynthetic rate of young Swiss stone pines (*Pinus cembra*) to be more closely related to soil temperatures than to ambient temperature. Also, transpiration depends more on soil temperature than on air temperature and moisture (Lyr and Hofmann, 1967). By contrast, in the Medicine Bow Mountains (southeastern Wyoming), considerable variation in photosynthesis occurred, dependent on a couple of factors (stomatal effects, summer drought, sky exposure) among different tree sites across an altitudinal timberline ecotone while the correlation between low soil temperature and photosynthetic performance was not significant (Brodersen et al., 2006).

Though many soil organisms have adapted to the extreme arctic and alpine environment (cf. Broll, 1998 and further references there; see also Withington and Sanford, 2007), it can be taken for a rule of thumb that in general a soil temperature <5°C will seriously impede biological activity (e.g., Retzer, 1974) and thereby nutrient supply. Experimental studies in mountain birch (*Betula tortuosa*) in northern Sweden, for example, revealed nitrogen balance and growth to be controlled by soil temperature rather than by nutrient supply and substrate (Karlsson and Nordell, 1996). The same was found in *Nothofagus solandri* at the upper timberline (1.300 m) on Craigieburn Range in southern New Zealand (Benecke and Havranek, 1980; Wardle, 1985c). Sveinbjörnsson et al. (1996) found mountain birch at tree-line to be more nutrient-limited than birch at lower elevation.

Recordings of soil temperatures at the upper timberline near Obergurgl (Tyrol) showed that soil temperatures close to the surface (5 cm depth) exceeded 5°C at 128 days. At 20 cm depth temperatures never rose above 15°C and did not exceed 5°C for half a year at 1 m depth (Aulitzky, 1961). Under snow cover, soil temperatures usually do not drop far below zero at timberline. Thus, organic matter may even decompose in winter, at least occasionally. On Niwot Ridge (Colorado Front Range), for example, microbial respiration still continues beneath winter snow cover at soil temperatures being as low as −5°C (Brooks et al., 1996; see also Withington and Sanford, 2007). On ridges and in gullies in the timberline ecotone on Stillberg (Dischma Valley, Switzerland), the decomposition outside the growing season amounts to 22% to 40% of the total annual decomposition (Reichstein et al., 2000).

In contrast to snow covered topography, soil freezes to several metres depth at sites sparsely covered with vegetation and lacking a continuous snow cover in winter. On wind-exposed ribs at timberline on Stillberg (2.000 to 2.300 m, Dischma Valley, Switzerland), for instance, soil stayed frozen from the end of November until the middle of April (Turner et al., 1975; Turner and Blaser, 1977; Blaser, 1980; Schönenberger and Frey, 1988). In summer, soil temperatures are lower on wind-exposed topography if compared to wind-sheltered sites with normal snow cover (Figure 17). However, even at relatively favourable snow-covered tree line sites near Obergurgl, temperatures (10 cm depth) between 0°C to 5°C prevailed in the beginning and in the end of the growing season (Figures 18 and 19). During summer, temperatures exceeding 10°C were only a little more frequent than temperatures below 10°C (Aulitzky, 1961; Havranek, 1972).

The presence of a plant cover considerably modulates diurnal and annual soil temperature fluctuations. From August 1998 to August 1999 we recorded soil temperatures at different sites on a wind-exposed knoll (Figures 20 and 21) in the timberline ecotone on Koahppeloaivi (420 m) in northern Finnish Lapland. At one site the sandy mineral soil (glacial till) is exposed by wind erosion, while the other site, a few metres downwind, is not eroded and covered by scattered mountain birch with dwarf shrub-lichen undergrowth (cf. Photo 36). At both sites, soil (2.5 cm depth) was frozen for about two-thirds of the year. At the eroded site (Figure 20), soil temperature did not rise above zero before the middle of May and ranged between 0°C and 10°C until early June, followed by further temperature increases. Maximum temperatures of 20°C to almost 30°C at 2.5 cm were recorded on clear days, in

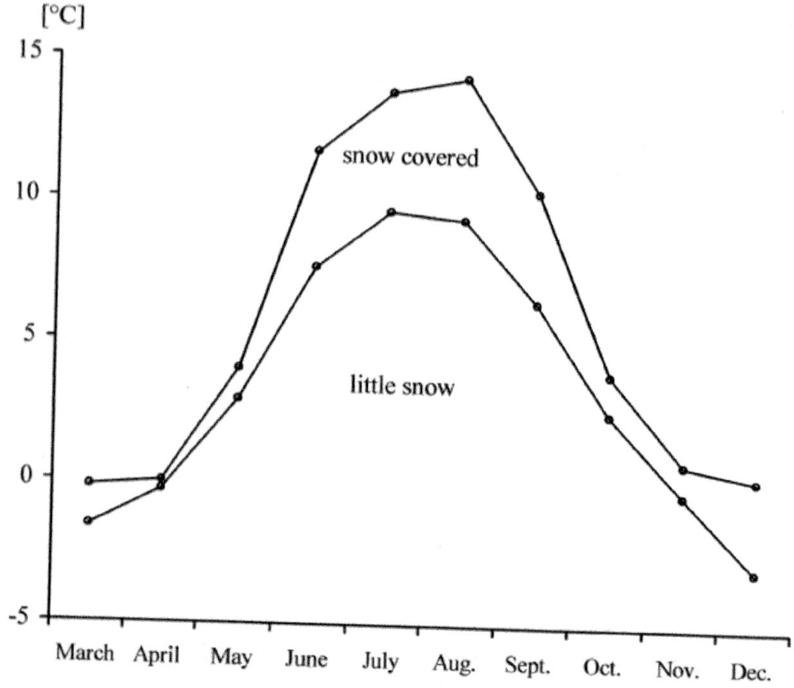

Figure 17. Monthly means of soil temperature (1957/1958) at 10 cm depth at a snow-covered and a snow-free site at the upper timberline (2.072 m) near Obergurgl (Tyrol, Austria). Based on data from Aulitzky (1961), modified from Havaranek (1972).

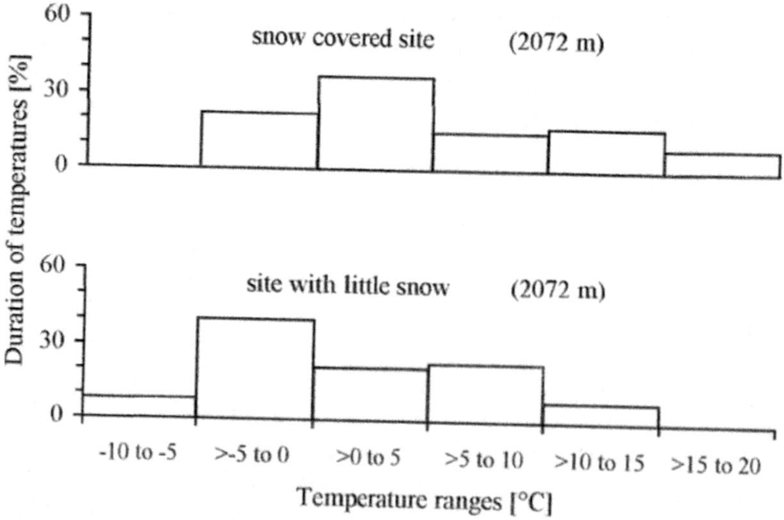

Figure 18. Duration of different temperatures (1958) at 10 cm depth at a site with little snow and a snow-rich site at the upper timberline (2.072 m) near Obergurgl. Based on data from Aulitzky (1961), modified from Havranek (1972).

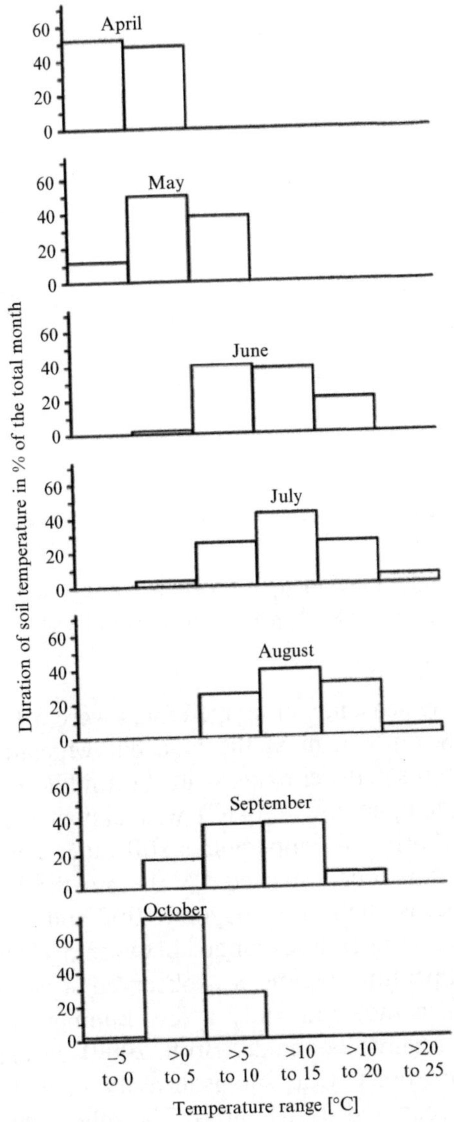

Figure 19. Mean monthly duration of soil temperature at 10 cm depth at a snow-rich site at the the upper timberline near Obergurgl. Based on data from Aulitzky (1961), modified from Havranek (1972).

spite of the loss of sensible heat through turbulent mixing. Soil temperature did not drop below 5°C from the middle of June to almost the end of July.

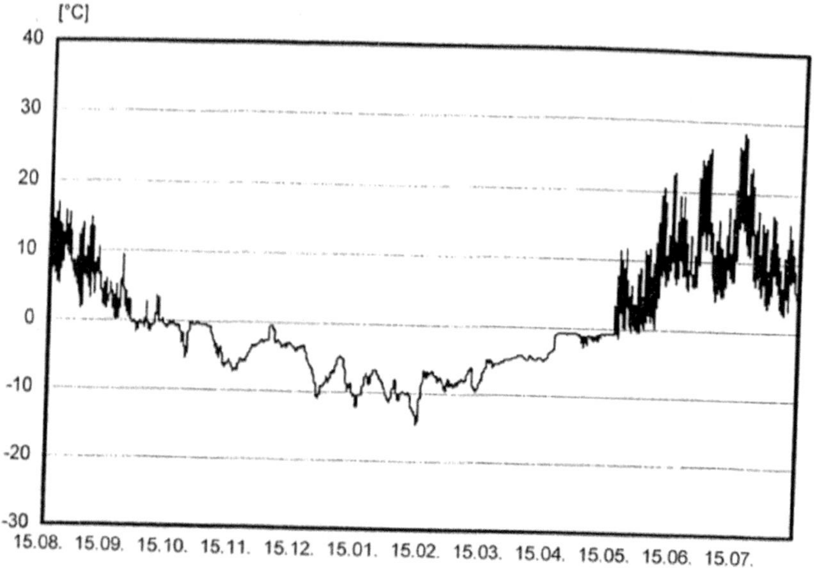

Figure 20. Soil temperatures (1998–1999) at 2.5 cm depth recorded at a wind-exposed site (310 m) with open mineral soil on Koahppeloaivi (northern Finnish Lapland). Data provided by Broll and Holtmeier.

During the growing season, soil temperatures were 5°C and more on 120 days, considerably warmer than at the adjacent vegetated site (Table 9). There, soil temperatures fell below zero in the middle of October (Figure 21). The minimum temperature (−14°C) was almost the same as on the knoll. In the end of April soil temperatures still ranged around the freezing point and did not continuously exceed 5°C before the beginning of June. Then temperatures occasionally rose beyond 10°C and reached 15°C at the maximum. In June soil temperatures ranged between 5°C and 10°C again.

A similar soil temperature regime was observed at two different sites on Jesnalvaara (320 m), a mountain only a few kilometres distant from the Koahppeloaivi sites (Figures 22 and 23; Broll, 2000). Jesnalvaara is exposed to wind from all directions. One site is heavily grazed by reindeer and sparsely covered by scattered dwarf shrub. The other site had been fenced about 30 years ago (exclosure) and was not grazed since. It is covered by closed dwarf shrub-lichen heath. In summer, soil (2.5 cm depth) at the grazed site was considerably warmer (Table 10) but stayed frozen (2.5 cm depth) about 1 week longer (until early May) than the ungrazed site. In winter, the minimum temperature dropped to −23°C, while the minimum temperature at the ungrazed site did not fall below −17°C (2.5 cm depth). Despite these relatively unfavourable conditions mountain birch and some

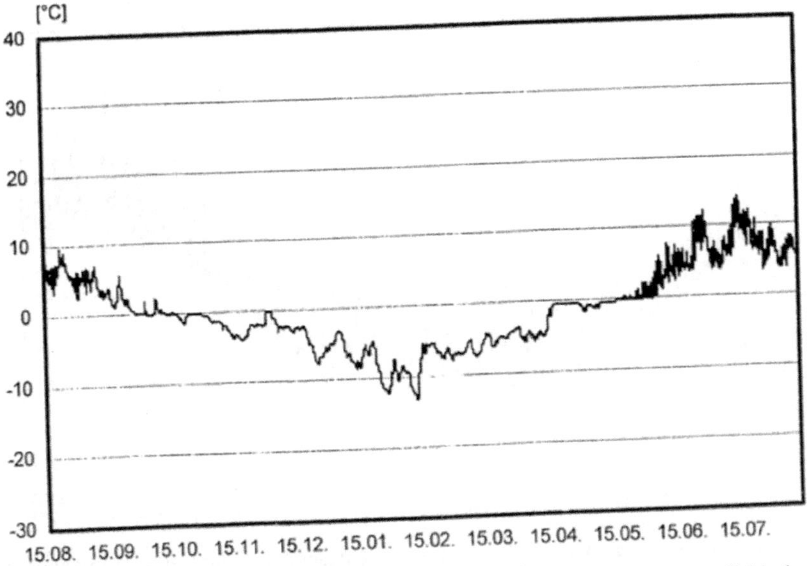

Figure 21. Soil temperatures (1998–1999) at 2.5 cm depth recorded in a small birch stand (310 m) with low dwarf-shrub undergrowth on Koahppeloaivi (northern Finnish Lapland). Data provided by Holtmeier and Broll.

Table 9. Mean temperatures of ≤ 0 °C, ≥ 5 °C and ≥ 10 °C (from 1 August 1998 to 31 July 1999) at two sites (at 310 m) in the forest-alpine tundra ecotone on Koahppeloaivi in Finnish Lapland

Site	Number of days		
	≤ 0 °C	≥ 5 °C	≥ 10°
windexposed knoll, open mineral soil	229	102	57
birch stands with dwarf shrub undergrowth	229	70	8

solitary Scots pines could establish themselves. These trees are crippled and only a few decimetres high (Holtmeier, 1974, 2002). The long-lasting frozen soil is likely to be more disadvantageous to the pines than to the birches, which cannot photosynthesize before being in leaf in the middle of June. In the timberline ecotone on Niwot Ridge (Colorado Front Range) Hansen-Bristow (1986) found that at sites where soil stays frozen until early July, the buds of *Picea engelmannii* and *Abies lasiocarpa* will not open before mid-July.

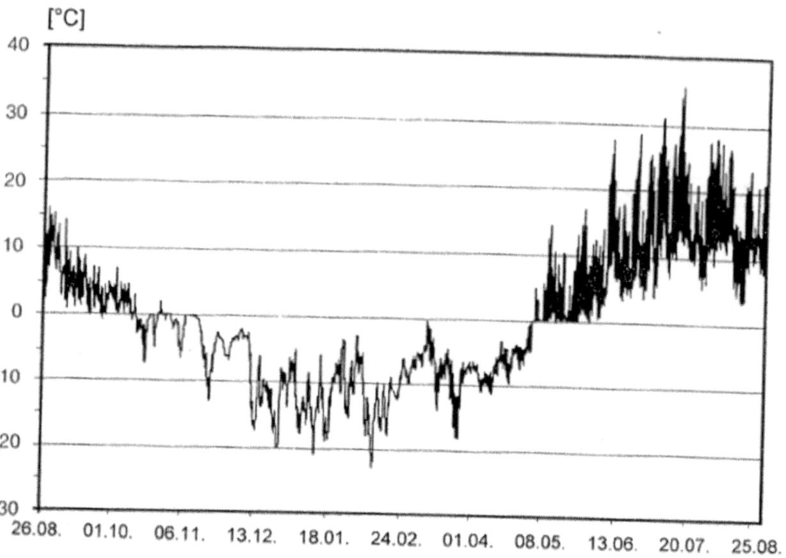

Figure 22. Soil temperatures (1996–1997) at 2.5 cm depth recorded at a wind-exposed site sparsely covered by dwarf shrubs and grazed by reindeer on Jesnalvaara (320 m, northern Finnish Lapland). Data provided by Broll.

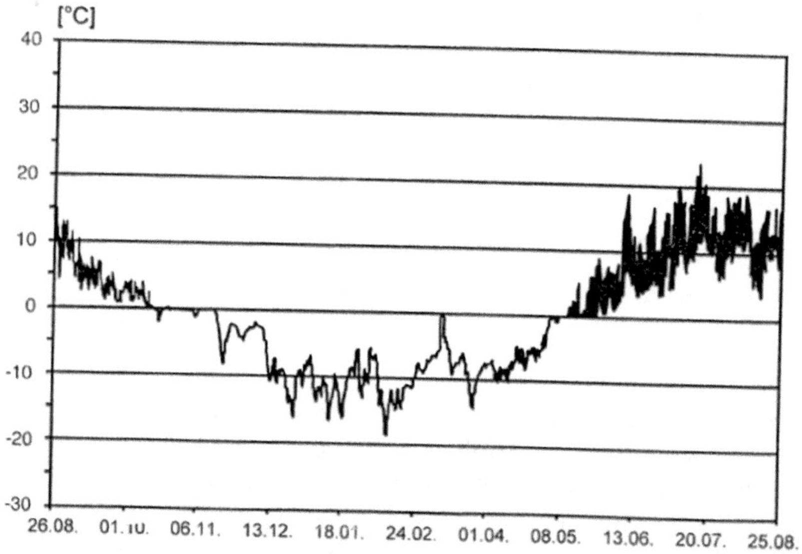

Figure 23. Soil temperatures (1996–1997) at 2.5 cm depth recorded on an ungrazed site covered by dwarf shrubs and lichens on Jesnalvaara (330 m, northern Finnish Lapland). Data provided by Broll.

Table 10. Mean temperatures ≤ 0 °C, ≥ 5 °C and ≥ 10 °C (from 26 August 1996 to 25 August 1997) at a site grazed by reindeer and an ungrazed site (exclosure) on Jesnalvaara (at 330 m) in Finnish Lapland. Data provided by G. Broll (Broll, 2000)

Site	**Number of days**		
	≤ 0 °C	≥ 5 °C	≥10 °C
grazed site	218	101	59
ungrazed site	227	91	45

Above upper timberline (4.000 m) on Pico de Orizaba (Mexico), occurrences of *Pinus hartwegii* young growth are obviously restricted to sites with relatively favourable soil temperatures (Lauer and Klaus, 1975a). The young trees grow mainly between bunch grasses (tussocks) that provide sufficient shelter from permanent strong and cold winds in the spring. Also, the topsoil does not cool down as much as on the open sites. Nevertheless, the pines cannot but develop low, cushion-like growth forms. After a couple of favourable years, or if roots can advance to greater depth, the pines are no longer affected by the extremely fluctuating temperatures and frequent lack of moisture in the topsoil and will be able to grow taller than the surrounding bunch grasses. The authors suppose soil temperatures to be the main factor controlling the establishment of pine seedlings and young growth ('cold-dry limit').

Though low soil temperatures basically impede root growth and dry matter production at timberline also other factors such as available moisture are involved as was evidenced by studies on young conifers (*Pinus cembra*, *Picea abies* and *Pinus mugo*) at timberline on Stillberg (near Davos, Switzerland). Root growth starts shortly after snow melt at soil temperatures between 2°C and 3°C and ceases at the same temperatures in the middle of October. However, at sites exposed to the east the period of root growth was twice as long as on the northern exposures. At a given soil temperature, conifer young growth on the north-exposed site developed more root tips during the growing season than young growth on the warmer east slope. After 6 years, however, dry matter was higher in the conifers on the east slope. Obviously, other factors had overridden the effects of soil temperature. Also, differences in the response of the conifer species to site conditions had become apparent.

In the Alps, soil temperatures at timberline are usually high enough during the growing season (June through September) to not seriously impair growth and dry matter production (Tranquillini, 1979a), occasional adverse conditions excepted (Havranek, 1972). Also, at timberline in New Zealand, low soil temperatures are not very likely to limit tree growth. Temperatures at 10 cm depth ranged above 10°C for 4 months, and the roots of *Nothofagus solandri* and also of the non-native *Pinus contorta* grow at least for 8 months per year (Benecke, 1972).

In the tropics, soil temperature regime at timberline is totally different from outside the tropics because of the diurnal climate. Referring to their studies in the Venezuelan Andes Walter and Medina (1969) hypothesized that all year round low mean soil temperatures (about 7°C) determine timberline by impeding protein synthesis in the roots. At timberline in East Africa, however, other authors found great spatial and temporal variations in soil temperatures that have to be ascribed to local site conditions and alternating rainy and dry seasons (Fröhlich and Willer, 1977; Winiger, 1979, 1980, 1981; Miehe and Miehe, 1994, 1996). At these conditions, mean temperature would disguise different site-dependent daily temperature cycles that have to be considered the true causal factors. Mean temperature might be considered a rough indicator only.

The daily cycle of soil temperature may also be more or less influenced by forest canopy structure, for example, as has been demonstrated by recordings in two different forest stands at the upper timberline (4.100 m) in Ecuador (Lauer, 1999, 2000). In both stands, soil at 10 cm depth did not warm up before early afternoon. Warming was delayed, however, for 1 h in the dense forest stand (*Baccharis*, *Gynoxis*), compared to the more open stand (*Gynoxis*, *Hersperomeles*, *Sarache*). The daily amplitude of soil temperature was small in both forest stands. In the dense forest, it ranged from 5.3°C (minimum) to 5.7°C (maximum), while the corresponding temperatures were 6.2°C and 7.8°C in the open stand. At a more open forest canopy, even higher soil temperatures had very likely occurred. Regrettably, the author does not provide information on soil properties, such as texture, humus content, bulk density, and water content that all influence thermal conditions such as heat conductivity, for example.

Whatsoever, trees at tropical treeline may obviously grow at lower soil mean temperature than reported (e.g., Walter and Medina, 1969; Miehe and Miehe, 1994) and supposed to be necessary for root growth at treeline on the Alps for instance (Körner, 1998a, b). Bendix and Rafiqpoor (2001), for example, found soil temperature (50 cm depth) under tree stands at the upper treeline in the Páramo of Papallacta (Eastern cordillera of Ecuador) significantly lower compared to other tropical mountains. Temperature in the rooting zone (10 cm depth) was about 4.8°C. This value corresponds to the soil temperature recorded at pine treeline on Mexican volcanoes (Iztacihuatl, 3.970 m; Pico de Orizaba, 4.020 m; Körner, 1998a) and at some other tropical treeline sites (Körner and Paulsen, 2004). A tree growth impeding effect seems likely at such low temperatures. With respect to the small differences of soil temperatures (50 cm depth) between the forest and bunchgrass sites in the Páramo de Papallacta, Bendix and Rafiqpoor (2001) hypothesize that the uppermost groves of *Polylepis* (about 4.100 m) indicates the potential climatic limit in this area. High soil temperatures measured by Bader et al. (2007) above the upper timberline in the tropical South American Andes probably reflect the anthropogenic origin of the present timberline in these regions. Furthermore, Richter et al. (2008) reported comparatively high soil temperatures from timberline in the Cordillera Real (Ecuador). The authors, however, do not mention any possible human impact on the existing timberline and assert that the soil temperature is not a decisive factor limiting tree growth in their research area. A mean soil temperature of 7°C (50 cm depth) should be expected here at altitudes from 400 m to over 800 m above the present timberline.

Although soil frost may occur at upper timberline in tropical high mountains throughout the years, particularly during the dry season, little information is available on its effects on vegetation (e.g., damage to roots). On the other hand, nocturnal frost at the surface and freezing of the uppermost topsoil are very likely to cause the so-called inverted timberline in many high-mountain valleys, as has already been demonstrated in a previous chapter (Chapter 3). Although soil does not freeze to greater depth (Furrer and Graf, 1978), seedlings may be seriously injured. According to Smith (1980), frequent soil frosts prevent open mineral soils at elevations above 3.800 m from being invaded by vegetation, even in the long-term.

Also, there exists only a little information on the effects of soil properties on soil temperature at tropical timberline. Lauer (1978, remark to a paper presented by Troll, 1978), for instance, estimated daily freeze–thaw events (night frost) in Mexican mountains to occur less frequently (40 to 50 days fewer) on block debris and rock sites compared to sandy substrate on Mexican mountains. Very likely he referred to surface temperatures. Anyway, the author supposed the block-rich and rocky sites to be considerably more favourable to woody plants. The same assumption was made as to the occurrences of *Polylepis* at similar sites. Also, in the Páramo of Venezuela strong frost temperatures are less frequent on block-rich substrate and mean temperature was 2°C to 3°C higher compared to the grassland on finer material (Azócar and Monasterio, 1979). However, recordings of soil temperature at different depth (5 cm, 12.5 cm, 17.5 cm) on open and forested block debris and grassland on fine soil at the upper limit of *Polylepis* groves in the Bolivian Andes did not provide any evidence of more favourable thermal conditions on block debris. In contrast, mean temperatures as well as minimum and maximum temperatures were even 1°C lower in the open block debris and 2°C lower in the rooting zone of the forested block debris compared to the temperatures in the open grassland (Kessler and Hohnwald, 1998). The relatively low minimum temperatures in the forest block debris were ascribed to cold air moving downhill inside the block debris. Cold air flowing out from the lower end of boulder fans was often mentioned in the literature (e.g., Ellenberg, 1963; Furrer, 1966). For physical reasons, however, cold-air flow inside the boulder fan cannot pump warm air from above the fan surface, as was supposed (Walter and Medina, 1969; Walter and Breckle, 1984) to be an effect favourable to tree growth at such sites.

In general, soil temperatures under shade-giving compact tree canopies are comparatively low and do not much fluctuate during the growing season compared to soil temperatures outside the shaded areas (e.g., Aulitzky, 1961; Holtmeier and Broll, 1992; Oke, 1995; see also Figures 20, 21 and 70). Shading of the ground by the trees and the resulting low temperatures in their rooting zone have been put forward to be the main cause of the climatic timberline (Körner, 1998a, b). In other words: taller growing trees would reduce their possible life span by themselves. Some circumstances, however, can hardly be brought in line with this hypothesis. Thus, it is questionable how trees have been able to grow for many decades, hundreds or even thousands of years (e.g. ancient *Pinus longaeva* and *Pinus aristata*; Currey, 1968; Brunstein and Yamaguchi, 1992; *Picea engelmanni*, Kullman, 2005a) though their shade-giving crown has been keeping soil temperature at a critical low level all the time. Even the uppermost age-old trees (tall trees and 'krummholz'), which often established in the timberline ecotone under more favourable climatic conditions than at present are still producing growth rings. Shading by the tree canopies may even result in a positive feedback. At timberlines in New Zealand or on tropical high mountains for example seedlings of radiation-intolerant tree species survive only under a closed forest canopy shading the forest floor (cf. Section 4.3.12). Bader et al. (2007) ascribe the occurrences of abrupt tropical timberlines at least partly to this effect. Regeneration of less shade-tolerant tree species inside dense forests may be impeded due to low light intensity (see also Cierjacks et al., 2008) rather than by the low soil temperatures. The authors consider the extreme minimum and maximum temperatures occurring in the Páramo at 15 cm above the ground to be more stressful to tree seedlings than low soil temperature. In the open stands of *Polylepis tarapacana* at timberline on the Sajama volcano (Bolivia; cf. Photo 21), for example, ground shading by the tree canopies is likely to mitigate severe moisture stress rather than affecting tree growth.

Dieback of mature trees is usually due to the normal aging process and decreasing resistance to climatic influences as well as to diseases and parasites and insects attacks. It cannot be excluded, however, that low soil temperatures may accelerate dieback. Thinning of the canopy would reduce shading of the ground and result in higher soil temperature (Holtmeier, 1986b), which would allow establishment of new trees, provided that no other factors prevent regeneration.

Under clear and calm conditions, open patches rapidly warm up and exhibit higher soil temperatures during the growing season than occur under compact trees and tree stands (e.g., Aulitzky, 1961; Holtmeier and Broll,

1992; Karlsson and Weih, 2001; Bader, 2007). Thus, within scattered timberline tree stands, tree root systems extending into the open patches will benefit from relative high soil temperatures. Tree seedlings would also profit from the warmer conditions as long as sufficient soil moisture is available from the upper soil horizons at these sites and solar radiation will not exceed the radiation tolerance of the seedlings. However, decrease of tree population close to the upper climatic tree limit and wider spacing of trees should not be taken as a 'natural adaptation' of tree growth to escape low soil temperatures. Scattered distribution may also have negative effects as isolated trees are more exposed to climatic injury than more densely grouped trees. Moreover, the supposed positive function of wide spacing 'mechanism' does not explain abrupt natural climatic timberlines.

Also, dwarfed growth widely occurring in the timberline ecotone is unlikely to be generally caused by low soil temperatures during the growing season as suggested by Körner (1998a). This can be concluded, for example, from several metres high erect and slightly flagged stems. These have emerged from suppressed mat- and wedge-like 'krummholz' growth forms (Section 4.3.11, Figure 58) while soil temperatures have remained as low as under low compact 'krummholz' of the same tree species only a few metres distant from the flagged trees (cf. Holtmeier and Broll, 1992). Consequently, dwarfed growth forms have to be attributed to factors other than low soil temperature.

Permafrost is widely distributed also in the world's high mountains (Gorbunov, 1978). The lower limit of continuous permafrost declines at increasing latitude. Patchy permafrost occurs at lower elevation where it is confined to specific localities (such as steep shaded slopes, beneath thick block debris). At the given climate, however, permafrost does not determine the altitudinal position of upper timberline, although it locally occurs at tree line. For example, evidence of patchy permafrost was found in the forest alpine tundra ecotone at 2.224 m (tree line 2.290 m) on Plateau Mountain (2.519 m, Highwood Range, 80 km west of Calgary, Canada). Above 2.305 m, continuous, 30 m deep permafrost exists. It has been supposed to be caused by the present climate that is characterized by extremely strong winds keeping the mountain plateau snow-free in winter (Harris and Brown, 1978). Also, on Niwot Ridge (Colorado Front Range), located about 1.300 km farther south, permafrost is said to occur at wet and snow-free sites at 3.500 m and above (Ives and Fahey, 1971; see also Ives, 1973a). Crippled trees and low wind-shorn tree islands (*Picea engelmannii* and *Abies lasiocarpa*; see Photos 51, 52, 53, 90, 91) are still common to this elevation. Curry (1962) reports 15 to 40 m deep permafrost beneath a *Picea engelmannii* stand at 3.360 m elevation near Climax (Colorado). In the Rocky Mountains, persistent ice is common in boulder fans and talus cones on northern exposures. Melt water supplies additional moisture to the trees at the lower rim of these sites through late summer (Arno, 1984). However, these comparatively small permafrost patches do not significantly influence the general spatial structure of the timberline ecotone in the middle and southern Rocky Mountains, because boulder fans and talus cones are usually treeless (orographic timberline) or only sparsely covered with trees. The same holds true for the Alps, where permafrost locally occurs above 2.300 m (Furrer and Fitze, 1970), that is below the potential timberline. By geoelectrical sensitivity measurements, Kneisel et al. (2000) found some evidence of permafrost lenses under the present forest cover even at 1.840 m on a north-facing slope in the Bever Valley (Upper Engadine).

In contrast, permafrost may strongly affect the spatial and temporal structure at northern timberline (Kryuchkov, 1973; Walter and Breckle, 1986, 1991; Larsen, 1989; Veijola, 1998; Kokelj and Burns, 2004) which may also occur as a more or less wide altitudinal ecotone (e.g., Hare and Ritchie, 1972; Holtmeier, 1974; Viereck, 1979). Permafrost impedes warming of the ground during the short growing season and thus impairs root growth and nutrient uptake (Larsen, 1989). Low temperatures and waterlogged active layers also hamper decomposition. Long-lasting seasonal soil frost at temperate mountain timberlines may have a similar effect, however. The varied mosaic of trees and tree groves alternating with open tundra vegetation in the ecotone is closely related to the micro relief on the surface topography and to different thaw depth of permafrost (Arno, 1984; Larsen, 1989). In dry regions, water supply in the active layer may favour rather than prevent tree growth (Kryuchkov, 1973, cited in Veijola, 1998). This might also apply, for example, to the situation in northwestern Mongolia (Charchiraa Mountains) described by Treter (1996, referring to Hilbig et al., 1989). Beneath larch forests on north-exposed slopes, soil stays frozen far into summer or even all year round, at least locally (Treter, 1996). Soil temperatures at 20 to 30 cm depth will not rise above 4°C in such places, while 10°C to 13°C were measured at the same depth on sun-exposed slopes covered by mountain steppe vegetation. Close to the surface, soil temperature sometimes even exceeded 17°C (Hilbig et al., 1989).

The interactions of thaw depth, plant cover and succession are features unique to the forest-tundra ecotone. Thaw depth closely depends on the insulating effect of the plant cover and litter layer and thus may change in response to succession. In compact *Pinus pumila* clumps, for example, increasing litter and moss cover prevent warming of the soil. Thaw depth does not exceed 20 to 25 cm. Most of the pine roots gradually die off. At the periphery of the pine clumps, however, where soil is warmed up by incoming solar radiation, pines expand progressively by layering (formation of adventitious roots). After several decades, the pine layers become independent from the 'mother tree', while the initial plants in the centre of the pine clump die, thus leaving a gap. In this opening, the soil can better be heated now, and increasing evaporation improves aeration of the soil. Thus, a new pine clump may establish itself. This is the way the patchy distribution of *Pinus pumila* might be explained (Kryuchkov, 1973; Figure 24).

In larch-alder stands (*Larix dahurica, Alnus* sp.) growing on low water sheds and similar convex topography at timberline in central and eastern Siberia, comparable changes can be observed. Due to increasing moss and peat layer, thaw depth is reduced to 30 cm and less. Larches, alders and even dwarf birches (*Betula nana*) gradually die off. Hence, wind blows the snow off and removes moss cover and peat. As a result, soil thaws to greater depth in summer and site conditions become more favourable to seedling establishing and tree growth. Larch-alder stands may develop again, and a new cycle will begin (Kryuchkov, 1973).

At high and middle latitudes solifluction (solifluction lobes, solifluction terraces, sorted steps, etc.) and other frost and freeze–thaw effects such as frost heaving (e.g., Zoltai, 1975) and needle ice formation may locally influence site conditions in the timberline ecotone. In the winter rain regions of California, for example, periglacial forms, micro-polygons included, can be found down to timberline (Höllermann and Poser, 1977). In the Alps, solifluction lobes occurring at the timberline level are relics of a cooler and probably warmer climate during the Holocene. In the High Tauern, the lower distribution limit of solifluction lobes is located at about 2.300 m, while active solifluction lobes do not occur below 2.650 m (Veit, 1993). The rather low limit of solifluction in the Alps has been supposed to be a consequence of timberline depression (Höllermann and Poser, 1977). Solifluction lobes may provide shelter to seedlings from strong winds (microsite facilitation).

Active solifluction lobes are common at the upper birch limit in the Norwegian fjordland (Holtmeier, 1974). Solifluction lobes, in particular if water-saturated, may collapse when entering the steep trough walls from the above trough shoulder. The loose masses slide down to the valley bottom destroying the birch forest and leaving deep furrows behind. In other places, such events may favour birches and other vegetation to colonize open block fields in the middle and lower parts of the trough walls by covering them with finer material (Figure 25).

Active turf-banked solifluction lobes do not generally exclude tree growth, while stone-banked lobes and solifluction terraces are usually treeless. In northern Finnish Lapland, the present author locally found such periglacial landforms even far below tree line. On the non-vegetated or only sparsely vegetated lobe or terrace surface, as also on other sites (wind-exposed, convex microtopography) with exposed mineral soil, frost heaving and also needle ice may occur. Needle ice forms at high soil moisture and air temperature a few degrees below zero. At upper timberline in Lapland, in the Rocky Mountains as well as in New Zealand and in the Alps such open sites

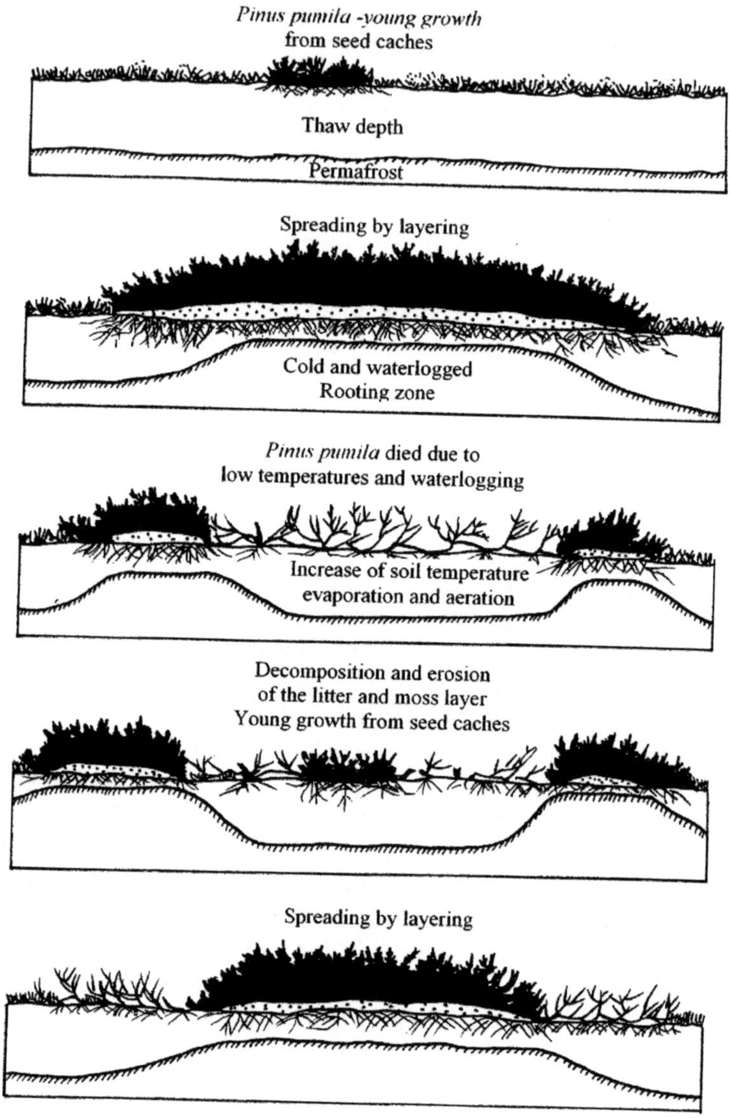

Figure 24. Interactions between *Pinus pumila* and site conditions in the forest-tundra ecotone in northern Siberia. Drawn by Holtmeier after the description by Kryuchkov (1973).

are extremely prone to deflation (Broll and Holtmeier, 1994; Holtmeier, 1996). Seedlings, already suffering from injurious climatic influences at such sites because not being sufficiently protected by snow in winter, are additionally impaired or even killed by these frost effects. Frost heaving and needle ice formation may locally destroy the fine roots of the seedlings and push them out of the soil (see also Black and Bliss, 1980).

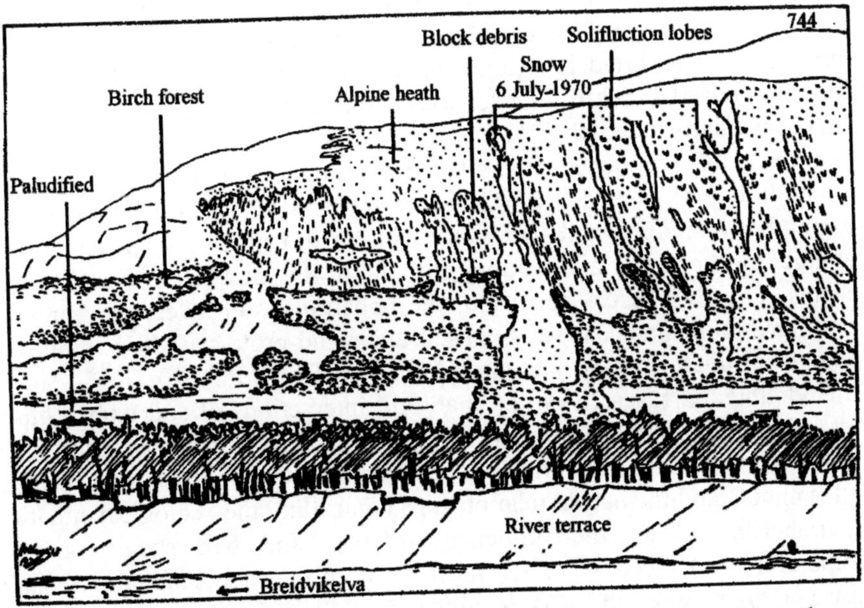

Figure 25. View from Brudal (Breivikeid, Troms, northern Norway) on the north–west-exposed slope of the Nakkefjeld. The steep blockfields are partly covered by fine material resulting from the collaps of solifluction lobes on the upper slope. Alpine heath (*from above*) and birch forest (*from below*) are invading sites that provide more favourable moisture and nutrient conditions than the boulder fields. Modified from Holtmeier (1974).

On the foregoing pages, the effects of low soil temperatures in the timberline ecotone have been considered. However, also extremely high temperatures occurring locally and temporally in the timberline ecotone may negatively affect germination, seedling establishment, growing conditions and decomposition. At the upper timberline in the Gurglertal (Tyrol), temperatures were recorded during a 12 days period of clear weather conditions (end of June/ beginning of July; Turner, 1958). On a south-west-exposed site (2.700 m) covered with dark and dry raw humus, soil temperature immediately below the soil surface exceeded 80°C for a short time and then ranged for several hours above 60°C, and stayed even longer above 40°C. The air temperature was about 13°C. Surface temperatures above 70°C were measured on dark substrate at clear and calm weather in the White Mountains (California). On light-coloured substrate (2.800 m) temperature still reached 67°C (Rien et al., 1998). Unfortunately, additional information on the site characteristics was not provided. On Pico de Orizaba (Mexico), the daily range of soil temperature was found to be ranging between 50°C and up to 70°C at extreme conditions (Lauer and Klaus, 1975a). At Mt. Wilhelm (Papua, New Guinea), soil temperature reached 60°C on a horizontal non-vegetated surface (3.480 m) while air temperature was only 15°C (Barry, 1978). At the present timberline (3.600 m) in the Ethiopian Highland, daily maximum temperatures may rise to 70°C on non-vegetated soil surface at intensive solar radiation (February through April), as was reported by Klötzli (1975). However, even at the northern timberline near Churchill (Hudson Bay), soil surface temperatures of 45°C and occasionally beyond 50°C were frequently recorded at clear weather conditions (July) on lichen-covered sites. At 6 cm depth, in the rooting zone, temperature still exceeded 35°C (Scott et al., 1987b). Soil temperatures, ranging between 40°C and 50°C for a while, will cause heavy physiological damage to seedlings and young growth of the timberline-forming tree species. Thus, excessive heating of the uppermost soil layers may prevent seedling establishment at sun-exposed and wind-protected sites within the ecotone at middle and low latitudes (Aulitzky, 1961, 1963a, b).

Undoubtedly, research on the manifold interactions of soil temperatures and other site factors as well as the effects of soil temperatures on germination, seedling establishment and tree growth should be intensified to better understand the causal relationships controlling the ecological situation at timberline. Also, the influence of trees and tree stands on soil temperatures (cf. Section 4.3.12) must be more intensely studied. After all, however, in view of the very locally varying soil temperatures (daily and annual temperature cycles, minimum and maximum temperatures, etc.) and the lack of data comparable to each other, it seems not very promising, at least in the ecological perspective, to focus on 'discovering' a mean soil temperature considered to be the factor controlling timberline worldwide (e.g., Daubenmire, 1954; Körner, 1998a, b).

4.3.6 Wind

At middle and high latitude, wind speed generally increases by elevation. At exposed high elevation sites, wind speed may be as high or even higher than recorded from stormy ocean coasts. In high mountain ranges at mid-latitudes influenced all year round by the westerlies, wind speeds are highest in winter. Then, at extreme weather conditions (e.g., Mt. Washington in New Hampshire; east slope of the Colorado Front Range), mean wind speeds reach 20 m s^{-1} at timberline and gusts may even exceed 70 m s^{-1}. Mean annual wind speed is about 15 m s^{-1} (Mt. Washington, 1.917 m) and 10.3 m s^{-1} (Niwot Ridge, 3.743 m). For comparison, mean annual wind speed on Säntis (2.440 m, Switzerland) and on Sonnblick (3.106 m, Austria) did not exceed 7.7 or 7.5 m s^{-1}, respectively (Braun-Blanquet, 1964). On Muottas Muragl (2.568 m, Upper Engadine, Switzerland) 4 m s^{-1} were calculated. All these stations are located on or close to the mountain top.

At timberline in the inner tropics wind speeds are by far not as high as on extratropical mountains. Hnatiuk (1994), for example, mentions generally weak winds in the mountains of New Guinea. On Carstenz Peak (4.720 m, New Guinea) mean wind speed (December through January) at 4.250 m was calculated 2 m s^{-1} (Allison and Bennet, 1976). Referring to different sources and to his own observations Hedberg (1964) emphasizes rareness of strong winds in the East-African Mountains. On the other hand, he also mentions comparatively strong and cold winds blowing downhill at night. The author considers the undisturbed growth form of several metres high giant groundsels and lobelias growing at high elevation an indicator of generally low wind speed. In the Bale Mountains (Ethiopia), however, Miehe and Miehe (1994) observed outliers of *Erica trimera* at about 4.200 m, the mat-like growth form of which they ascribed to strong eastern trade winds. Richter (2001) presents observations on wind-shearing at exposed sites on mountain passes and ridges in Ecuador. Average wind speed (2 m above the ground) was about 4 m s^{-1}, maximum gusts reached about 15 m s^{-1} on the mountain crest (Richter et al., 2008). Bendix and Rafiqpoor (2001) report occasional strong winds in the Páramo of Papallacta (Ecuador). On El Misti (4.760 m, Peru), far above timberline, mean wind speed was estimated at 5 m s^{-1} while gusts were about 16 m s^{-1} (Bailey, 1908). In the outer tropics, wind speed increases as the zonal westerlies move to lower latitude. Thus, in the Himalayas strong winds from the west prevail from October through May and will be replaced by comparatively weak eastern air currents a month later (Barry, 1992). The effects of wind on timberline spatial pattern and growth form of trees are by far less conspicuous compared to the Rocky Mountains, for example (Schickhoff, 2005). However, there are some reports on wind-influenced timberlines from Nepal, for example, where the uppermost *Betula utilis* stands occur at wind-protected sites with depressed timberline on south-facing and wind-exposed slopes compared to shady lee slopes (Miehe, 1982, 1984, 1989).

In mountainous terrain, wind speeds and directions are controlled by the local topography and thus usually are more or less different from air currents above the mountains. Ridges and gullies with a relief of 5 to 12 m can modify the wind-speed by ±60% when the wind direction is perpendicular to the ridges (Nägeli, 1971). The resulting mosaic of wind-exposed sites alternating with wind-protected topography strongly influences the ecological conditions at timberline, particularly in high mountain ranges with prevailing strong and permanent winds from one direction, such as the north–south oriented Rocky Mountains (cf. Photos 48, 50, 51, 52 and 53), the Scandinavian mountains or the New Zealand Alps. Wind recordings are rare at timberline and often hardly comparable because of different instrumentation and methods of measurement (e.g., height of the anemometers above soil surface, continuous or occasional measurements, different seasons, vegetated or non-vegetated sites, etc.). In open terrain wind speed usually increases rapidly above the surface. At 2 m height, it may be already twice or three times as high as at the surface, according to the surface roughness (e.g., Geiger, 1961; Caldwell, 1970; Kind, 1981; Barry, 1992; Oke, 1995).

Wind affects tree growth physiologically and mechanically (wind pressure, abrasion) and by influencing the site conditions (air and soil temperature, soil moisture, height and duration of the snow cover, etc.). Most direct wind effects have already been considered in a previous chapter (Section 4.3.4). In addition, the effects of heavy storms such as breakage (crowns, stems), removal of foliage, wind throw, and uprooting of the trees must be mentioned. Heavy loads of rime ice and hoar frost, particularly at the wind-exposed side of the trees (cf. Photo 79), may cause breakage to crowns and branches. Such effects contribute to the development of asymmetric growth forms reflecting the locally varying wind speeds and directions at windy timberline sites (Section 4.3.11). Such tree deformation was used as an indicator of wind conditions in mountain areas because of extreme data scarcity for constructing detailed maps of wind directions and also of wind speeds (e.g., Lawrence, 1939; Troll, 1955b; Krivsky, 1958; Yoshino, 1966, 1973; Rudberg, 1968; Holroyd, 1970; Holtmeier, 1971b, 1978, 1996; Yoshimura, 1971; Wade and Hewson, 1979, 1980; Robertson, 1986; Wooldridge, 1989). Wind also influences conditions at timberline by dispersing pollen and seeds (Section 4.3.10.1). Needle loss due to strong winds during the growing season does not only reduce photosynthetically active tissue but also causes nutrient loss, because premature loss of green needles prevents resorption of nutrients. Sveinbjörnsson et al. (1996) report that early season foliage loss caused by a snow storm had reduced seasonal carbon gain probably by more than 90%. Moreover, deflation is common at wind-exposed sites in the timberline ecotone, in particular on sandy, rapidly draining substrate and if intense grazing and trampling by wild or domestic ungulates have destroyed the plant cover. By removing the humus layer and the top soil the wind not only reduces the waterholding capacity but also nutrient supply.

The influence of wind on site conditions must be considered one of the most important agents in the timberline ecotone. Increasing wind velocity enhances turbulent flow and thus vertical fluxes of heat and moisture (evaporation). Such effects are particularly pronounced in high mountains characterised by high mean wind speed close to the surface, as in the Rocky Mountains, for example, where evaporation is additionally increased by advection of comparatively dry air from western directions (LeDrew, 1975; Isard and Belding, 1986; Barry, 1992).

Wind also reduces temperature differences between plants and ambient air and between different exposures to solar radiation. While over-heating of the upper soil layers can be occasionally observed at sun-exposed and simultaneously wind-protected topography it will usually not occur at the windward sides. Under clear weather conditions at night, however, wind-exposed sites are usually warmer due to turbulent mixing than wind-protected, less ventilated sites, such as depressions or valley bottoms that may experience relatively low temperatures (cold air pockets, inverted timberline).

At temperate timberlines, the relocation of snow by winter wind is of paramount importance to the site conditions. Relocation depends on wind velocity, turbulent flow and consistency of the snow (coherence of the snow particles, hardness of snow surface, water equivalent, density; Dyunin, 1967; Flemming, 1969; Formozov, 1969; Kobayashi, 1971; Tabler and Schmidt, 1973; Tabler, 1975; Berg, 1986). At a wind speed of 3 m s^{-1} already (measured at 1 m above the snow surface, which corresponds to 6 m s^{-1} at the normal anemometer height; Flemming, 1969; Formozov, 1969) snow moves like thin haze along the snow surface. At a wind speed of about 10 m s^{-1}, drifting snow may strongly reduce visibility. A certain amount of wind-blown snow vaporizes either partly or sometimes completely during transport by wind (Schmidt, 1972; Tabler and Schmidt, 1973; Tabler, 1975); most of it will be deposited at sites favourable to accumulation. Wind-driven ice particles cause abrasive damage to the plants (Section 4.3.4). Moreover, accumulation of wind-transported detritus in snow banks, depressions, on the leeward side of compact tree islands or inside of open tree groups may considerably increase nutrients and water retention characteristics of soils (cf. Wilson, 1958; Teeri and Barret, 1975).

4.3.7 Snow cover

Microtopography and plant cover (e.g., low dwarf shrub and grass vegetation, tree groves alternating with glades, etc.), which both determine surface structure and roughness, strongly influence wind direction and velocity and thus relocation of the snow (Friedel, 1961; Kronfuss, 1967, 1970; Holtmeier, 1971b, 1978, 1980, 1987a, 1989, 1996, 1999c; Billings, 1973; Cairns and Fonda, 1974; Wooldridge et al., 1996). Since the duration of the snow cover at a given site depends primarily on the amount of snow, almost the same snow-patch pattern recurs every year, particularly in the alpine zone where vegetation does not affect surface-near windflow as much as it does compared to the timberline ecotone (e.g., Friedel, 1965, 1967; Kronfuss, 1967, 1970; Prutzer, 1967; Burns, 1980; Burns and Tonkin, 1982; Minnich, 1984).

4.3.7.1 Distribution and characteristics of snow cover

Usually, convex topography becomes snow-free earlier because wind has partly or completely removed the snow in winter already. Leeward slopes of small ridges or concave topography, on the other hand, where big snow masses pile up, may be covered with snow until early summer. This holds true even for sun-exposed leeward slopes. Compared to the windward northern slopes much more snow is accumulated. Although solar radiation load is much higher on the south-facing slopes, it does not provide enough energy for melting the snow masses as rapidly as the lower and often patchy snow cover on windswept northern sites. The situation is quite different, however, in case southern slopes are exposed to the wind (Figure 26; see also Turner, 1961; Neuwinger, 1972; Holtmeier, 1974, 1985a, 1987a; Rychetnik, 1987). On the sunny slopes snow will go very early, maybe in late winter already, while snow melt will be extremely delayed on the shaded adjacent leeward slope. Snow deposition on a mountain slope also depends on wind direction. With prevailing upslope winds the snow drifts behind obstacles are shorter than with downslope winds. In addition, slope inclination affects relocation

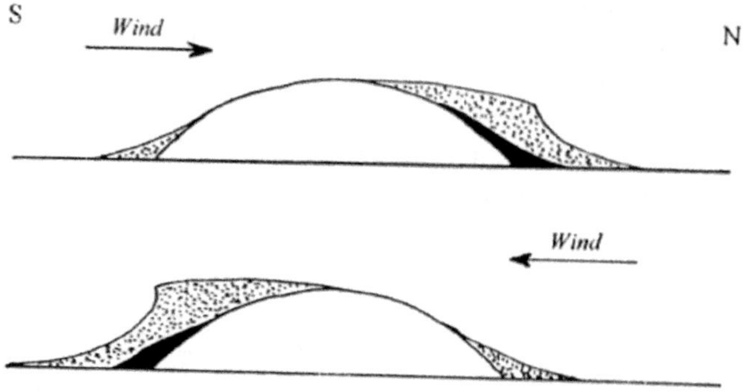

Figure 26. Influence of wind and radiation on the duration of snow cover. Modified from Aulitzky (1961).

of snow. On steep windward slopes snowdrifts are shorter than on gentle wind-exposed slopes, while length of snow drifts increases at increasing steepness on leeward slopes (Figure 27; Schneider, 1962; for the physical reason see Schmidt, 1970; Martinelli, 1975). In the timberline ecotone these effects of microtopography are partly smoothed or exacerbated by the influence of trees and tree stands on windflow, relocation of snow and also on incoming radiation (Section 4.3.11; see also Walder, 1983). Thus, the local situation always requires specific consideration.

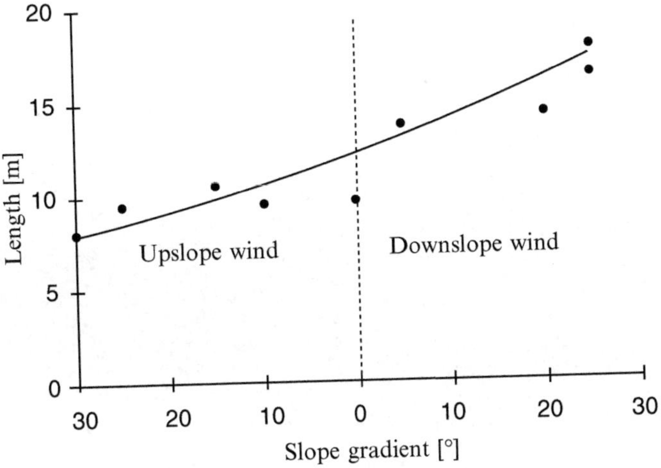

Figure 27. Maximum length of snow drifts formed by upslope and downslope winds as a function of slope inclination. After Schneider (1962), from Holtmeier (1996).

In the course of studies on precipitation and snow cover at timberline in the Gurglertal (Turner, 1961), snow melt started 1 month earlier in the forest than in the above treeless zone. This difference was supposed to be an effect of the forest canopy heated by solar radiation. The present author observed the opposite in the Colorado Front Range. On extensive wind-swept treeless areas above the forest-alpine tundra ecotone, winter snow disappeared much earlier than in the ecotone itself where big snow masses had accumulated. The mosaic of scattered trees and tree islands alternating with open glades increases surface roughness and thus snow accumulation (Photo 32; see also Photos 51, 52, 88 and 90). Big snow masses pile up particularly behind the leeward end of compact tree islands and other massive obstacles (e.g., blocks, solifluction lobes and aprons, terraces). Much snow is deposited also within scattered tree stands and in the glades, where it may last until the beginning or middle of July (e.g., 1979, 1990), occasionally even until early August (1984) (Holtmeier, 1986a, 1987a, 1996; Holtmeier and Broll, 1992; Broll and Holtmeier, 1994). In the closed forest below, however, it is the forest canopy that intercepts much snow. The intercepted snow partly vaporizes or is blown downwind. As a result, comparatively little snow covers the forest floor.

In wind-exposed treeless areas, snow cover, if not rapidly removed, gets increasingly compacted and hardened by the wind. After a while the snow surface may be hard enough to carry even a man. Consequently, higher wind speeds are needed to pick up snow particles from the snow surface. Gradually,

Photo 32. The forest-alpine tundra ecotone (*Picea engelmannii, Abies lasiocarpa*) on the west slope of Rollins Pass (Front Range, Colorado) at about 3.470 m (view SE). Because of the surface roughness (mosaic of clonal conifer groups and open glades) which enhances snow accumulation, the lower and the middle part of the ecotone are still covered with snow. In the closed forest, where most of the falling snow is inter-cepted by the forest canopy and in the Alpine, where the snow is blown off, snow has already gone. F.-K. Holtmeier, 8 July 1979.

snow surface will exhibit characteristic erosional features such as sastrugi, for example. On the other hand, snow usually remains loose within tree groves and scattered tree stands (Holtmeier, 1974, 1987b, 1996).

Mean snow depth may be used to characterize the climate character of a region. It does not provide, however, any information on the real locally varying snow pattern in the ecotone. Snow depth is usually greater at timberline in maritime mountains than in more continental regions. However, in areas with an average snow depth of 40 to 60 cm (e.g., Finnish Lapland), for example, the differences in depth between wind-exposed and wind-sheltered topography may easily exceed 2 or even more metres (Holtmeier, 1974; Autio and Heikkinen, 2002). On the other hand, even under extremely snow-rich conditions, such as reported from Paradise Valley (Mt. Rainier, Cascades; Blüthgen, 1980) with a mean annual snow fall of almost 15 m, convex topography may regularly be blown free of snow in winter.

In timberline ecotones characterized by alternating convex and concave topography the percentage of snow-covered surface (grooves, hollows and similar depressions, leeward slopes) is usually higher compared to small knolls or crests, which are snow-free or sparsely covered with snow. On extended wind-swept plain surfaces and less sculptured gentle slopes, however, snow may be completely relocated downwind to better wind-protected areas – unless scattered trees and more or less widely spaced tree clumps increase surface roughness and thus accumulation (Section 4.3.12).

4.3.7.2 Effects of the snow cover on sites

Positive and negative effects of winter snow are described in almost every treatise on high-mountain vegetation outside the tropics (e.g., Frey, 1977; Körner, 1999, see for further references). The following pages will thus refer only to those effects considered to be relevant to tree growth and to the general ecological conditions and spatial structures in the upper timberline ecotone (Table 11).

Soil moisture conditions are closely related to microtopography. The same substrate provided, soil moisture is usually higher at concave topography as on plain surfaces than on well-drained knolls, ridges and crests (cf. Photos 46, 47, 88 and Figures 32, 41 and 42). At concave sites, melting snow (run-off, seepage) provides additional water supply that may considerably increase soil moisture (e.g., Billings and Bliss, 1959; Holway and Ward, 1963; Billings, 1973; Cairns and Fonda, 1974; Holtmeier, 1996). In the field, however, each site requires specific consideration.

Table 11. The influence of snow cover on the trees at timberline

Positive effects	Negative effects
Protection from frost frost drought ice particle abrasion grazing, browsing Moisture supply (important on Rapidly draining substrate)	Short growing season Delayed rise of soil temperature → Germination impeded → Root growth delayed → Decomposition impeded → Nutrient uptake impeded Mechanical damage caused by → Snow break → Snow creep → Snow slides → Avalanches Snow fungi infections

In case of impermeable substrate, for example, melt water from the snow cover in situ will increase long lasting soil moisture. Also, different humus content and plant cover (composition, density, root biomass, root penetration) affect soil moisture conditions. On coarse debris or block fields, melt water rapidly percolates to greater depth. On less permeable ground at the lower, distal end of slope debris, seepage outflow may cause waterlogging (cf. Figure 43). High soil moisture or even waterlogging also occurs along the lower rim of persistent or long-lasting snow banks and snow patches typical of leeward slopes and behind other topography similarly favourable to snow accumulation. Particularly on sun-exposed leeward slopes additional water supply until summer from snow banks may be of paramount importance to the vegetation that otherwise might suffer from drought (cf. Figure 26; Holtmeier, 1996). However, long-lasting melt water supply may also cause leaching and impoverishment of the soils. On the other hand, fine aeolian material ('alpine loess') that was trapped by the rough snow surface or inside and at the leeward side of tree islands during the winter is accumulated at the soil surface (Thorn, 1978) and may become an important factor (fine texture, relatively high calcium content) to soil forming processes at such sites (see also Holtmeier and Broll, 1992; Broll and Holtmeier, 1994).

In winter, a closed and thick snow cover protects plants from injurious climatic influences (frost, ice particle abrasion, intense solar radiation). Under such conditions, even plants susceptible to frost and drought may survive at altitudes they would not be able to invade if they were not protected by snow. Seedlings and young growth still enjoying protection by sufficient snow cover are in the same situation as the chamaephytes. In the wind-swept upper part of the timberline ecotone on the Rocky Mountains, for example, some species such as *Ribes montigenum* and *Vaccinium myrtillus* (cf. Figures 65 and 74), usually growing in the more or less closed conifer forest, are restricted to the wind-protected and snow-rich leeward side of the wind-shaped tree islands, while grass and herbaceous vegetation prevails in the sparsely snow-covered or snow-free open sites. *Kobresia myosuroides*, for example, which is intolerant of deep, late-lying snow, is restricted to such locations (Holtmeier, 1978, 1996; Komarkova, 1979; Billings and Bliss, 1959; Burns and Tonkin, 1982; Walker et al., 2001).

Young trees growing higher than the average winter snow cover get increasingly exposed to adverse climatic effects. Under clear weather conditions extreme daily temperature cycles occur close to the snow surface. Needles and shoots warm up far above ambient air temperature by absorbing re-radiation from the snow surface, while nocturnal temperatures at the snow surface are very low due to radiative cooling. Under such conditions stressed foliage and shoots are prone to be damaged by frost and winter desiccation. Additionally, ice particles driven along the snow surface abrade unprotected shoots and needles. For several decades it thus depends on the weather conditions whether young growth and wind-shorn trees may adapt normal phanero-phytic growth (cf. Figure 56 and Photos 72–75). However, plants usually protected by snow in winter will also seriously suffer from climatic injuries if suddenly exposed by deflation or off-sliding snow, since they are less frost hardened than plants already projecting above snow cover.

Contrary to upright growing trees, krummholz species (in its original sense; Chapter 3) such as *Pinus mugo* and *Pinus pumila* are normally completely covered by snow in winter. The height of *Pinus pumila* stands, for example at timberline in the Japanese mountains, largely corresponds to the locally varying snow depth (Hultèn, 1974; Wardle, 1977; Okitsu and Ito, 1984; Nakashinden, 1994). At extremely wind-blown sites, these conifers exhibit mat-like growth forms not exceeding 10 to 20 cm in height (Wardle, 1977; Okitsu and Ito, 1984). In view of these observations *Pinus pumila* has been supposed to need a protective snow cover in winter (Wilmanns et al., 1985) which would not be in accordance with its very high frost tolerance (−70°C, Table 5), however. Nevertheless, Wardle (1977) found exposed needles slightly desiccated while lethal winter desiccation did not occur. In general, the needles of *Pinus pumila* are usually not strongly affected by winter desiccation because of being buried under the winter snowpack (Maruta et al., 1996; Takahashi, 2003).

On the other hand, *Pinus pumila* obviously does not tolerate being covered too long by snow, as is the same with *Pinus mugo*. Both conifer species suffer from parasitic snow fungi at such sites. In contrast, deciduous *Alnus maximoviczii* scrub does not become infected and thus may settle also snow-rich terrain (Wardle, 1977). Apparently, growing conditions are most favourable to *Pinus pumila* at more wind-protected sites characterized by moderate snow cover.

The length of the growing season corresponds only roughly to the snow-free period of the year. Young growth of evergreen conifers, for example, starts photosynthesis in spring though still being snow covered. At timberline in the Alps, young Swiss stone pines (*Pinus cembra*) 50 cm beneath the snow surface were able to compensate for respiration loss and exhibited even a positive net balance at decreasing snow cover (Tranquillini, 1959). Young trees that had become snow-free relatively late in the season usually started growth shortly after snowmelt, while at sites where the snow had already gone the beginning of growth was considerably delayed relative to melt-out (Tranquillini and Unterholzner, 1968). The same was observed in young growth of mountain pine (*Pinus montana*) and European larch (*Larix decidua*) at timberline on Stillberg (near Davos, Switzerland) (Figure 28; Turner et al., 1982; Turner and Streule, 1983).

Shoots of *Picea engelmannii* at timberline in the Colorado Front Range reached their maximum length 55 days after the snow cover had disappeared (1st May). If the spruces did not become snow-free before 22nd May, shoot growth continued for 45 days. At a melt-out on 1st June, however, shoots had already completed growth after 35 days (Wardle, 1968). In subalpine firs (*Abies lasiocarpa*) and Engelmann spruces (*Picea engelmannii*) distinctly projecting beyond the snow surface, the present author found the development of needles and annual shoots to be delayed for about 1 week at the still

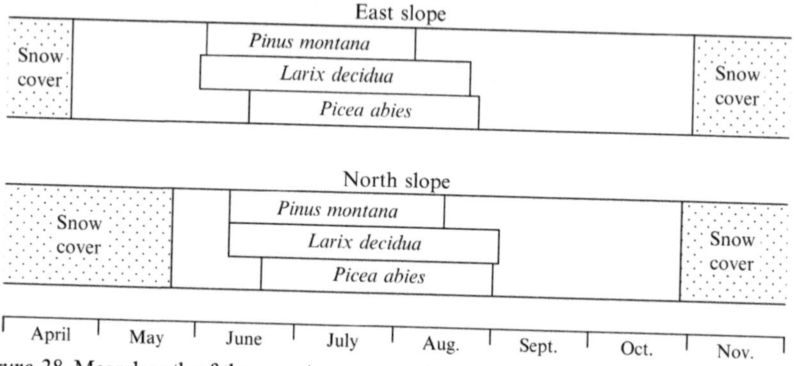

Figure 28. Mean length of the growing season of young trees (*Pinus montana, Larix decidua, Picea abies*) and duration of the snow cover on east- and north-exposed slopes (inclination 40°) on Stillberg (2.180 m, Dischma Valley, near Davos, Switzerland) in 1960–1965. Modified from Turner et al. (1982).

snow-covered base of the trees compared to the tree tops that had become snow-free much earlier. Also, in low wedge-like growing spruces and firs at extremely wind-exposed sites in the timberline ecotone (3.500 m) the phenological development of the snow-free windward foliage and also of the still snow-covered leeward needles was considerably delayed if compared to the better wind-protected but already snow-free foliage (Figure 29). To a certain extent increased growth rate may compensate for a short growing season, favourable weather conditions provided. Cold spells, however, often accompanied by late frost, may considerably delay phenological development, which additionally depends on age and height of the trees. Smaller young growth is usually covered longer by snow than taller plants, which also stored more energy. For example, in subalpine spruces and subalpine firs at the wind-exposed side of 'ribbon forests' (3.335 m, Colorado Front Range) that had been almost free of snow in the end of July/beginning of August (Photos 93 and 94), shoot elongation was almost completed in the second third of July, while in those young trees that had recently emerged from the snow needles were right flushing. Shoot elongation had not even commenced. In some plants, still partly covered with snow, buds did not open until the first week of August (Holtmeier, 1987b). Under such conditions, young growth and seedlings in particular will not be able to fully develop during the short growing seasons. As seedlings get older and taller their vitality gradually increases and they may survive even strong winters with extremely long-lasting snow cover.

The effects of the length and thermal conditions (as overlapping factors) of the growing season on tree growth and survival are different under different site conditions. At extremely windy and sparsely snow-covered timberline sites in northern Patagonia, for example, growth of *Nothofagus pumilio* is primarily controlled by temperature during the growing season while duration of the winter snow cover is the critical factor at snow-rich sites (Villalba et al., 1997).

Late-lying snow and resulting high soil moisture keep the soil temperature low until early summer. After soil has dried temperature rapidly increases (cf. Figures 76 and 77). Low soil temperatures until early summer hamper photosynthesis, decomposition, root growth and nutrient uptake as has been demonstrated in a previous chapter (Section 4.3.5). Such conditions usually prevent seedling establishment and sites will remain treeless (Photo 33, see also Photos 50–54).

On the other hand, thick snowpack insulates effectively the ground beneath, helps to conserve soil heat and minimizes frost penetration into the ground, contrary to sites sparsely covered with snow or even without snow. At snow-rich sites in the timberline ecotone on Stillberg (near Davos, Switzerland), for example, frost advanced to 15 cm depth in the beginning of the cold season. Then soil temperature gradually increased and varied closely around the freezing point throughout the winter (Turner et al., 1975; Turner and Blaser, 1977; Blaser, 1980; Schönenberger and Frey, 1988). Thus, roots of taller trees usually reach deeper than the ground is frozen.

Open sites, however, may freeze to several metres depth. Thus, the whole rooting zone may be affected. Although the upper soil thaws relatively early in the season, soil at greater depth may stay frozen until summer, particularly on shaded sites. In this connection we refer back to the heavy needle damage that occurred in spruce and pine in the Swedish Scandes during the extremely cold winter 1986/1987 (Section 4.3.3). Little snow on the ground let the soil freeze to great depth. In forest gaps the soil was still frozen just

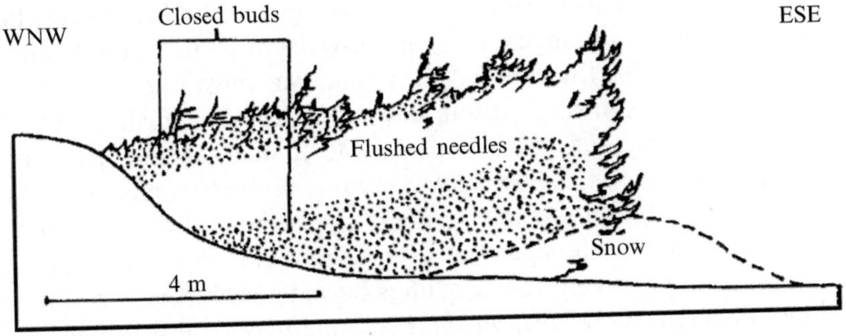

Figure 29. Phenology (18 July 1984) of a clonal group (*Picea engelmannii*) on Niwot Ridge (Front Range, Colorado) at about 3.450 m. The windward edge and the canopy of the group were already snow-free in the beginning of June, while the lower leeward part did not become snow-free before the last week of June. Needles flushed first in the relatively wind-sheltered part not too long buried in the snow.

Photo 33. Because of late-lying snow cover and snow-fungi infection (e.g., *Herpotrichia juniperi* and *Phacidium infestans*) conifers will not be able to invade this shallow depression (west slope of the Front Range, Colorado, at about 3.470 m). At the lower rim of the concavity, meltwater and seepage promote willows (*Salix plani-folia*). F.-K. Holtmeier, 2 September 1977.

below the surface after the snow had gone. Under forest cover soil remained frozen at 30 to 40 cm depth until the end of June. At these conditions, winter desiccation was supposed to have caused the needle damage (Kullman, 1989a). In warm summers, snow-rich sites are favourable to germination because of relative high soil moisture supply. This was evidenced, for instance, by the invasion of conifers into subalpine meadows close to timberline (cf. Photo 120) in many mountain ranges in the north-western United States since the beginning to the 1950s (Section 5.4). Invasion was most intense in cool and moist sites. A short growing season, however, strongly impaired growth of seedlings and saplings (cf. Photo 121). On Mt. Rainier, for example, seedling density (*Abies lasiocarpa, Tsuga mertensiana*) is highest on moderate slopes, covered with snow for 3 to 4 months. These slopes are well-drained, however, and carry subalpine heath vegetation (*Phyllodoce empetriformis, Vaccinium deliciosum*). Comparatively few seedlings became established on warm and dry slopes covered by *Festuca viridula* communities. The seedlings experienced much better growth, however (Table 12; Henderson, 1973; Franklin and Dyrness, 1973).

Also, the effects of the thermal conditions that have become more favourable compared to the past centuries ('Little Ice Age') are different on the west and east slope of Mt. Rainier. While trees have invaded the subalpine meadows on the snow-rich west slope almost continuously since the 1930s, invasion on the dryer east slope was restricted to a few short periods. In other words: warming and thus earlier snow melt has favoured advance of trees into the meadows on the west slope, while it impeded seedling establishment on the dry east slope (Rochefort and Peterson, 1996). Similar change occurred in the Olympic Mountains (Washington). *Tsuga mertensiana* growing at normally snow-rich, cool and moist sites (south-west slope) regenerated most intensely during warm and dry summers, while regeneration of *Abies lasiocarpa* that grows mainly at the drier sites on the north-east slope was favoured by snow-rich winters and humid summers. In areas with more balanced conditions such differences did not occur (Woodward et al., 1995).

In the Bernardino Mountains (California), seedlings of *Pinus contorta* and *Pinus flexilis* could invade normally snow-rich sites only temporarily during years with little snow cover, as was the case between 1945 and 1965. After a few decades most young growth died due to topographically enhanced accumulation of big snow masses. This happened after 1969 and between 1978 and 1980 (Minnich, 1984). On sun-exposed leeward slopes, particularly in dry regions such as the Rocky Mountains, melt water may considerably improve moisture supply to the plants and also locally enhance germination and seedling growth.

Table 12. Tree invasion, distribution, growth and growth forms of *Abies lasiocarpa* and *Tsuga mertensiana* at different sites on Mt. Rainier (Modified from Henderson, 1973; Franklin and Dyrness, 1973)

Meadow community	**Site conditions**	**Intensity of tree invasion**	**Spatial pattern**	**Growth rate**	**Growth form**
Carex nigricans	Late-lying snow, growing season <3 months	Very low	Singly	Very slow	Distorted
Phyllodoce empetriformis-Vaccinium deliciosum	Gentle slopes, moist, moderately to well drained soils, 3–4 months snow-free	High	Widespread	Slow	Distorted
Valeriana sitchensis-Veratrum viride	Steep and fresh slopes, avalanche chutes	Moderate	Singly or small groups	Moderate	Straight
Festuca viridula	Dry and warm, intensely disurbed by pocket gophers	Low	Singly or copses	Moderate	Straight

At extremely snow-rich sites, evergreen conifers are heavily endangered by parasitic snow fungi such as the brown snow felt fungus (*Herpotrichia juniperi, Herpotrichia coulteri*) and snow blight (*Phacidium infestans* with different races and varieties; Roll-Hansen, 1989) which develop best within the snowpack at high humidity and temperatures around freezing (e.g., Gäumann et al., 1934; Björkman, 1948, 1962, 1963; Petrak, 1955; Bazzigher, 1956, 1976, 1978; Donaubaur, 1963; Butin, 1996; Nierhaus-Wunderwald, 1996; Table 13). Wet and late-lying snow favours fungus development and infection of the needles buried in the snow. Infection causes more or less serious needle loss (Photo 34, see also Photos 85 and 89). In case of heavy needle loss seedlings and young growth usually die after having been infected the first time. Young trees will be safe from snow fungus infection only when having grown beyond the average winter snow cover. However, at the commonly slow growth in the timberline ecotone it my take a young tree 50 or even more years to grow higher than a 150 cm deep snow cover.

Table 13. Influence of temperature on the growth of three snow fungi (after different sources, from Nierhaus-Wunderwald, 1996)

Growth	*Herpotrichia juniperi*	*Herpotrichia coulteri*	*Phacidium infestans*
Minimum growth in culture	-3 °C to -5 °C	-5 °C	-3 °C to -5 °C
Limit of growth in culture	15 °C	10 °C	15 °C
Maximum growth in culture	30 °C	24 °C	27 °C
Growth at high humidity under deep snow cover	~0 °C	~0 °C	~0 °C

Snow fungi are very likely the factor most adverse to high elevation reforestation (e.g., Aulitzky, 1961, 1963a, b; Holtmeier, 1974; Schönenberger and Frey, 1988; Schönenberger et al., 1990; Senn and Schönenberger, 2001). During the wind shelter experiments on the wind-swept Pru del Vent experimental site (see Section 4.3.4), snow accumulation at the leeward side of artificial windbreaks prevented winter desiccation but, however, caused a snow fungus infection by *Phacidium infestans* and *Gremeniella abietina* to the young Swiss stone pines (Streule and Häsler, 2006). Also natural regeneration at timberline may experience high losses due to snow fungus infection, even on wind-exposed topography. As the seedlings grow taller they increasingly influence local windflow, and as a result snow accumulation and prolonged duration of the snow cover (see also Photo 85).

Snow blight attacks trees regardless of their condition. Another snow fungi, *Gremeniella abietina* (=*Ascocalyx abietina*), which is common to subalpine conifer forests from about 1.500 m up to tree line, causes shoot dieback in young trees and thus growth disturbances and reduced increment growth (Heiniger and Kanzler, 1988; Schönenberger and Frey, 1988; Nierhaus-Wunderwald, 1996). This fungus is common to subalpine conifer forests. Mainly young conifers are attacked. In high-altitude afforestations (between 2.080 and 2.230 m) on Stillberg (near Davos) about 60% of Swiss stone pines and almost 50% of prostrate mountain pine fell victim to *Gremeniella abietina*, while only 2% of larch trees were killed (Senn, 1999). Weakened trees in particular are highly prone to infection by *Gremeniella abietina*. Short, rainy and cool summers (formation of fruiting bodies, infection) reduce tree resistance and increase infection risk. The pathogenous stage of shoot dieback begins as soon as the fungus penetrates into the shoots buried under snow. Consequently, the highest losses by shoot dieback occur at sites where the winter snow remains longest (Figure 30).

Photo 34. Young Swiss stone pine (*Pinus cembra*) infected by *Phacidium infestans* in a snow-rich forest gap in the lower part of the forest-alpine tundra econe (ca. 2.120 m) on the northwest-facing slope of the Upper Engadine main valley (Switzerland) below Muottas da Schlarigna. F.-K. Holtmeier, 8 September 1996.

High snow cover also protects seedlings and young growth from animals such as red deer, ibex, chamois and other free ranging ungulates as well as from ptarmigans during the winter. On the other hand, small herbivorous mammals (mice, voles, pocket gophers, etc.) enjoying relatively favourable living conditions under the snow cover (protection from cold and predators) may cause heavy damage to buds, shoots and roots (girdling) of seedlings and young growth of the trees (Holtmeier, 1999c, 2002). Locally, sliding snow, snow creep and settling snow cover may cause mechanical damage to the trees (Section 4.3.11), which then are more easily attacked by parasitic fungi and insects. In particular, young trees completely buried by snow, are affected (Freiray and Schweingruber, 1994). Snow cover settles rapidly under warm weather and wet snow conditions. Branch-breakage usually occurs due to excessive winter snowpack. Snow breakage of stems and branches resulting in considerable loss of foliage cause growth reductions, as Kajimoto et al. (2002), for example, found in *Abies mariesii* at timberline on Mt. Yumori (northern Japan). Branches were lacking up to 4 to 6 m height from the ground or 2 to 3 m above the maximum level of snowpack (2 m) during normal winters. As changes in snowpack occur at almost regular intervals, they influence population dynamics in *Abies mariesii* forests. During extremely snow-rich winters, snow breaks limit canopy development of old trees and cause gaps in the forests, thus enhancing growth of previously suppressed nearby trees and facilitating regeneration (Kajimoto et al., 2002).

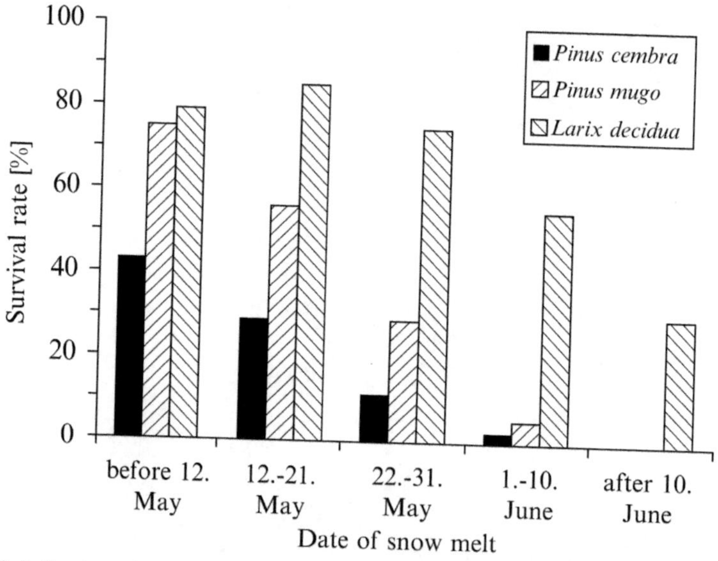

Figure 30. Influence of snow-cover duration on survival of *Pinus cembra*, *Pinus mugo* and *Larix decidua* in the forest-alpine tundra ecotone of Stillberg (Dischma Valley, near Davors, Switzerland) 20 years after planting. Modified from Senn (1999).

The presence or absence of tree species at timberline appears frequently to be a matter of enough but not too much snow (Cox, 1933, cited by Arno, 1984). This holds true in most cases but the ecological situation does not always follow the rule. For example, in the very windy timberline ecotones on many mountain ranges in the western United States, also Cox has worked there, by the way, snow cover would be lacking if low tree groups had not caused snow accumulation. Seedlings first established themselves behind stones, grass tussocks and willow shrubs that provided protection from the strong winds, and after the seedlings had grown taller they influenced windflow near the surface and thus increased snow deposition (cf. Photo 87, 90 and 91).

Snow conditions mainly controlled by the surface structure (microtopography, arborescent vegetation) not only influence growth, growth forms and distribution of the trees but also of the low-lying plant communities (dwarf shrubs, lichens, herbs, grasses, etc.) in the timberline ecotone. The patchy distribution of these communities clearly reflects snow depth and duration of the snow cover and thus the need of the plants to be protected by snow in winter. Frequently a pronounced topographic gradient is apparent. Wind-exposed knolls, crests and similar convex topography are usually covered with plants highly resistant to wind, frost and winter desiccation because thin snow does not provide sufficient protection from these influences. In the Alps, for example, dwarf azalea (*Loiseleuria procumbens*) growing espalier-like close to the soil surface, a few centimetres high mats of *Empetrum nigrum* and *Vaccinium uliginosum* and extremely hardy lichens (*Alectoria ochroleuca, Thamnolia vermicularis, Cladonia* species) are typical of such sites. Between the dwarf shrub-lichen patches mineral soil often is exposed and, at the best, sparsely vegetated with scattered *Juncus trifidus* and cushion plants such as *Silene acaulis*. Towards the better wind-sheltered downwind sites, continuously covered with snow in winter, the height of the dwarf shrubs increases and less hardy species such as *Vaccinium myrtillus* and *Rhododendron ferrugineum* (*Rhododendron hirsutum* on carbonate soil) prevail. Willow-scrub, rush mats, sedge communities and even snowbed vegetation occur in the snow-rich and often waterlogged hollows and groves adjacent to convex topography.

In the timberline ecotone in Lapland we observe similar topographic gradients reflecting the more or less great hardiness of the plants (Holtmeier, 1974; Broll et al., 2007). Wind-exposed topography is covered with espaliers of dwarf azalea, dense cushions of *Diapensia lapponica* and low *Empetrum hermaphroditum* mats, interfused or alternating with other dwarf shrub species such as *Betula nana, Arctostaphylos alpina* and *Vaccinium vitis idaea* and many lichen species (e.g., *Ochrolechia frigida, Cetraria nivalis, Cetraria crispa, Cladina* species, *Alectoria ochroleuca, Alectoria nigricans Thamnolia vermicularis* and others). At patches with the mineral soil exposed *Juncus trifidus* is common. Approaching the snow-rich sites, coverage and height of the dwarf shrubs increase while lichens are less frequent. Dwarf birch (*Betula nana*) is most flexible and exhibits a great variety of growth forms under different conditions. While usually mat-like at wind-swept sites, it grows 1 m high or even higher on snow-rich concave topography and leeward slopes.

Also in the timberline ecotone of the Rocky Mountains the distribution pattern of the low vegetation follows the same principle, although the species are different. Wind-exposed topography with little or no snow in winter are covered with alpine dry meadow (*Kobresia myosuroides, Carex rupestris*) and cushion plants such as *Selaginella densa, Trifolium dasyphyllum, Silene acaulis*) while herbaceous alpine meadows (e.g., *Acomastylis rossii, Deschampsia caespitosa, Trifolium parryii*) dominate snow-rich sites. Snow-bed communities replace this vegetation at sites covered with snow less than 75 days (cf. Figure 65). Wet meadows and willow thickets are typical of moderately snow-covered but often waterlogged shallow hollows and grooves (cf. Billings, 1973, 1979; Burns, 1980; Holtmeier, 1978, 1996).

Sequences of plant communities controlled by snow cover conditions along a microtopographical gradient (transect) as described above reflect almost a rule that became represented in so-called ecograms used in high-altitude re-forestation (Aulitzky, 1963a; Turner et al., 1975). These ecograms show a transect through a small crest and the adjacent downwind groove with the plant communities corresponding to the different snow cover conditions (depth and duration), relative wind velocity, radiation load, soil moisture and risk of snow-fungus infection along the topographical gradient. Thus, the overlapping effects of the different ecological factors become apparent 'at first sight', and the plant communities are used as biological indicators. These ecograms were done to make it easier for the workers involved in re-forestation of abandoned high-altitude to assess local site conditions by just visually comparing the ecogram with the given situation in the field and to apply then the appropriate planting techniques.

4.3.8 Soils

Because of its varying physical and chemical properties and its many functions soil belongs to most complex site factors, also in the timberline ecotone. Studies on soils at upper timberlines are relatively rare, except for the Alps. In Swiss and Austrian research programs comparatively many investigations on soils have been carried out for creating a sound scientific base for re-forestation of abandoned alpine pastures and other areas at high elevations from which man had removed the forest during history (e.g., Aulitzky, 1961; Neuwinger, 1967, 1970, 1972, 1978, 1980, 1986; Blaser, 1980; see also

Section 4.3.14.2). Subalpine and alpine soils were also studied in connection with basic research for prevention of torrential washes and avalanche catastrophes (e.g., Czell, 1967, 1972; Neuwinger and Czell, 1959; Österreichische Akademie der Wissenschaften, 1980, 1985; further references there). Little is known about soils at tropical mountain timberlines, and the existing information is very general.

Most investigations on mountain soils refer to the alpine and periglacial (subnival) belt. Very locally varying mosaics of different soils characterize these altitudinal belts and also the timberline ecotone (e.g., Blaser, 1980; Neuwinger, 1984). Thus, for example, Regosols, Cambisols, Podzols and hygromorphic soils may form such a mosaic corresponding to the given micro-topography (cf. Figures 32 and 42). No single soil type can be considered typical of the timberline ecotone, either in the temperate or in the tropical mountains. Although soil forming processes and also the effects of the soils on vegetation (temperature, water supply, decomposition and plant available nutrients) depend partially on the altitudinal change of climatic conditions, the influence of microtopography, different parent material, microclimates and plant cover on soils in the ecotone is by far more important than the effect of elevation (Friedel, 1967). Moreover, the physical quality and chemical properties (e.g., evergreen or deciduous, amount and C/N ratio of the litter), structure (open, dense, height) and coverage (low, high) of the vegetation influence pedogenesis. From a study on decomposition of organic matter at timberline on Stillberg (Dischma Valley, Swiss central Alps; Bednorz, 2000) it became apparent that decomposition on ridges with mor (humus form) depends primarily on temperature while at groove sites with Rhizomulls and higher decomposition rates the litter quality (C/N ratio) controls decomposition.

Human use, such as grazing cattle and lumbering, has persistently disturbed soil development in the present timberline ectone and also above (Section 4.3.14). Probably, there is no other factor besides tree rings reflecting site history (climate, use, fires, slides, etc.) as clearly as soils. Altogether, at the given great variety of soil conditions it does not make much sense to expect any altitudinal gradient (e.g., pH) in soil-ecological conditions that might be related in any way to timberline.

Though results of local studies on soil conditions can usually not be applied to other timberline areas, some general principles common to soil at timberline will be considered. Most soils in the ecotone are shallow and exhibit a relatively high percentage of skeletal material and humus. Mixing of mineral and organic matter by bioturbation has occasionally been considered unimportant (e.g., Bochter et al., 1983). However, in the Cascades and in the Rocky Mountains, for example, bioturbation obviously has a great effect on pedogenesis (Bockheim, 1972; Holtmeier, 1982, 1987b, 1999c, 2002; Schütz, 1998).

Ranker or Rendzina of different quality are widely distributed on mountain slopes, in steep gullies, in avalanche tracks (winter-cold mountains) and similar locations subjected to erosion. Also, unstable slope debris impedes soil development (Photo 35). Initial soils are also common on wind-exposed topography due to wind-erosion (Photo 36; Figure 31). Badly drained sites such as former melt water channels and shallow wet depressions between convex topography often exhibit organic soils or Gleysols as can frequently be observed at timberline in the Alps, in northern Europe or in the Rocky Mountains, for example (cf. Photos 46 and 47; Figures 32 and 42). Most soils in the timberline ecotone are relatively young, particularly if the ecotone extends up to cirques that were repeatedly covered with temporally advancing glaciers during the Holocene, as was the case, for example, in the Colorado Front Range (Benedict, 1973, 1981) and other high mountain ranges in the western United States. On the other hand, soils on interfluves that never were glaciated are comparatively old.

Photo 35. Forest is almost unable to invade the unstable dry debris on the south slopes of Mt. Tukuhnivatz (La Sal Mountains, Utah), where almost no soils have developed. F.-K. Holtmeier, 21 July 1994.

In the Alps, in the Scandinavian Mountains, the Rocky Mountains and many other temperate high mountains, Podzols have developed under coniferous forests. In the timberline ecotone of the Colorado Front Range, for example, such 'forest soils' alternate with soils more typical of the

Photo 36. Organic layer and top soil eroded on this wind-exposed site (327 m) on the northern slope of Koahppeloaivi (northern Finnish Lapland). Lack of nutrients and soil moisture deficiency prevent birches (*Betula tortuosa*) from invading this formerly forested site. F.-K. Holtmeier, 8 September 1996.

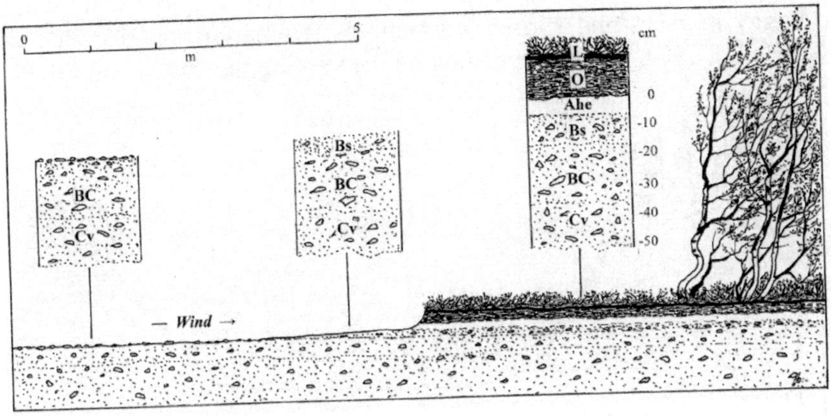

Figure 31. Soil profiles along a transect through the wind-eroded site (Photo above). Left of the wind scarp (*middle* profile) the organic layers and the upper mineral soil horizon of the original Podzol (on the *right*) are missing while further upwind (profile on the *left*) the Bs horizon also eroded.From Holtmeier et al. (2004).

alpine belt, such as Cryochrepts and Cryumbrepts (Komarkova, 1976, 1979; Burns, 1980; Litaor, 1987; Holtmeier and Broll, 1992; Broll and Holtmeier, 1994), shallow soils exhibiting already (visible) distinct horizons (Cambisols,

International Society of Soil Science Working Group RB, 1998; Inceptisols, Soil Survey Staff, 1998). They developed at moderately to well-drained sites. Previously, they were called 'alpine turf soil' (Retzer, 1956; Johnson and Cline, 1965) that may be compared to 'alpine brown soil', which also is an old term out of use now. Holtmeier and Broll (1992) found distinct leaching (Cryoboralfs) under dense conifer tree islands above the closed forest. In a few cases initial podsolization was evidenced (see also Broll and Holtmeier, 1994). Also Komarkova (1976, 1979) emphasizes sharp contrasts between soils under conifer islands and surrounding alpine vegetation. Similar conditions were reported from other timberlines in North America (Bliss and Woodwell, 1965; Nimlos and McConnell, 1965; Harries, 1966; Bockheim, 1968, 1972).

In the lower part of the timberline ecotone in the Front Range, Dystric Cambisols are the most common soils on gently sloped old surfaces (Retzer, 1974; Burns, 1980), while shallow leached soils occur under coniferous scrub at timberline on steep valley sides. In high-lying glacially moulded valley heads and on the floor of cirques, the timberline ecotone is characterized by a very locally varying mosaic of different soils that is closely related to microtopography (cf. Benedict, 1973, 1981; Holtmeier, 1978, 1979a; Haase, 1983, 1987). Dystric Cambisols are to be found on roche moutonnées covered with conifers, while different wet mineral and organic soils have developed in poorly drained hollows and grooves where snow accumulates and stagnant water is present for most of the growing season (Figure 32; see also Burns, 1980). The wet sites are usually covered with low willow thickets, boggy marshes and spring communities (Komarkova, 1976; Benedict, 1981; Haase, 1983, 1987), depending on the varying moisture conditions.

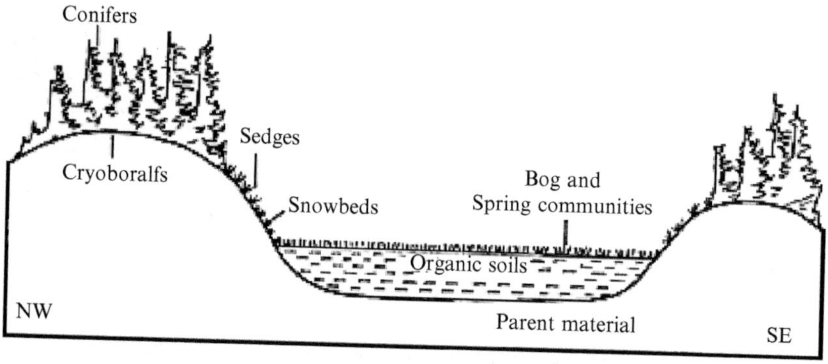

Figure 32. Relationships of soils and vegetation to local topography in the forest-alpine tundra ecotone on the floor of high elevation cirques (schematic, based on the situation in the Front Range, Colorado).

Similar locally varying soil mosaics, mainly caused by alternating convex and concave microtopography, are obvious also at timberline in the Alps and in the Scandinavian mountains. In the Alps, for example, alpine Cambisols have originated under alpine turf, while Podzols were formed under dwarf-shrub vegetation on crystalline substrate. Poorly drained concave topography is characterized by wet or even water-saturated mineral and organic soils covered with marshy meadows or bog communities. A special type of Rendzina characterized by an acid organic layer overlying the calcareous soil occurs under stands of spruce and mountain pine with acidophilic dwarf shrubs. It is interlocked with other calcareous Rendzina typical of the alpine grassland.

Mature Rendzina soil profiles result from a long and undisturbed soil development, which is usually an exception in the timberline ecotone. In many sites we observe truncated or buried soil profiles. Such disturbances are common to the soils of extensively grazed alpine pastures in the Upper Engadine (Switzerland), for example ('alpine pasture soil' in the sense of Pallmann and Haffter, 1933; colluvial Podzols in the sense of Neuwinger, 1970). Podzols and remains of Podzols occurring above the closed forests are often considered to indicate ancient forest that had been removed by human impact or receded due to unfavourable climate. Thus, for example, remains of eroded Podzols in the upper, wind-swept ecotone on Niwot Ridge (Colorado Front Range) which is only sparsely covered with stunted wind-shaped trees and tree groves at present, provide evidence of a former forest that declined probably because of adverse climatic conditions (Broll and Holtmeier, 1994), as has been evidenced also in other Rocky Mountain areas (Andrews et al., 1975; Carrara et al., 1984; Beaudoin, 1989). With the given dry regional climate, active podzolization is unlikely in the alpine zone of the southern Rocky Mountains (see also Johnson and Billings, 1962; Johnson and Cline, 1965; Burns, 1980).

Remains of former forest soils can also be found above the present closed forest in the Alps. On the Gurgler Heide (above Obergurgl Village, Gurglertal, Tyrol), for example, Podzol remains occur up to an elevation of 2.800 m under the humus layer of alpine Cambisols and also at the upper limit of alpine dwarf-shrub vegetation. The original Podzol was very likely formed during the Holocene climatic optimum (Neuwinger, 1970). Also in the Upper Engadine Podzol relics are common under alpine pastures and dwarf-shrub vegetation (e.g. Müller, 1983). The Podzol remains are partially buried under 10 to 30 cm of colluvial material that contains charcoal. They also give some evidence of former forest reaching to a greater elevation than the present forest. Rübel (1912), for instance, supposed the upper limit of the dwarf azalea (*Loiseleuria procumbens*) to almost coincide with the position of the timberline at the postglacial climatic optimum. In the Valaisian Alps, podzolization occurs up to 2.570 m. Fully developed Podzols were found up to 2.550 m indicating former coniferous forest and/or heath communities (Carnelli et al., 2004). In the Upper Engadine, Hiller (1996, 2001) found evidence of podzolization even beneath snow-bed communities at about 2.600 to 2.800 m, far beyond the present anthropogenic timberline. However, the uppermost occurrences of Podzol remains can certainly not be ascribed to former forest cover. According to Neuwinger (1970) active podzolization (Iron-Podzols with a fully developed humus profile) can only be expected below the potential timberline at sites covered with moss-rich dwarf-shrub vegetation and also in *Pinus cembra* stands with undergrowth of *Rhododendron ferrugineum*.

In northern Europe, Podzols developed not only under coniferous forest but also under mountain birch forest. However, in this case it is the litter of ericaceous dwarf shrubs on the forest floor that favours podzolization. Shallow Podzols are typical of dwarf-shrub-lichen-heaths above the closed mountain birch forests, where podzolization goes on also with the present climate (communication G. Broll; see also Broll et al., 2007). In northernmost Finnish Lapland we found eroded relict Podzols far above the present birch forest limit. Together with wood remains found mainly in eroded peat layers (Holtmeier and Broll, 2006) and peaty hummocks they give evidence of the existence of former birch forests at greater altitude than at present (Holtmeier et al., 2003).

Podzolization, generally supposed to be restricted to coniferous forests at middle and northern latitudes, also occurs in many tropical mountains from a certain altitude up to timberline (Hardon, 1936; Jenny, 1948; Askew, 1964; Harris, 1971; Burnham, 1974; Smith, 1977b). Above timberline, however, relatively dry climatic conditions and diurnal freeze–thaw cycles prevent podzolization (Zeuner, 1949; Coe, 1967; Agnew and Hedberg, 1969).

Also at the tropical mountain timberline conditions for soil formation vary very locally depending on microtopography, substrate, microclimate, plant cover and soil moisture, although winter-snow cover and its effects on site conditions is missing. Thus, one should desist from any generalisation (Askew, 1964; Smith, 1977b; Rehder et al., 1988).

Soils that originated from lava, pyroclastic falls and flows, volcanic ashes and debris need to be specifically considered. They are particularly common to many high mountains of the Tertiary mountain belt, particularly in the tropics and subtropics and also on the volcanoes of East Africa and many oceanic islands. Due to very different chemical and mineralogical properties (e.g., allophane content due to weathering of volcanic glass), texture, conditions during ash deposition (wind direction, wind velocity) and weathering different soils (Andosols) developed (Shoji et al., 1993a, b; Kimble et al., 1999, further references there).

Andosols (IUSS Working Group WRB, 2006) or Andisols (Soil Survey Staff, 1998), for example, widely occur at high elevation where they developed under grass and/or forest cover on well-drained tephra not too much eroded. These volcanic soils have been intensely studied. Comparatively little information, however, is available on their influence on sites and growing conditions at timberline. Although exhibiting different properties due to the different local parent material, most Andosols are characterised by a several decimetres thick humus-rich A_h-horizon covering the underlying tephra. Moreover, the bulk density of tephra is usually low whereas permeability is very high. On Kilimanjaro, for example, such dark Andosols of about 60 cm depth occur in the ericaceous belt rich in giant groundsels (3.000 to 3.500 m). At the upper limit of this zone the dark Andosols are interlocked with high-desert soils that also developed on volcanic ash (Franz, 1979). On Mt. Kenya, Andosols prevail in the *Hagenia abyssinica*-zone between 2.700 and 3.400 m (Lange et al., 1997).

Andosols are also widespread at timberline in Middle and South America. Lauer and Klaus (1975a) report Andosols on Pico de Orizaba (Mexico). At an elevation of 3.000 to 4.000 m scattered forest and scrub grow on these dark soils that cover a light-coloured substrate. Soil depth often exceeds 1 m. Bader et al. (2008b) mention deep dark humic Andosols that developed on old and more recent volcanic ashes at treeline sites (3.500 to 3.600 m) in the Ecuadorian Andes. Frei (1958) describes 'black Andean soils' from the upper montane forest in this region. Andosols of 70 to 90 cm depth, not much differentiated, are to be found under dwarf forest (3.900 to 3.950 m, upper limit of the closed forest 3.600 to 3.800 m) in the Columbian Central Cordillera: 'Humic Cryandepts' on the east slope, 'Dystric Andepts' on the more humid west side (Salomons, 1986; Verweij, 1995).

The acid and comparatively nutrient-poor organic layer (Moder) is about 50 cm thick (Thouret, 1989). Hildebrand-Vogel et al. (1990) report relatively young Andosols developed under *Nothofagus pumilio*-forests (timberline 1.300 to 1.400 m) on Choshuenco volcano in eastern Patagonia: High permeability prevents waterlogging and reduces surface run-off although precipitation amounts to 4000 mm. At low elevation in warm and humid regions, Andosols have developed on young ashes. At high elevation, however, Andosols occur mainly on older tephra because of slow weathering (Martini and Palencia, 1975).

In the following, the effects of volcanic soils on timberline will be considered exemplarily. Generally, recent volcanic ash blankets have caused a depression of timberline below its potential climatic limit. It depends on the given climate, soil conditions and distance to the seed trees how far and how rapidly forest may advance again to higher elevation. The depth of volcanic ash seems to be a key factor for natural succession after ash was deposited. After the eruption of the Hudson volcano (Chile) in 1991, for example, all forest stands at timberline were destroyed, and in many areas timberline receded for 100 to 300 m (Vogel, 1996, 2000). Also, on the sides of Antillanca Valley (Puyehue National Park, southern Central Chile) volcanic scoria and ash deposition caused a similar depression of the timberline (*Nothofagus pumilio*). Climate and, on steep slopes, avalanches prevent forest from advancing to greater altitude. In addition, instability (steep slopes) of the loose and coarse material, particularly of scoria, hamper natural reforestation. Moreover, fine material that would be favourable to soil development is blown off by permanent strong winds. Thus, soil in the proper sense usually does not exist in the *Nothofagus pumilio* scrub zone. During summer, moisture conditions may become critical to invading trees because of high temperatures occurring on the scoria surface (Veblen et al., 1977).

After the eruption of Ksudach volcano (southern Kamchatka) in 1907, a pumice layer more than 1 m thick covered the land surface, even at a distance of 20 km from the volcano. The plant cover was completely destroyed, and only lichens and scattered occurrences of herbs and shrubs have returned as of now. The depth of the ash varies because it was repeatedly relocated by wind and erosion processes. In areas with ash deposits not deeper than 70 cm, birch stands (2 m high) interspersed with dwarf shrub and lichens have developed. Locally, even older birch trees are to be found that have survived the catastrophe. A few stunted Siberian dwarf pines (*Pinus pumila*) that originated from seed caches of the nutcracker (*Nucifraga caryocatatces*) are admixed to these stands. Also *Alnus kamtschatica* is represented. Acid substrate, drought and very low nutrient supply are adverse factors to invading trees. In addition, the great instability of the substrate impairs forest advance, as can be concluded from the many roots that were shorn off and from birches that were uprooted though having an extended root system (>500 m^2). At such places, only initial soils have developed. Even birches the roots of which penetrated the ash layer and advanced to the humus layers of the buried old soils are heavily suffering from nutrient deficiency. Because of the unfavourable substrate and also due to harsh climatic conditions, soil formation goes on very slowly. Grishin et al. (1996b) suppose that on dazitic ash deposits the development of a 'soil-vegetation-system' that would be comparable to the system prior to the eruption, will take more than 2.000 years at least. Succession on basaltic lava, on the other hand, is proceeding faster (Grishin and Del Moral, 1996).

However, 8 to 10 years after the eruption of Shiveluch volcano (central Kamchatka) herbaceous vegetation, willows, poplars (*Populus suaveolens*), alder (*Alnus kamtschatica*), young growth of larch (*Larix kamtschatica*) and spruce (*Picea ajanensis*) had already established themselves on the light dazitic pumice deposits. In contrast, succession is comparatively slow in those areas that were located in the main direction of the explosive eruption. On pyroclastic flows, most favourable conditions for re-colonisation by vegetation are provided if fine organic matter accumulates between the pumice particles thus increasing soil moisture in the uppermost ash layers. Succession and pedogenesis depends on how rapidly closed alder stands will establish themselves, providing sufficient litter and nitrogen supply. This process takes approximately 50 to 100 years (Grishin and Del Moral, 1996).

The substrate properties may also be responsible for the altitudinal position of timberline. For example, on the tephra-covered south-eastern slope of Mauna Loa volcano (Hawaii) timberline (*Sophora chrysophylla*) is located at about 2.900 m while single trees occur 50 and more metres higher on the lava flows that project the ash deposits a little. The upper limit of *Sophora chrysophylla* stands is lower on the dry northern side compared to the more humid slope areas. Henning (1974) attributed this difference to the low water-holding capacity of the ashes and to slow soil formation above the zone of orographic rainfall (trade winds). On the lava flows, fine material originating from weathering of the substrate or relocated ashes accumulates in cracks and ruptures and increases moisture. In addition, within the cracks evaporation is lower because of reduced solar radiation (shade). Thus, on the lava flows, trees enjoy relatively favourable growing conditions if compared to the dry tephra deposits. Altogether, the altitudinal limit of tree growth is likely caused by low precipitation above the trade wind inversion and, which is as important, limited plant-available soil moisture (Henning, 1974).

When the Hoei (a secondary crater of Mt. Fuji) erupted in 1707, ash deposition caused a timberline depression of approximately 400 m. The present timberline ecotone extends up to 2,400 m and then merges into an altitudinal zone dominated by herbaceous vegetation. The upper 50 to 90 m broad 'dwarf forest' zone (0.5–1.5 m high 'trees'; Masuzawa, 1985) with alders (*Alnus maximoviczii*), willows (*Salix reinii*), birches (*Betula ermanii*) and larches (*Larix leptolepis*) borders high-stemmed larch stands (*Larix leptolepis*) that are replaced by fir-spruce forest (*Abies veitchii, Picea jezoensis*) with some admixed hemlocks (*Tsuga diversifolia*) at lower elevation. The oldest larches in the upper part of the ecotone are pioneers of the re-invading mountain forest. These stunted trees are about 65 years old now, while the oldest individuals in the closed forest below exceed 200 years. A few trees are even older than 300 years (Saito, 1971, cited by Ohsawa, 1984). Larch is a pioneer tree in this area mainly because of its ability to reproduce and propagate by root suckers. While the ashes are poor in nutrients they meet the ecological requirements of larch, however, because they are well aerated (Takei, 1995). Though almost 300 years have passed since the eruption of Hoei, soil formation has not yet advanced beyond the initial stage (Photo 37). The still low nitrogen content, lack of organic substances and favourable seed beds, instability of the substrate, high snow cover and insufficient seed dispersal impede vegetation from invading the ash-covered areas (Masuzawa, 1985; Maruta, 1994; Del Moral and Grishin, 1999).

Photo 37. Larch (*Larix leptolepis*), dwarf willows (*Salix reinii*) and alder (*Alnus maximoviczii*) are pioneers on the volcanic ashes of Mt. Fuji (Japan). Though almost 300 years have passed since the last volcanic eruption, soil formation is still at an initial stage. G. Broll, 26 August 1990).

Nevertheless, litter accumulation between the stone fragments on the surface gradually augment carbon and nitrogen content and thus increase water-holding capacity of the highly permeable substrate. The accumulated organic matter does not originate from the local vegetation only but also comes from lower elevation by upslope winds. Obviously, nitrogen-fixing plants such as alders that produce large amounts of litter play an important role in nitrogen increase and soil development (Ohsawa, 1984; Masuzawa, 1985; Masuzawa and Kimura, 1987; Chapin III et al., 1994; Walker, 1999).

The timberline zone on Tenerife is also strongly influenced by the volcanic substrates. Different types of Andosol prevail. While comparatively dry Andosols have developed on southern exposures, less dry types are common on the northern slopes (Caldas and Delgado, 1971). Also topography and age of the volcanic substrate affect soil development. Thus, depending on substrate age and slope gradient, different stages of Andosol development occur, ranging from simple Lithosols to initial and fully developed Andosols (Höllermann, 1978, 1982). Obviously, the geological age of the substrate is the factor controlling the altitudinal position of the upper timberline. Timberline is highest (2.400 m) on old phonolithic substrate with advanced soil formation, while it is comparatively low (1.800 to 2.000 m) on young basalts and trachyte where still Lithosols prevail. As already demonstrated from the Hawaiian volcanoes, the surface-near layers of the usually very permeable substrates (lava, pumice) are very dry due to rapid draining. The deeper layers are protected from evaporation and thus are relatively moist. However, seedlings with still shallow root systems are not able to use this moisture. Basaltic sands that are not protected from solar radiation and evaporation by pumice or stone layers get completely dry to 30 or even more centimetres depth during the long dry summer season above the trade-wind inversion (upper cloud surface at 700 to 800 m, i.e., below timberline). For many weeks, available soil moisture falls far below the wilting point. Severe conditions of the undeveloped soils negatively affect regeneration on high altitude sites (Höllermann, 1978; Šrutek and Lepš, 1994; see also Chapin and Bliss, 1989). The properties of the volcanic substrate influence the altitudinal position of timberline more than exposure, for example (Höllermann, 1978). However, also thermal influences interfering with the edaphic conditions in a complex way play an important role at timberline.

Also on Popocatepetl the combined effects of climatic factors and recent volcanic deposits determine the altitudinal position of the upper timberline. Drought and instability of the substrate affected by deflation and surface runoff seem to prevent the establishment of trees (Beaman, 1962).

Mycorrhiza is another important edaphic-biotic factor at timberline. This symbiosis of roots and fungi enables the trees to take up sufficient nutrients as the fine mycelium better traps nutrients than the roots proper. The fungi are also able to metabolise phosphorus and other minerals thus making it available to the trees. In addition, the fungi stimulate root growth and produce substances that protect the roots from infections and pathogenic organisms (cf. Slankis, 1973; Marx, 1973; Allen, 1991). From a global view about 5.000 ectotrophic mycorrhizal fungi exist (mainly Basidiomycetes and Ascomycetes). On the other hand, about 50 fungi only form endotrophic mycorrhizae. Endotrophic mycorrhiza is particularly common to tropical tree species and not that strongly tied to specific host tree species. In the temperate zones of both hemispheres upper timberlines are formed mainly by tree species growing with ectotrophic mycorrhiza (e.g., *Abies*, *Picea*, *Pinus, Larix*, *Tsuga*, *Nothofagus*, *Podocarpus* and others). Also, in the other climatic zones, ectotrophic mycorrhizae are common or even prevail (Moser, 1967; Trappe and Fogel, 1977). There is little information on the influence of the tropical mycorrhizae on trees. This holds particularly true with respect to the timberline zone (Allen, 1991). VA-mycorrhizae (endotrophic) have been found in *Polylepis*, for example (Hensen, 1993). Probably this mycorrhiza enables *Polylepis* to grow at very high altitude and to invade successfully initial soils (Kessler, 1995).

The fact that in winter-cold mountains outside the tropics ectotrophic tree species advance to much higher elevation than endotrophic species or species devoid of mycorrhizae seems to indicate that ectotrophic mycorrhizae are an advantage to the trees at the given harsh climatic conditions at timberline (short and often unfavourable growing season). Endotrophic mycorrhizae enable the trees to acquire sufficient resistance to adverse climatic effects. The ectotrophic mycorrhizal fungi are able to exploit the major nutrients nitrogen and phosphorus contained in the organic substances. In addition, the fungal root mantle makes the roots less susceptible to attacks by pathogens (Read, 1998). Mycorrhiza may also act as a nutrient reservoir from which nutrients can be obtained on demand (Haselwandter, 2007).

In the opinion of Moser (1967) the natural climatic timberline would be located several hundred metres lower (in the Alps at about 1.500 m) if it were formed by endotrophic species or species lacking any mycorrhiza. The map of Gasteiner Valley (northern slope of the High Tauern, Austria; Figure 33; Schinner, 1978) showing the distribution of endotrophic and ectotrophic tree species seems to confirm this hypothesis. Environmental conditions

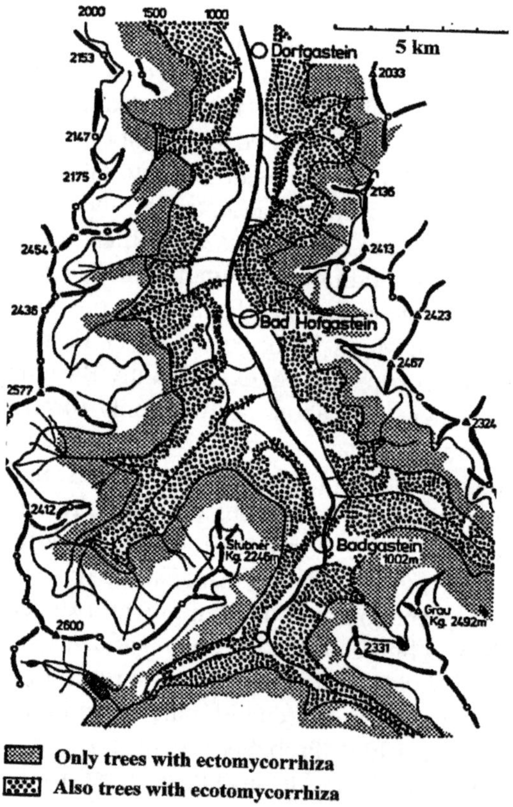

Figure 33. Distribution of tree species with (finely dotted) and without (coarsely dotted) ectotrophic mycorrhiza in the mountain forest of the Gasteiner Valley (Austria). From Schinner (1978).

appear to be limiting the altitudinal distribution of ectomycorrhizal fungal taxa. Thus, a great number of ectomycorrhizal fungal species occurring in the sub-alpine forest have never been found in the timberline ecotone (Moser, 1982). Also, in the eastern Canadian Rocky Mountains, richness and diversity of ectomycorrhizal fungi decrease with elevation (Kernaghan and Harper, 2001).

Ectomycorrhizal fungi already existing on pioneer plants may encourage the growth of other plants that become established concurrently or later. A study on Mt. Fuji (Nara and Hogetsu, 2004), for example, suggests that ectotrophic mycorrhiza on willow shrubs (*Salix reinii*) which became established on the volcanic scoria from the eruption in 1707 facilitate the neighbouring seedlings of other woody plant species such as planted willow seedlings, birch (*Betula ermanii*) and larch (*Larix kaempferi*). Increased growth of birch seedlings near willows on eroded sites in Iceland must possibly be attributed to infection by the ectomycorrhizal fungi existing on the roots of willows nearby (Magnússon and Magnússon, 2001).

The role of mycorrhiza is of practical concern for the success of high-altitude afforestation. In the Alps, for example, afforestation of abandoned alpine pastures, often covered by ericaceous dwarf shrub heath, with Swiss stone pine (*Pinus cembra*) may be more successful when the seedlings are inoculated with appropriate ectomycorrhizal basidiomycetes already in the nursery (Moser, 1964; Haselwandter, 2007).

4.3.9 Topography/geomorphology

Though the altitudinal position of climatic timberline depends on the zonal and regional climates, topography, and microtopography in particular, are the key factors controlling the locally varying site conditions and patchiness of vegetation and thus the spatial structure and physiognomy of the timberline ecotone (Figure 34; e.g., Holtmeier, 1974, 1996, 2000, 2005a, b; Holtmeier and Broll, 2005; Butler et al., 2004, 2007; Resler et al., 2005; Broll et al., 2007). The effects of microtopography on radiation load and wind completely override the influence of elevation (Friedel, 1967). Exposure to wind and solar radiation controls temperature and moisture (precipitation, relocation of snow, evaporation, etc.) and thus cause more or less different microclimates that affect other site factors and ecological processes such as mineralisation, for instance.

Because of surface run-off and seepage in particular the top of convex topography is usually characterized by drier conditions than the sloping sides and adjoining concave micro-relief. In winter-cold mountains relocation of snow exacerbates or smoothes these contrasts in soil moisture conditions, as has already been demonstrated in the previous chapters (Sections 4.3.6 and

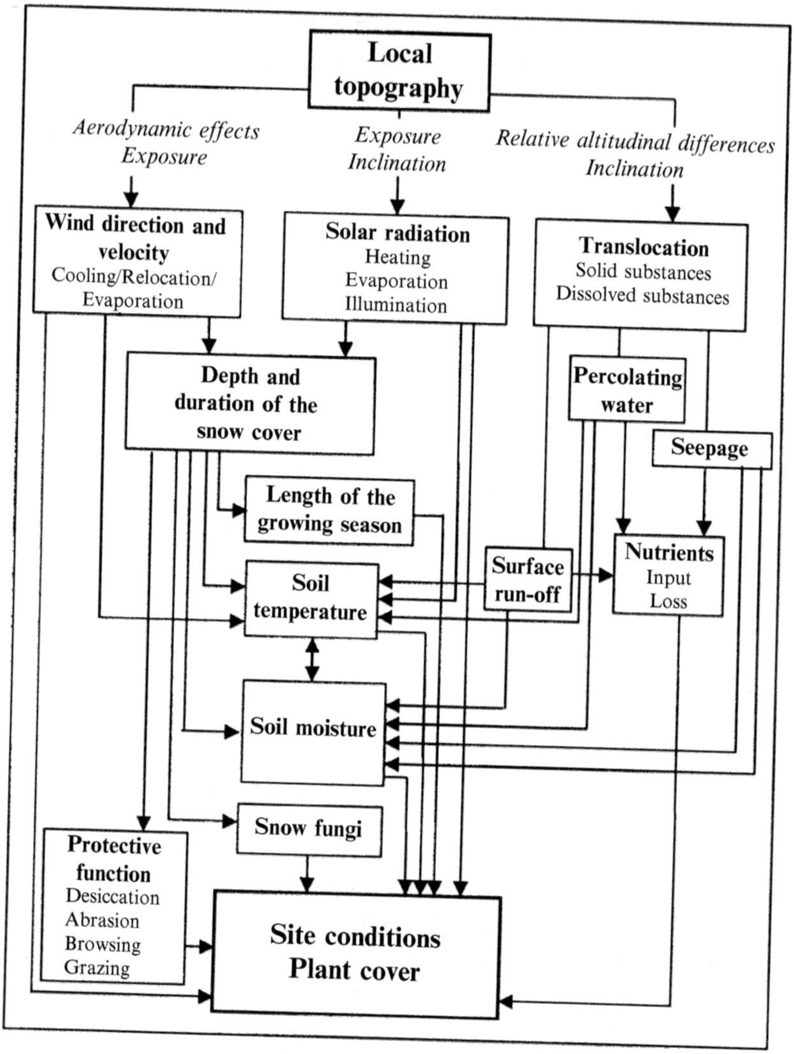

Figure 34. Influence of local topography on site factors, site conditions and plant cover. The site factors have direct and indirect effects on site conditions and vegetation.

4.3.7). Thus, only some additional information will be given in the following if necessary for understanding. Not least, topography (steepness, accessibility, avalanche zones, etc.) often controls human impact on altitudinal position and structure of the upper timberlines.

4.3.9.1 Slope gradient and geomorphic structure

In heavily dissected mountains (such as the European, New Zealand and Japanese Alps, many ranges of the Rocky Mountains and the Himalayas) the upper timberline is located on more or less steep mountain slopes often exhibiting interspersed rocky cliffs and more gentle terrain such as relics of old valley bottoms and trough shoulders. In some cases, however, the timberline ecotone extends across old uplifted, gently sloping land surfaces as in the Colorado Front Range (Photo 38) or on Beartooth Plateau (Montana) and in some other Rocky Mountain areas, for example. In many mountain valleys the timberline ecotone extends also into the relatively gently sloping trough shoulders.

Photo 38. Timberline ecotone (about 3.400 m) on the east slope of Mt. Audubon (Front Range, Colorado). Such wide ecotones are typical of elevated, gently rolling, old land surfaces in the Rocky Mountains. F.-K. Holtmeier, 10 August 1977.

Obviously, orographic timberlines are more widely represented than climatic timberlines. However, also in those areas where the forest reaches its upper climatic limit the spatial structure of the ecotone is strongly related to the varying topography. When comparing such timberlines certain rules become apparent that frankly require typification.

On steep mountain sides, downslope mass movements by debris slides, debris avalanches and snow avalanches as well as the surface structure caused by the geological conditions are important factors affecting site conditions in the timberline ecotone. Moreover, small ridges, grooves, rills and gullies caused by postglacial slope erosion, screes and talus cones create an almost regular microsite pattern of more or less wide forest strips alternating with

Photo 39. Rib and groove topography on the west-facing slope (2.000–2.220 m) of the upper Roseg Valley (Upper Engadine, Switzerland). Avalanches prevent upright growing conifer forest (*Pinus cembra, Larix decidua*) from invading the grooves successfully. Green alder (*Alnus viridis*), relatively tolerant of avalanches, occur at the lower rim of the scree. On the south-facing slopes of the small ribs, tree growth is very likely impeded by dryness and denudation. F.-K. Holtmeier, 6 October 1997.

almost treeless terrain oriented perpendicular to the slope gradient. Particularly in high-lying valleys with forest only on the footslope the forest is intensely dissected mainly by avalanche chutes down to the valley floor, and timberline is depressed (Photo 39).

Trees and tree groves are usually restricted to small ridges, little knolls and outcrops not affected by avalanches, whereas within avalanche chutes the high-stemmed forest is replaced by prostrate, scrub-like species such as alders, dwarf pines and willows and tree species that are able to regenerate from basal shoots or layering after breakage, such as birches or aspen, for example (Section 4.1.2). They may be considered typical substitutes of high-stemmed coniferous forest at such sites.

In the dry northwestern Himalayas, for example, a birch belt (*Betula utilis*) extends above the conifer forest on humid northern slopes. It is overtopped by willow scrub (depending on the region *Salix denticulata, Salix wallichiana, Salix karelinii*) that merges into the alpine zone. From there birches and willow thickets follow the avalanche chutes and gullies far into the conifer forests at lower elevation (Figure 35; Photo 40; Troll, 1939, 1964; Nüsser, 1998). A similar situation can be observed in Alaska and in some ranges of the Rocky Mountains, as for example in Glacier International Peace Park

(Montana) and in Glacier National Park (British Columbia), where Sitka alder (*Alnus sitchenis* = *Alnus sinuata*) settles the avalanche-affected slopes. Upright-growing conifers and conifer groups are usually restricted to rocky cliffs and ledges that provide protection from avalanches (cf. Photo 43 and Figure 38).

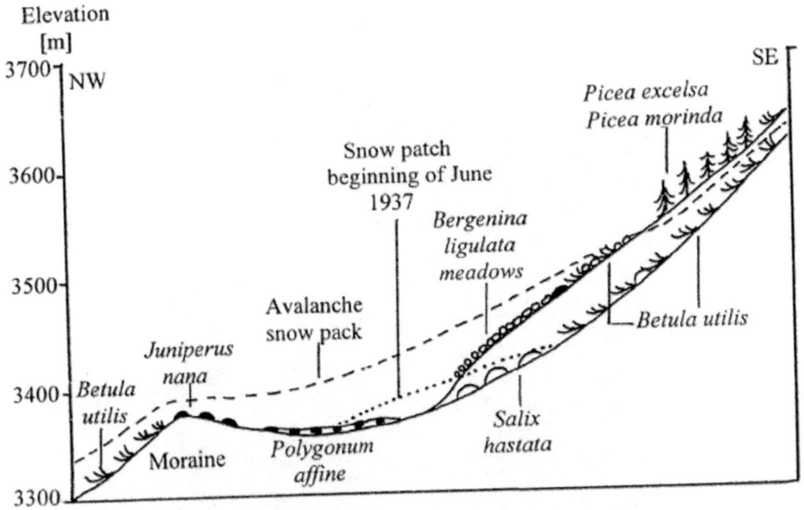

Figure 35. Vegetation on a north-facing slope in the Nanga Parbat area (Northwest-Himalaya, Pakistan). Modified from Troll (1939).

Photo 40. Betula utilis in an avalanche chute at the upper timberlines on a NNW-facing slope in the Nilt Valley (north slope of Raksposhi, Pakistan) at 3.650 m. U. Schickhoff.

Photo 41. Norway spruce (*Picea abies*) invading an avalanche chute on the east-exposed slope oft he Stubai Valley (Tyrol, Austria). F.-K. Holtmeier, 20 April 1981.

Figure 36. Larch (*Larix decidua*) in an avalanche chute (Upper Engadine, Switzerland). Great flexibility enabled the tree to survive at such a site (drawing after a photo by the author). From Holtmeier (1965).

In avalanche chutes young growth of the conifer species forming the high-stemmed forest stands regularly occur (Photo 41), but it usually does not survive the next avalanche. Still, flexible young plants such as young larches, for example, are relatively tolerant of avalanches and are likely to survive such events. Stems are pressed to the ground by the snow masses and may emerge again when the snow has gone (Figure 36). *Larix lyallii* (Rocky Mountains is obviously particularly tolerant in this respect. Arno and Habeck (1972) report larch individuals over 6 m in height and 13 cm in diameter that gradually turned in an upright position again after snow melt without having been damaged. *Abies mariesii* less than 5 m high and 50–100 years old behave similarly (Kajimoto et al., 2004). Also, young spruces (e.g., *Picea engelmannii, Picea abies*) as well as sub-alpine fir (*Abies lasiocarpa*) and lodgepole pine (*Pinus contorta*) are relatively resistant (see also Johnson, 1987; Pattern and Knight, 1994). However, as soon as trees have grown to a certain diameter they will be broken or up-rooted by the avalanches. Broken spruces may recover by turning plagiotropic branches into an upright position (see also Figure 59) and also by intense layering of the branches in contact with the ground (Section 4.3.10.2). Frequently, winter-snow pressure alternating with orthotrophy leads to butt-sweep of the stems. In the long-term, however, mostly stunted growth forms develop that do not project above the surrounding scrub and 'krummholz'. Thus, it depends on the frequency of recurrent avalanches whether conifers may ever develop tree-like habitus at such conditions. At a recurrence of less than 15 to 20 years this appears to be impossible (see also Johnson, 1987). Young growth of Swiss stone pine, which is very sensitive to mechanical damage, does not have any chance of survival at such sites.

In the Craigieburn Range (New Zealand, South Island) where the upper timberline is formed by evergreen *Nothofagus solandri*, scrub (true krummholz) of *Podocarpus nivalis*, *Phyllocladus aspleniifolius* var. *alpinus* and *Hebe* species instead of *Nothofagus* trees grows within avalanche-prone slopes (Norton and Schönenberger, 1984), as it were the southern hemisphere counterpart to the thickets of dwarf mountain pine and green alder in the European Alps. On the other hand, *Nothofagus* recovers from breakage due to heavy snow loads and avalanches by thriving basal shoots: Thus, contrary to conifers, southern beech may persist at such sites (Schönenberger, 1984). Once eliminated, however, *Nothofagus* is almost unable to invade again because its seedlings need protection from intense solar radiation and strong frost by the forest canopy (Wardle, 1974). The almost impenetrable even-aged *Nohofagus pumilio* thickets that established themselves within avalanche chutes on Choshuenco volcano in southern Central Chile can be considered a parallel (Veblen et al., 1981).

The lower the slope gradient and the greater the distance between the footslope and the upper tree limit, the smaller is the timberline zone affected by avalanches, screes and slope debris. On gentle topography, as for example on high-lying (uplifted) old land surfaces (cf. Photo 38), rolling peneplains and gently sloping monadnocks (e.g., northern Finland), the timberline pattern is quite different from steep valley sides in high mountains. Forest and tree groves usually reach their highest position in small valleys and grooves, providing better moisture conditions and shelter from strong winds (Holtmeier, 1974). Avalanches do not occur at these sites.

In many mountain valleys, the forest, more or less dissected by avalanches and erosion rills, covers only the talus cones and slope debris at the foot of steep rocky slopes. The distribution pattern of trees, tree groups and small forest stands follows the geomorphic structure of the valley sides. Structural benches, ledges and similar almost level (flat) topography give trees limited places to grow (Photo 42; see also Photo 43).

On slope debris and talus cones, avalanches and soil moisture conditions determine the spatial distribution pattern of trees, tree stands and other vegetation. Soil moisture varies due to geomorphic microfeatures, different substrate and humus content. Rills and gullies, often identical with avalanche chutes, are usually richer in snow and moisture compared to the adjoining small ridges. In the Alps, for instance, the moist sites are normally covered with green alder thickets and willow scrub or by tall grasses (e.g., *Calamagrostis villosa*) and tall perennial herbs. At such snow-rich sites, dwarf mountain pine (*Pinus mugo*), although tolerant of avalanches, would be at a disadvantage compared to green alder (*Alnus viridis*) because of being highly susceptible to infection by the brown snow felt fungus (*Herpotrichia juniperi*). Thus, in the Central Alps, extended stands of dwarf mountain pine are usually restricted to early snow-free, avalanche-prone southern exposures, whereas green alder prevails on shaded avalanche-endangered and more humid slopes (cf. Photo 115).

Occasionally, however, dwarf mountain pine and green alder thickets occur in close vicinity on north-facing valley sides, both species related to specific microsite conditions. On the northeast-facing, extremely avalanche-prone slope of the Bernina Valley (Upper Engadine, Switzerland), for example, dwarf mountain pine is confined to comparatively dry ridges and other convex microtopography, whereas green alder thickets extend alongside wet gullies. However, green alder also grows on convex sites just below steep rocks in the upper slope, where surface-runoff provides ample moisture supply, which is important to green alder, as the silicaceous substrate is slope debris, characterised by high permeability.

High-stemmed *Pinus cembra* and *Larix decidua* could advance 200 m higher on the ridges that are less frequently affected by avalanches than in

Photo 42. East-facing slope (2.000–2.200 m) of the Roseg Valley below Piz Rosatsch (Upper Engadine, Switzerland). The distribution of the tree stands (mainly *Pinus cembra* and a few *Larix decidua*) follows the geomorphic structure of the steep, rocky mountain slope. F.-K. Holtmeier, 5 August 1963.

the outlet of the avalanche-prone gullies at the lower part of the footslope (Figure 37). Obviously, this distribution pattern of both krummholz species is primarily controlled by the different moisture conditions related to the microtopography as both species are relatively tolerant of avalanches. At the given continental climate the higher moisture in the gullies and below the rocks in the upper slope enable alder to successfully compete with dwarf mountain pine. In more humid ranges such as the northern and southern Alps, dwarf mountain pine is usually more typical of carbonatic (dry) substrate whereas green alder prevails on silicaceous substrate. Therefore, dwarf mountain pine and green alder are often considered to be vicarious species. Schweinfurth (1966) describes characteristic topographically controlled distribution patterns of woody species from steep slopes in the Alexander Range (Taramaku Valley, New Zealand, South Island) and from the Franz-Josef-Glacier area. *Hoheria glabrata* low forest (coppice forest) grows within gullies and ravines, whereas *Metrosideros umbellata* stands cover the ridges. At its upper limit the low forests give way to scrub (up to 900 m) that borders high elevation tussock grassland.

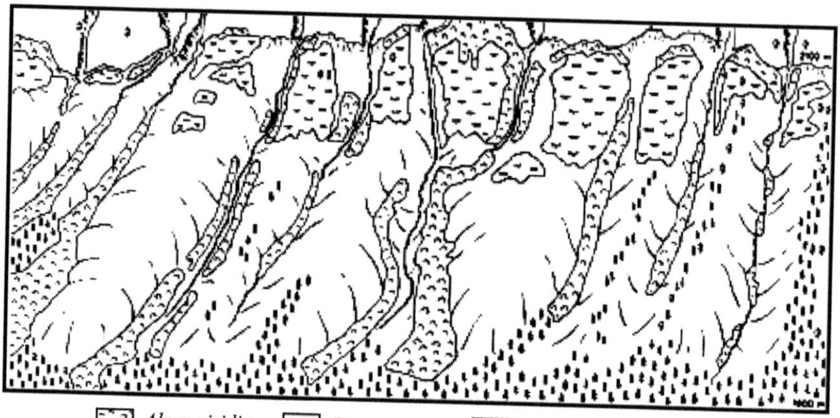

Figure 37. Distribution of *Alnus viridis*, *Pinus mugo*, *Pinus cembra*, and *Larix decidua* on the northeast-facing slope oft he Bernina Valley (Upper Engadine, Switzerland). *Pinus mugo* is restricted to relatively dry ridges whereas *Alnus viridis* grows on wet avalanche-prone gullies and also just below steep rocks (surface runoff) in the upper slope. *Larix decidua* and *Pinus cembra* are confined to sites not endangered by avalanches. From Holtmeier (1967b).

Steep slopes formed by almost horizontal or slightly dipping sedimentary strata are often characterised by narrow forest strips alternating with comparatively broad debris zones perpendicular to the slope gradient (Photo 43, Figure 38). Forest strips and tree groves are restricted to the structural benches and other rocky outcrops that provide protection from avalanches and mass wasting, while the dry and avalanche-prone debris zones are almost devoid of trees and are locally covered with dispersed grass and herb communities. True krummholz (e.g., alders) also occurs. Avalanches and also drought prevent the forest from invading these areas.

Even in the humid Norwegian fjordland, for example, slopes covered by coarse and permeable debris exhibit relatively dry conditions that do not allow birch forests to establish themselves. Just above the structural benches, however, out-flowing seepage increases moisture (Figure 39; see also Figure 38). If not regularly affected by avalanches, such sites are comparatively favourable to tree growth, particularly on dry southern exposures. At the bottom of rocky cliffs, surface-runoff may improve moisture conditions and thus encourage tree growth. Occasionally, however, the uppermost outliers of tree growth can be found in rock-cliffs, if fine mineral and organic matter that accumulated in cracks provides sufficient moisture and nutrients.

In glacially moulded valley sites, conditions abruptly change if crossing the trough rim and entering the relatively gentle trough shoulder. In many valleys of the Scandinavian mountains and particularly in the Norwegian fjordland, for example, comparatively flat trough shoulders are often covered with boggy vegetation interspersed with single birches and birch clumps that are confined to slightly convex microtopography, while birch forest, more or less dissected by avalanches, extends on the steep and better drained trough walls (Figure 40). The same effect is caused by geological structures separating steep and flat topography (cf. Photo 108). At such conditions the forest may normally not advance to its upper climatic limit.

Geomorphic landscape structures different from those in the high-mountain valleys and fjords, cause also different site patterns in the timberline ecotone. In northern Finland, for example, the timberline ecotone extends on rolling old land surfaces or on gently sloping fjelds and monadnocks that rise for some hundred metres above the undulating landscape. Except for the outermost northwestern region of Finland, where the overthrust of the Scandes (superimposed on the crystalline bedrock) causes steep topography (Photo 44), avalanches do not significantly influence the timberline. In contrast to high-mountain

Photo 43. South-exposed slope on Logan Pass (Glacier National Park, Montana). The slope is characterized by structural benches alternating with slope debris. Upright growing trees are restricted to subtle convex topography not affected by avalanches. Willows (*Salix* sp.) and alder thickets (*Alnus sinuata*) grow on the upper slope, which is only thinly mantled by debris, and also below the structural benches, where surface runoff and seepage increase moisture supply (cf. Figure 38). The snow patches provide melt water far into the summer. F.-K. Holtmeier, 5 August 1998.

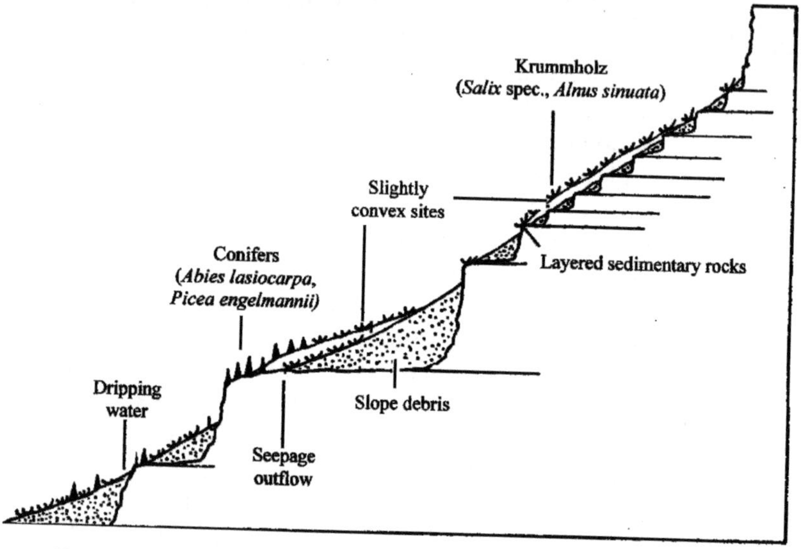

Figure 38. Schematic cross section through the slope shown in Photo 43.

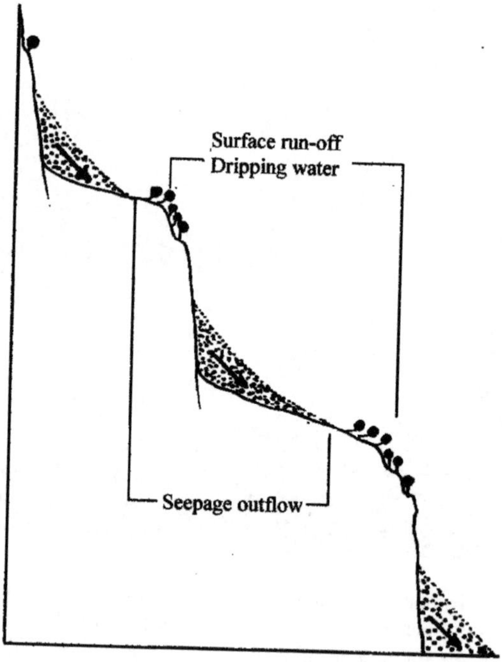

Figure 39. Schematic cross section through a steep slope with cliffs in the fjordland of northern Norway.

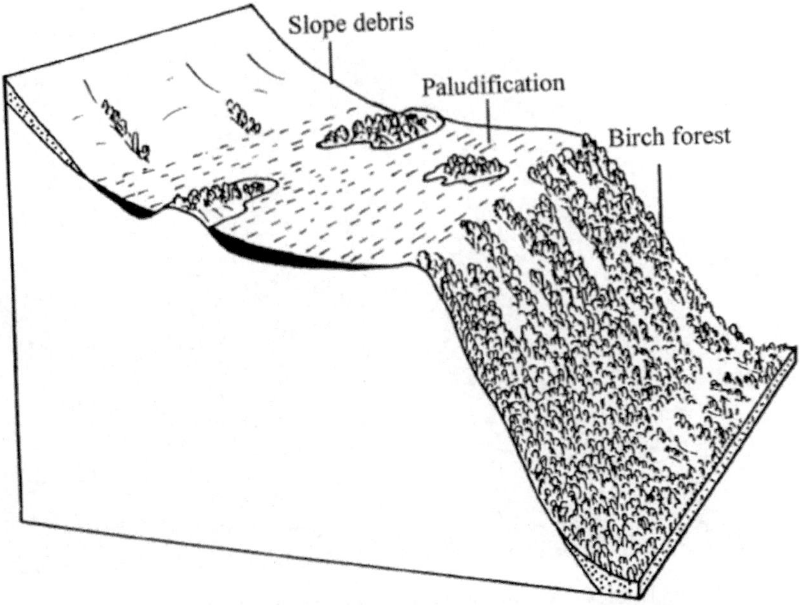

Figure 40. Birch forest limit (*Betula tortuosa*) at the upper rim of a trough wall. (schematic, based on observations in the fjordland of northern Norway).

Photo 44. Southwest-facing slope of the Saana Fjeld near Kilpisjärvi (Finnish Lapland). Numerous avalanche chutes enter the upper rim of the mountain birch forest (*Betula tortuosa*). F.-K. Holtmeier, 4 August 1999.

Photo 45. The birch forest (*Betula tortuosa*) on the northeast-facing slope of Rodjanoaivi (Finnish Lapland) reaches its highest position in the shallow valleys, whereas the convex topography (light grey areas) is covered with dwarf shrub-lichen heath. The light spots are heavily wind-eroded. F.-K. Holtmeier, 5 August 1998.

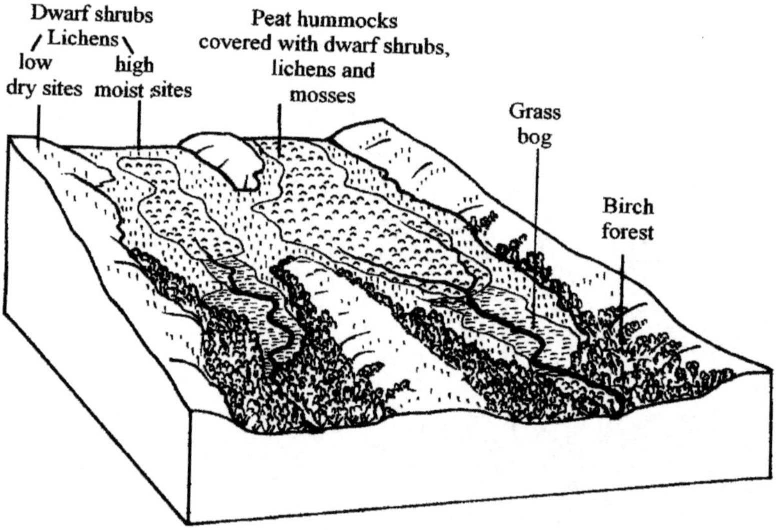

Figure 41. Gentle, paludified slope in Lapland. Birchs stands (*Betula tortuosa*) are restricted to slightly convex topography (schematic).

valleys and to the fjordland, the ecotone reaches higher in small valleys and gullies compared to ridges and similar convex topography. However, within these depressions trees and tree groups are usually restricted to slightly convex and not too-wet sites (Photo 45; Figure 41; see also Broll et al., 2007). On wind-exposed slope areas with little snow cover or even missing snow in winter, the birch forest gets already scattered at relatively low altitude, and the uppermost outposts of tree growth are usually restricted to more protected sites such as swales and other shallow depressions or behind low terraces and stone blocks (Figure 42).

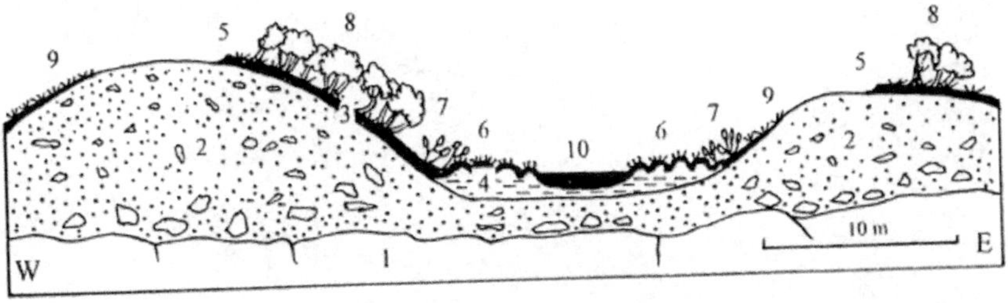

Figure 42. Site pattern (schematic) at the upper limit of birch forest (*Betula tortuosa*) on Koahppeloaivi (northern Finnish Lapland) at about 310 m. *1* – bedrock, *2* – till (sandy-skeletal), *3* – organic layer, *4* – peat, *5* – wind scarp, *6* – hummocks, *7* – willows, *8* – birch stand on Podzol, *9* – dwarf-shrub-heath, *10* – pond.

Photo 46. View (to the southeast) from Arapaho trail (ca. 3.580 m) to the locally varying site mosaic in the lower area of the Fourth of July cirque (ca. 3.420 m, Front Range, Colorado). F.-K. Holtmeier, 22 August 1991.

Photo 47. Site mosaic (ca. 2.200 m) on the northwest-facing slope of the Upper Engadine main valley (Switzerland). Bogs and sedge vegetation occur in the water-logged shallow depressions while dwarf-shrub heaths grow on the convex topography. The trees are also confined to the latter sites. F.-K. Holtmeier, September 1968.

Particularly variegated site mosaics related to alternating microtopography occur at the uppermost section of glacially moulded high-lying valleys and cirques in the Colorado Front Range, for example, and also in other ranges of the Rocky Mountains. The effects of roche moutonnées, glacial spill ways (drainage channels), rock bars and alluvial flats on microclimates and soil conditions (Section 4.3.8) control the distribution pattern of small forest stands, tree clumps and non-arboreous vegetation. Conifers (*Picea engelmannii*, *Abies lasiocarpa*), for instance, are usually confined to dry and mostly nutrient-poor topography, such as rocky outcrops, moraines, etc., whereas willow thickets, boggy and marshy vegetation covers the adjoining wetter, often waterlogged sites (Photo 46). Similar conditions can be found, by the way, on trough shoulders and comparably glacially sculptured slope areas in the Alps, as in the Upper Engadine, for example (Photo 47).

The oblique air photo (Photo 48) and the terrestrial photos 49 and 50 showing the Blue Lake Valley in the Colorado Front Range may provide a little more detailed example of the topographically controlled varying patchiness in the timberline ecotone. The valley is located on the eastern flank of the Front Range and gradually drops from the continental divide (west) to the east. The relatively broad valley bottom (about 3.350 m) was intensely sculptured

Photo 48. Detail of an oblique air photo of the Blue Lake Valley (Front Range, Colorado), copied from a coloured air photo of the US Geological Survey (F16CN 08013 277449), 2 October 1974. Scale at the valley bottom about 1:14.500. North is on the right. Further information in the text.

by the Pleistocene glacier. The great geomorphic variety is mirrored by the distribution of the vegetation. Also in this valley, snow-rich, wet or even waterlogged microtopography (alluvial flats, drainage channels and other depressions) is mostly devoid of trees and is covered instead by willow thickets and bog or marsh communities, whereas sharply contoured conifer stands occur on the convex sites (roche moutonnés and other rocky outcrops, ridges, rock bars, structural benches, etc.). The spatial structure of the timberline ecotone in the valley head (top of Photo 48) is different from the lower valley section from which it is separated by the cirque threshold (Photo 49). Roche moutonnés and spill ways on and above the cirque threshold are oriented approximately from west to east, i.e., more or less parallel to the valley, as is obvious from the distribution of tree stands and conifer scrub (dark grey colour). In the lower valley section (in the middle of the air photo), however, the geomorphic situation, rock steps alternating with relatively flat topography partly covered with debris, is determined by geological bedrock structures oriented almost perpendicular to the valley. Also here, ribbon-like forest stands and conifer scrub on the prominent sills and rocky outcrops and almost treeless glades clearly reflect the geomorphic situation and resulting site conditions. The tree stands, acting as natural snow fences, enhance snow accumulation at the leeward side of the rocky steps and thresholds where the snow cover often lasts far into summer (cf. Photo 50). Trees are unable to invade such sites because of the considerably shortened growing season. In case these sites are

Photo 49. View west into Blue Lake Valley. In the background is the Continental Divide. From the left to the right: Mt. Toll (3.956 m), Paiute Peak (3.989 m), Mt. Audubon (4.030 m). F.-K. Holtmeier, October 1974.

covered with coarse and permeable debris (light grey colour in the air photo) moisture deficiency prevents vegetation from colonizing such sites, except for the lower end of the debris cover where out-flowing seepage increases soil moisture (Figure 43).

Photo 50. Melt-out pattern in the middle section of Blue Lake Valley (View SE) controlled by the influence of the geomorphic structures and ribbon-like tree stands on snow relocation. F.-K. Holtmeier, 18 July 1984.

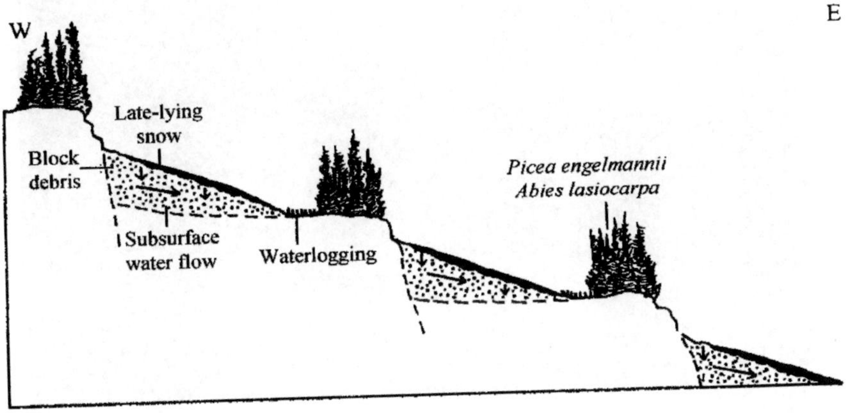

Figure 43. Schematic transect through the middle section of Blue Lake Valley (cf. Photo 50) showing the influence of topography, substrate, and tree vegetation on snow accumulation and moisture conditions.

Forests could also not establish themselves on the steep south-exposed footslope of Mt. Audubon (cf. Photos 48 and 49, right), which is covered for the most part by a mantle of rock waste. Very likely, the slope is too dry for tree growth and if trees invaded for a while avalanches would soon eliminate them. Exception is made for some convex microtopography slightly higher than the surrounding debris and less affected by avalanches, where tree stands have grown up. At the relatively flat cirque exit and also on the gently sculptured cirque floor (cf. Photo 49, right) patches of stunted scrub-like growing conifers (same species as in the forest on the valley bottom) could become established. On the steep southern side of Blue Lake Valley (cf. Photo 49) trees and forest stands can survive only on small ridges, rock spurs and other outcrops providing protection from the frequent avalanches.

On high mountain ridges that never were glaciated, so-called interfluves, the spatial structure of the timberline ecotone is somewhat different. Niwot Ridge may be considered a typical example (Photos 51–53; see also Holtmeier and Broll, 1992). The microtopography was primarily formed from periglacial processes such as solifluction (solifluction lobes, solifluction terraces, etc.). Consequently, the surface (above 3.000 m) is only gently sculptured and is characterised by slightly convex topography alternating with shallow depressions. A few shallow north–south trending valleys are cut into the southern

Photo 51. Shallow, southwest-trending valley on the south slope of Niwot Ridge (Front Range, Colorado). This valley is also visible in Photo 52 (in the background, right). Long-lasting snow cover and waterlogging exclude trees from the valley bottom. As elevation increases, the conifers (mainly clonal groups of *Picea engelmannii* and *Abies lasiocarpa*) adapt wind-trimmed growth forms. F.-K. Holtmeier, 24 July 1989.

Photo 52. Relocation of snow in the forest-alpine tundra ecotone (about 3.500 m) on Niwot Ridge during a storm (view east). F.-K. Holtmeier, 7 April 1989.

Photo 53. On the wind-sheltered side of this waterlogged solifluction terrace (foreground) on the east-facing slope of Niwot Ridge (at 3.335 m), low tree clumps (*Picea engelmannii*, *Abies lasiocarpa*) grow. The bottom of the SW-NE-oriented valley is treeless because of long-lasting winter snow. Wind-trimmed clonal groups grow on the wind-exposed slope in the background. F.-K. Holtmeier, 22 August 1990.

flank of the ridge. As these valleys are oriented perpendicular to the prevailing winds from the west, big snow masses accumulate on the leeward valley sides, on the valley bottom and in the lower part of the wind-exposed slopes. The convex topography between the small valleys is scoured snow-free in winter (Photo 52; Sections 4.3.6 and 4.3.7). Normally, avalanches and larger snow slides do not occur because of the relatively gentle topography. Thus, on the gentle slopes of shallow valleys the upper limit of tree growth advances to higher elevation compared to wind-exposed convex areas. Tree stature is gradually reduced as the upper limit of tree survival is approached (cf. Photos 51–53; see also Section 4.3.11). Almost west–east oriented parallel conifer hedges and rows of wind-shaped tree islands, usually expanding downwind (upslope in this case) by layering (Section 4.3.10.2; see also Holtmeier, 1982; Holtmeier and Broll, 1992; Broll and Holtmeier, 1994), cause a striped vegetation pattern on the wind-exposed upper valley sides (Photo 52). This structure of the ecotone determines the distribution of the winter snow cover (cf. Section 4.3.12). On the valley bottoms, melt water and seepage cause waterlogging mainly during and after snow melt (Photo 51). Moreover, late-lying snow cover shortens the growing season. Temporary water-saturated soils and a too short growing season prevent conifers from invading these sites.

Photo 54. The influence of local topography on the wind and translocation of snow near Caribou Lake (3.396 m, west slope of the Front Range, Colorado) is reflected in the physiognomy and distribution of the tree vegetation. Clonal groups of *Picea engelmannii* and *Abies lasiocarpa* are restricted to the wind-exposed slopes of the moraine (*right*) and to the moraine crest, whereas the leeward slope (*left*) and the shallow depressions between the morains have remained treeless because of the late-lying winter snow cover. F.-K. Holtmeier, September 1977.

On prominent knolls and other exposed topography, wind-trimmed, mat-like or wedge-like conifers form the uppermost limit of the timberline ecotone. Usually, but not as a rule, these outposts of tree growth are restricted to the leeward edge of low solifluction terraces, solifluction lobes and rocks or to shallow hollows providing sufficient protection from the permanent strong winds. However, long-lasting snow cover on the leeward sides of low ridges, terminal moraines, swells and other convex topography may restrict conifer stands to the strongly wind-swept sites less rich in snow (Photo 54).

Avalanches entering the high-altitude forest, and restriction of tree stands to ridges and other topography that is comparatively safe from avalanches are peculiar to the timberline ecotone on high-mountain valley sides outside the tropics. Troll (1959) considered this type of timberline to be representative of the winter-cold climates and contrasted it in a schematic sketch with a tropical type of timberline that is related in a different way to topography than extra-tropical timberline and is not influenced by avalanches. He referred mainly to his observations and studies in the eastern Bolivian and Peruvian cordillera, where he observed timberline reaching its uppermost position in valleys, gullies and grooves while being located at comparatively low elevation on ridges on other convex topography. Troll took this for a rule and ascribed this pattern to more favourable moisture conditions and to reduced heat loss through long-wave radiation at the valley sites. This scheme has been adopted by many textbooks and reference books (e.g., Price, 1981; Klink and Mayer, 1983; Leser et al., 1991). However, as has been evidenced by many modern studies, tropical upper timberline does not always follow this rule (Fries and Fries, 1948; Hedberg, 1951; Walker, 1968; Paijmans and Löffler, 1972; Corlett, 1984, 1987; Stadtmüller, 1987; Young, 1993; Miehe and Miehe, 1994). As was demonstrated on the previous pages, also timberline in the higher latitude mountains exhibits a much greater variety than can be expected from the strongly generalised schematic sketch drawn by Troll.

In many temperate and tropical high-mountains the upper timberline gradually declines towards the valley heads (Fries, 1913; Fries and Fries, 1948; Holtmeier, 1965, 1974, 1994a; Friedel, 1966, 1967; Schiechtl, 1967). Fries (1913) who described this feature first from the Torne Lappmark (northern Sweden) called it 'valley phenomenon' (Figure 44). He supposed that decreased temperature and increased wind velocity at the timberline level were the cause. Later Fries and Fries (1948) reported the 'valley phenomenon' also from Mt. Kenya and from the Aberdare Mountains in East Africa and explained them as the result of cold airflow following gullies and gorges. Wardle (1962) observed the 'valley phenomenon' in the Tararua Mountains (New Zealand, North Island). In this area, the upper timberline formed by silver beech (*Nothofagus menziesii*), declines for about 200 m towards the heads of the trough valleys, which might be attributed to the short and cool growing season in the valley heads. Also, in the mountain ranges of the South Island the 'valley phenomenon' occurs.

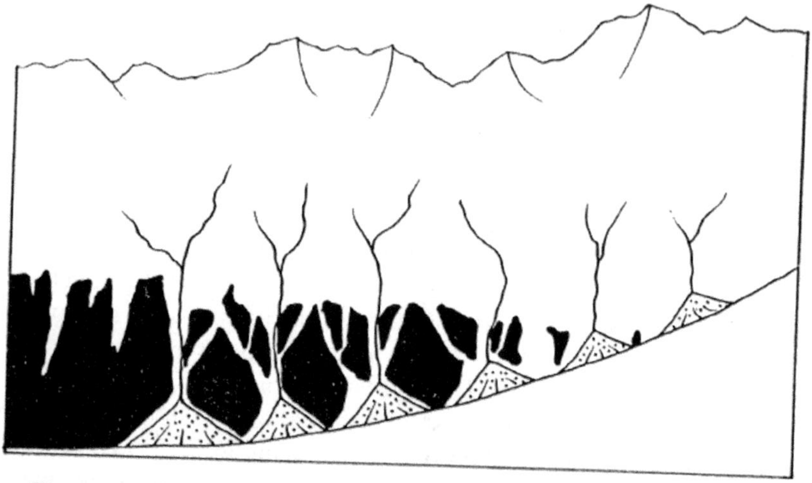

Figure 44. 'Valley phenomenon'. Timberline declines towards the valley head.

The present author's observations in the European Alps and in the Rocky Mountains provide some evidence that the decline of the upper timberline towards the valley heads is primarily due to reduction of global radiation on the valley floor by the steep shade-giving valley sides. Moreover, the 'warm slope zone' (Geiger, 1961; Aulitzky, 1968) peters out approaching the valley head thus resulting in cooler conditions. Unfavourable climatic conditions in the valley head such as frequent late and early frost, could impede larch (*Larix decidua*), for example, from entering the old Swiss stone pine stands in the upper Scharl Valley (Lower Engadine, Switzerland). Late frost often destroys buds and thus hampers regeneration of larch, which is highly susceptible to freezing temperatures when flowering and thriving needles (see also Holtmeier, 1995b). If larch thrives a second time in summer after having been defoliated by the caterpillars of the larch bud moth (*Zeiraphera diniana*) early frost may easily destroy the second needle generation formed in the same year (Zuber, 1995). Not least, orographic influences such as screes, mass-wasting and talus cones split up the narrow forest belt and lower the timberline in the upper valley sections. In the Alps and also in many high-lying valleys of the Rocky Mountains, intense anthropogenic impact (such as grazing cattle and sheep, mining, etc.; see Section 4.3.14.1) has caused or at least contributed to the depression of timberlines in the valley heads.

Even more complicated is the situation in still glaciated valley heads. Towards the upper valley section the upper timberline not only declines but the forest also retreats upslope as the glaciated area is approached (Figures 45, 46). The upslope retreat of the forest, the so-called 'glacier valley phenomenon' (Friedel, 1967), has long been attributed to the effects of the glacier wind (e.g., Schlagintweit and Schlagintweit, 1854; Holtmeier, 1965; Friedel, 1967). Certainly, the effects of the glacier wind have been over-estimated (Holtmeier, 1974). Glacier winds are not very strong and normally die out already within a few hundred metres distance from the glacier tongue because they are slowed due to the friction with the valley floor and the opposing valley wind. Thus, glacier winds affect vegetation (e.g., phenology) in the outwash plain close to the glacier front only. Consequently, apparent wind effects such as slight flagging of the trees or wind-scarps are likely caused by strong winds that cross the mountain crest and blow down-valley on the leeward mountain side, overriding the effects of the glacier wind (Holtmeier, 1974). Compared to unglaciated valley heads, however, climatic conditions close to a glacier front are more unfavourable to tree invasion because of the cooling by the glacier ('ice box effect').

Also, the upslope retreat of the forest stands close to the glacier front cannot be attributed to the direct effect of the glacier wind but rather is the result of the generally cooler conditions close to a glacier, as is evidenced by the rapid colonisation of the outwash plain by trees and good growth of these trees (Holtmeier, 1974, 1994a). In the end, however, the unstable fresh lateral moraines, which cover the lower slopes and are only very slowly invaded by trees, turn out to be the primary cause of the 'glacier valley phenomenon'.

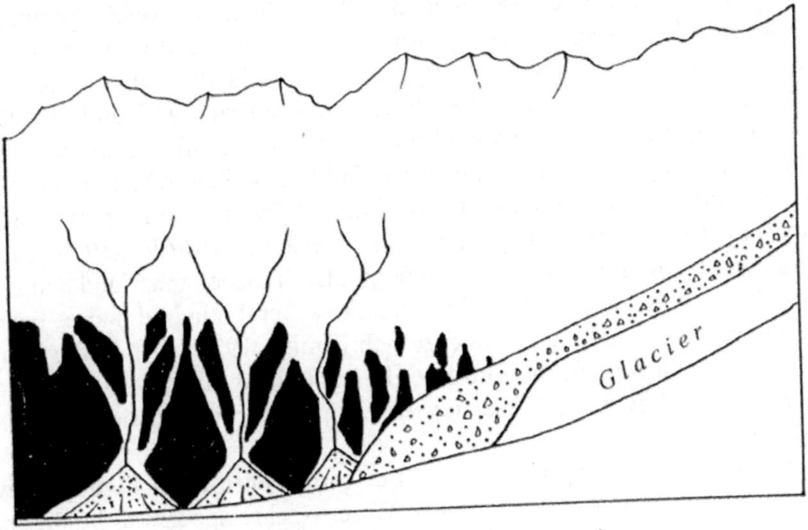

Figure 45. 'Glacier-valley phenomenon'.

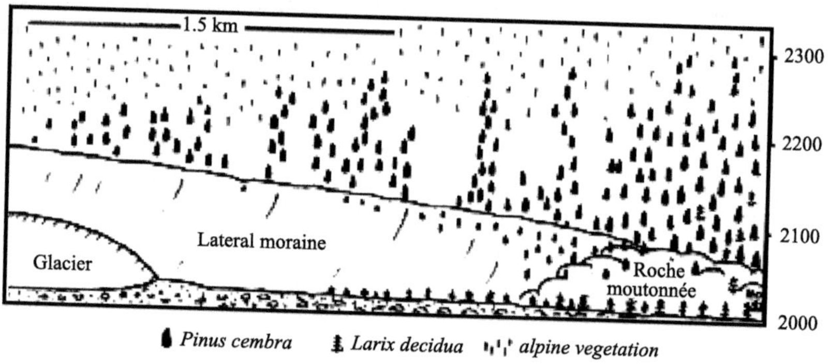

Figure 46. Invasion of the forefield and lateral moraine of a retreating glacier (schematic, based on the situation at Morteratsch glacier in the Upper Engadine, Switzerland). From Holtmeier (1994a).

Thus, the term 'glacier valley phenomenon' would stand for a lower also edaphically-caused timberline, which should not be confused, however, with the so-called inversion timberline or bottom timberline due to frequent frost and/or water-saturated soils on the valley floor (cf. Table 3).

Also, in the tropics, topography causes a great variety of timberline, mainly caused by steepness, substrate and moisture conditions, while avalanches do not occur in the timberline ecotone. Timberlines in the upper glacially moulded sections of high-lying valleys of New Guinea are particularly suitable for comparison with the upper timberlines on higher latitude mountains (cf. Photo 10). The southern flank of the Mt. Wilhelm massif, for example, is characterised by extremely steep-sided valleys. On some slopes, the bedrock is exposed, while on other slopes, even at a slope gradient of exceeding 45°, soil and a closed plant cover (mostly tussock grassland) could develop. Frequent land slides and earth flows have left deep rills and gullies. Often, water-soaked peat that lost its hold on the bedrock or till was the triggering factor. In the uppermost section of the Imbuktum Valley, which was not that much disturbed, forest climbs up to almost 3.760 m. Solitary tree stands formed by 10 to 15 m high *Dacrycarpus compactus* (Löffler, 1979) can be found even 50 m higher on convex topography. Tussock grassland and scrub that follow the wet gullies and rills extend to several hundred metres beyond the uppermost occurrences of tree growth (Smith, 1977a, b). At the bottom of the steep valley sides, the forest borders grassland that covers the flat and waterlogged valley floor (inversion timberline or inverted timberline, see Table 3).

The grassland is interspersed with tree ferns confined to slightly convex sites such as moraines and thresholds. Waterlogging appears to be the main factor preventing forest growth. Seedlings that may just establish themselves on the temporarily dry surface succumb to light nocturnal frosts regularly occurring at this elevation (3.500 m; Smith, 1975, 1977b; Corlett, 1987). Very likely, man-caused fires (hunting-fires; Corlett, 1987) have encouraged the expansion of grassland. Natural fires, on the other hand, are very rare because thunderstorms are not very common at this altitude and are usually accompanied by heavy rains (Smith, 1980). Smith (1980) is very sceptical if natural fires occur in this area at all. Anyway, the expanding grassland may have increased the frost risk on the valley floors. Below the glacially sculptured, relatively broad valley heads, the valleys narrow and get V-shaped (cross section). The steep and better-drained sides are completely covered with forest up to timberline.

Young (1993) reports similar conditions from the Rio Abiseo National Park (northern Central Peru). The upper limit of closed forest is located between 3.200 and 3.600 m. Solitary tree groves can be found up to 3.700 m, whereas the flat glacially moulded valley bottoms (3.200 to 3.500 m) are covered with grass vegetation. Water-saturated soils, cold air accumulation and man-caused fires (set to increase pasture) are considered the main factors keeping the valley floor devoid of forest. Also, in the *Polylepis* belt of the western Bolivian cordillera cold air and/or waterlogging prevent *Polylepis* from establishing in concave topography (Jordan, 1983). In the Manu National Park (southern Peru) again, forest climbs to greater elevation in the V-shaped valleys than on the steep slopes (Jordan, 1983).

In the valleys on the western flank of Mt. Kenya and in the Aberdare Mountains west of Mt. Kenya, timberline and the bamboo belt below are depressed for about 100 to 150 m compared to the adjacent valley sides (Fries and Fries, 1948). Hedberg (1951, 1964) mentions a relatively low position of the upper limit of the *Philippia* and *Erica* stands in the valleys. Likewise, in the Bale Mountains (southern Ethiopia) species-rich *Hagenia-Hypericum* forests and *Erica* dwarf forests climb to greater elevation on slopes and ridges than on broad, flat valley floors and on plateaus prone to occasional waterlogging and formation of cold air layers (Miehe and Miehe, 1994). However, timberline is often higher in wind-protected, well-drained (water and cold air) and relatively fireproof valleys and gullies than on the slopes (cf. *Polylepis* gallery forests in the Páramo of Ecuador, Section 4.1.3). The isolated stands of *Hagenia-Hypericum*, which occur up to 200 m above the closed mountain forest, are very likely remains of a previous higher-reaching forest belt. These stands are restricted to sites not only protected from fire and other anthropogenic disturbances but also exhibiting relatively favourable thermal (wind-sheltered) and hygric conditions. Moreover, competition between the woody species *Hagenia*, *Hypericum* and *Erica* plays an important role, as was described by Miehe and Miehe (1994) from the northern slope of the Bale Mountains. In this area, *Erica trimera* grows on steep, fireproof cliffs of the basaltic plateau (3.400 to 3.500 m), whereas the uppermost outliers of the *Hagenia* forest are confined to the better wind-protected and warmer sites. At the bottom of the steep slopes, water dripping of the rocks surface improves soil moisture conditions.

Also, occurrences of *Hypericum* in the ericaceous belt above the closed *Hagenia-Hypericum* forests are normally restricted to block-rich sites where the fire risk and competition by *Erica trimera* are obviously reduced. At the lower rim of slopes covered by block debris moisture is higher due to outflowing or surface-near seepage. In this connection Miehe and Miehe (1994) draw parallels to the *Polylepis* problem (Section 4.1.3). Miehe and Miehe (1994) observed timberline structures similarly controlled by alternating substrate on Ruwenzori (Uganda), where *Erica* stands occur on rocky sites with shallow soils, while waterlogged sites are treeless. This timberline is considered untouched by human impact for the most part.

In Simen (Ethiopia), the mountain forest gets increasingly scattered above 3.600 m, which is attributed to extremely wet conditions on flat topography during the rainy season (Klötzli, 1975). In the dry season, however, no moisture is available to the plant because the soils completely dry out to 65 cm depth. At increasing waterlogging the *Erica arborea* bush retreats to rocky well-drained sites, such as knoll and gully sides. *Hypericum revolutum* stands occur at the periphery of the waterlogged sites and replace *Erica arborea* stands on block debris. On block debris with humus pockets, however, *Erica arborea* stands with interspersed single *Lobelia* plants can be found up to an elevation of 3.900 m.

On the southern slope of Kilimanjaro, the upper limit of the ericaceous belt is higher on small ridges and crests than in the valleys (Klötzli, 1958). In other areas, however, the mountain forest (*Hagenia*, *Podocarpus*, *Hypericum*) gradually merging into *Erica* bush (height ≥ 6 m) at its upper limit, reaches its highest position in the valleys (Walter and Breckle, 1984). Also, on Mt. Elgon (Kenya/Uganda) the highest timberline (3.000 to 3.450 m; *Philippia excelsa*) is located in humid valleys where the tree stands are relatively well protected from the frequent fires. From the air a dendritic branching of the timberline is apparent related to the valleys and their tributaries (Hamilton and Perrott, 1981). It is only this type of tropical timberline that is represented by the schematic sketch of Troll (1959).

4.3.9.2 Exposure

Exposure to solar radiation and wind is an important factor influencing the altitudinal position of timberline and the distribution pattern of the tree species represented at timberline. Solar radiation increases exponentially with altitude due to the concentration of water vapour in the lower atmosphere. Wind speed also increases with elevation. Consequently, the contrasts between sunny and shaded and between windward and leeward slopes exacerbate by increasing altitude. In temperate mountains and also in the outer tropics, timberline is usually higher on sunny exposures than on shaded slopes, if no orographic or anthropogenic factors prevent the forest from reaching its climatic altitudinal limit. In the European Alps, for example, the altitudinal position of timberline on sunny and shaded slopes differs for about 100 m. In the mountain ranges of Oregon and Washington the difference is about 150 m (Bailey, 1936; Arno, 1984). The upper timberline in the Brooks Range (Alaska) is located at 700 to 800 m on the southern slope. Stunted trees occur even at 950 m. On the northern side, however, the foot-slopes are already covered with treeless tundra (Billings, 1990). In Norway, Odland (1996) found great regionally varying differences in the altitudinal position of birch timberline caused by different exposure to solar radiation. The most pronounced differences (150 m) occur in the interior fjordland whereas they are smaller in the coastal region and in the more continental eastern area. However, the greatest differences between sunny and shaded slopes occur in subtropical high-mountains.

As sunny slopes are not only warmer but also drier than shaded slopes, a change in exposure often causes a change of the tree species, which may be demonstrated by many examples. Sun-exposed avalanche-endangered slopes in the Alps, for instance, are often covered with dwarf mountain pine (*Pinus mugo*), while comparatively narrow green alder stands (*Alnus viridis*) are restricted to the rills and gullies providing better moisture conditions. In contrast, avalanche-prone northern exposures are completely covered with almost impenetrable green alder thickets. In the more continental mountain ranges in the western United States, whitebark pine stands (*Pinus albicaulis*) with sparse undergrowth occur on sunny exposures whereas subalpine larch with luxuriant heath vegetation (*Larix lyallii*) cover shaded slopes (Franklin and Dyrness, 1973; Arno, 1984; Arno et al., 1995; Peterson and Peterson, 1995). At timberline in the White Mountains of California, open bristlecone pine forests (*Pinus longaeva*) growing mainly on northern and western exposures are replaced by mountain steppe on the south- and east-exposed slopes. At the given dry climatic conditions in this area the field capacity of different substrate is an additional factor controlling the distribution pattern of the vegetation. Pine stands climb to greater elevation on dolomite (Sheep Mountain) than on more acidic sandstone and granite (Campito Mountain). The soils on dolomite although poor in nutrient provide higher soil moisture compared to the much drier acidic soils on sandstone and granite (Wright and Mooney, 1965).

On many mountains in the climatically dry regions of Central Asia, dense forests and other hygrophilous vegetation are often restricted to moist northern exposures, whereas steppe vegetation covers the southern aspects. Locally scattered arborescent juniper occurs forming the upper treeline (cf. Photos 13 and 14; Schäfer, 1938; Troll, 1939, 1964; Von Wissmann, 1960, 1961,

1972; Jacobsen and Schickhoff, 1995; Winkler, 1997; Miehe et al., 1998; Richter, 2001).

Also, in the dry northwestern Himalayas, the species represented in the 'krummholz belt' above the coniferous forests differ by exposure of the mountain slopes. Willow scrub (*Salix denticulata, Salix wallichiana, Salix karelinii*) with admixed *Rododendron anthopogon* var. *hypenanthum* prevails on humid northern slopes, while a mosaic of mat-like growing junipers (*Juniperus communis* var. *alpina, Juniperus squamata*) and alpine vegetation is characteristic of the sunny exposures (Troll, 1939; Schickhoff, 1993; Nüsser, 1998). Depth and duration of the winter snowpack vary considerably in exposure. In the Karakoram, for example, southern exposures (above 3.000 m) receiving extremely high radiation loads may become snow-free already in January while winter snowpack lasts until May on the northern slopes at same altitude (Cramer, 1997a). Consequently, juniper trees growing on the sun-exposed slopes are facing severe climatic stress, such as excessive solar radiation, photoinhibition and drought stress whilst coniferbirch forest on the northfacing slopes enjoy prolonged moisture supply from the late-lying snow cover. Recently, it has been doubted whether the natural conditions or human disturbances prevent the forest from invading the sun-exposed slopes in these dry regions. Probably, heavy grazing is the decisive factor (Miehe et al., 1998; Miehe and Miehe, 2000). In any case, on the Tibetan Plateau anthropogenic impact has been present for several thousand years (Frenzel, 1994; Weiwen, 1994). Troll (1941) had already considered human disturbances to be an important factor in this region and supposed a forest steppe as natural vegetation.

Also, in the comparatively humid Himalayas of Nepal, exposure to solar radiation and its effects on site conditions (mainly moisture) is the factor controlling the distribution pattern of forests. Juniper stands (*Juniperus recurva, Juniperus communis, Juniperus wallichiana*) are restricted to dry slopes, while subalpine forests of *Betula utilis* and *Rhododendron campanulatum* are common at moist sites (Schmidt-Vogt, 1990a, b). Snow cover lasts relatively long under the shade-giving forest canopy on the shaded slopes thus increasing humidity and soil moisture. Under these conditions a closed plant cover may develop which in turn prevents erosion and loss of soil that retains moisture. In contrast, on hot and dry sunny aspects only open forests with scattered shrub and field layers are left, and the permeable slope debris is exposed. Deforestation and extensive grazing have exacerbated the situation. Increased evaporation and lack of melt water, previously provided by the long-lasting snow cover under the forest canopy, enhanced soil drought after the forests had been removed or had become scattered by humans.

In other high-mountain ranges, the direct effects of exposure to solar radiation are also partially overridden by the influence of other factors. In New Zealand, for example, the upper timberline, which is partly formed by *Nothofagus*, reaches its highest position on ridges and spur, disregarding exposure to radiation (Wardle, 1993). Instead, the lower risk of deep minimum temperatures on convex topography at clear and calm nights seems to be the controlling factor. This restriction of the upper forest stands to convex topography seems to confirm the hypothesis that finally the length of the growing season determines the upper limit of southern beech forests (Wardle, 1993).

In the inner tropics, differences in the altitudinal position of upper timberline that can be attributed to the direct effect of exposure on solar radiation have long been supposed to not exist. Instead, hygric and thermal differences caused by the daily cycle of cloudiness and exposure to moisture-carrying air masses have been considered to be the decisive factors leading to different altitudinal position of climatic timberline on the mountain slopes (Troll, 1941; Salt, 1951; Coe, 1967; Van Steenis, 1972; McVean, 1968, 1973; Smith, 1977a, 1980; Lauer, 1979b; Löffler, 1979; Sarmiento, 1986; Corlett, 1987; Barry, 1992; Winiger and Menz, 1993). Probably, global radiation reflected from the low-lying morning cloud cover can increase radiation load at higher elevation on the mountain slopes. In any case, solar radiation load 70% higher than below the cloud cover has been recorded (Flenley, 1992, 1995). Based on the results of their studies on tropical timberlines in the Andes Bader and Ruijten (2008) recently set up the hypothesis that, at a smaller scale, the comparatively low position of timberline on east-facing slopes within the eastern flanks of the Andes might be due to excessive radiation loads causing photooxidative stress and photoinhibition. These effects are strongest when working together with cold temperatures as they regularly occur in the morning after cold nights.

Approaching the outer tropics the effects of different exposure gets increasingly conspicuous. In the western Bolivian cordillera, for instance, the upper limit of *Polylepis* groves climbs higher (200 to 300 m) on the north-facing sun-exposed sides of the west–east or east–west trending valleys compared to the southern exposures. The fact that in the arid southern part of the cordillera *Polylepis* groves are mainly confined to north-facing slopes indicates unfavourable thermal conditions on southern exposures to be the decisive factor (Jordan, 1983).

Besides exposure to solar radiation also exposure to the prevailing winds affects the altitudinal position of timberline (Photo 55), by direct physiological and mechanical influences of the wind on tree growth or by relocation of the winter snow and resulting effects on the site conditions (Section 4.3.7 and 4.3.11).

At finer scales (cf. Figure 2) microtopography may cause a highly varying pattern of sites receiving different solar radiation loads. Exposure of the microsites on a valley side varies in relation to the orientation of the valley. In west–east oriented valleys, for example, the strongest contrasts of irradiation are between the north- and south-exposed valley sides, while the difference caused by microtopography are less pronounced on both of the opposite slopes. In north–south trending valleys the situation is different. On both valley sides the rib and groove topography causes a corrugated transverse surface (cf. Photo 39) with a 'regular' pattern of sites receiving high radiation loads (sun-exposed sides of the ribs) alternating with less irradiated sites (north-exposed side of the ribs). In continental climates extreme temperature amplitudes, frequent freeze-thaw events and insufficient soil moisture may locally impede tree establishment on the southern exposures, and closed forest may even climb to higher elevation on the northern slopes of the ribs.

Photo 55. View on Arthur's Pass (1.008 m, New Zealand, South Island). The strong winds blowing from the north (*left*) cause a depression of timberline on the pass. Only in niches and similar wind-protected sites, does the forest (*Nothofagus solandri* var. *cliffortioides*) reach higher elevation. Downwind from the pass (*right*), the upper timberline climbs to higher altitude, while the trees are becoming more and more dwarfed as elevation increases. F.-K. Holtmeier, 26 November 1979.

The effect of similar microtopography on wind-mediated snow relocation and its influences on tree establishment vary according to the particular situation. On the wind-exposed slope of a west–east oriented ridge, for

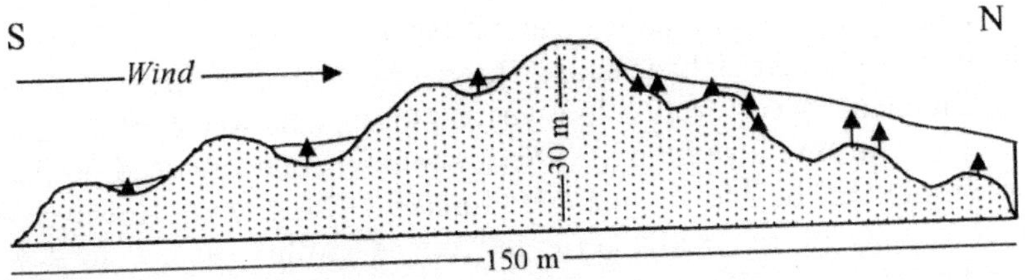

Figure 47. Transect through a west–east oriented ridge (schematic) showing the different effect of similar slope microtopography on snow relocation and tree establishment on the wind-facing slope and on the lee side of the ridge. On the wind-exposed slope of the ridge with little or no winter snow on convex microtopography, trees can be found only in the wind-sheltered, relatively snow-rich concavities. On the snow-rich lee slope, trees are restricted to the early snow-free sites whereas deep and long-lasting snow impedes tree establishment in the slope concavities.

example, with mainly southerly winds (Figure 47), snow accumulating in concavities and on the lee side of convex topography would facilitate tree establishment.

Contrary to that, on the snow-rich lee side of the ridge trees would be restricted to the early snow-free sites.

4.3.10 Regeneration

If not depressed by orographic influences or anthropogenic disturbances, the position of the upper timberline depends in the long-term on physiological resistance of the trees to mountain climate, and on natural regeneration. While most reviews of timberline focus on the influence of environmental factors on tree growth and survival, the effects on natural regeneration, particularly on seedlings and saplings during the first years, are not sufficiently considered. This seems a little strange because it depends on successful regeneration whether the present forest stands will persist and timberline advance to greater elevation or higher latitude due to global warming (Holtmeier, 1985b, 1989, 1993a, 1994a, 1995a, 1996; Rochefort et al., 1994; see also Chapter 5).

4.3.10.1 Seed-produced regeneration

Regeneration can be expected only if viable seeds are produced at not too-long intervals. Moreover, seeds must be dispersed to suitable seedbeds and microsites where a bundle of interacting factors (microclimate, illumination, soil moisture, soil acidity and others) allow germination and seedling surviv-al. In this respect conditions get increasingly unfavourable by elevation. Norway spruce (*Picea abies*), for example, produces good seed crops at intervals of 3 to 5 years. At timberline, however, good seed production (great quantity and high quality) can be expected every 9 to 11 years only (Tschermak, 1950). At high elevation, *Picea engelmannii* and *Abies lasiocarpa* produce abundant seeds every 3 to 6 years (Oosting and Reed, 1952; Franklin et al., 1971). In Swiss stone pine (*Pinus cembra*) abundant seed production occurs at intervals of 7 to 10 years (Holtmeier, 1974, 1999a). However, seed production may be regionally different (e.g., Oswald, 1963, 7 to 8 years; Rohmeder, 1941, 3 to 10 years on an average). On the other hand, complete failure in seed production is rare. Exception is made for European larch (*Larix decidua*) the flowers of which are often destroyed by late frost. Larch produces good seed crops every 10 years. Ripening of seeds is delayed with increasing altitude. At timberline in the Alps, larch seeds do not fully develop until the beginning of October. A cool summer or a wet and cold autumn may considerably impair the morphological differentiation of seeds.

Approaching upper timberline, amount and quality of seeds decrease. Thus, germination capacity in *Picea engelmannii* seeds, for example, at timberline on the Colorado Front Range, was practically zero (Dahms, 1984; Table 14). The same we found in *Pinus sylvestris* at its upper limit in northernmost Finnish Lapland (Holtmeier, 2005a). According to Henttonen et al. (1986; see also Harju et al., 1996) good seed maturation in northern boreal forests is likely to occur twice in a century. Sveinbjörnsson et al. (1996; see also Sveinbjörnsson, 2000) mention a high percentage of empty seeds at the upper birch treeline in the Torneträsk area (Swedish Lapland). Also, the amount of light filled seeds increased with altitude. Small seeds normally provide little energy. Kullman (1984) noticed a drastic decline of regeneration above the closed mountain birch forest in Central Sweden. Moreover, he found only empty seeds at sites without any occurrences of birch seedlings. Seed production at the upper timberline (1.350 m) in the Craigieburn Range (New Zealand, South Island) was checked to be about 5% compared to lower elevation (1.000 m), and the germination rate of viable seeds even decreased from 20% to 5% at the same altitudinal difference (Wardle, 1970; Norton and Schönenberger, 1984). Also, the survival rate of the 1-year-old seedlings was much lower at timberline than below. On the other hand, a narrow band of young growth established itself along the upper margin of the closed southern beech forest during the second half of the 20th century. The trees show good growth, and losses occur in young seedlings only (Wardle and Coleman, 1992).

In the long-term, the upper limit of production of viable seeds is usually located below the physiological limit of tree growth (Figure 48). If the climate deteriorates this limit will drop to lower altitude and even in the closed forest production of viable seeds will occur at greater intervals. At

Table 14. Percentage of empty seeds and germination capacity of seeds of *Picea engelmannii* in the forest-alpine tundra ectone (Niwot Ridge, Colorado Front Range) (After Dahms, 1984)

	Subalpine belt 3.350–3.500 m	**High montane belt 3.150–3.350 m**
Endosperm and embryo missing	60.3%	52.9%
Endosperm present, embryo missing	32.7%	18.3%
Necroses, embryo not viable	6.6%	4.8%
Germination capacity	0.4%	24.0%

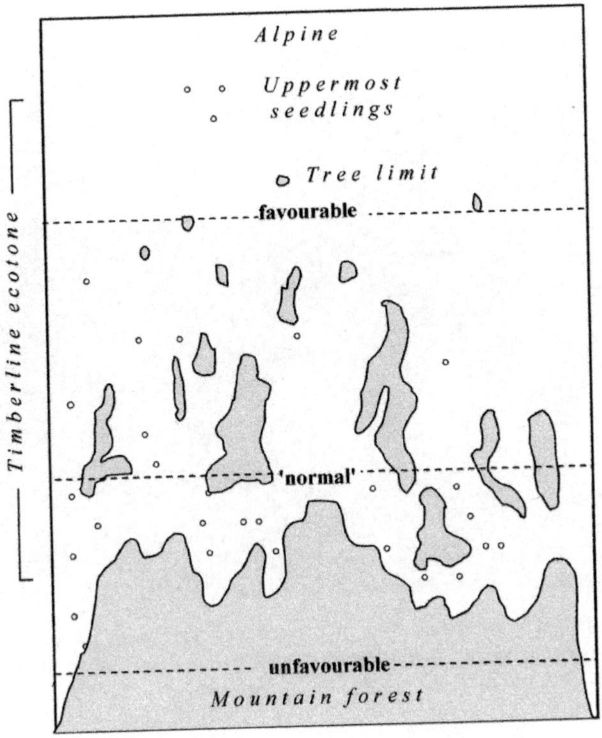

Figure 48. Altitudinal position oft he upper limit of production of viable seeds under 'normal', favourable and unfavourable conditions.

favourable climatic conditions, however, even the uppermost stunted outliers of the mountain forests produce cones and often large amounts of seeds (Photo 56). Arno and Habeck (1972) reported dense even-aged young growth of subalpine larch (*Larix lyallii*) invading subalpine glades in the Bitterroot Range (Montana). In this area such 'reproduction glades' are typical of the vegetation and site mosaic in the timberline ecotone. 'Seedling meadows' of different age and also small older larch groves provide evidence that the production of viable seeds is high enough to allow sufficient regeneration at irregular intervals.

Mostly, however, the seeds produced at timberline are empty or did not fully mature. Consequently, regeneration from seeds is extremely critical in the timberline ecotone, which partly explains the very low number of seedlings and young growth, especially close to treeline (Holtmeier, 1993a, 1995a, 1999a; see also Chapter 5). Some authors (e.g., Marr, 1977), however, suppose the supply of viable seeds to be normally sufficient for an effective natural regeneration, even in the uppermost zone of the timberline ecotone, and ascribe the low number of seedlings primarily to lacking favourable sites. This cannot be confirmed by the present author.

Photo 56. Abundant cone production in the leeward part of a wedge-like spruce (*Picea engelmannii*) growing in the forest-alpine tundra ecotone on Niwot Ridge (Front Range, Colorado) at about 3.500 m. F.-K. Holtmeier, 27 July 1989.

Also, at timberline in eastern Patagonia, which is formed by *Nothofagus pumilio* and locally also by *Nothofagus antarctica*, good seed years are an exception. Kalela (1941a), however, attributed the lack of seedlings outside the forest to moisture deficiency rather than to seed supply. Wardle (1998) reports from the Chilean Andes seedlings that invaded glades within the scrub of stunted *Nothofagus pumilio* and *Nothofagus antarctica* and also advanced to higher elevation. Considering the fact that layering and thriving basal sprouts is the normal way of regeneration at this timberline, the great number of seedlings is rather surprising. The author believes recent warming and resulting shorter length of the winter snow cover to be the factors having triggered increased sexual regeneration. He does not provide, however, more detailed information on the possible different effects on the reproductive process. Thus, warming could mean better seed supply, more favourable conditions for germination and seedling growth or both. Cuevas (2002) provides evidence from the *Nothofagus pumilio* forest in the Chilean part of Tierra del Fuego that regeneration may happen at all altitudes about the same time, while seedling density, however, was lower at treeline (690 m).

In our study areas on the Colorado Front Range, young growth is normally represented by only a few generations (age classes) in the timberline ecotone, whereas seed-produced regeneration is frequent in the upper montane forest at a little lower elevation. Thus, successful generative reproduction seems to be unusual at the given climatic conditions in the ecotone. If lack of suitable sites were the only factor impeding sexual regeneration at timberline, low numbers of seedlings could be expected rather than many age classes missing.

In contrast to the sporadic seedlings of Engelmann spruce (*Picea engelmannii*) and subalpine fir (*Abies lasiocarpa*), seedlings and young growth of limberpine (*Pinus flexilis*) are abundant at sites preferred by the Clark's nutcracker to establish seed caches. Nutcrackers harvest the nutrient-rich seeds mostly in the lower forest. Thus, they exhibit a comparatively high germination capacity and may give rise to seedling clusters. Moreover, the seeds cached just beneath the surface enjoy better conditions for germination than wind-mediated spruce and fir seeds that are deposited on the surface (Section 4.3.13.3).

Finally, insufficient seed qualities as well as lack of sites favourable to germination and seedling establishment are the main reasons of almost episodic regeneration in the timberline ecotone. When we studied regeneration of mountain birch in northernmost Finnish Lapland (1996 to 2002) we found conspicuously high numbers of a few years old seedlings (>25 seedlings/m^2) alongside little streams (about 380 m; Photo 45; see also Holtmeier et al., 2003; Broll et al., 2007). Seedlings concentrated at sites covered by low grass, sedge and herbaceous vegetation, while comparatively few seedlings of the same age occurred either in dwarf birch and willow thickets or on the wind-sheltered sides of adjoining convex topography rising 2 to 3 m above the stream sides (cf. Figures 42, 88 and Photo 45). Older young growth, however, was completely missing at both streamsides and convex topography, which means that seedling establishment failed for many decades.

At timberline, most tree species reproduce at a relatively old age. *Pinus cembra*, for example, reaches its reproductive age at 70 or 80. *Larix lyallii* takes even 100 and more years before producing seeds, and good seed years can be expected in subalpine larch not before the trees are 200 to 300 years old (Arno and Habeck, 1972). *Pinus sylvestris*, on the other hand, produces seeds at an age of 15 to 20 years and *Pinus mugo* after only 10 years. A single reproductive cycle extends over several years. Climatic, biotic and mechanical influences may interrupt the process at any time (Figure 49).

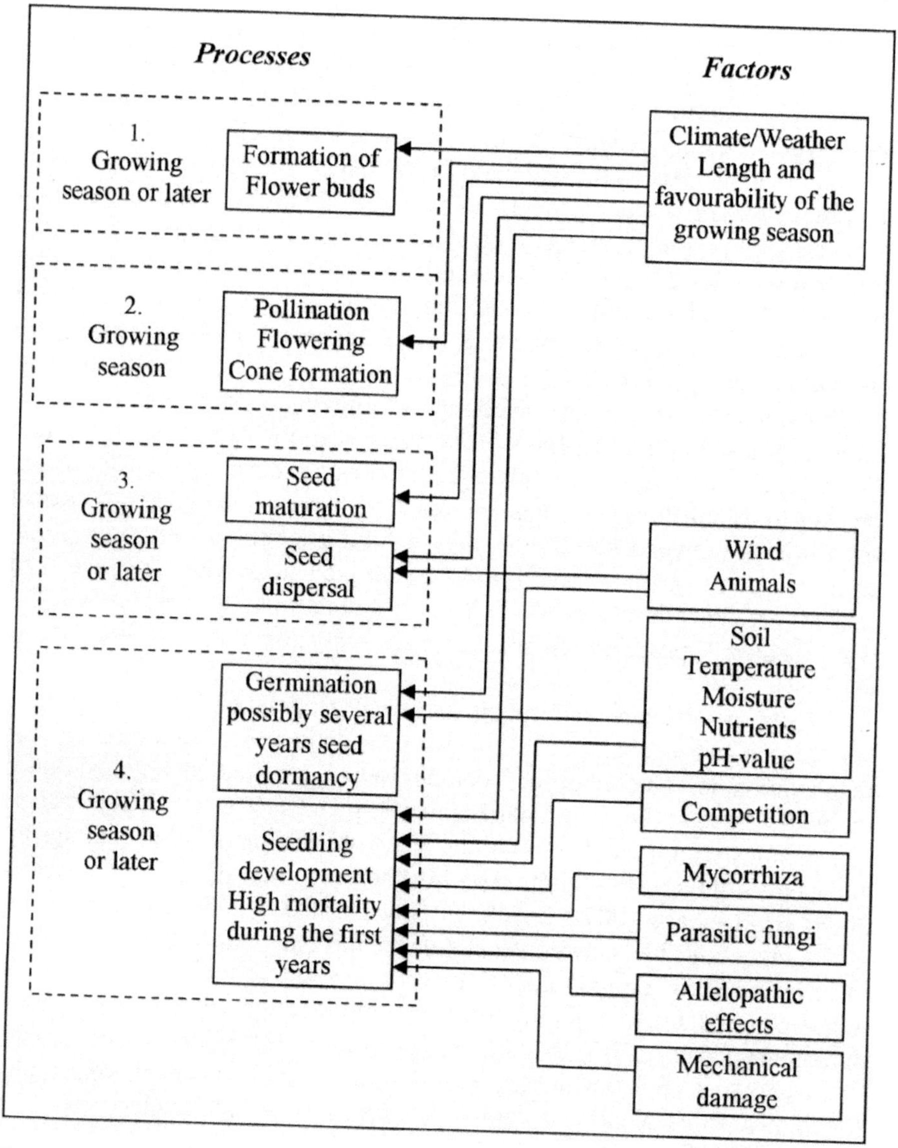

Figure 49. Factors and processes controlling seed-based regeneration at timberline. Modified from Holtmeier (1993a).

However, even in case this process from bud formation, flowering and cone production to release of mature seeds was not disturbed, successful regeneration will not be guaranteed. First, seeds must reach a suitable seed bed and germinate. Then seedlings have to develop at usually harsh environmental conditions.

Germination is influenced by many factors (temperature, soil moisture, soil acidity, air humidity, illumination, moulds and others) and also depends on the specific properties of the seeds. The germination rate of birch, alder and larch seeds is almost 100% at temperatures between 20°C and 30°C, sufficient light conditions provided. Low temperature (5°C to 15°C) reduce germination rate. However, if seeds were exposed to freezing temperatures, germination at low temperatures would increase (Farmer, 1997). The wingless seeds of *Pinus cembra*, *Pinus albicaulis*, *Pinus pumila*, *Pinus flexilis* and some other stone pines are comparatively heavy and characterised by a hard seed coat. They will germinate only after long stratification at freezing temperatures, while conditions of light obviously do not affect germination. In contrast, seeds of *Pinus sylvestris* and other anemochorous pine species normally germinate shortly after release from the cones (Granström, 1987). However, they will lose viability after only 10 to 16 months. Elliott (1979) emphasizes that seed banks do not exist at the polar treeline formed by black spruce (*Picea mariana*). Also, regeneration of *Picea engelmannii* and *Abies lasiocarpa* from seed banks is not very likely. Thus, tree species the seeds of which do not survive for a prolonged period in seed banks may only successfully regenerate if a good seed crop and a sufficiently warm growing season will coincide, as was hypothesised by Kearney (1982), for example, in view of the reproduction of both Engelmann spruce and subalpine fir in Jasper National Park (Alberta, Canada).

Seeds of stone pines, on the other hand, may overlay for several years (Tomback et al., 1993; McCaughey, 1994; Kajimoto et al., 1998). The germination rate of *Pinus albicaulis* seeds is highest after 2 years but seeds also germinate later. Consequently, abundant seed production is not necessarily identical with intensive regeneration. In young growth clusters of *Pinus albicaulis*, for example, which had originated from seed caches of the Clark's nutcracker, the age of the individuals differed by up to 7 years, although the nutcracker had cached all seeds simultaneously (Tomback et al., 1993). Askawa (1957) found similar disparities of age in *Pinus pumila* stands, and the present author found such age differences in young growth clusters of *Pinus cembra* at timberline in the Upper Engadine (Switzerland).

Birch seeds remain viable for 2 to 3 years. As seed years occur more frequently than in timberline forming pines, viable birch seeds are practically always available for regeneration. Seeds of rowan (*Sorbus aucuparia*), a tree species occurring at the highest treelines in the European mountains and also in some regions of the central Himalayas (Miehe and Miehe, 2000), may survive for more than 5 years (Granström, 1987). Thus, the fluctuating fructification of rowan may be partly compensated for. Experimental studies on seedling growth showed faster growth in seedlings that had originated from seeds collected at high elevation than the seedlings that had been grown from low elevation seeds, though size and nutrient content of the seeds decreased by altitude. Faster growth might be an adaptation to the shorter growing season at high elevation (Barclay and Crawford, 1984).

Conditions favourable to germination do not necessarily also encourage seedling growth. For example, germination may possibly be enhanced by long-lasting snow cover and high soil moisture supply, while seedling development is normally impeded by resulting low soil temperatures, shortened growing season, and eventually also by parasitic snow fungi (evergreen conifers only; Section 4.3.7.2; Hiller et al., 2002). This has also become evident from experimental studies in the Swedish Scandes (Jämtland) on the development of seeds that were sewn in permanent plots (795 to 930 m). Snow cover was different while sufficient soil moisture was provided at all plots. Seeds of the previous years germinated shortly after snow melt, whereas height growth and survival rate of the seedlings clearly corresponded to the length of the growing season (Kullman, 1984). At sites exposed to solar radiation and characterised by little snow cover, drought and high soil temperatures often hamper germination (Turner, 1958; Noble and Alexander, 1977). However, seedlings once established may show better growth performance (cf. Table 12; Franklin and Dyrness, 1973; see also Holtmeier, 1985b).

Dense dwarf shrub and grass cover may impair regeneration by preventing seeds from reaching a suitable seed bed, by competition for moisture and nutrients and by too much shading of the tree seedlings (see also Schönenberger, 1975; Weih and Karlsson, 1999). Germination of larch seeds, for instance, will be impeded by even a thin (<2 cm) humus or moss layer (Table 15), because seedlings will normally parch before their fine roots have reached the mineral soil, as was evidenced by experiments of

Table 15. Germination of larch seeds (*Larix decidua*) and survival rate of seedlings as affected by the plant cover (Modified from Auer, 1948)

Vegetation in experimental pots, favourable light and moisture conditions	**Germination rate (%)**	**Survival rate (%) of seedlings after 105–170 days**		**Survial rate (%) of the sown seeds**
Mineral soil	64	77	73	47
Moss cover	55	83	3	45
Mosses and herbs	45	91	0	0
Dense herb layer	43	2	0	0

Auer (1947). On the other hand, dense dwarf shrub heath or grass vegetation protects seedlings from injurious climatic influences. In stands of *Polylepis pauta* and *Polylepis incana* (Photo 113) at timberline in the Ecuadorian Andes a thick litter layer resulting from all year round leaf shedding may impede seedling establishment (Cierjacks et al., 2008).

Sufficient moisture and warmth provided, open mineral soil is particularly favourable to seedling establishment (Photo 57), except for seeds that need humus-rich substrate for germination. On exposed mineral soil, seedlings at high numbers may form true 'seedling meadows' (e.g., Holtmeier, 1995b). At dry sites, however, exposed to wind or solar radiation and characterised by permeable substrate and very low humus content, germination rate is normally extremely low. Germination rate of larch and spruce seeds, for example, closely depends on constant soil moisture, since the seeds are coated with wax and thus swell up very slowly. Mechanical damages caused to the seeds while being dispersed by wind along the snow surface, for example, may improve germination (Auer, 1947).

Photo 57. Intense regeneration of *Picea engelmannii* on open mineral soil (boulders in a loamy-sandy matrix) in the forest-alpine tundra ecotone on Wheeler Peak (Great Basin National Park, Nevada) at 3.360 m. F.-K. Holtmeier, 30 July 1994.

Birch timberline sites (330 m) in northern Finland, where strong winds had removed the organic layer and exposed permeable sandy substrate (glacial till), were almost devoid of birch seedlings, whereas seedlings occurred at great numbers in shallow hollows, gullies and similar snow-rich depressions that provide higher soil moisture (Section 5.4). Successful germination does not necessarily mean sustainable regeneration. The first year is most critical, as the initial seedling growth closely depends on the nutrient content of the seeds, which is different in different tree species. Small and light seeds (e.g. *Larix*, *Betula*, *Populus*, *Salix*) are by far less rich in nutrients than larger seeds (e.g. stone pines, oaks, beeches). Thus, it is not by chance that the comparatively heavy seeds of *Pinus cembra*, *Pinus albicaulis*, *Pinus pumila* and other stone pines (Table 16) are an important food for many animals such as nutcrackers, woodpeckers, mice and bears (Holtmeier, 1999c, 2002, further references there).

Table 16. Average number of seeds per kilogram of some American conifers (Data from McCaughey et al., 1986)

Tree species	**Number of seeds (×1.000)**
Picea glauca	498.2
Larix lyallii	313.1
Picea engelmannii	297.6
Tsuga mertensiana	251.5
Picea mariana	890.1
Abies lasiocarpa	76.1
Pinus flexilis[1]	10.8
Pinus albicaulis[1]	5.7

[1]Wingless seeds.

Cui and Smith (1991) studied mortality rates of *Abies lasiocarpa* seedlings at high elevation sites (2.672 to 2.950 m) in the Medicine Bow Mountains (southern Wyoming). The authors observed very high losses during the first 2 years after germination. About 60% (shaded) to 90% (sun-exposed) of the seedlings died. Later, however, almost no losses occurred. So it seems very likely that, after the first needles had flushed and roots had been infected with ectotrophic mycorrhiza, water and nutrient uptake and thus dry matter production considerably increased. In the timberline ecotone on Beartooth Plateau (Montana, Wyoming), losses in whitebark pine seedlings and seedling clusters were also high during the first years after germination. After 4 years, the annual mortality of the remaining individuals declined to less than 1% (Mellmann-Brown, 2002, 2005; Figure 50).

Many seedlings succumb to climatic injuries and root competition or are destroyed by frost heaving, heat-girdling (Noble and Alexander, 1977), cattle and wild game (grazing, trampling). Experimental studies at Toolik Lake (approximately 50 km north of the polar treeline at the footslope of the Brooks Range, Alaska) showed that artificial warming could not stimulate seedling growth (*Betula papyrifera*, *Picea glauca*, *Populus tremuloides*), while growth increased after the competing tundra vegetation had been removed (Hobbie and Chapin III, 1998).

Thus, nutrient-deficiency or, better, competition for nutrients may be a critical factor in seedling performance and survival, which also became obvious from experiments in *Nothofagus solandri* seedlings in New Zealand

(Wardle, 1985a). Seedlings only survived on nutrient-rich soils and if root competition was eliminated. Also, allelopathic effects of the associated vegetation may seriously impair germination and seedling growth, as is known from *Empetrum hermaphroditum*, for example, that negatively affects birch and pine seedlings (Nilsson et al., 1993; Nilsson, 1994; Farmer, 1997; Weih and Karlsson, 1999).

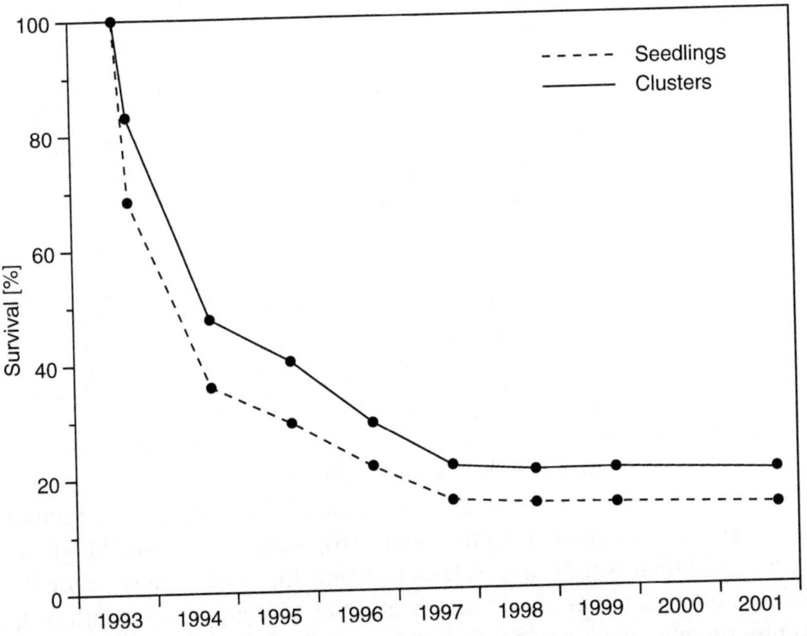

Figure 50. Survival of whitebark pine (*Pinus albicaulis*) regeneration (n = 543 seedlings and 175 seedling clusters on 19 experimental sites in the timberline ecotone on Beartooth Plateau (Montana/Wyoming) in 1992. From Mellmann-Brown, 2002).

Occasionally, lichens also impair seedling growth by exuding toxic compounds that impede mycorrhiza development. The strongest effects come from *Cladonia stellaris* (=*Cladonia alpestris*) followed by *Cladonia arbuscula* and *Cladonia rangiferina*, whereas *Stereocaulon paschale* is not that adverse. Toxic exudations also negatively affect saprophytic fungi and thus indirectly reduce decomposition and nutrient supply (Brown and Mikola, 1974). So, moderate reindeer grazing as well as fire may improve conditions for regeneration.

Young growth that has survived the seedling stage enters a critical phase as soon as it starts projecting above the winter snow cover. Thereupon it will be determined whether young growth will be able to display its genetically predetermined 'normal' tree form or will be continuously pruned by injurious winter climate (Sections 4.3.3, 4.3.4 and 4.3.11).

Seed-based regeneration in the timberline ecotone can be well compared with a very difficult 'hurdle race' (Figure 49). In case of a general cooling, sexual reproduction fails first. As to the supply of the timberline ecotone, especially of its uppermost zone, with germinable seed, seed dispersal from lower elevation is an important factor. In this respect, the situation at timberline in high mountains is relatively favourable compared to the polar timberline. At a cooling climate, for example, the upper limit of production of viable seeds would decline for several 10 m, maybe about hundred metres. At least occasional seed supply from lower elevation could be expected. In the forest-tundra ecotone, however, a cooling of the same magnitude would result in a southward shift of the northern limit of viable seed production for at least several tens if not hundreds of kilometres. Consequently, viable seeds would hardly reach northern treeline or even tundra (e.g. Holtmeier, 1974; Nichols, 1975a, b, 1976; Elliott and Short, 1979; Black and Bliss, 1980; Larsen, 1989).

With respect to successful seed-based regeneration, the way of seed dispersal, either by wind (anemochorous) or animals (zoochorous), is an important factor. Many timberline forming conifers such as *Picea*, *Abies*, *Larix*, *Tsuga* and some pine species produce relatively light winged seeds that are dispersed mainly by wind. Also, dispersal of seeds of *Betula*, *Alnus*, *Populus*, *Salix* and other deciduous trees at timberline depends on wind.

Wind-mediated seed dispersal is influenced by a bundle of factors (cf. Table 18) such as seed weight (Table 16), seed-wing size, height above surface at which seeds are released from the tree, wind velocity and direction, other weather conditions (wet, dry), topography (smooth or gently sloping, rugged, upslope, downslope) and plant cover. Low seed weight, large wings, high wind velocity, smooth topography and low plant cover are favourable to seed dispersal by wind, dry weather conditions provided. Energy content and germination capacity are positively correlated with seed size (see also Sveinbjörnsson et al., 1996). In general, regeneration from small seeds is often less effective than from large seeds.

High surface roughness, for example, caused by uneven forest canopy or dense scrub or krummholz, impedes anemochorous dissemination from forest at lower elevation up to and beyond treeline. In general, most seeds reach the ground already within a distance two to three times the height of the seed tree (Hesselmann, 1934, 1938; Lehto, 1957; Kuoch, 1965; Shiyatov, 1966; Holtmeier, 1974, 1993a; McCaughey et al., 1986).

In northern Lapland, for example, the present author found young growth of Scots pine (*Pinus sylvestris*) being usually restricted to close proximity of old seed-bearing trees (Holtmeier, 1974, 2005a). The same was observed at the pine tree limit in the southern Swedish Scandes (Kullman, 2007a) and in the timberline ecotone in the polar Urals, for example. In this region, Siberian larch (*Larix sibirica*) releases seeds usually in June or July when the winter snow cover is melting. The wind carries the seeds no farther than 40–60 m (Shiyatov, 1966). This concentration of young growth around the seed source might also be attributed to the facilitation of seedling establishment and survival by the tall trees as is generally observed in boreal forest (see also Slot et al., 2005). Even on windthrow areas, the more or less great distance from the seed source is notable. The longer the distance the smaller is the number of seeds on the ground. Thus, high seedling and young growth density usually occur on the smallest windthrow areas (Schönenberger and Wasem, 1999). A few seeds only are carried over greater distances and dispersed fairly irregularly.

Seeds released from high trees have a greater chance of being dispersed over longer distances compared to seeds from low trees. Thus, ample seed supply from lower elevation can be expected only within a comparatively narrow zone just above the closed forest (see also Cuevas, 2000). Wind-mediated seed dispersal beyond timberline is very sparse, as was also found by Kuoch (1965) who studied seed supply at timberline (1.970 to 2.010 m) in the Sertig-Valley (Switzerland) by using seed traps. The author concluded from the results that a natural reforestation of abandoned alpine pastures only by wind-borne seeds were not to be expected within the foreseeable future. Contrary to that, Schmidt and Shearer (1995) found seeds of *Larix occidentalis* distributed by warm and dry upslope winds up to 250 m and supposed that major frontal winds would carry seeds even much farther. In the birch timberline ecotone in northern Finnish Lapland, the present author came across 20 to 40 years old Scots pines (*Pinus sylvestris*) that must have originated from seeds of a several kilometres distant and 200 m lower seed source.

On fjelds in northern Finland, Luoto and Seppälä (2000) found pine young growth to about 10 km distant from the nearest pine forest. The present author found some spruces (*Picea abies*) on the forefields of Roseg- and Tschierva-glacier (Upper Engadine, Switzerland). The oldest spruce trees are about 60 to 80 years old and more or less deformed. They must have originated from seeds that had been carried over long distances either from solitary spruce stands at the mouth of the Roseg-Valley or from the neighboured Upper Engadine main valley west of the Roseg-Valley. In case the seeds came from there, they would have covered a horizontal distance of about 5 km and an altitudinal difference of 700 to 1.000 m (Holtmeier, 1974). In the main valley, however, the upper limit of spruce trees is located at about 2.000 m and no young growth of spruce occurs in the timberline ecotone (2.300 to 2.400 m) where only *Larix decidua* and *Pinus cembra* are present. Thus, it does not seem to be very likely that the seeds that gave origin to the solitary spruce in the Roseg-Valley had been carried over the mountain crest.

However, strong cyclonal winds may disperse conifer seeds over even much longer distances. Aario (1940), for example, reported solitary pine occurrences about 40 km north of the northern limit of closed pine forest in Petsamo-Lapland that originated from wind-borne seeds. Also, the northernmost tree islands of black spruce (*Picea mariana*) 80 km beyond the northern treeline emerged from seeds that were carried with the wind from a far distant seed source at treeline or even farther south (Payette and Delwaide, 1994). Obviously, regeneration in the timberline ecotone depends on seed quality rather than quantity, other factors influencing regeneration disregarded (see also Figure 49).

Wind-mediated seed dispersal is very irregular due to the influences of changing weather, topography and plant cover. Seeds carried along the snow surface may accumulate at great numbers at wind-sheltered sites such as hollows, terraces and craters caused by wind scouring or ablation around old trees and at the edges of compact tree clumps. Many dense tree clusters have originated from such seed deposits. Generally, however, it depends on chance whether wind-borne seeds will reach a suitable seedbed within the microsite pattern in the timberline ecotone or even beyond the present treeline. Thus, the chance decreases considerably approaching the alpine zone. Anyway, also in the timberline ecotone, microsites favourable to germination and seedling establishment are rare. Often, dense plant cover prevents wind-borne seeds from reaching the ground. Moreover, extremely long-lasting snow cover, waterlogged soils or drought impede seedling establishment. As a rule, regeneration is most successful on convex topography, sufficiently warm and moist conditions provided.

In general, regeneration from animal-mediated seeds seems to be more effective than from wind-borne seeds (Holtmeier, 1999c, 2002). This holds particularly true for some subalpine pines species such as *Pinus cembra*, *Pinus albicaulis* and *Pinus flexilis*, which produce heavy wingless (or almost wingless) seeds that are dispersed mainly by nutcrackers (*Nucifraga caryocatactes*, *Nucifraga columbiana*). Seed dispersal by these birds is a key factor in stone pine forest ecosystems and in timberline dynamics. Because the nutcrackers' role is so important, it will be considered in detail in an extra chapter (Section 4.3.13.3).

Little information is available on sexual regeneration at tropical timberlines, with a few exceptions. Paijmans and Löffler (1972), for example, mention abundant regeneration of *Dachrycarpus compactus* and *Papuacedrus papuanus* at timberline on Mt. Edward (New Guinea). They ascribe the great amount of seedlings and young growth to better conditions of light compared to the closed forests. Byers (2000) reports intensive natural regeneration of relictic *Polylepis* in the upper Pisco-Valley (Huascarán National Park, Peru) and invasion of pastures by trees after these areas had been protected from any use. Information by Laegaard (1992) and Kessler (1995) on regeneration of *Polylepis* in the Bolivian and Ecuadorian Andes does not provide an idea of the situation at the upper distribution limit of this species. Normally, we are dependent on rather incidental observation or on indirect information. Smith (1977b), for example, reports in his studies on the high mountains of New Guinea low forest (coppice forest) invading burn areas. However, the author does not provide more detailed information about how succession proceeds: by seedlings, basal shoots and/or root suckers of groves that had survived the fire? In respect of the adverse present climatic conditions, forest is supposed to not resettle those areas from which humans had removed it. It is not clearly said, however, how the present climate prevents natural reforestation. Thus, we depend on speculation. Besides climatic influences that would impede tree growth, an insufficient supply of viable seeds could be the factor preventing re-colonisation of the burn areas by forest. Competition of grasses and herb vegetation could be an additional agent keeping the burns devoid of trees. Klötzli (1975) mentioned missing seedlings of *Erica arborea* and *Hypericum revolutum* in the high elevation grassland of northern Ethiopia, and supposed lack of soil moisture and competition by the grass vegetation were the reason. Thus, one could speculate that seed-produced regeneration would have occurred if site conditions had been more favourable to germination and seedling establishment, which cannot be taken for sure, however. Wesche (2002) reports abundant seed production in *Erica arborea* in eastern Africa. At timberline on Mt. Kenya, successful seed-based regeneration of *Hagenia abyssinica* and thus persistence of the forests seem to be contingent on fire at moderate frequency, while light intensity has no effect, for example (Lange et al., 1997).

There is evidence from the tropical Andes that severe microclimates as well as pasture management (recurrent burning) prevent seed-based regeneration at the present timberline. While tussock grasses rapidly resprout after having been burned (e.g., Beck et al., 1986; Laegaard, 1992; Hofstede et al., 1995; Wesche et al., 2000) shrub vegetation and occasional young growth of trees that became established in the Páramo will recover very slowly, if ever (Di Pasquale et al., 2007). Radiation-tolerant tree seedlings, saplings and ramets may occur at great numbers near the forest limit. Bader et al. (2008b) suppose that tree seedlings benefit from shading by the surrounding Páramo vegetation rather than being suppressed by competition. Seedlings fully exposed to excessive solar radiation and extreme fluctuations of temperature in the Páramo usually become seriously damaged. Discolouration of leaves and subsequent loss of foliage often cause seedling death. Moreover, over-heating may result in drought stress and desiccation. In addition, photooxidation and photoinhibition may affect tree seedlings. The adverse effects of the Páramo environment to seedling establishment seem to be substantiated by transplantation experiments (Bader et al., 2008b). Mortality increased when the seedlings were fully exposed to solar radiation, while artificial shading reduced seedling mortality (Section 4.3.12). On the other hand, shade-in-tolerant *Polylepis pauta* and *Polylepis incanca* regenerate more vigorously at the forest edge of the Ecuadorian timberline than in closed forest stands (Cierjacks et al., 2008). Besides the harsh Páramo climate also insufficient seed dispersal is likely to impede timberline advance as suggested by the number of seedlings rapidly decreasing from the forest edge (Bader et al., 2008b; see also Wesche et al., 2008). Altogether, precise information on regeneration at tropical timberlines is scarce. Thus, we refer mainly to temperate timberlines in the following.

4.3.10.2 Vegetative reproduction

Tree species that are able to reproduce by layering (formation of adventitious roots), basal shoots (stump sprouts) and root suckers are at an advantage compared to species reproducing only by seeds, as vegetative reproduction is by far less impeded by unfavourable climatic conditions than sexual regeneration and still continues at low temperatures that would prevent any production of viable seeds (see also Larcher, 1980a). In addition, clonal groups act as nutrient supply units as long as the clonal members have not become independent (Kuoch and Amiet, 1970).

Although vegetative reproduction is more common in broad-leaved trees (Smith et al., 1997) it also occurs in some conifer species such as spruce (e.g., *Picea engelmannii*, *Picea abies*, *Picea mariana*, *Picea glauca*), fir (*Abies lasiocarpa*, *Abies balsamea* and others; Photo 58), hemlock (*Tsuga mertensiana*) and Alaska cedar (*Chamaecyparis nootkatensis*). These species reproduce vegetatively mainly by layering. Also larch species (e.g., *Larix decidua*, *Larix lyallii*) are able to layer (Photo 59), though layering is much less common in larch than in the other conifers mentioned above (Arno and Habeck, 1972). At some timberlines, however, larch reproduces more effectively by layering and also by root suckers than by seeds, as is typical of *Larix gmelinii*, for example, in central Kamchatka (Okitsu, 1997).

In most pine species vegetative reproduction is exceptional. However, a few pines such as *Pinus mugo* (Hafenscherer and Mayer, 1986) and *Pinus pumila* (Okitsu and Ito, 1984; Wilmanns et al., 1985; Kajimoto, 1992) regularly reproduce by layering. This is advantageous to these pines as low light intensity and thick litter layers impede seedling establishment in dense pine shrub (see also Michiels, 1993). Layering was observed also in hedge-like growing *Juniperus excelsa* at wind-exposed sites on Dinar Pass (3.250 m) in the southern Zagros Mountains (Pontecorvo and Bokhari, 1975). In the Himalayas, Miehe and Miehe (2000) found multi-stemmed, bush-like junipers that had originated from stumps of felled arborescent junipers.

Photo 58. Clonal group of *Picea abies* in the forest-alpine tundra ecotone on Yllästunturi (Finnish Lapland) at about 480 m. F.-K. Holtmeier, 1 September 2000.

Approaching timberline and treeline, vegetative reproduction increases also in deciduous tree species. Red beech (*Fagus sylvatica*), for example, layers frequently (Fanta, 1981). Mountain birch (*Betula tortuosa*) reproduces mainly through thriving shoots from the root stock. Vegetative reproduction is also common in the deciduous *Nothofagus* species (*Nothofagus pumilio*, *Nothofagus antarctica*) at timberline in South America (Cuevas, 2002). In *Nothofagus solandri* at timberline in New Zealand, vegetative reproduction is even more important than seed-based regeneration, in particular after heavy snow loads, avalanches, winter desiccation, land slides or occasional fires damaged the trees. Southern beech usually recovers rapidly by thriving new shoots from the remained rootstock (Norton and Schönenberger, 1984; Schönenberger, 1984).

Photo 59. Clonal group of *Larix decidua* on the extremely wind-swept Pru dal Vent (Puschlav, Switzerland) at 2.170 m. The group has been extending downwind (*left*) by layering. F.-K. Holtmeier, 24 September 1968.

Layering often occurs on steep valley sides and in avalanche chutes after the stems has been partly covered by downhill mass movement or if the treetops were broken by avalanches or sliding snow. The damaged trees normally elongate by layering at their lower (downhill) end (see also Stimm, 1985, 1987). If growing on level ground, the plan view of a clonal group is usually almost circular, because the groups gradually expand at their periphery. Clonal groups exposed to strong permanent winds from almost exclusively one direction, may elongate at their leeward end thus extending upslope or downslope according to the prevailing wind direction (cf. Figure 60; Holtmeier, 1993a).

Clonal groups are by no means rare at lower elevation. However, they occur more frequently approaching timberline (see also Cooper, 1911, 1931; Mayer, 1976) as seed-produced regeneration declines and damages (climatic injuries, breakage, etc.) triggering vegetative reproduction increase considerably.

At tropical timberlines, vegetative reproduction, mainly by root suckers and stump sprouts, seems to be of particular importance, as can be concluded from the sparse available information concerning timberline dynamics and regeneration (e.g., Gillison, 1970). Thus, some authors repeatedly mention a pronounced ability of the tree species forming timberline in eastern Africa (*Hypericum*, *Erica*, *Philippia* and others) to rejuvenate from stump shoots, which guarantees tree survival after anthropogenic disturbances and frequent fires (Klötzli, 1975; Beck et al., 1986; Miehe and Miehe, 1996, 2000; Hertel and Wesche, 2008). Also, *Polylepis* may effectively reproduce by thriving new shoots from the stump (e.g., Koepke, 1961; Rauh, 1988; Kessler, 1995; Sturm, 1998), and vegetative reproduction increases according to the frequency of fires.

Trees regenerating only from seeds will die after a maximum individual age. Clonal groups, on the other hand, may continue to exist almost indefinitely, if not destroyed by fire, insects, fungi or deteriorating climate that would prevent the trees from gaining sufficient tolerance of environmental stress and injuries. Under a 5 m tall Norway spruce (*Picea abies*) growing near the treeline (910 m) in Central Sweden, Kullman (2005a) found "generations" of genetically identical subfossil wood remains. The oldest was 9.000 years old indicating the possible long live-span of clonal groups. With regard to the survival of the uppermost forest stands, reproduction by layering, root suckers and stump sprouts is much more effective than seed-based regeneration, in particular at a cooling climate (e.g., Kihlman, 1890; Bryson et al., 1965; Larsen, 1965; Tolmachev, 1970; Nichols, 1974, 1975a, b, 1976; Payette, 1976, 1983; Elliott, 1979; Payette and Gagnon, 1979; Payette et al., 1982; Elliott-Fisk, 1983; Holtmeier, 1985b; Kullman, 2002; Laberge et al., 2000).

In the following, the present author refers mainly to his own studies and observations in the Rocky Mountains, in northern Europe and in the Alps (see also Holtmeier, 1986a, 1993a, 1999b). Obviously, clumping of trees due to layered branches is the rule rather than the exception at timberline on the Rocky Mountains. Clonal tree islands are more common at the timberline in the high mountains of North America and also in the northern forest-tundra ecotone than at timberline on the Alps. This may be ascribed to greater ability of the North American tree species (e.g., *Picea engelmannii*, *Abies lasiocarpa*) to layer if compared with *Picea abies* in the Alps, as Lüdi (1961) already assumed. In view of the many North American timberline ecotones the physiognomic aspect of which is virtually characterised by the prevalence of climatically shaped clonal tree island, we would not hesitate to agree (see also Larsen, 1965, 1980, 1989; Ives, 1973b, 1978; Nichols, 1974, 1975a, b, 1976; Elliott, 1979; Hansen-Bristow, 1981; Ives and Hansen-Bristow, 1983). In the timberline ecotone on the Colorado Front Range, for instance, more than 70% of the tree islands of *Picea engelmannii* and *Abies lasiocarpa* have originated from layering (Holtmeier, 1999b). However, if comparing timberlines in this respect it must be considered that in the Alps high-elevation spruce forests were mostly removed to create alpine pastures (Section 4.3.14.1). Surely, spruce stands at the climatic timberline had a higher percentage of clonal groups than the present spruce forest.

Under severe conditions, formation of adventitious roots is usually triggered by damages at the apical shoots caused by climatic (frost, winter desiccation) and/or mechanical (breakage, abrasion, browsing) influences. Thus, the plagiotropic branches, no longer hormonally controlled by the leader, will elongate vigorously, and the lowest ones will contact the ground and get gradually imbedded in the litter (Figure 51; see also Kuoch and Amiet, 1970; Zimmermann and Brown, 1971; Fanta, 1973; Brown, 1974; Schönenberger, 1978, 1981, 1986; Fink, 1980; Stimm, 1985; Holtmeier, 1993a, b). Adventitious roots mainly form at the lower side of the imbedded branches, usually in the O_f-horizon of the organic layer that exhibits higher soil moisture than the surface litter. Very likely, humic acids have a stimulation effect (Fanta, 1973).

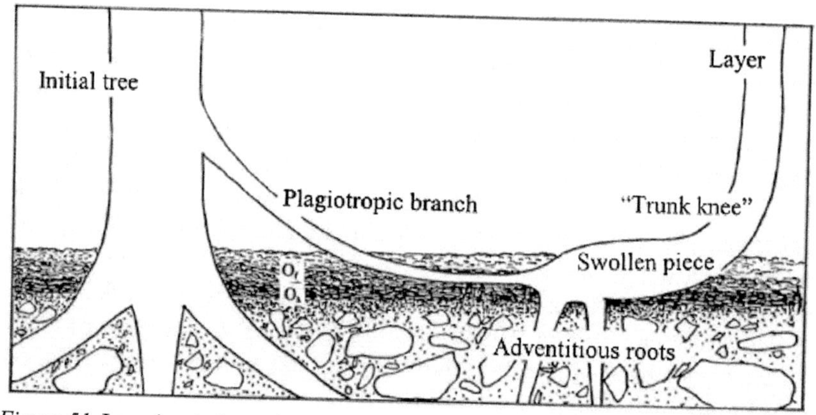

Figure 51. Layering (schematic). Adventitious roots usually form in the organic layer.

In the timberline ecotone, climatic influences cause damage at the terminal shoots of relatively young trees already, particularly as soon as they start growing above the protecting snow cover. Obviously, this is the main reason for the increase of clonal tree islands from lower elevation to treeline. Continuous layering produces a more or less great number of clonal generations consisting of several clonal stems each. The age of the clonal generations and clonal members decrease from the parent trees to the periphery of the tree islands. Thus, layering gives origin to compact and sharply contoured tree islands (Photos 60, 61). At timberline in the Colorado Front Range and in many other mountain ranges in western North America, the present author found old clonal tree islands consisting of more than 70 clonal stems (Holtmeier, 1999b). At the northern timberline, Laberge et al. (2000) discovered a 1800 years old clonal group of black spruce (*Picea mariana*) composed of more than 80 stems. Occasionally, clonal colonies have originated from seedlings of Engelmann spruce and subalpine fir that had established themselves around old pine trees. The pines provided protection to the seedlings, which grew up and then propagated by layering (see also Patten, 1963b).

Photo 60. Excavated connections of a clonal group of *Abies lasiocarpa* at a wind-exposed site (at about 3.400 m) in the forest-alpine tundra ecotone on Niwot Ridge (Front Range, Colorado). Under the influence of permanent strong winds from the west (*left*) this group had gradually expanded downwind by layering. F.-K. Holt-meier, 5 August 1987.

Photo 61. Compact, sharply contoured clonal groups (*Abies lasiocarpa*) on Hurricane Ridge in the Olympic Mountains (Washington) at about 1.900 m. F.-K. Holtmeier, 5 August 1995.

Alder, aspen, birch, willows, southern beech and other broad-leaved species exhibit a high ability of thriving sprouts from stump and stem. Reproduction from stump sprouts, for example, guarantees survival after breakage caused by avalanches, heavy snow loads or debris slides. Aspen regenerates also from root suckers after fire or mechanical damage (see also Photo 19). *Larix leptolepis* at timberline on Mt. Fuji, for example, reproduces mainly in this way. Recovery from stump sprouts is typical of mountain birch in northern Europe after defoliation by the autumnal moth (*Epirrita autumnata*; for further references see Holtmeier, 1999c, 2002; see also Sections 4.3.13.4 and 5.4). Complete recovery takes many years and is impeded by many factors such as reindeer grazing and voles. Also root rot spreads gradually from the decaying rootstock into the new shoots and impair recovery. The cause of root rot are fungi such as *Armillaria borealis* and *Cerrena unicolor* (communication Y. Mäkinen). Probably, also some other fungi species are involved (*Piptorus betulinus*, *Fomes formentarius* and *Daedalopsis septentrionalis*; communication J. Lehtonen). *Armillaria* attacks healthy trees as well as stressed trees and also is a saprophytic decomposer of dead trees (Wargo and Shaw, 1985; Gregory et al., 1991; Guillaumin et al., 1993) Birch trees respond to the infection by excreting compounds that prevent spreading of these saprophytic fungi into the new stems (so-called compartmentalization; Photos 62 and 63; see also Shigo, 1985). Thus, living stems may even exist 200 years or longer (Holtmeier et al., 2003; Kullman, 2005a). The oldest birch we found at timberline was 225 years old The stump, however, from which such living stems were released may be 500 years old or even older (cf. Kullman, 2005a). Generally, longevity of mountain birch is supposed to be relatively short (Karlsson and Wielgolaski, 2005).

4.3.11 Influence of site conditions on growth form

Response of tree growth to high-altitude climate and site conditions is partly reflected in tree physiognomy. At lower elevation, physiognomy depends mainly on the specific ‘plan of construction’ (tree architecture) which is genetically predetermined. Conifers, for example, are characterised by a monopodial branching system (Brown, 1974; Strasburger et al., 1991) which is particularly obvious in the spruce genus. An orthotropic and radialsymmetrical trunk forms the main axis (monopodium) from which the plageotropic branches turn off ramifying monopodially again. The terminal shoot controls growth of the plageotropic branches, as long as the leader has not been destroyed. Also, plageotropic branches of first order control those of second and higher order. Due to apical dominance, spruce and many other

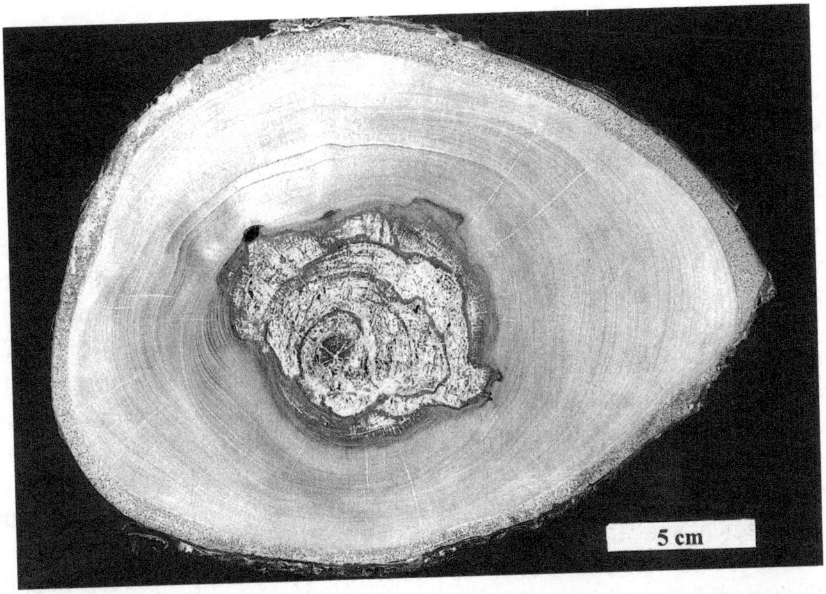

Photo 62. Compartmentalization in the stem of a 200-year-old mountain birch (*Betula tortuosa*) which had been growing in the forest-alpine tundra ectone on Staloskaidi (northern Finnish Lapland) at about 250 m. F.-K. Holtmeier 1999.

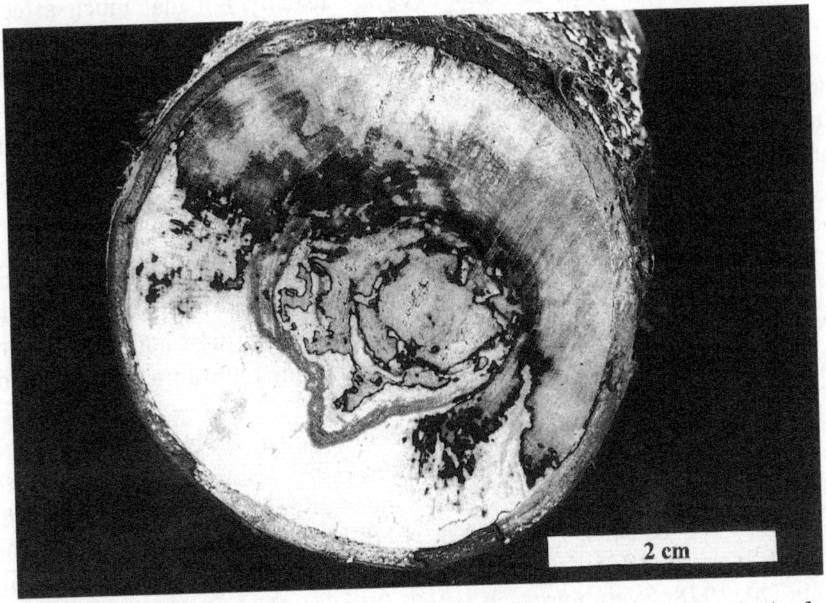

Photo 63. In this basal sprout of a mountain birch (*Betula tortuosa*) growing in the forest-alpine tundra ecotone on Koahppeloaivi (northern Finnish Lapland) at about 310 m compartmentalization could not stop root rot, and the sprout died. F.-K. Holtmeier 1999.

conifers normally exhibit pointed conical growth forms. Particularly in many widely distributed species (e.g., of the genera *Picea*, *Abies*, *Larix* and *Pinus*) many subspecies, races and varieties evolved that may be distinguished by their morphological characteristic, coarse or fine twigs, a ramification system, curved or straight boles (e.g., subalpine larches; Wettstein, 1946; Rubner and Reinhold, 1953) and other peculiarities.

Additionally, growth forms change with aging of the trees. Height and radial growth decreases, and the ability to produce roots declines. Branches die off, the apical dominance ebbs away and instead lateral branches may take control. Moreover, external agents (climate, insects and other animals, fire) leave their marks. The effects accumulate during the lifetime of a tree (maybe several hundred years) and, in total, may influence its growth form in one or the other way. Later, however, it is rather difficult if not impossible to identify single events (e.g., strong frost, drought, defoliation, breakage etc.) and to assess their relative importance to growth form performance. Dendrological analyses, however, may provide some information.

Finally, site conditions such as nutrient supply, temperatures, moisture, illumination, wind, biotic factors (herbivores, parasitic fungi) and slope gradient effectively influence the development of growth forms. In closed forests and dense tree stands, competition for water, nutrients and light also plays an important role. Solitary trees are usually not that much affected, except at seedling stage when still competing with grasses, herbs and shrubs.

With respect to a wealth of different growth forms, it seems hard to identify growth forms that would be specific to the timberline environment. Thus, for instance, curved stems caused by soil creep or heavy snow pressure, multi-trunk growth forms resulting from layering or coppicing (stump sprouts) are not unique to the timberline ecotone. This also holds true for asymmetric, more or less intensely flagged trees. They not only occur at timberline but also on wind-exposed sites at lower elevation, along the ocean coast and in wind-swept lowlands (e.g. Weischet, 1955; Barsch, 1963). Although often producing similar growth forms at wind-controlled high and low elevation sites, wind acts in a different way in the timberline environment (e.g., winter desiccation, ice particle abrasion).

Approaching the climatic temperate or tropical timberlines, height and annual radial growth decrease (Däniker, 1923; Beaman, 1962; Clausen, 1962/1963; Hueck, 1962; Mark and Sanderson, 1962; Oswald, 1963, 1969; Holzer, 1967; Wardle, 1970, 1971, 1978, 1993; LaMarche and Mooney, 1972; Ellenberg, 1975; Lauer and Klaus, 1975a; Höllermann, 1978; Smith, 1980; Ott, 1978; Sveinbjörnsson, 1993; Šrutek and Lepš, 1994; Šrutek et al., 2002; Flenley, 1995; Gallenmüller et al., 1999; Kronfuss and Havranek, 1999). In the short-term, radial growth may depend more on the inter-annual thermal differences during growing season than on the general decrease of

Photo 64. Several-hundred-years-old limber pine (*Pinus flexilis*, diameter 100 cm, height about 500 cm) at a wind-exposed site in the lower part of the forest-alpine tundra ecotone on Trail Ridge (Rocky Mountain National Park, Colorado) at about 3.350 m. F.-K. Holtmeier, 22 September 1974.

temperature by elevation (see also Paulsen et al., 2000). Also, microsite effects on tree growth override the influence of the altitudinal thermal gradient, in particular if young growth is concerned.

As height growth is more hampered than growth in diameter, old trees in particular often look very compact (Photo 64, see also Photo 119). Schweinfurth (1962), for example, noticed enormous diameters compared to tree height in *Libocedrus bidwillii* and *Weinmannia racemosa* growing at timberline (about 1.000 m) on the eastern slope of Mt. Egmont (New Zealand, North Island). Canary pines (*Pinus canariensis*) at timberline (2.000 m) on Pico de Teide (Tenerife), similar in diameter to those at lower elevation, exhibit a substantially shorter stature (Šrutek et al., 2002). The same disproportion between tree height and diameter has been reported also from tropical timberlines (Junghuhn, 1852; Hueck, 1962; McVean, 1968; Hope, 1976; Smith, 1977b, 1980; Miehe and Miehe, 1994; Bruijnzeel and Proctor, 1995). Grubb (1971) assumed that slow growth and low tree height at the upper limit of tropical cloud forests have to be attributed to low temperatures impeding decomposition and thus nutrient supply (cf. Wardle, 1971, 1978).

Trees growing at timberline are much more exposed to climatic influences (wind, radiation, frost and others) than trees in the closed mountain forest. Solitary trees will be more affected by the climatic agents than trees growing

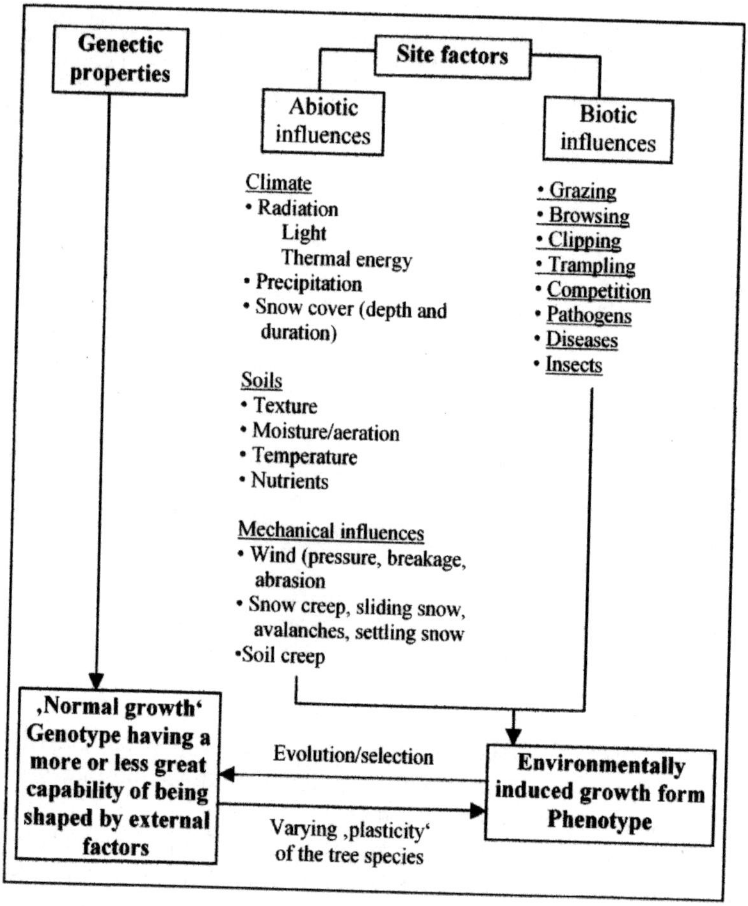

Figure 52. Genetic properties and site factors influencing the growth form of trees at the upper timberline in temperate mountains.

within groups. Thus, climatically stunted growth forms more or less different from the 'normal growth' of the trees are by far more frequent in the ecotone than in the forest below (Figure 52).

The more or less great ability of the tree species to repair such damages (e.g., breakage, drought, defoliation, etc.) by thriving stump sprouts and by layering or by second order branches taking the function of destroyed leaders is of paramount importance to growth form development (see also Khutornoy et al., 2001). Thus, the locally varying site conditions in the ecotone are reflected in a great variety and in the spatial distribution pattern of the environmentally controlled growth forms. Growth forms and their site-related distribution in the ecotone may serve as indicators of the timberline environment and can help to differentiate the ecotone from physiognomic and ecological aspects. Additionally, growth forms and the shape of the tree crowns in particular mirror the influence of fluctuating climate (Holtmeier, 1969; Kullman, 1986a; Scott et al., 1987a; Payette et al., 1989; Payette and Lavoie, 1992; Payette and Delwaide, 1994; see also Figure 58). Also, defoliation by leaf-eating insects (see Holtmeier, 1999c), fire, breakage of treetops, stems and branches (storms, heavy wet snow) have often left their marks. Evaluation of such marks, in particular if supported by tree ring analyses, allows conclusions as to events having affected growth form during the lifetime of a tree. Besides varying ring-width and density of the tree rings, frost cracks and compartmentalization (cf. Photos 62, 63), the pattern of tension and compression wood and the more or less pronounced eccentricity in tree diameter provide instructive information (cf. Bannan and Bindra, 1970; Schweingruber, 1980; Kienast, 1985; Kienast and Schweingruber, 1986; Kontic et al., 1986; Schweingruber et al., 1986; Mattheck, 1991; Schönenberger et al., 1994; Holtmeier, 1999b), although the events that caused the peculiarities to the individual tree ring pattern cannot always be unambiguously identified.

Information on growth forms at the tropical timberlines is comparatively scarce. In general, observations refer to altitudinal decrease of tree stature and to the local type of timberline, mainly whether timberline occurs as a line or a more or less wide ecotone characterised by gradual transition from closed high-stemmed forest to scrub and high-elevation grassland.

Dome- or umbrella-shaped tree crowns with very dense and relatively small and hard foliage restricted to the superficial part of the umbrella, which occur in tree species of different taxonomic orders, are typical of many tropical timberlines. Troll (e.g., 1955a, 1959, 1968, 1973), in particular, repeatedly referred to this peculiar feature ('evergreen umbrella-shaped trees'). Climatically shaped growth forms are rarely mentioned. Jordan (1983), for example, reported compact, heavily stunted growth forms of *Polylepis* at its upper limit and ascribed these growth forms to unfavourable thermal conditions and wind. Miehe and Miehe (1994) consider growth forms of *Erica arborea* in the relic 'dwarf forests' (Chapter 3 and Section 4.1.3) on the Bale Mountains (Ethiopia) to be caused by the harsh high-elevation climate. The uppermost *Erica* individuals, exhibiting a globular shape, are rarely higher than 1½ m. The authors do not provide information on the development of these growth forms. However, they emphasize a physiognomic similarity to dwarf mountain pine (*Pinus mugo*), that also display sabre-like, curved stems trailing on the ground and rising at the tips. This comparison is weak, however, as down-lying growth form of dwarf mountain pine is inherent and should not be confused with climatically induced low growth (Holtmeier, 1973, 1981a). Probably, decumbent growth of *Pinus mugo* results from evolutional adaptation to regular heavy snow load and permanent mechanical stress due to avalanches, sliding snow and snow creep, as Wilmanns et al. (1985) assumed, for example.

This hypothesis sounds plausible but is not stringent however, as stands of mountain pines are not restricted to snow-rich sites. They are also common on rocky steep valley sides and outcrops almost snow-free in winter. Thus, the occurrences of dwarf mountain pine at such sites might be attributed to other factors than snow pressure. For example, limited nutrient supply, and thus reduced competition with more nutrient-demanding high-stemmed conifer species, might be favourable to *Pinus mugo*, which is not very competitive at less extreme sites (Holtmeier, 1967b, 1974).

In contrast to the inherent multi-trunk and decumbent growth of *Pinus mugo*, *Pinus pumila* and *Alnus viridis*, environmentally dwarfed growth forms result from the response of the individual tree to climatic, biotic or mechanical impact at a given site (Holtmeier, 1981a; cf. Figure 52). Undoubtedly, such stunted trees would grow tree-like at more favourable conditions. Thus, the multi-stemmed stunted growth of *Erica trimera* within the 'dwarf forests' might be caused by environmental influences and in this case should not be paralleled with dwarf mountain pine, except low growth where hereditary. The same holds true for *Podocarpus compactus* and *Rapanea vaccinoides* in New Guinea. At lower elevation, these species grow as high-stemmed trees, while they gradually adapt shrub-like habitus as a treeline is approached. Low-growing individuals advance to the altitudinal limit (4.480 m) of the 'subalpine scrub zone' (Robbins, 1970). With respect to these most advanced outliers and many floristic similarities the 'scrub zone' is considered by some authors (e.g., Brass, 1964) to be a part of the upper mountain forest.

Twisted, gnarled and often polycormic growth forms virtually similar to those displayed by mountain birch and *Polylepis* (Photo 65; see also Photo 21), for example, are typical of tropical timberline (see also Paijmans and Löffler, 1972; Seibert and Menhofer, 1991). Due to layering and, after the apical meristems were repeatedly destroyed, shrub-like growth forms develop quite similar to the true hereditary krummholz (Hueck, 1966; Ellenberg, 1975). Therefore, Ellenberg (1975) called these dwarfed forests 'sclerophyllous krummholz forests'), which may be misleading again as the growth forms are environmentally controlled.

Dome- or umbrella-shaped crowns, typical of many tropical timberlines, also occur at timberline in New Zealand (Wardle, 1968; Troll, 1973) which might reflect the influence of comparable climatic conditions. In New Zealand, however, dome-shaped crowns forming a close even canopy are restricted to the mixed mountain forest on the wind-exposed slope where the highly competitive southern beech does not occur (see also Schweinfurth, 1966; Wardle, 1973). Particularly, tall shrubs such as *Olearia colensoi* display dome-shaped crowns. Southern beech, if growing at wind-swept sites, also exhibits this kind of crowns. Dome-shaped crowns have been supposed to be an adaptation that would enable trees to use solar radiation more effectively at the given cool and cloudy climate, as their shape would seem to provide maximum exposure to light in a foggy, light-deficient environment (Wardle, 1965b, 1978). On the other hand, energy loss is reduced by the more or less hemispherical crown surface, which is relatively small compared to the volume of the crown. Moreover, heat loss to the atmosphere by turbulent mixing is comparatively low because of the smooth crown surface, which might also be advantageous in a windy climate. Dome-shaped crowns prevail also at timberline on other maritime subantarctic island such as Tierra del Fuego, Île Amsterdam and on Auckland Islands (Troll, 1959).

Photo 65. Gnarled *Polylepis* ssp. in the Puna on Huascarán (Peru) at about 4.500 m. F. Klötzli.

In view of the corresponding life forms of different species and genera a common ecological principle might be supposed behind it (e.g., Troll, 1955a, 1973; Schweinfurth, 1978). However, one should be aware that tropical mountain climates are distinctly different from subantarctic maritime climates, a few correspondences disregarded. This holds particularly true for the New Zealand mountain climate when compared to smaller islands and tropical timberline. New Zealand mountain plants exhibit a strict seasonal cycle of shoot elongation and winter dormancy, even if their phenology (New Zealand *Nothofagus* is evergreen) is less striking than in the north temperate zone (Wardle, 1963, 1973). Winter is not that strong, and the growing season is longer but cooler than at temperate timberline in the northern hemisphere.

The effects of true winter conditions on growth and physiognomy of trees at timberline are obvious, however (Mark et al., 2000). Schweinfurth (1962, 1966), for example, considered ice particle abrasion a possible factor among others shaping the wind-trimmed growth of *Libocedrus bidwillii* at timberline on Mt. Egmont (New Zealand, North Island). This author also argued that, besides permanent strong winds, also snow was an important agent influencing the development of the dense canopy of the mountain forest. Wardle (e.g., 1978, 1985c, 1991) mentions winter desiccation in wind-exposed *Nothofagus solandri* at timberline, which has very likely been enhanced by ice particle abrasion (see also Schönenberger, 1984). Typically, this kind of damage is restricted to stems and shoots projecting above the winter snow cover (Section 4.3.4). As a result, flagged and mat-like growth is very common at wind-exposed sites, although the erect flags typical of *Picea engelmannii*, *Abies lasiocarpa* and other strongly monopodial conifers do not occur (Wardle, 1978). Also the present author's observations in the Craigieburn Range (South Island) give evidence of the strong influence of winter climate on growth forms of *Nothofagus* and *Libocedrus* at timberline (cf. Photo 66).

Comparing upper timberlines of the southern Andes with tropical timberlines might be even harder (Eskuche, 1973; Veblen et al., 1977) as the southern Andes are characterised by a seasonal climate with clear thermal contrasts of winter and summer (Wardle, 1977). Contrary to New Zealand southern beech the subalpine beech species (*Nothofagus pumilio*, *Nothofagus antarctica*) are deciduous. Also, both species are more prone to develop stunted growth forms if compared to the southern beech at timberline in New Zealand (Wardle, 1973). Even *Pinus hartwegii* at timberline on the Mexican volcanoes does not show extremely climatically stunted growth forms, as are typical of the temperate timberlines (Beaman, 1962; Lauer and Klaus, 1975a). However, wind-trimmed scrub-like juniper (*Juniperus monticola*) occurs at wind-exposed and block-rich sites (Wardle, 1965b; Klink, 1973; Lauer, 1975) and cushion-like *Pinus hartwegii* were found above timberline (Lauer and Klaus, 1975a). Thus, the physiognomic-ecological affinities that were repeatedly emphasized by Troll (e.g., 1959, 1973) should be reconsidered in view of recent studies on tropical and southern temperate timberlines. Very likely a finer differentiation will be needed.

As a general rule, low growth form is less affected by the altitudinal and longitudinal decrease of temperature, for different reasons. Turbulent heat loss is reduced due to low wind velocity near the ground and relatively smooth surface of low growth forms, which thus are almost decoupled from the free atmosphere (e.g., Grace, 1980). At a positive radiation balance, tissue temperature is higher in low prostrate growth forms, young growth and seedlings than in several metres high trees (cf. Figure 54; Wegener, 1923; Wardle,

1974; Dahl, 1986; Wilson et al., 1987; Grace, 1988, 1989). Warmer temperatures may enhance survival rate and reproduction, if water and nutrient supply is not limited and any damages due to extremely high radiation loads and excessive heat will not occur.

In temperate high mountains, winter snow cover normally, but not necessarily, protects low growing plants. Also, the uppermost outliers of the mountain forest have often developed mat-like growth forms that are enjoying the relative favourableness of the climate near the ground (cf. Photos 68, 69 and 87). If able to reproduce by layering they may persist even at deteriorated climatic conditions. In case of a general warming, these outliers may produce viable seeds (cf. Figure 48) and thus will become an important factor to potential advance of timberline (Lavoie and Payette, 1992; Lescop-Sinclair and Payette, 1995; Weisberg and Baker, 1995). Birch seedlings at upper timberline in northern Europe were considered real 'opportunists' that keep dwarf-shrub-like 'suppressed' growth as long as climate is adverse to developing 'normal growth'. At improving climate, however, they will grow up to tree size (Kallio and Lehtonen, 1973; Kullman, 1984). The same holds true for the low cushion-like young *Pinus hartwegii* beyond the upper timberline on Pico de Orizaba (Mexico). After a couple of subsequent favourable years or if the roots have penetrated to greater depth they may assume macrophytic growth (Lauer and Klaus, 1975a).

Injuries caused to the trees by climatic elements increase abruptly above the closed forest where they influence tree physiognomy more or less effectively. Resulting growth forms are conspicuously similar at temperate timberlines in the northern hemisphere and northern timberline. This similarity may be partly ascribed to the relatively close floristic relationship of the timberline forming tree species in the northern hemisphere (Section 4.1.1). Moreover, the effect of winter snow cover on growth forms is common to the upper and northern timberline environments. Under severe conditions, the different tree species, despite their specific type of branching, may develop quite similar growth forms. As to survival and persistence in the long-term, trees that are able to develop growth forms closely adapted to particular environmental situations are at an advantage. Among the conifers, spruces, firs and larches are most flexible in developing habitat-adapted growth forms, not least because they are able to regenerate and propagate by layering.

Growth forms controlled by strong prevailing winds are particularly common to the timberline ecotone on the Rocky Mountains and many other mountain ranges in western North America (e.g., Griggs, 1946; Billings, 1963; Marr, 1977; Holtmeier, 1978, 1980, 1981b, 1985a, b, 1996; Arno, 1984; Alftine and Malanson, 2004) as well as on the Japanese and New Zealand Alps and at the northern timberline. In the European Alps, on the other hand, such extremely wind-shaped growth forms are not abundant and are usually restricted to specific microsites. This might be due to the less windy climate if compared with the Rocky Mountains, for example. However, high-altitude forests in the Alps were widely cleared in consequence of alpine pasturing and other human impact during history (Section 4.3.14.1). Consequently, the original zone where wind-trimmed growth forms would occur under undisturbed conditions does not exist anymore. The width of the present timberline ecotone is mainly anthropogenic. Nevertheless, recent invasion of abandoned pastures appears to be strongly affected by the present climate and microclimates (Holtmeier, 1965; Müterthies, 2002).

Frost, winter desiccation, ice particle abrasion and other wind effects (cooling, evaporation, banging the branches and twigs, removal of foliage, breakage, etc.) injure needles and shoots projecting above the snow cover. Thus, many trees cannot grow higher than the average snow cover, while others may be successful, but usually not for several decades. Their physiognomy, however, is more or less different from 'normal growth'. At sites exposed to strong prevailing winds, asymmetric growth forms prevail, particularly in tree species that are normally controlled by apical dominance. The resulting growth modification range from flagging of upright growing trees, almost devoid of needles and twigs at their windward side (Photo 66, 67), to flagged skirted trees ('supra nival skirted', Lavoie and Payette, 1992; Photos 72, 73, 74 and 75) and strong wind-shearing resulting in table- and mat-growth (Photos 68, 69) at sites lacking snow or only occasionally covered with snow in winter (Holtmeier, 1980). As growth forms gradually develop in response to environmental constraints, many transitional forms between different types exist. Accordingly, terminology is differentiated and not always unambiguous. In the following, only the main types will be considered (Figure 53).

At extremely wind-controlled sites, trees may develop only mat growth forms that, decoupled from the free atmosphere, take advantage of the more favourable climate near the ground (cf. Photos 68 and 69; Figures 54 and 55). Thus, meristem temperatures of dwarfed pine (*Pinus sylvestris*) growing within the timberline ecotone on the Cairngorm Mountains (Scotland) are significantly higher than ambient air temperature, the largest difference being 10°C while the mean difference is 4.3°C (Wilson et al., 1987; Figure 54). Also, close proximity to the ground is crucial for avoiding ice particle abrasion and for conserving tissue water via snow burial during severe winter periods of desiccation (Hadley and Smith, 1986; Smith and Knapp, 1990). Mat growth is the most severe deformity of the growth. Mat size ranges from a small stunted shrub only a few centimetres high and less than 1 m in length to compact mats 15 m across and 1 m high, again representing one conifer (Arno, 1984). Normally, the thin and often patchy snow cover protects these low growth forms only partly and temporarily from the adverse winter climate. In the long-term, however, conifer mats may gradually turn into wedge-like growth forms expanding by layering at their higher leeward end. The upwind surface usually appears to be clipped. Occasionally, single stems that start projecting above the average snow cover will get flagged (Photo 70). Stem height generally increases from the windward to the leeward edge of the clonal groups while age decreases.

Incidentally, such growth forms are inconsistent with the hypothesis (Körner, 1998a, b) that short tree stature at the climatic timberline is related to low soil temperature during the growing season. Soil temperature in the rooting zone does not vary much under the tree canopy (cf. Figure 70). Moreover, soil temperature in the lee of the clonal groups will decrease when this area covers with the trees due to expansion by layering. The resulting decrease in soil temperature, however, will not cause growth limitation or further downwind migration. Release of erect stems from the dwarfed growth

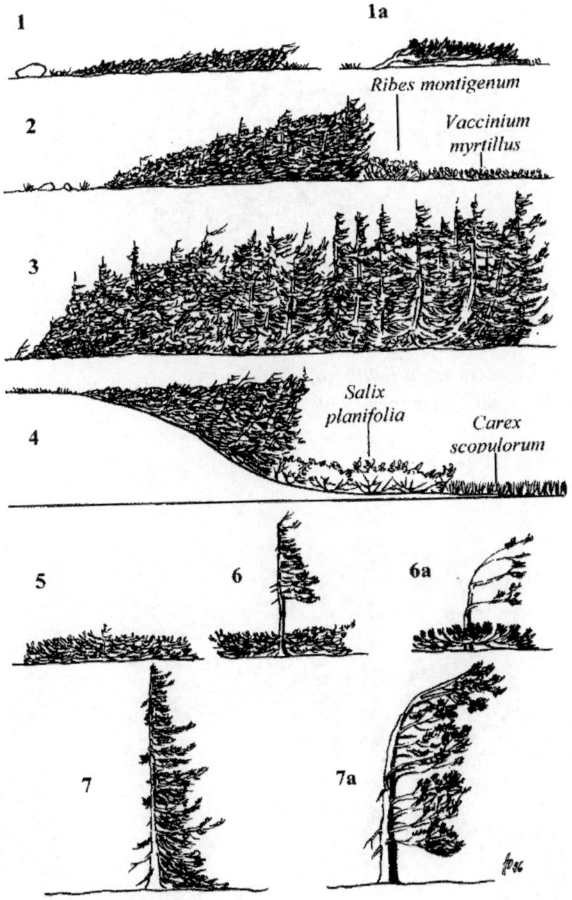

Figure 53. Wind-trimmed growth forms at timberline. *1* – matgrowth, Ia – matgrowth (*Pinus* species), *2* – wedge, *3* – hedge, *4* – cornice, *5* – table, *6* – flagged table tree, *6a* – flagged table tree (*Pinus* species), *7* – flag tree, *7a* – flag tree (*Pinus* species). From Holtmeier (1996).

Photo 66. Flagged *Libocedrus bidwillii* on Arthur's Pass (New Zealand, South Island) at 915 m. F.-K. Holtmeier, 26 November 1979.

Photo 67. Flagged *Picea engelmannii* and *Abies lasiocarpa* in the timberline ecotone on Niwot Ridge (Front Range, Colorado) at about 3.420 m. The depth of the snow cover is about 120 cm. F.-K. Holtmeier, 7 April 1989.

Photo 68. Downwind-migrating mat-like *Picea engelmannii* at an extremely windy site above Diamond Lake (Front Range, Colorado) at about 3.465 m. In front of the tree island are wind scarps. F.-K. Holtmeier, 18 August 1977.

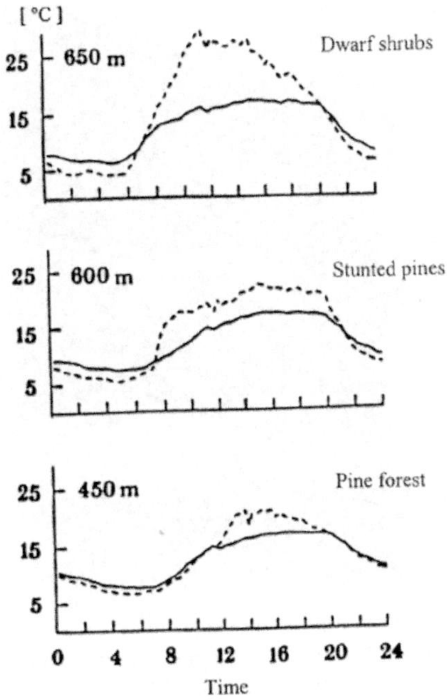

Figure 54. Mean daily temperature (27 May to 30 June 1985) in the forest-alpine tundra ecotone and in the dwarf-shrub belt in the Cairngorm Mountains (Scotland). Solid line means air temperature, dashed line means temperature of terminal shoots. Modified from Wilson et al. (1987).

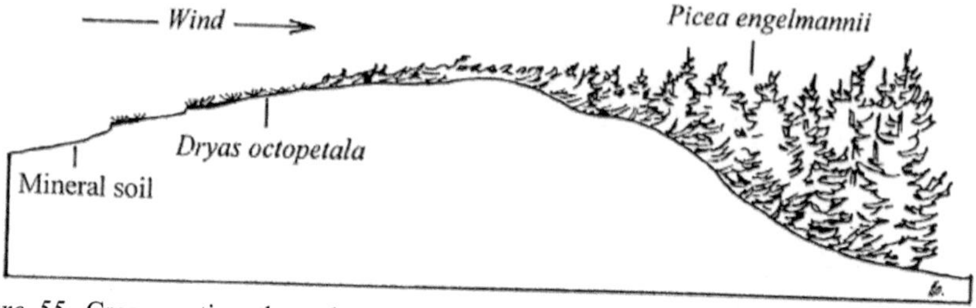

Figure 55. Cross section through the site shown in Photo 68. Though the spruce (*Picea engelmannii*) displays mat growth on the wind-exposed, convex topography it develops upright stems at the lee side.

form will continue as long as the climatic conditions allow undisturbed height growth beyond the previous low tree canopy. As the erect stems become increasingly coupled to the cooler atmosphere and apical meristem temperatures are close to ambient air temperature most of the time, stem-release and accelerated height growth is likely to be related to reduced wind action and winter injury.

Continued downwind layering may result in more or less long 'tree hedges' oriented parallel to the prevailing wind direction (Photo 71; Holtmeier, 1978, 1981a, 1982, 1986a, 1996; Alftine and Malanson, 2004). Roder (1895) reported such 'hedges' also from the polartimberline on Kola Peninsula.

Photo 69. Mat- and wedge-like bristlecone pines (*Pinus aristata*) at the upper, wind-exposed edge of the forest-alpine tundra ecotone on Kingston Peak (Front Range, Colorado) at 3.360 m. F.-K. Holtmeier, 18 August 1987.

Photo 70. Clonal groups of Subalpine fir (*Abies lasiocarpa*) on the wind-exposed west slope oft he Front Range (Colorado) near Devil`s Thumb Pass at about 3.590 m. The leeward flagged stems grow higher than the windward shoots. F.-K. Holtmeier, 2 September 1977.

Photo 71. Hedge-like clonal *Picea engelmannii* and *Abies lasiocarpa* extending down- wind (to the *left*). Some of the 'hedges' became established in the lee of big blocks. Lower, wind-swept area (strong winds from the NW, *right*) of the Fourth of July circque (east slope of the Front Range, Colorado) at about 3.420 m. F.-K. Holtmeier, 27 July 1977.

Mats, wedge-like growth forms and 'hedges' may considerably increase snow accumulation at their sites (Section 4.3.12); cf. Photos 87 and 90). On the leeward slope of convex topography, growth forms may develop, the clipped surface of which is almost identical with the snow surface of the cornice (Figure 53). Frequently, extremely wind-trimmed growth forms occur close to less deformed trees of the same species at the same site. Thus, for example, low mats can be found between 'hedges' running parallel to the prevailing wind direction. This pattern of different growth form standing side-by-side results from wind funneling between the 'hedges' that increase wind velocity near the ground. This effect is also reflected in wind-scarps (Holtmeier, 1978; Broll and Holtmeier, 1994; cf. Figures 74 and 75; Photo 92).

So-called table trees and flagged or skirted table trees are widely spread as well at the altitudinal timberline in temperate mountains and at the polar timberline (Figure 53, Photos 72–75). They usually occur at sites regularly covered with deeper snow and exhibit comparatively luxuriant foliage. The almost level 'table surface' corresponds to the average snow depth because shoots growing higher than the snow surface will normally succumb to climatic influences within a short time.

Photo 72. Flagged table tree (*Pinus longaeva*) in the timberline ecotone on Sheep Mountain (White-Inyo Mountains, California) at about 3.475 m. F.-K. Holtmeier, 4 August 1992.

Photo 73. Flagged table tree (*Larix decidua*) on Muottas da Schlarigna (Upper Engadine, Switzerland at 2.260 m. F.-K. Holtmeier, August 1971.

Photo 74. Flagged table tree (*Pinus sylvestris*) on Mustavaara (Finnish Lapland) at about 480 m. Pallastunturi in the background. F.-K. Holtmeier, 7 June 1969.

Photo 75. Flagged table tree (*Betula tortuosa*) on Jesnalvaara (northern Finnish Lapland) at 320 m. In general, flagging is less spectacular in mountain birch compared to conifers. F.-K. Holtmeier, 9 September 2000.

However, a couple of subsequent favourable years may allow almost undisturbed growth above the extremely dangerous zone close to the snow surface (ice particle abrasion, wide amplitudes of tissue temperatures due to intense solar radiation and nocturnal cooling). In the following a usually more or less flagged crown may develop having foliage on the leeward side of the tip of the stems and a band almost without foliage below the flagged tip. The degree of flagging depends on wind-exposure and tree species. Annual short shoots thriving on the upwind side of the stems during summer will normally be killed in winter. Often, trees gradually lose all their needles and shoots at their wind-facing side within a few decimetres wide zone just above the snow surface. In this zone, ice particle abrasion and also re-radiation from the snow is most intensive. The tip of the stems, however, gradually escapes this endangered zone and foliage may normally develop at least on the leeward stem sides (Holtmeier, 1969, 1974, 1985a; Scott et al., 1987a, 1993). Not only conifers but also broad-leaved trees such as mountain birch (*Betula tortuosa*) or southern beech (*Nothofagus solandri*; Norton and Schönenberger, 1984) growing under similar severe conditions may display such growth forms, even though apical dominance is not as important as in coniferous tree species.

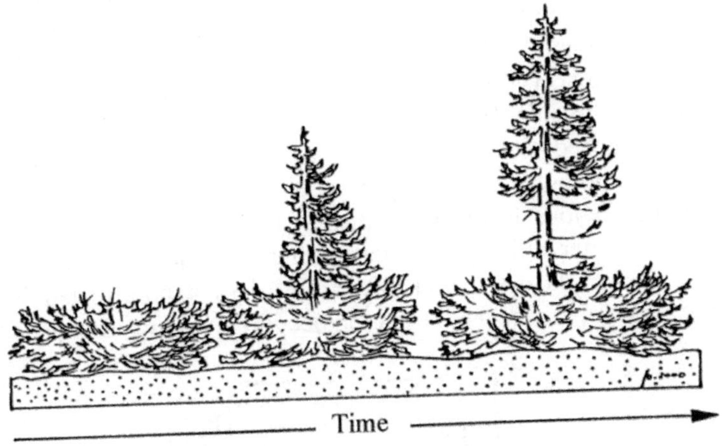

Figure 56. Development of a flagged table tree. According to Holtmeier (1969).

At extreme sites, those trees that are able to display growth forms closely adapted to the peculiar environmental situation are at an advantage (Löve et al., 1970; Holtmeier, 1981b). That is even true for very resistant tree species such as *Pinus aristata, Pinus longaeva, Pinus flexilis* and *Pinus cembra*, which are generally considered less susceptible to the kind of deformation described above than spruce and fir species (e.g., Wardle, 1965b; Tranquillini, 1974). The great abundance of flagged and other climatically shaped growth forms of spruce and fir occurring already in the lower parts of the timberline ecotone seems to substantiate this hypothesis. In the hardy pines, we rarely observe asymmetric, flagged tree crowns with reduced growth of shoots and needles at the windward side of the trees, despite pronounced apical dominance. Instead, downwind-leaning and/or curved stems and particularly leeward-trained treetops more frequently develop (cf. Figure 53). At extremely wind-swept sites in the timberline ecotone, all tree species display mat growth or similar growth forms training along the ground (cf. Photos 67, 68). Spruces, firs and larches are most flexible in developing growth forms in adaptation to habitat conditions, not least because of their ability to regenerate and propagate by layering. Thus, these species in particular are able to colonize sites where upright growth would be impossible. Schröter (1898) already considered spruce to be the most variable of all tree species (see also Holtmeier, 1980, 1981b, 1989; Lavoie and Payette, 1992; Laberge et al., 2000). The author should have mentioned, however, that fir and larch do not come after.

The physiognomy of the growth forms described in the foregoing pages reflect mainly the effect of the prevailing winds in winter, as is obvious from the lower snow-protected parts that have remained undisturbed (table trees, flagged skirted trees). On the other hand, living foliage and shoots at the wind-exposed side of the tree stems projecting above the snow cover are usually shorter than at the leeward side (Figure 57; Bernbeck, 1907, 1911; Martin and Clements, 1953; Wilson, 1959; Grace, 1977; Wade and Hewson, 1979; Holtmeier, 1980, 1981a; Dahms, 1992) and often incompletely developed (Holtmeier, 1981b; Dahms, 1992). This again has to be attributed to the influence of wind during the growing season which more or less delays the phenological development of wind-exposed needles and shoots, as has become obvious from our studies at timberline on the Colorado Front Range (Holtmeier, 1980). Ten days after bud burst the needles on the leeward side of the trees and also at the less wind-affected base of the flagged trees were twice as long as the needles on the wind-exposed side.

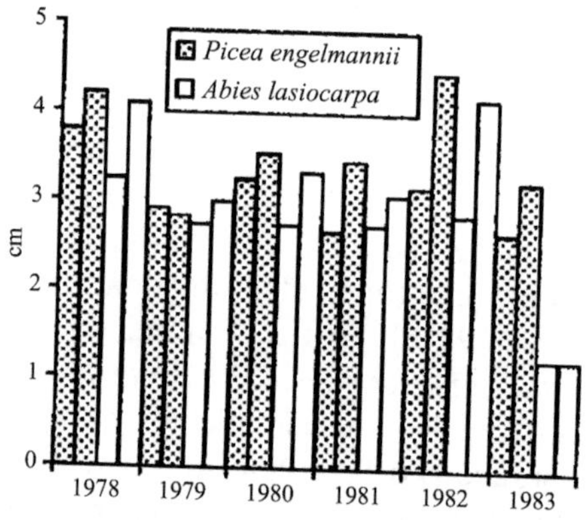

Figure 57. Length of needles at the wind-ward side (left bar) and leeward side (right bar) of *Picea engelmannii* and *Abies lasiocarpa* in the forest-alpine tundra ecotone on Niwot Ridge (Front Range, Colorado). Data from Dahms (1992).

In addition, studies on the seasonal distribution of wind directions in the upper Fourth of July Valley (cf. Photo 46; Holtmeier, 1978; Ruwisch, 1983) provide evidence that also winds during the growing season contribute to flagging. In this study, anemometers recorded the distribution of wind directions that were also mapped using flagged trees and other wind-shaped growth forms as well as wind scarps and long-lying snow patches. Average wind direction (300°) derived from the wind-trimmed growth forms (*Abies lasiocarpa* and *Picea engelmannii*) differs more from the recorded wind direction in winter (320° to 325°) than in summer (293°). This clearly indicates that wind during the growing season plays an important role in growth form development.

Strong prevailing winds during the growing season bend away exposed twigs and still elastic branches. Gradually they get fixed in this forced downwind position (cf. Figure 53; see also Holtmeier, 1971b; Yoshimura, 1971; Yoshino, 1973; Wade and Hewson, 1980). Also compression wood may reflect the influence of prevailing winds during the growing season. Thus, conifers exhibit narrow growth rings on their wind-exposed side and wider rings with conspicuous compression wood at their downwind side. In deciduous trees it is just the opposite (see also Mattheck, 1991).

Discussion is still going on as to whether the development of these growth forms is genetically predetermined (Clausen, 1963; Löve et al., 1970; Grant and Mitton, 1977; Mitton, 1985). According to Löve et al. (1970) this holds particularly true for species that are able to display different growth forms whereas other species cannot. Wardle (1974), for example, referring to a personal communication of Godley considers both the prostrate and arborescent growth of *Nothofagus antarctica* growing side by side at the same sites to be inherent. Eskuche (1973), on the other hand, ascribes the prostrate beeches to snow pressure (settling snow), as also does Wardle in a later paper (1998) relating to *Nothofagus pumilio*, however. Electrophoretic analyses in *Picea engelmannii* and *Abies lasiocarpa* at timberline in the Colorado Front Range gave evidence that those different growth forms of the same tree species result from different genetic properties of the individual tree (Grant and Mitton, 1977; Mitton, 1985).

However, the present author's observations in the same area do not support this hypothesis (Holtmeier, 1981a). In wind-trimmed 'tree hedges', for example, that originated from leeward elongation of one conifer species, a great variety of more or less wind-shaped growth forms may be represented: mat- or wedge-like growth at the wind-facing edge, flagged stems projecting above the snowpack a little more downwind, and several metres high stems with almost undisturbed apical-controlled growth at the better wind-protected leeward end. Consequently, in this case the wind effect obviously overrides the influence of possible different inherited character. Furthermore, recovery of hitherto suppressed growth forms in *Picea engelmannii* and *Abies lasiocarpa* by release of erect stems in response to changing environmental conditions (cf. Section 5.4) are inconsistent with the theory of genetically controlled prostrate growth of these species in harsh environments (cf. Figure 58). Also, prostrate growth of *Pinus longaeva* close to treeline on Mt. Washington (Nevada) cannot be considered a genotype as LaMarche and Mooney (1972) emphasized. Contrary to that, prostrate and arborescent (6 m, ±0.5 m) *Pinus aristata* at treeline (3,360 m) on Mt. Evans (Colorado; cf. Photo 69) exhibit morphological and physiological features that might be ascribed to different genetic properties (Schoettle, 1993).

On extremely wind-exposed topography at timberline in North American high mountains, the present author occasionally came across whitebark pines (*Pinus albicaulis*, Sierra Nevada, California) and limberpines (*Pinus flexilis*, Wheeler Peak, Colorado Front Range) that show prostrate growth like *Pinus mugo*, which is typical of avalanche chutes. The limberpines grow solitarily displaying more or less wind-trimmed growth (cf. Photo 104), while the whitebark pines form relatively dense thickets (Photo 76). In the latter case, all pines, ranging in height between 1 and 5 m according to wind-exposure, are multi-stemmed. The tips of their decumbent stems turn into an upright position. Asymmetric wind-shaped growth, however, is restricted to the very windward edge of the pine thickets. Thus, we might assume inherent prostrate growth, particularly as all individual pines within the groups show almost the same habitus. On the other hand, an extreme frost event (Section 4.3.3.1) that seriously damaged the pines at young age might be the responsible factor. After the controlling apical shoots (leaders), not yet projecting far above the ground, had been destroyed, the tips of the plagiotropic branches would have gradually turned into a vertical position thus resulting in an almost 'basket-like' epicormic growth form of the individual. However, we could not substantiate this hypothesis, as we were not allowed to sample tree rings or trunk disks.

In the upper timberline ecotone on the eastern and northern slopes of Mount Olympus (2.917 m, Greece), *Pinus heldreichii*, normally growing as big trees (up to 20 m high and measuring 200 cm in diameter) e.g., Habeck and Reif, 1994 at the upper limit of the high-stemmed forest (about 2.000 m) display growth forms (Photo 77) that could easily be taken for *Pinus mugo* although being shaped by environmental factors (communication R. Brandes).

Also, wind-trimmed spruces and firs trailing along the surface and developing into 'normal' trees if protected from the wind by artificial constructions (snow fences, wind breaks) suggest that external factors may

Photo 76. Pinus albicaulis on a wind-swept ridge above St. Mary`s Pass (Sierra Nevada, California) at about 3.180 m. The whitebark pines are displaying growth similar to *Pinus mugo*. F.-K. Holtmeier, 27 July 1994.

Photo 77. Environmentally-shaped *Pinus heldreichii* in the timberline ecotone on the eastern slope of Mt. Olympus (Greece) at about 2.300 m. Frost and/or winter desiccation are likely to be controlling the growth form. R. Brandes, 10 May, 2000.

override hereditary growth. Structural changes in the timberline ecotone modifying the local windflow pattern near the ground may have the same effect. On the other hand, stems that were able to grow beyond their well protected mat- or table-like basal part may die back if environmental conditions deteriorate, while the part encased in the snowpack will mostly remain undisturbed (Figure 58). Such die back occurred in *Larix laricina* and *Picea glauca* near Churchill (Manitoba), for example, after the eruption of Pinatubo volcano that was followed by a cool unfavourable summer in 1992 (Scott et al., 1997). Such changes of habit repeatedly occurred during site history and thus may provide useful information for reconstruction of timberline dynamics.

Wind combined with heavy snow loads, particularly after wet snow fall (Photo 78), cause breakage to tree crowns, branches and stems mainly in the lower zone of the timberline ecotone where big snow masses accumulate due to increased surface roughness (Section 4.3.12). A possible increase of moisture-carrying air masses may enhance the formation of heavy snow loads increasing the risk of stem and crown breakage at the altitudinal timberline. Resistance to stem breakage caused by heavy snow loads differs by the tree species. *Pinus sylvestris*, for example, is more vulnerable than *Picea abies* or *Betula tortuosa* (Jalkanen and Konôpka, 1998). The risk of breakage is highest when low temperatures reduce wood flexibility.

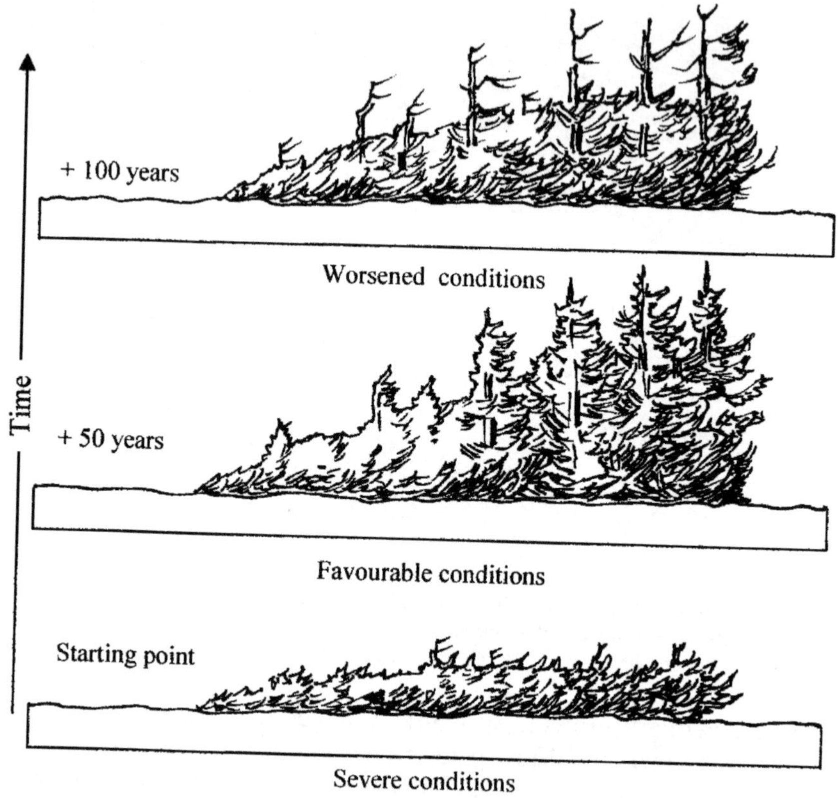

Figure 58. Change in physiognomy of a clonal conifer group under changing climatic conditions. After observations at timberline in the Front Range (Colorado).

In the wind-swept open terrain above the scattered tree stands, glaze storms causing heavy ice and rime loads (Photo 79) may cause high losses of foliage on the wind-exposed side of the trees (e.g., Yoshino, 1973; see also Troll, 1955b). At timberline on extremely windy Mt. Washington (New Hampshire) more than 2 m of rime frost may form at the wind-facing sides of the trees during a single night at such weather conditions and cause serious damage (Billings, 1990). Rime frost may also increase the risk of stem and crown breakage.

Settling snow pulls down and breaks branches and twigs encased in the snow (Photos 80 and 81). While tall trees are affected only at their base, young growth is often completely destroyed (Freiray and Schweingruber, 1994). Such damage is very common on snow-rich leeward slopes of convex topography, in depressions and at the downwind edge of compact tree islands and 'ribbons'. Damage by settling snow occurs as well within small forest gaps acting as snow traps (Photo 81; Section 4.3.12). Some species such as mountain birch or southern beech may recover by forming basal sprouts resulting in multi-stemmed growth, which is common to these species.

Photo 78. Heavy snow load (‚tykky') at timberlinbe (460 m) on Vasalaki (Aakenus Tunturi, near Kittilä in Finnish Lapland. T. Tasanen, beginning of February, 1999.

Photo 79. Abies mariesii with hoar frost at its windward side (*left*) after 2 days of strong winter monsoon at timberline on Mt. Azuma (Central Japan) at about 2.200 m. M. Yoshino.

Photo 80. Deformation of a birch (*Betula tortuosa*) by settling snow in a snow-rich hollow on the west-facing slope of Lommoltunturi (Pallastunturi area, Finnish Lapland) at about 500 m. F.-K. Holtmeier, 2 June 1970.

On slopes, particularly at the lateral edges of avalanche tracks and rills, snow creep and snow slides abrade bark and remove plant cover. Trunks are thrown and often show strong deformation (Photo 82; Figure 59; Schönenberger et al., 1994). In conifers, compression wood is formed on the downslope side of the lower butt-log or at the lower side of decumbent stems (vice versa in broad-leaved trees). In some cases the stems would later turn up into a vertical position (negative geotropism) by formation of compression wood at the lower sides facing the soil. Broad-leaved trees are pulled in upright position by forming tension wood at their upper sides facing away from the soil (Mattheck, 1991). Continued one-sided pressure may cause cracks particularly to the curved stem base thus allowing access to saprogenous organisms. On fallen but still rooting trees, branches on the upper side of the trunk gradually develop into upright stems thus adapting harp-like growth (cf. Figure 59). Often, all branches on the upslope side of the lower trunk get lost. Missing foliage,

Photo 81. Settling snow has pulled down the lower branches of the trees (*Abies lasiocarpa*). The trees increase accumulation of blowing snow by reducing wind velocity. Needle loss has been caused by the brown snow-felt fungus (*Herpotrichia juniperi*).Trail Ridge (Front Range, Colorado) at about 3.450 m. F.-K. Holtmeier, 27 July, 1989.

Photo 82. The curved stems oft he birches (*Betula tortuosa*) reflect the influence of creeping snow. Southwests-facing slope of Sörtind (Kvalöya, northern Norway) at about 250 m. F.-K. Holtmeier, middle of February, 1971.

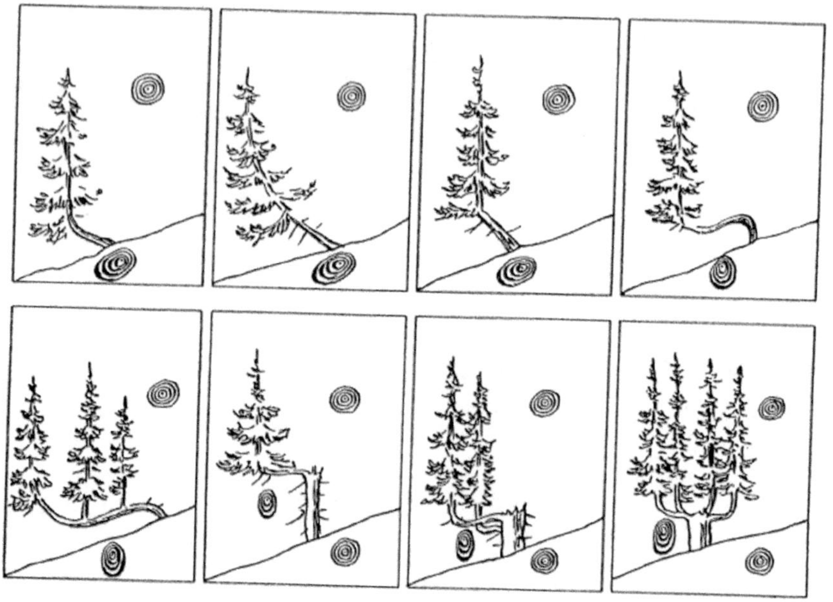

Figure 59. Growth forms of trees typical of sites influenced by moving snow cover (avalanches, gliding snow, snow creep). The cross sections of the tree trunks show the position of compression wood. Curved stems (on the *top left*) may also result from soil creep. Modified from Schönenberger (1994).

dead branches and tips at the trunk base may also result from pathogenic snow fungi. However, it is often hard to find clear evidence because the damaged zone gets increasingly vague as the tree grows taller.

The struggle for existence is impressively mirrored in the physiognomy of the several hundreds of years old trees at timberline. They look more or less stunted and often exhibit several tops originated from branches that erected after the main trunk had been broken. Since none of these branches became a leader, candelabrum-like growth is the result (cf. Figure 59).

In case the branches just below the point of breakage took the apical control so-called bayonet-growth developed (cf. Figure 59). These are the most conspicuous growth forms caused by mechanical impacts. Others were influenced by a bundle of simultaneous or subsequent different kinds of damages, the specific effects of which can now hardly be identified.

Within forest stands at the present timberline in the European Alps, old environmentally shaped trees occur that are several hundred years older than the other trees. These veterans are relics of the previous high-elevation forests that were removed for the most part by human activities (Section 4.3.14.1). As grazing prevented almost any natural regeneration, these trees could mature without competing with other trees and develop their specific old-age growth physiognomy.

Thereby, response of tree growth to the many external factors, storms, snow load, lightning, fire and not last human impact, played a more or less important role. For example, mountain people cut and broke off branches and twigs of Swiss stone pines to harvest the energy-rich seeds that were an important food. When direct human impact and grazing ceased, new forest grew up around the old broad and often multi-stemmed trees (Photo 83; see also illustrations in Rikli, 1909; Bichsel, 1995; Bischoff, 1995; Zuber, 1995).

Photo 83. This multi-stemmed Swiss stone pine (*Pinus cembra*) close to upper timberline has grown for many centuries in open terrain without competing with other trees. East-facing slope of the Bernina Valley below Muottas da Schlarigna (Upper Engadine) at about 2.190 m. F.-K. Holtmeier, 14 September 1990.

Vegetative reproduction and propagation of many timberline-forming tree species effectively influence the habit of dendroid vegetation at timberline (Section 4.3.10.2). The physiognomy of dense and sharply contoured clonal tree islands is mainly influenced by the slope gradient and wind-exposure (Figure 60). On level terrain, the plan view is almost circular (cf. Photo 61) whereas it is rather longish or elliptical on steep topography. If the initial trees within the clonal tree islands die, so-called 'timber-atolls' develop (Griggs,

1938; Figure 60Ad). Tree islands permanently influenced by strong winds from one direction display asymmetric growth with an almost triangular (wedge-like) plan view (Figure 60Ab, Bb, Bc). Also, more or less parallel 'hedges' may develop. At extremely wind-swept sites, clonal tree islands often gradually 'migrate' downwind. By layering they elongate at their leeward end, while dying off at their wind-facing front due to winter desiccation, ice-particle abrasion and other mechanical effects (Photo 84; cf. Figure 60Ac and Figure 75; see also Marr, 1977; Holtmeier, 1978, 1980, 1986a, 1999b; Benedict, 1984). Leeward migration will come to an end if the tree islands die off faster at their wind-exposed edge than elongating downwind. This may happen, for example, to upslope moving clonal groups reaching an altitude at which adverse climate will prevent any growth (see also Marr, 1977). Downwind migration will also cease if the leeward edge of the tree island enters snow-rich terrain such as the leeward slope of convex topography where big snow masses usually pile up and last until early summer.

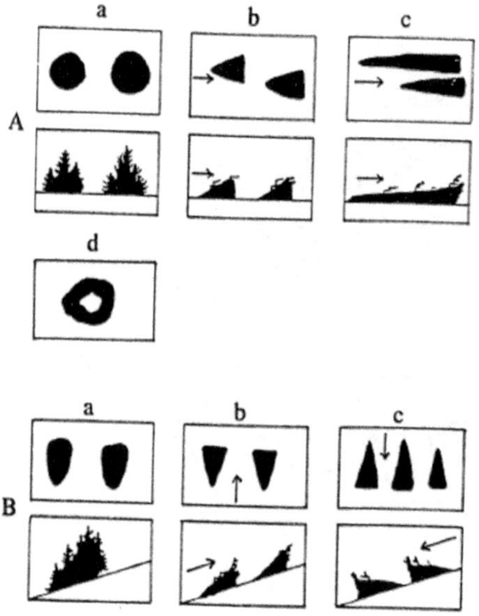

Figure 60. Physiognomy of clonal groups (plan and side view).

A – On horizontal terrain
a – No wind influence
b – Strong wind influence
c – Very strong wind influence
d – Atoll

B – on slope
a – No wind influence
b – On a wind-exposed slope
c – On a lee slope

The growth form of the single clonal members depends mainly on their position within the tree islands. Almost circular clonal groups growing on flat and wind-protected terrain normally show crowded apical-controlled upright stems at more or less great numbers (cf. Photo 89). In general, the younger layers at the margin of such tree islands do not grow as high as the older ones in the centre. In strongly wind-trimmed clonal groups, stem height usually increases from the upwind edge to the better-protected leeward side. Often, wind has forced the stems in a downwind leaning position (cf. Photos 59, 89 and Figure 53). Within the almost horizontal section of the layering branches (cf. Figure 51), comparatively wide growth rings with compression wood develop at the layer's lower side imbedded in the litter, while narrow rings are produced on the upper side away from the surface. Increased diameter growth at the lower side of the horizontal layering branches results from the weight of the branches and also from snow pressure (tree islands acting as snow traps).

Photo 84. Dowind-migrating clonal groups (*Picea engelmannii, Abies lasiocarpa*) in the forest-alpine tundra ecotone on Trail Ridge (Rocky Mountain National Park, Colorado) at about 3.420 m. F.-K. Holtmeier, 24 July 1987.

Although prevailing winds may shape the entire tree islands, growth of the clonal stems is also controlled by many group-internal factors, the intensity of which may change in the course of time, such as water and nutrient supply, illumination, varying snow loads and competition after the connections to the mother tree have decayed. Also, exposure of the clonal members to wind may considerably change to gradual or sudden (e.g., breakage, windthrow, fire, insects) changing structures of the colonies resulting from irregular growth. We found such disturbances reflected in the growth rings sampled from different clonal stems with tree islands of *Picea engelmannii* and *Abies lasiocarpa* at timberline on the Colorado Front Range (Holtmeier, 1999b). However, the only feature common to a number of stems were occasional reductions in diameter growth that obviously had occurred simultaneously. They might have been caused by strong late or early frost events, windstorms or extreme snow load overriding the irregular effects of the internal growth controlling conditions.

Multi-stemmed growth is very common at upper timberline and does not only occur in clonal groups proliferating by layering. Although a single tree is better protected from adverse climatic influences within a group, clumping of trees should not be considered a 'useful (precautionary) strategy' because it results from the response of the trees to injuries caused by a great number of factors. Multi-stemmed growth, for example, may be a consequence of a tendency toward side-branching, with lack of hormonal dominance by the leader. It may also be caused, however, by substitute shoots from the trunk base (Section 4.3.10.2) or lower branches turning into a vertical position after the main stem was broken. Not least, multi-trunk growth may result from fusing of several stems at the base that originated from a nutcrackers seed cache, for example (Section 4.3.13.3; see also Tomback et al., 1993; Tomback and Schuster, 1994). Larger multi-stemmed groups often result from vegetative proliferation. Occasionally, multi-stemmed groups that originated from basal sprouts or root suckering may be easily confused with clonal groups. Consequently, they have been called 'false' or 'pseudo layers' (see also Phares and Crosby, 1962; Fanta, 1973, 1981). Likewise, many seed produced tree stands can hardly be distinguished from clonal groups at first sight, if not looking for the rooting system (Figure 61).

Figure 61. Multi-trunk conifer groups (spruce, fir) of different origin. Clonal group on the *left* side, seed tree with seedlings at its periphery on the *right* side.

4.3.12 Influence of trees and tree stands on site conditions

In the previous chapter, the influence of site conditions on tree growth has been treated in a manner that is traditional in presentations on timberline and mountain vegetation. As to understanding the great physiognomic, structural and ecological variety of timberline, however, the effects of trees and tree stands on site conditions and on the immediate surrounding area are hardly less important. These effects result mainly from the distribution pattern of trees (and different tree species) and stand structures, which again depend on the way of reproduction (sexual, vegetative) and seed dispersal (wind- or animal-mediated seeds) and on primary site conditions. The less broken the terrain, the more wind, snow cover, radiation transfer and balance are directly influenced by the distribution pattern of the tree stands and more or less wide openings (e.g., Geiger, 1961; Billings, 1969; Hare, 1971; Buckner, 1977; Holtmeier, 1978, 1982, 1987b, 1993a, 1996, 2005b; Daly, 1984; Scott et al., 1993; Oke, 1995; Ott et al., 1997; Hiemstra et al., 2002; Kullman, 2005e; Geddes et al., 2005). In the following, these effects will be considered by way of example from the temperate and northern timberlines (European Alps, Rocky Mountains, Northern Europe), which are well documented, while only little information is available from tropical timberlines.

While microclimates in the treeless alpine zone result from the influence of microtopography on solar radiation and windflow near the surface (e.g., Turner, 1980), local climate in the timberline ecotone is additionally controlled by the effects of the tree stands, glades and smaller openings on the climatic elements. These effects may facilitate or impede the establishment of additional trees. Increasing tree population (Figure 62) gradually reduces the number of sunlit sites, for example, where seedlings and saplings profit from warmer conditions (higher air and soil temperatures) compared to shaded sites. On the other hand, reduced exposure to intense solar radiation by shade-giving trees may have a positive effect on seedlings (e.g., Germino and Smith, 2002; Slot et al., 2005), particularly during periods of drought and in dry regions. Moreover, diminished exposure of seedlings and young trees to intense solar radiation following cold nights may reduce the risk of summer frost injury (e.g., Lundmark and Hällgren, 1987; Örlander, 1993). The abrupt limit of closed *Nothofagus solandri* forest in New Zealand (cf., Chapter 3; Photo 3), for example, appears to be due mainly to the intolerance of mountain beech seedlings to high radiation loads and desiccation in the alpine tussock grassland (Wardle, 1965b, 1971, 1973; see also Norton and Schönenberger, 1984; Schönenberger, 1985). Normally, seedlings only survive under a closed shade-giving forest canopy. The same holds true for many abrupt altitudinal timberlines in the tropics (Bader et al., 2008a, b). Intense solar radiation probably prevents the establishment of shade-demanding tree seedlings beyond the current forest limit. Removal of shade-giving plants next to tree seedlings transplanted to the Páramo resulted in decreased seedling survival from 44% to 19% (Bader et al., 2008b).

At treeline in the Kazbegi region (Central Greater Caucasus Mountains, Georgia), Akhaltkatsi et al. (2006) found most birch seedlings (*Betula litwinowii*) growing beneath the canopy of dense *Rhododendron caucasicum* shrub. This

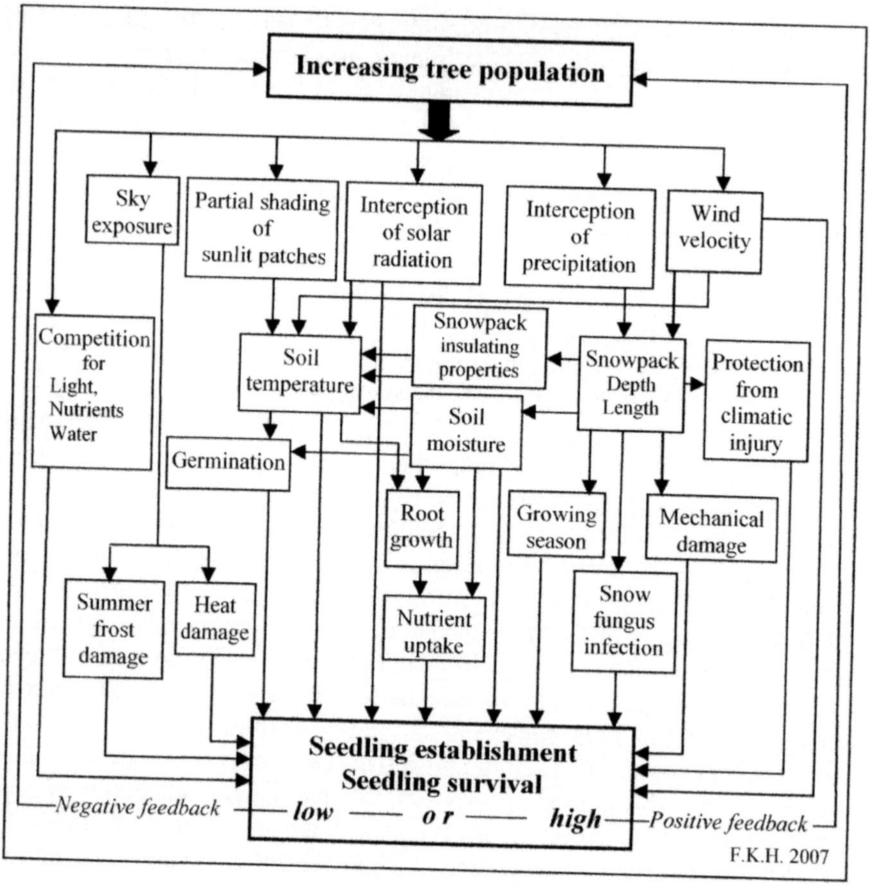

Figure 62. Influence of increasing tree population on site conditions and seedling establishment and survival. – mofied from Holtmeier and Broll (2005)

site produced the greatest number of cold days and the lowest mean minimum air and soil temperature during the growing season (mid July to the end of October). Birch seedlings rarely become established beyond the *Rhododenron* shrubline. Thus, the authors conclude that the benefits from reduced sky exposure (see also Germino and Smith, 2000; Smith et al., 2003) to incident sun light outweighed the disadvantages of low cold air and soil temperatures under the *Rhododendron* shrub canopy. Older but stunted birches occur just below the shrub line.

In the timberline ecotone in Finnish Lapland, we found the population of mountain birch seedlings (*Betula tortuosa*) to be at its highest within willow thickets. The willows have probably facilitated establishment of the birch seedlings, though temperatures (air and soil temperatures) under the shrub canopy remained relativly low during the growing season.

In the altitudinal timberline ecotone outside the tropics and in the northern timberline ecotone, the influence of trees and tree stands on snow relocation is probably the most spectacular effect and has far-reaching consequences (e.g., Holtmeier, 1987a, 1993a, 1996, 2005b; Geddes et al., 2005). In wind-swept terrain, where snow cover is normally very patchy, young growth of trees starts to definitely influence the surface winds as soon as the tops begin to project above the alpine grass and dwarf shrub vegetation or tundra (Holtmeier, 2005b; Holtmeier and Broll, 1992; Broll and Holtmeier, 1994). While singly standing trees exert relatively little influence, the effects of tree clumps are rather conspicuous and persistent. Bekker (2005), for example, describes such a feedback between vegetation pattern and process in the timberline ecotone on Lee Ridge (Glacier National Park, Montana) where an upslope, windward to leeward pattern of older trees (*Pinus contorta* var. *latifolia*) facilitates the establishment of young growth downwind (=up-slope) of the older pines by deposition of blowing snow. The extent of these effects depends on size, density, structure and distribution of the tree stands alternating with open patches. In case of loose stands, with low canopy cover more snow usually accumulates inside than outside the tree stands and in compact clumps with closed canopy. These differences in snow accumulation are most spectacular in wind-swept terrain. Thus, Swiss stone pine clusters, for example, growing at such sites, may over-compensate the primary influence of micro-topography on windflow and snow by increasing snow deposition and thereby length of snow cover to a degree that the trees may be heavily damaged or even killed by *Phacidium infestans*. In particular, the less competitive and generally slow-growing younger pines will be affected (Section 4.3.13.3; Figure 63; Photo 85; Donaubaur, 1963; Holtmeier, 1965, 1987a; Bazzigher, 1976). The older individuals already projecting above the snow, they had triggered increased snow accumulation, lose only the foliage encased in the relatively long-lying snowpack. Such effects are particularly common in tree stands that became scattered due to competition. The upper limit of snow fungus infestation marks the average depth of the snowpack and thus may be used as a snow gauge (cf. Photo 81, 85 and 89; Figure 63).

Conversely, a decline of existing snow-trapping tree stands will result in increased removal of snow by the wind and exposure of plants in the understory to climatic injury, particularly on wind-swept topography. Complete defoliation of the subarctic mountain birch forest, for example, in northernmost Finland during a mass outbreak of the autumnal moth (*Epirrita autumnata* in the mid-1960s caused dieback of birch tree stands in the timberline ecotone (Section 4.3.13.4; Kallio and Lehtonen, 1973, 1975: Holtmeier, 1974, 2005a, b; Lehtonen and Yli-Rekola, 1979; Seppälä and Rastas, 1980; Holtmeier and Broll, 2006). Consequently, juniper (*Juniperus communis*)

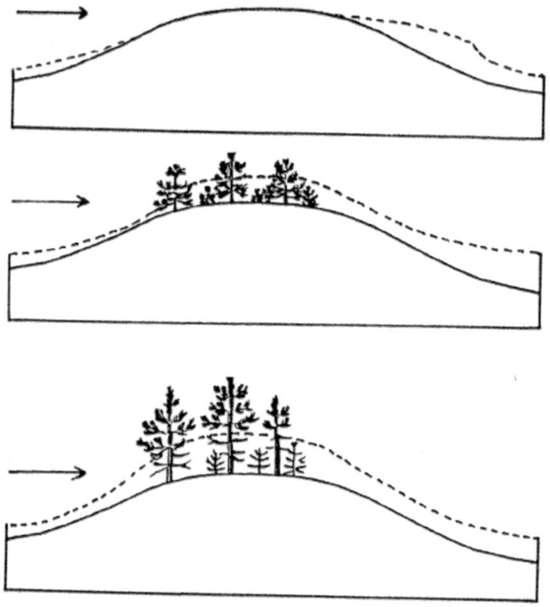

Figure 63. Influence of a *Pinus cembra* cluster, originated from a seed cache of the European nutcracker on snow accumulation. After a couple of years, the younger pines will be killed by snow blight (*Phacidium infestans*). Modified from Holtmeier (1974).

Photo 85. Cluster of *Pinus cembra* heavily damaged by *Phacidium infestans* due to the increased snow accumulation within the tree stand. Northwest-facing slope of the Upper Engadine main valley at about 2.200 m. F.-K. Holtmeier, 23 September 1968.

which was growing in the understory, thus being well-protected by the snow cover, became increasingly exposed to ice particle abrasion and frost; Holtmeier, 2005b). Since juniper is intolerant of such influences its canopy surface died back parallel to the now shallower winter snow cover (Photo 86; Figure 64).

Photo 86. After a mass outbreak of the autumnal moth (*Epirrita autumnata*) the mountain birch stands near Oadašamjavvrik (about 330 m, Utsjoki, northern Finnish Lapland) during the 1960s the mountain birch forest (*Betula tortuosa*) declined. Juniper (*Juniperus communis*) was no longer protected by snow and suffered from the consequent adverse climatic conditions. F.-K. Holtmeier, 21 August 2003.

The influence of compact tree stands such as wind-trimmed clonal groups is somewhat different. The wind blows around and over them, causes scouring effects and often clears the snow completely from the ground on their windward sides and sometimes round their lateral edges, too (blow outs; Photo 87). The snow intercepted by the canopy partly evaporates or is blown downwind, while comparatively little snow is trapped inside the clonal groups where it remains relatively loose (Holtmeier, 1987a, 1989). Big snowdrifts, length three to five times the height of the tree stand, pile up on their leeward side (snow-fence effect). The snow surface is compacted and will easily carry a person walking on it.

Crowded tree islands increase surface roughness compared to treeless terrain thus enhancing accumulation of snow in the whole area covered by the tree islands (Photo 88). Thus, snow may last even longer than on the lee sides of solitary clonal groups in the timberline ecotone (cf. Photos 87, 89). Because of the great depth of the snowpack, scouring effects usually do not expose the ground as is typical on less snow-rich sites. In areas with long-lying

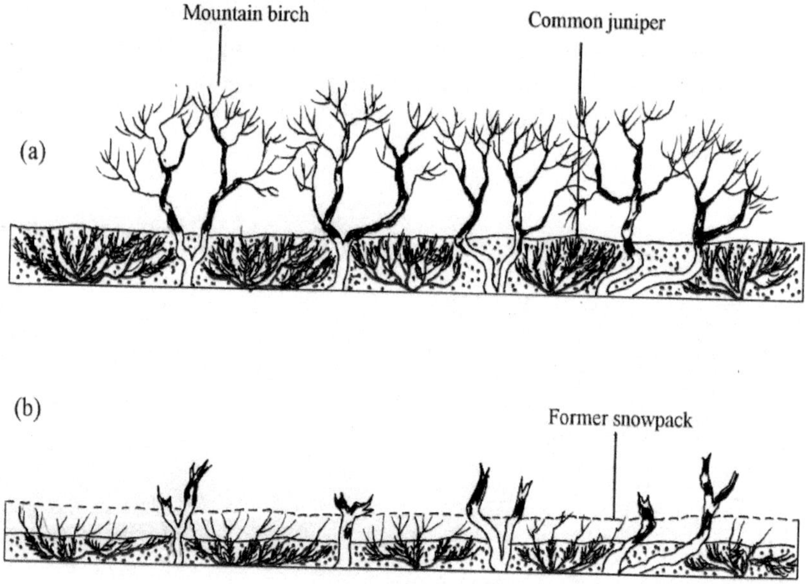

Figure 64. The effect oft the destruction of the mountain birch forest (*Betula tortuosa*) on the winter snow cover and common juniper (*Juniperus communis*) in the understorey (schematic). From Holtmeier (2005b).

Photo 87. Distribution of snow influenced by compact clonal groups (*Abies lasiocarpa, Picea engelmannii*) on Niwot Ridge (Front Range, Colorado) at about 3.450 m. F.-K. Holtmeier, 7 April 1989 .

Photo 88. Air photo of a shallow, north–south-oriented valley on the south slope of Niwot Ridge (Front Range, Colorado). The leeward slope (*left* side) is still covered by snow, except for some convex topography. On the wind-exposed slope (*right*), snow has persisted only at sites where snow accumulation had increased because of the greater roughness of the surface caused by the wedge- and hedge-like clonal groups of *Picea engelmannii* and *Abies lasiocarpa* alternating with narrow treeless corridors and gaps. H.-U. Schütz, 19 June 1990.

snow, many clonal groups show heavily needle loss due to the brown snow felt fungus (*Herpotrichia juniperi*; Photo 89; see also Photo 81). Obviously, fungus must have occurred after the clonal tree islands already projecting above the snowpack, increased snow accumulation to critical extends. Otherwise, the conifers would not have survived. At present, the brown snow felt fungus would kill occasional seedlings and young growth that became established during favourable less snow-rich years because they are completely buried under the snowpack. Also, snow pressure may destroy the young trees at these islands.

Photo 89. As they grow taller, these clonal groups (*Picea engelmannii, Abies lasiocarpa*) gradually enhance snow accumulation to the extent that the needles near the ground were killed by *Herpotrichia juniperi*. Fourth of July cirque (Front Range, Colorado at about 3.420 m). F.-K. Holtmeier, 19 July 1992.

The compacted snow drift on the leeward side of dense clonal tree islands often persists until early summer (Photo 90; Holtmeier, 1982, 1987a, b, 1996; Holtmeier and Broll, 1992; Broll and Holtmeier, 1994), thus resulting in a short growing season, long-lasting low soil temperature and high soil moisture which, in turn, control water and nutrient uptake, root growth, and decomposition. Also, the risk of snow fungus infection is increased, in particular as needles and twigs with adherent hypha and spores are accumulated (often as narrow strips) together with relocated snow, as it happens to wind-blown detritus in general (cf. Wilson, 1958; Teeri and Barett, 1975). As the snow is melting, this allochtonuous litter will stick on the living foliage and may become an additional source for fungal attacks (see also Simms, 1967).

Photo 90. Clonal group (*Picea engelmannii*) with several-metres-long leeward snow drift. West-facing slope of a shallow valley (see Photo 88) on the south slope of Niwot Ridge (Front Range, Colorado) at about 3.450 m. H.-U. Schütz, 7 June 1990.

At such conditions tree seedlings can hardly establish themselves. In case the snowpack does not last too long, a closed grass- and herbaceous cover will develop, in contrast to the windward side of the tree islands where coverage is usually sparse. Also dwarf shrubs such as *Vaccinium myrtillus* and *Ribes montigenum* that need protection by the snow cover in the winter are common in the leeward snowdrift area (Figure 65). Extremely late-lying snow, however, may prevent any dwarf shrubs and grass vegetation.

At the leeward edge of clonal groups, usually not exceeding 1-m height in extremely wind-swept terrain, little snow accumulates compared to the large snowdrifts behind taller and better wind-protected tree islands. Consequently, snow melts earlier and the growing season is prolonged. In case germination was not prevented by too rapidly draining soil, sporadic seedlings might establish themselves growing fairly well for decades as protected by the upwind clonal group (Photo 91). The treetop will usually be deformed or even destroyed by wind when growing as high or higher than the windward group. In the following, the tree, if spruce or fir, will elongate downwind by layering. In case 'hedges' alternate with narrow treeless corridors, wind-channeling may prevent deposition of snow in these corridors and also cause soil erosion (Holtmeier, 1978). However, strong winds blowing occasionally from other than the 'normal' direction may cause an over-lapping drift accumulation (Figure 66) burying these 'hedges' and corridors completely.

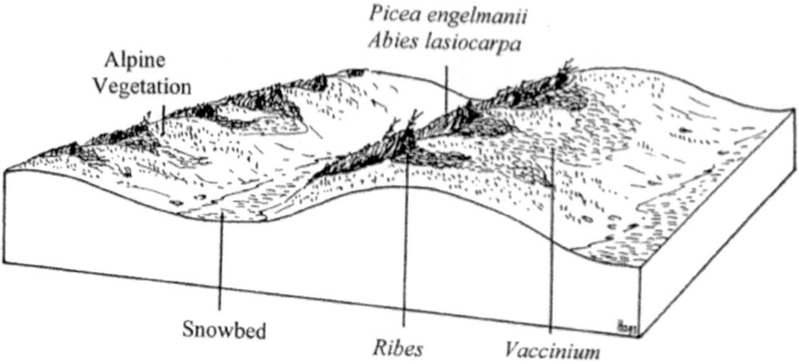

Figure 65. In the very windy environment close to the treeline, *Ribes montigenum* and *Vaccinium myrtillus* are restricted to the areas in the lee oft he conifers, where snow drifts piles up in winter.

Photo 91. Clonal group (*Abies lasiocarpa*) at an extremely windy site on Niwot Ridge (Front Range, Colorado) at about 3.490 m. Behind the leeward edge of the group and the low willows (*left side*), a young subalpine fir has become established where the snow drift builds up in winter. F.-K. Holtmeier, 18 August 1990.

When the wind switches back to the 'normal' direction, the deep snow cover will reduce wind-channeling between the 'hedges' and thus decrease wind erosion (Holtmeier, 2005b).

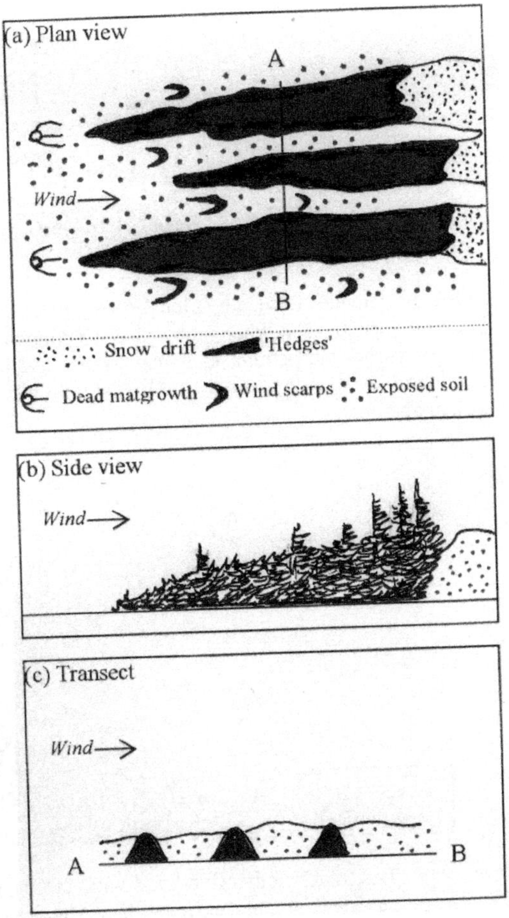

Figure 66. 'Hedges' and their effects on snow-relocation at different wind directions: (**a**) Prevailing wind direction parallel to the 'hedges'. Wind channeling keeps the narrow 'corridors' snow-free. Large snow masses accumulate at the leeward sides of the 'hedges'. (**b**) Side view of (a). (**c**) Wind from other than the normal direction. Snow may bury the normally snow-free 'corridors' and most part of the 'hedges'. From Holtmeier (2005b).

Many of the 'hedges' that might be taken for a single clonal group at first sight are later seen to consist of several clonal groups (spruce and fir) in line (Figure 67; Holtmeier, 1993a). Often, pines (*Pinus flexilis*) that originated mainly from seed caches of Clark's nutcracker can also be found at the leeward end edge of clonal tree islands. On wind-exposed topography above the mountain forest on Hokkaido (Japan), snow accumulation enhanced by Siberian dwarf pine (*Pinus pumila*) supports the gradual propagation of this true krummholz by layering (Okitsu and Ito, 1984).

At better wind-protected sites, the tree stands do not just influence the action of wind and snow, they also influence heat distribution in the timberline ecotone. The dark surfaces of the tree clumps absorb most of the solar radiation

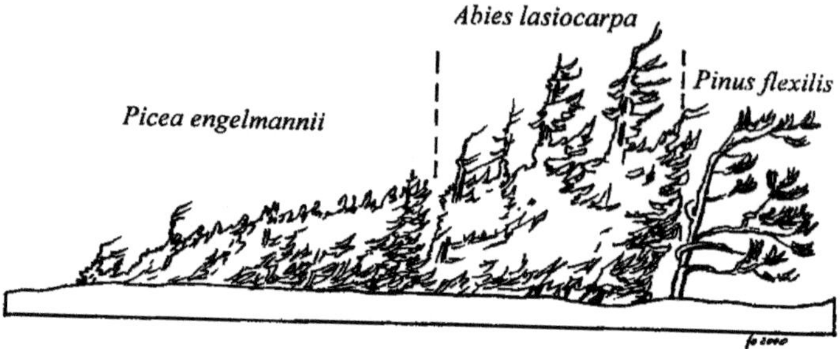

Figure 67. 'Hedge' formed by clonal *Picea engelmannii* and *Abies lasiocarpa*. The subalpine fir became established where snow that accumulates in the lee of the spruce provides some protection from the permanent strong winds. The limberpine (*Pinus flexilis*), which became established latest, enjoys the shelter by the both windward conifers. After a field sketch of the present author.

Figure 68. The 'black-body effect' and wind scouring cause early snow melt at the sun- and wind-exposed side of a compact tree group. After a field sketch of the present author.

to pass it on to their environment in form of sensible heat ('black-body effect'; Brink, 1959; Swedberg, 1961; Franklin and Mitchell, 1967; Habeck, 1969; Holtmeier, 1987a). On the sides directly exposed to solar radiation the snow melts much faster than in the centre of the tree islands or on their shaded sides. The melting produced by the 'black-body effect' in the snow cover is especially intense when the direction of radiation and that of the prevailing wind are identical (Figure 68). In this case grass and dwarf shrub vegetation at the windward side of the tree islands very soon emerge from the snow, warm up by absorbing more solar energy and thus accelerate snow melt (Holtmeier, 1986a). Such sites are relatively favourable to seed-based regeneration of the conifers because of the prolonged growing season and melt-water supply. On the other hand, if during the winter the direction of solar radiation lies opposite to that of the prevailing wind, this can lead to over-compensation of the 'black-body effect' by large snow masses accumulating at the wind-protected sunny sides of the tree clumps (see Figure 26; Sections 4.3.6 and 4.3.11). Also in loose tree stands this heat radiation, which would promote melting of the snow, remains almost ineffective due to the high accumulation of snow within most of them.

In the centre of clonal groups it is usually more humid compared to their surroundings, at least during dry weather conditions, because of interception of solar radiation by the dense canopy and less turbulent mixing. In addition, the thick raw humus layer (20 to 30 cm) holds moisture (Figure 69).

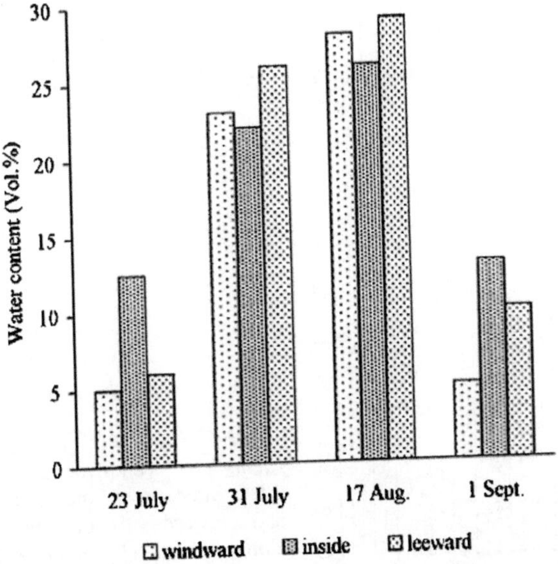

Figure 69. Mean soil-moisture content oft he A-horizon (0–4 cm depth) at the windward side, interior and leeward side of three clonal groups (*Picea engelmannii*, *Abies lasiocarpa*) on Niwot Ridge (Front Range, Colorado) at about 3.450 m during dry (23 July, 1 September) and humid (31 July, 17 August) weather conditions. From Broll and Holtmeier (1992).

Moreover, soil temperatures are comparatively low and amplitudes are smothered (Figure 70). Probably, root activity and root growth may be impeded at such conditions as has been speculated by Körner (1999) and Paulsen et al. (2000), for example. On the other hand, roots, in particular those of clonal members at the margin of the tree islands, usually extend beyond the shaded area thus enjoying higher soil temperatures that may positively affect root growth and nutrient uptake, sufficient soil moisture supply provided. In dry regions, shading within the tree islands might be even favourable to site conditions because of the resulting delay of snow melt, reduced temperatures and low evaporation. Thus soil moisture deficiency is prevented until early summer when soil moisture outside the tree islands is already exhausted (see also Van Miegroet et al., 2000).

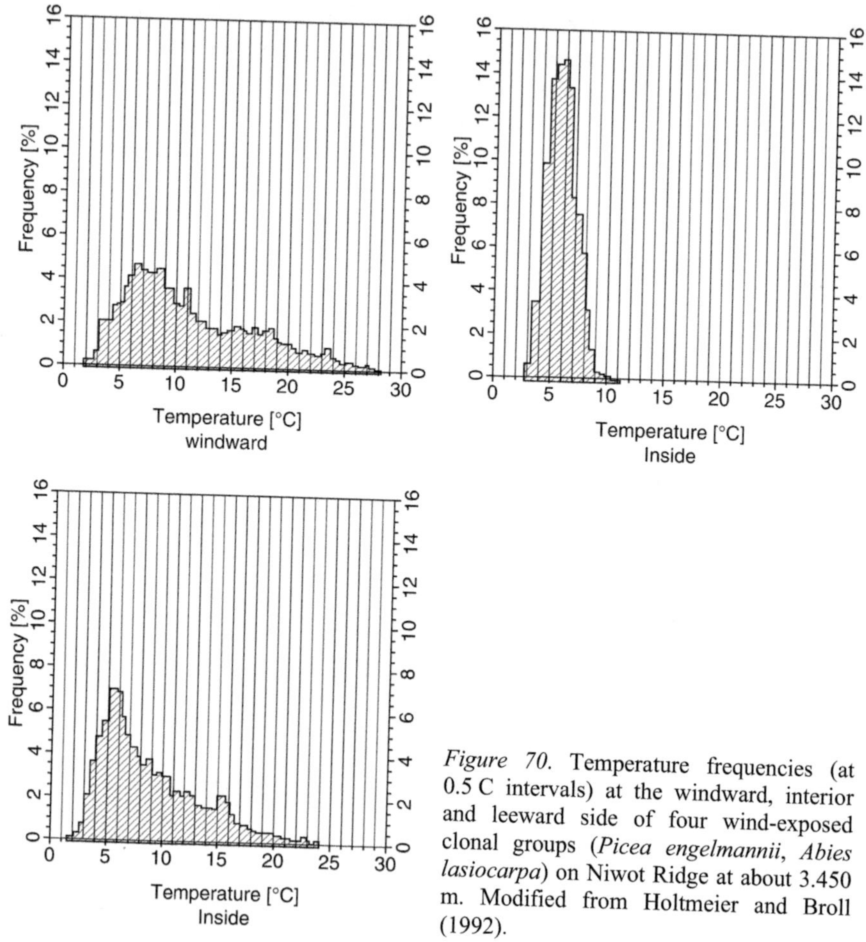

Figure 70. Temperature frequencies (at 0.5 C intervals) at the windward, interior and leeward side of four wind-exposed clonal groups (*Picea engelmannii*, *Abies lasiocarpa*) on Niwot Ridge at about 3.450 m. Modified from Holtmeier and Broll (1992).

Inside the compact tree islands, ground vegetation is very scarce if not missing, in particular because of low light intensity. Thus, the litter consists mainly of slowly decomposing needles, remains of broken twigs, bud scales and cones. Site conditions gradually changing under the influence of the tree islands as well as the different amount and quality of the litter at the windward side, inside and in the snowdrift area have affected and still influence pedogenesis, thus causing distinct differences at the micro scale (Holtmeier and Broll, 1992; Broll and Holtmeier, 1994). Obviously, tree vegetation is the factor affecting soil forming processes and thus nutrient conditions in our study sites. Strangely enough, Bekker et al. (2001) refer to this paper (Figure 71; Holtmeier and Broll, 1992) to prove that soil fertility may be a factor influencing the alpine treeline ecotone, which would be just the opposite.

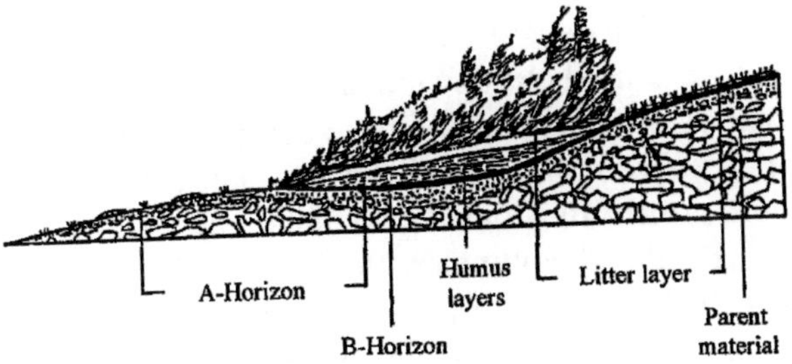

		windward	inside	leeward
C_{org}	[%]	4.8 (0.4)	5.3 (0.5)	7.5 (0.5) [a]
N	[%]	0.35 (0.03)	0.35 (0.03)	0.55 (0.03) [a]
C / N		13.6 (0.5)	15.3 (0.6) [a]	13.5 (0.3)
P	[ppm]	2.0 (0.2)	2.2 (0.2)	2.7 (0.2) [b]
K	[ppm]	150 (16)	120 (13)	210 (19) [a]
CEC	[meq/100g]	8.5 (0.6)	10.3 (0.9)	10.5 (1.1)
Ca	[meq/100g]	5.0 (0.8)	5.5 (0.9)	7.0 (1.0)
H / Al	[meq/100g]	2.5 (0.6)	4.0 (0.6)	1.5 (0.3) [c]
pH	[$CaCl_2$]	4.0 – 4.5	3.5 – 4.5	4.5 – 5.0
	[H_2O]	4.5 – 5.0	4.0 – 5.0	5.0 – 5.5

(a) Significantly different from other two sites (t test, $n = 10$, $P \leq 0.05$)
(b) Significantly different from windward (t test, $n = 10$, $P \leq 0.05$)
(c) Significantly different from inside (t test, $n = 10$, $P \leq 0.05$)

Figure 71. Mean and standard error of chemical soil properties (A-horizon) at the windward side, interior and leeward side of a clonal group (see Photo 90) at a wind-exposed site on Niwot Ridge at about 3.450 m. Modified from Holtmeier and Broll (1992).

In case the structure of dense clonal groups changes, for example by death of the initial trees, snow deposition will also change. More snow is accumulated within the gap (Figure 72). In addition, shading by the other clonal trees preserves the snow thus increasing the risk of snow fungus infection. Young growth may establish itself during a few subsequent years with little snow in winter but usually succumbs to fungal damage or snow pressure after a while.

Local climatic conditions, primarily controlled by the local topography, may fundamentally change due to the influence of clonal tree islands on the wind flow near the surface and resulting side effects (depth and length of the snow cover, soil temperatures, soil moisture, etc.; Figure 73). On a wind-facing slope, for example, or on rounded topography that never accumulates any appreciable snowpack, compact tree islands may create a very locally varying mosaic of sites with different depth and length of the winter snow cover (cf. Photos 88, 90).

On snow-rich leeward slopes, clonal tree islands additionally enhance deposition of big snow masses resulting in prolonged snow cover. By their influence on wind flow near the ground compact tree islands often become a factor initiating severe wind erosion exposing the mineral soil and creating wind scarps at their upwind and lateral sides (Figure 74). Tree islands gradually moving downwind have conspicuously influenced microtopography leaving exposed mineral soil, small 'plateaus' (10 to 20 cm high remains of the soil profile that developed under tree island cover) and wind scarps behind (Broll and Holtmeier, 1994; Figure 75; Photo 92).

So-called 'ribbon forests' are probably one of the most impressive examples of the interrelationships of site conditions and distribution pattern of the vegetation (Photo 93, Figure 76). 'Ribbon forests' are known only from the Rocky Mountains, so far (Billings, 1969; Buckner, 1977; Holtmeier, 1978, 1982, 1987b, 1996; Schütz, 1998, 2005). The spatial structure of 'ribbon forests',

Figure 72. Gaps inside a clonal group enhance accumulation of snow. Interception of incoming solar radiation and almost unimpeded outgoing long-wave radiation cause low temperatures and delay snow-melt.

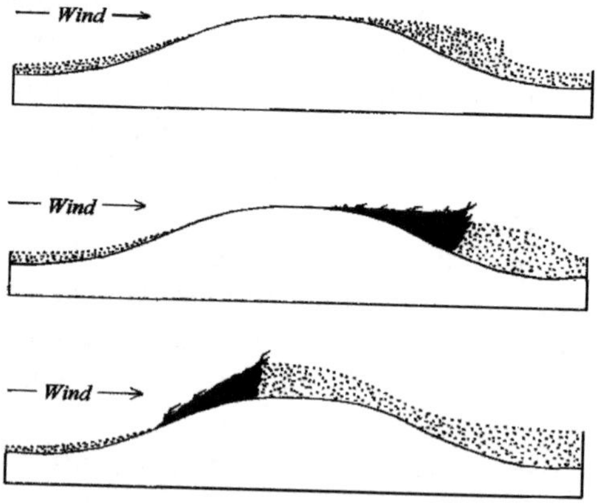

Figure 73. The relocation of snow, originally controlled by topography, is changed by the influence of tree vegetation.

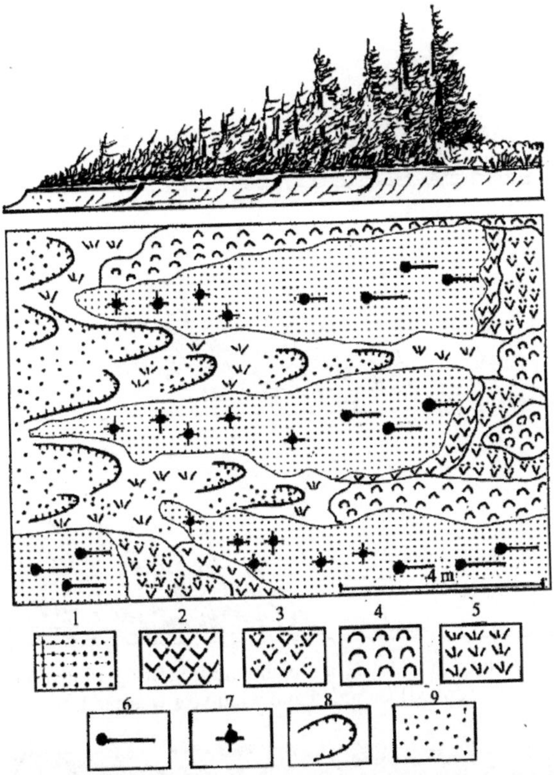

Figure 74. Wind-influenced site pattern (plan and side view) around strongly wind-influenced clonal groups (schematic, based on observations in the forest-alpine tundra ecotone in the Rocky Mountains).

1 – clonal groups (*Picea engelmanni, Abies lasiocarpa*), *2* – *Ribes montigenum*, *3* – *Vaccinium myrtillus*, *4* – *Salix planifolia*, *5* – *Kobresia myosuroides*, *6* – flagged tree canopies, *7* – dead terminal leaders, *8* – wind-scarps, open mineral soil.

which are oriented perpendicular to the prevailing winds, is caused by tree stands measuring up to a hundred or more metres in length and between 10 to 30 m in width, alternating with almost treeless areas ('snow glades') up to 100 m wide (cf. Figure 76; Photos 93, 94). The 'ribbons' are formed by *Picea engelmannii* and *Abies lasiocarpa*, while the glades are covered with often-luxuriant subalpine meadows. Normally, the 'ribbons' consist of several more or less extended sections separated by narrow gaps that increase wind velocity ('wind channel effect') thus removing the snow. 'Ribbon forests' are to be found in the Rocky Mountains mainly on the snow-rich leeward slopes dominated by strong and permanent downslope winds from the west relocating the snow from the extended alpine area to the leeward forest below (Holtmeier, 1978, 1982, 1987a, b, 1996).

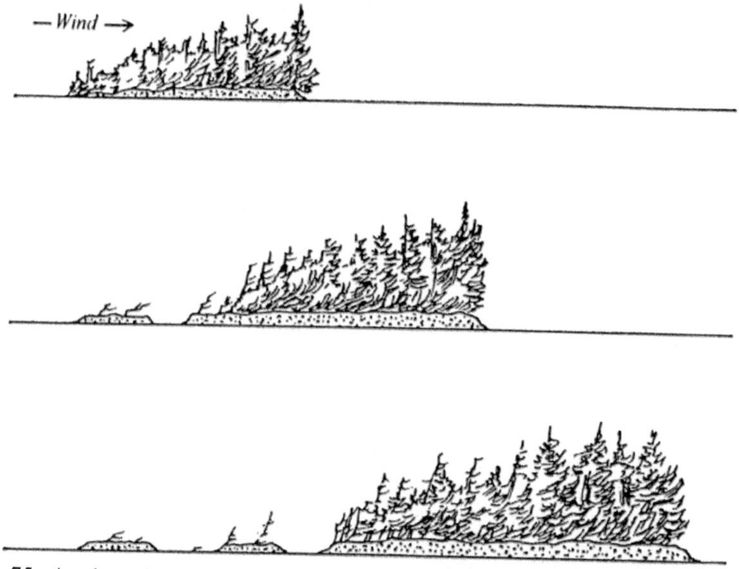

Figure 75. At the windward side of a downwind-migrating clonal group, the soil that developed under the trees (dotted) is eroded and wind-scarps are left.

Photo 92. Wind-scarps and wood remains at the windward side of a clonal conifer group that migrated downwind. Forest-alpine tundra ecotone on Niwot Ridge (Front Range, Colorado) at about 3.450 m. F.-K. Holtmeier, 31 July 1989.

Photo 93. 'Ribbon forest' on a north-facing slope above Lefthand Reservoir (Front Range, Colorado) at 3.350 to 3.430 m. The initial trees became established along solifluction terraces and low edges providing shelter from the strong downslope winds (see also Figure 77). F.-K. Holtmeier, 27 July 1989.

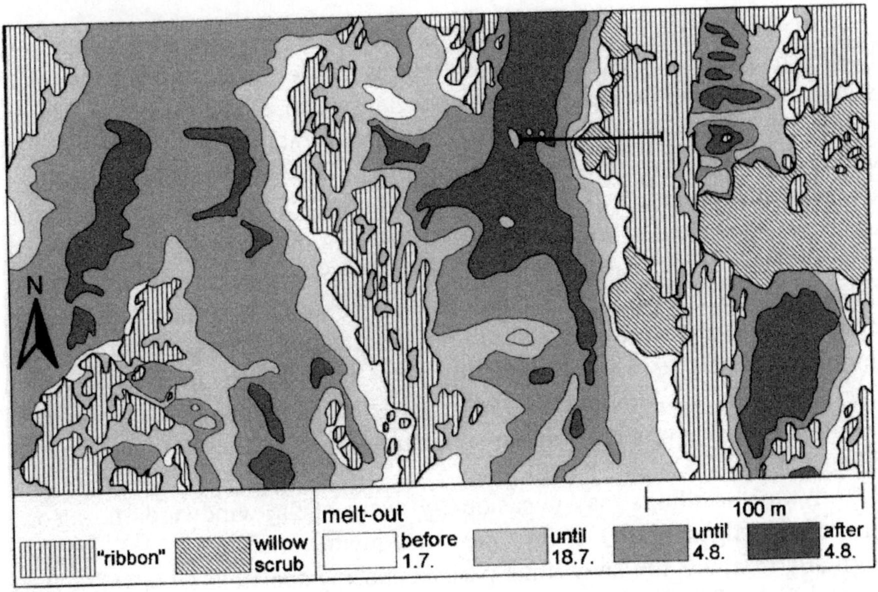

Figure 76. Melt-out in a 'ribbon forest' on the east-facing slope of Mt. Audubon (Front Range, Colorado) at 3.300 to 3.355 m (cf. Photo 94). Soil temperature was measured along the transect line (Figure 78). Modified from Holtmeier (1987b).

Photo 94. View downslope of a 'ribbon forest' (about 3.350 m) on the east slope of Mt. Audubon. Along the upwind rim of the 'ribbons'. where the snow melts relatively early, some young growth (*Picea engelmannii, Abies lasiocarpa*) has come up. F.-K. Holtmeier, 1 July 1984.

Usually, the 'ribbons' originated from seedlings of spruce and fir that established themselves along solifluction terraces and low edges giving shelter to young growth from the strong downslope winds. Also, a little snow accumulates behind this microtopography providing additional soil moisture. Growing taller and elongating downwind mainly by layering, these tree stands increasingly enhance accumulation of snow and thus produce large leeward snowdrifts the water equivalent of which exceeds the amount of local precipitation by far. Finally, these snow masses, often persisting until early or even mid summer (Photo 94; see also Marr, 1967), prevent the 'ribbons' from further expanding downslope. A short growing season, fungal infection (*Herpotrichia juniperi*) and snow breaks usually do not allow trees to establish themselves within the 'snow glades'. Although this snow fence effect gradually decreases with distance from the upwind 'ribbon' it usually affects the entire 'snow glade' (Figure 77).

Anyway, seedlings may occasionally occur at the windward and also at the lateral edges (gaps) of the next downwind 'ribbon' where scouring normally removes the snow thus partly exposing the field layer (cf. Figure 76 and Photo 94). During extremely snow-rich winters, however, with late-lying wet snow, most seedlings and also young growth will be eliminated by the snow conditions.

Inside the 'ribbons', long-lasting low soil temperatures (Figure 78) and low intensity of light do not allow any seed-produced regeneration (see also Patten, 1963a). In small gaps only, resulting from snow break or death of old

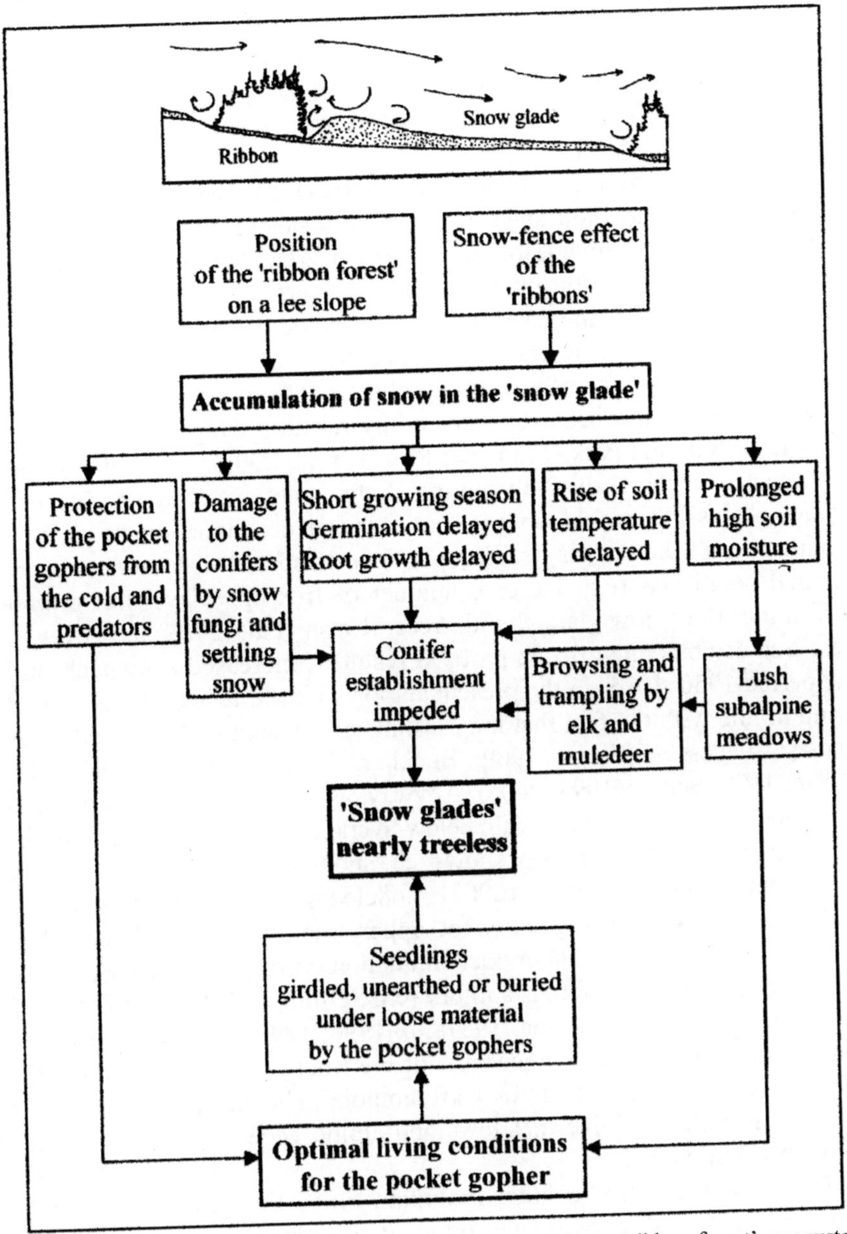

Figure 77. Interactions of biotic and physical factors in a 'ribbon-forest' ecosystem. Modified from Holtmeier (2002).

trees, seedlings may occasionally grow up provided that not too much snow accumulates that would shorten the growing season, lower soil temperature and increase the risk of fungal infection.

Mule deer (*Odocoileus hemionus*) and Wapiti (*Cervus elaphus*), which graze the luxuriant energy-rich subalpine meadows in the 'snow glades', are additional agents keeping the 'glades' almost devoid of trees (cf. Figures 76 and 77). Moreover, pocket gophers (*Thomomys talpoides*) enjoying optimal conditions of life within the 'snow glades' impede successful tree establishment (Section 4.3.13.1).

The position of the 'ribbons' which survived mainly by layering for hundreds and even thousands of years does not seem to have essentially changed during this time (Holtmeier, 1987b). Also, the mosaic of clonal tree islands alternating with open patches appears to have been relatively persistent throughout the centuries, even if some tree islands have migrated downwind. It is open to question if and how these spatial structures will change in consequence of a warming climate (Chapter 5). Disappearance of a 'ribbon forest' has been documented by repeat photography at timberline in Glacier National Park (Montana; Roush et al., 2007). This 'ribbon forest' still existed in the early 20th century. Later, trees were able to invade the 'snow glades'. It would appear that reduced accumulation of snow and lengthening of the growing season possibly combined with more favourable thermal conditions (e.g. increase number of frost-free days) has promoted infilling of the 'snow glades' with trees. Roush et al. (2007), however, suggest tree establishment to be partly a result of increased snowpack during the period 1950–1975. In the present author's view this hypothesis is hard to bring in line with the fact that long-lasting snowpack usually keep the 'snow glades' treeless (Billings, 1969; Buckner, 1977; Holtmeier, 1978, 1982, 1987b, 1996; Schütz, 1998, 2005). Anyway, as snowpack shows considerable interannual variation, winters with below-average snowfall could have initiated tree invasion into the normally snow-rich 'snow glades'.

Recently, Hiemstra et al. (2002) modeled snow redistribution by wind and interactions with vegetation at upper treeline in the Medicine Bow Mountains (Wyoming). The model simulation corresponds basically to field observations, which different authors (e.g., Billings, 1969; Buckner, 1977; Holtmeier, 1978, 1980, 1986a, 1987a, b, 1996, 1999b; Broll and Holtmeier, 1994) had made before in other parts of the Rocky Mountains and confirms the well-known effects of trees ('krummholz', 'hedges', 'ribbon forests'), alternating with subalpine meadows and alpine vegetation, on wind-driven snow distribution pattern.

Probably, there are no other mountain timberlines shaped in a similar way by climate and, in particular, by wind, as those in the wind-swept high-mountain ranges of western North America. Also, at the polar timberline wind effects on tree physiognomy are mostly not that spectacular, and are almost unimportant in this respect at tropical mountain timberlines.

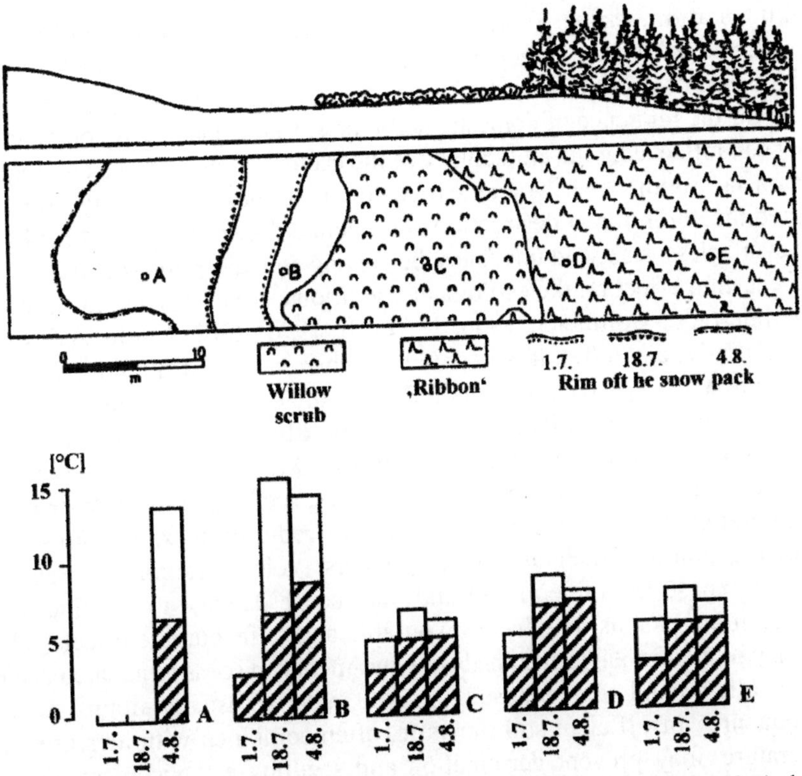

Figure 78. Soil temperatures at depths of 5 cm (white bar) and 20 cm (hatched bar) along the transect in the 'ribbon forest' on Mt. Audubon (cf. Figure 76 and Photo 94), July and August 1984. From Holtmeier (1987b).

As was already mentioned, at timberline in the Alps such natural spatial structures caused by the interaction of topography and tree islands got lost when the upper mountain forests were cleared in order to extend the alpine grazing area to lower elevation. Moreover, in the comparatively low but steep outer mountain ranges and also in other steep rugged mountains, the distribution pattern of trees is primarily controlled by orographic influences (mass wasting, slope erosion, avalanches, etc.). Consequently, the effects of trees and tree islands on their close environment are by far less conspicuous if compared to the timberline ecotone on the Rocky Mountains. On more gentle slopes, however, groupwise planting of trees in high-elevation re-forestation, thus mimicking natural structures, has proved to be an effective practice to increase stability of the timberline (Schönenberger, 1986; Schönenberger et al., 1990).

4.3.13 Influence of animals on timberline

Animals are among many other factors (cf. Figure 1) influencing the ecological conditions at timberline. While a wealth of literature has been published on their effects on forests in general (Holtmeier, 2002, there further literature) the influence of animals in the timberline ecotone has been widely disregarded, except for the influence of domestic herbivores. Cattle and sheep, for example, have been influencing the altitudinal treeline ecotone in many mountain regions of Eurasia, South America and Africa for thousands of years. Pastoral use resulted mostly in forest destruction and treeline decline (Section 4.3.14.1).

Nevertheless, also wild ungulates, borrowing animals, birds and insects may have a lasting effect on the ecological conditions, structures and dynamics of timberline (e.g., Holtmeier, 1966, 1967a, b, 1974, 2002; Cairns and Moen, 2004; Cairns et al., 2007). It is mainly by herbivory, seed dispersal, trampling, and bioturbation that wild animals influence the ecological conditions in the timberline ecotone. Herbivorous insects may affect trees mainly by feeding on leaves as well as on the xylem and phloem. Mammals affect trees mainly by consuming above ground phytomass (leaves, seeds) and by bark stripping, fraying, trampling, pawning and scraping (e.g., red deer, ibex, bighorn sheep). Burrowing animals such as pocket gophers, voles, ground squirrels and marmots expose the mineral soil and may cause damage to the fine roots. Exposed mineral soils may have favourable and unfavourable effects. Thus, they may provide suitable seed beds to wind-mediated seeds (e.g., larch seeds). On the other hand, the exposed mineral material usually drains rapidly. Consequently, insufficient soil moisture, often combined with increased soil temperature, may prevent germination and seedling establishment. For the sake of completeness grizzly bears must be mentioned in this context. They may locally influence site conditions in the timberline ecotone by digging for bulbs and tubers (e.g., Butler, 1995). In the Rocky Mountains of Montana, for example, digging for the roots of the yellow sweetvetch (*Hedysarum sulphurescens*) and the bulbs of the glacier lily (*Erythronium grandiflorum*) is very common in the ecotone zone. Moreover, the grizzlies excavate for pocket gophers, voles, marmots, ground squirrels and invertebrates (e.g., ladybirds and army cutworms; Chapman et al., 1955; Mattson et al., 1991) thus covering many square metres of the surface with loose mineral soil. It may be that grizzly bear soil disturbance favours tree seedling establishment (e.g., Cairns et al., 2007).

Birds may affect the tree vegetation directly by consuming seeds and buds and, what is more important, by dispersal of the seeds from some plant species, among them some subalpine tree species. Thus, they influence timberline dynamics. With regard to effects of the changing environment in the treeline ecotone (Section 5.4) at the landscape and local scales

(Holtmeier and Broll, 2005) the role of wild animals should not be disregarded. Compared to the effects of climate, topography and human impact the animal's effects are more of a local and regional importance. The following examples, mainly from Europe and North America, are presented to stimulate broader interest in the role of wild animals as agents involved in the complex interaction of timberline-influencing physical, biological and anthropogenic factors (cf. Figure 1).

4.3.13.1 Large herbivorous mammals

In addition to consumption of phytomass (foliage), food selection and foraging behaviour of the herbivores play an important role. The central questions are: which plants and which parts of the plants will be eaten at which time of the year or stage of forest succession. Energy uptake by consumption of fully developed leaves, for example, affects a tree or a tree stand in a way different from the uptake of the same quantity of energy by browsing terminal buds, fresh shoots or seeds. Moreover, the effect of phytomass consumption is also different in spring, summer or fall (e.g., Weis et al., 1982) and depends also on the successional stage. After a forest fire or a wind throw, for example, even comparatively little phytomass consumption has usually stronger effects than in undisturbed forests.

Ungulates, on the other hand, benefit from the larger quantity of nutrient-rich food available after such disturbances. This again may positively influence their reproduction and survival rate. Mammalian herbivores browse evergreens usually in winter and prefer broadleaves in summer. As conifers retain reserves mainly in needles and deciduous trees store reserves in stem and roots, damage occurs when it is most detrimental to the trees. In general, herbivores influence ecosystems more as regulators than by foliage consumption and through the transfer of energy (Chew, 1994; Holtmeier, 2002).

Wild ungulates (e.g., red deer, roe deer, reindeer, moose) have coevolved with the mountain forests ecosystems and ungulate density was in balance with the natural carrying capacity of these ecosystems. Natural factors such as availability of food, winter starvation, mortality and reproduction rate, diseases and losses to predators controlled ungulates and kept them at a comparatively low level. In the original forests of Middle Europe, for example, the density of wild ungulates is supposed to have been about ten times lower than nowadays (Widmann, 1991; Danilkin, 1996). Although ungulates affected tree vegetation by browsing and trampling they were not a threat to the existence of the mountain forests. The situation changed when ungulate population densities began exceeding the carrying capacity of the forests. In the present context carrying capacity means the maximum population density of ungulates that may live in the forests without preventing tree regeneration.

Altogether, wildlife population and damage control is not a question of forestry or wildlife management but requires an approach at the ecosystem level considering animal population and its habitat as a complex system of many interacting factors and processes operating over multiple scales of space and time (landscape ecological approach). Animal population dynamics as controlled by internal and external factors (e.g., Holtmeier, 2002, there further literature) need to be considered and managed simultaneously in the changing timberline environment (see also Bugmann and Weisberg, 2003). In most cases the magnitude of effects of wild ungulates in the timberline ecotone can only be understood if considered in combination with the changes in the timberline environment caused by historical and current human impact (Holtmeier, 2002).

Red deer: In most of its distribution areas, red deer was not a problem to the forest ecosystem as long as its population density was limited by the natural factors and did not exceed the natural carrying capacity. In most parts of Europe, red deer density has considerably increased since feudalistic times. Red deer profited from strict protection and support (e.g. winter feeding) by the sovereigns and noble landowners who had the privilege of hunting big game and therefore were interested in great numbers of red deer on their property. In the Alps, for example, winter feeding places should keep red deer herds from traditional seasonal migration to the low-lying mountain forelands. Thus, grazing pressure in the mountains increased. Moreover, regulators such as wolf (*Canis lupus*), bear (*Ursus arctos*), and lynx (*Lynx lynx*) were killed whenever possible.

Compared to chamois, big horn sheep, ibex and others red deer (*Cervus elaphus*, Europe) and elk (*Cervus canadensis*, North America) regularly occur at comparatively large numbers in and above the altitudinal treeline ecotone. Red deer affect tree growth in mountain forests and in the timberline ecotone by browsing fresh annual shoots, terminal buds, and twigs and by bark stripping. Moreover, red deer cause damage by rubbing their antlers against the tree stems to get rid of the velvet. During the rutting season, the males, being very aggressive at this time of the year, beat their antlers against the tree stems thus destroying the bark and making the trees more susceptible to fungus infection.

During the winter, red deer, if not supported by feeding, frequent the treeline ecotone and the alpine zone where strong winds remove the snow from exposed topography, thus making food more accessible. Seedlings and young growth that have invaded these areas are damaged by browsing and trampling. At very windy conditions, however, red deer retreat to the upper closed forest stands (Schmidt, 1993). Young growth may be completely destroyed in such places.

At timberline seedlings and saplings are particularly vulnerable by wild ungulates for mainly two reasons: First, the trees are growing close to their climatic limit, where any additional damage may reduce their vitality or even kill the trees. Second, as young trees in the treeline ecotone are usually growing slowly they are exposed to browsing for a longer period of time than more rapidly growing young trees at lower elevation. Some tree species (e.g. spruce, larch, birch, and rowan) are relatively tolerant to browsing. However, if heavily browsed, they also will produce distorted growth forms (reduced height growth, clipped surface, loss of terminal leaders, candelabrum-like growth, etc.).

Trampling does not only destroy seedlings but may also initiate soil erosion (Photo 95). Nutrient and moisture supply decrease after the removal of the organic layers (cf. Holtmeier et al., 2003, 2004; Holtmeier and Broll, 2005). On the other hand, erosion may also have a positive effect as wind-mediated tree seeds may reach a seedbed in exposed mineral soil more easily than in dense dwarf shrub or grass vegetation. Dispersal of tree seeds by ungulates within the treeline ecotone and beyond the treeline is not very effective, however.

Photo 95. Wind-scarps developed after elk (*Cervus canadensis*) and mule deer (*Odocoileus hemionus*) had destroyed the plant cover. Wind comes mainly from the west (*left*). Forest-alpine tundra ecotone on the east slope of the Front Range (Colorado) at about 3.450 m. F.-K. Holtmeier, 21 August 1979.

In many areas of the European Alps, too large populations of red deer resulted in over-aging of the forest and in a decline of newly established tree regeneration in the timberline ecotone and beyond the tree limit. In the Rocky Mountain National Park (Colorado), large numbers of elk (*Cervus canadensis*) are destroying tree seedlings and saplings as well as the willow shrub (*Salix glauca, Salix brachycarpa*), which is very common within and above the timberline ecotone. The decline of willows has fatal consequences to ptarmigan (*Lagopus leucurus*), for which willow buds are an essential food source (Holtmeier, 2002).

In the Scottish Highlands, red deer, in combination with sheep and human-caused fires, have prevented regeneration of Scots pine (*Pinus sylvestris*) until the 1950s (Watt and Jones, 1948). Since the red deer population declined and sheep grazing and fires decreased, Scots pine is recolonizing its former area (French et al., 1997). Apparently, the effects of red deer would have been less critical if not excessive pastoral use of the subalpine and alpine zone had negatively influenced the high-elevation forests and tree growth in the timberline ecotone already for centuries.

Maintenance of high-altitude forests, particularly of high-altitude afforestations above the present man-caused forest limit would benefit from an effective reduction of red deer density (Pfister et al., 1987). Natural young growth in the timberline ecotone is usually less affected by red deer (and also chamois) than are high-altitude afforestations (Schönenberger et al., 1990; Senn, 1999). Disturbances, e.g., by skiers, snow boarders and snow mobiles, must be reduced in order to avoid a progressive shrinking of red deer habitat, particularly of refuges. Fugitive red deer, when fleeing in deep snow or upslope, need up to 60 times the energy than at rest or when grazing (Holtmeier, 2002). Red deer must compensate for the loss of energy by food intake. Food, however, is rare in the winter. Consequently, damage in the remained (refuge) habitats is likely to considerably increase despite reduced red deer population (Reimoser et al., 1987; Reimoser, 1999).

Reindeer: Reindeer are semi-domestic animals. That means that they are neither purely wild animals nor domestic animals. Thus, their present impact on vegetation and timberline cannot be explained without considering the history of reindeer herding (Sections 4.3.14.1 and 5.4). Reindeer impact on the timberline ecotone is very complex. Reindeer directly affect tree seedlings and saplings as well as dwarf shrub-lichen heath by consuming living organic matter and by trampling. Reindeer feed on leaves of mountain birch and willows all summer preferring the tips of basal sprouts (see also Haukioja and Heino, 1974; Tenow, 1996). Reindeer affect mountain birch mainly at the seedling and sapling stage. Slow growth of seedlings and sprouts makes them available to reindeer for many years. Although mountain birch is very resistant against browsing and capable of partial compensatory growth, browsing for a couple of years, however, may destroy seedlings and also prevent regeneration from basal shoots (Kaitaniemi et al., 1995; Holtmeier, 2002; Neuvonen and Wielgolaski, 2005). Lehtonen (1987), in contrast, did not consider reindeer grazing to be a factor limiting shoot growth. Reindeer browsing of willows may reduce food availability for willow grouse (*Lagopus lagopus*; Den Herder et al., 2008), just as red deer limits forage to ptarmigan (*Lagopus leucurus*) in the timberline ecotone of the Rocky Mountains. In and above the birch-timberline ecotone in northernmost Finland, we found also about 60% of the pine saplings (*Pinus sylvestris*) damaged by reindeer. Usually, the terminal leaders were destroyed.

Moreover, reindeer initiate and intensify soil erosion thus influencing soil ecological conditions (e.g., Holtmeier, 1974, 2002, 2005a; Broll, 2000; Holtmeier et al., 2003, 2004; Anschlag, 2006; Holtmeier and Broll, 2006; Broll et al., 2007). Soil erosion is most spectacular on mountain tops and other windy places with substrates highly susceptible to deflation such as sandy till, eskers and sand dunes (cf. Photo 36). During summer reindeer usually move to late-lying snow patches and wind-exposed topography to escape molesting insects, particularly nose bot flies (*Cephenemyia trompe* L.) and warble flies (*Hypoderma tarandi* L) (Hagemoen and Reimers, 2004). Though consumption of lichens by reindeer is relatively low during summer, reindeer rapidly destroy dry lichen cover by trampling. Removal of the lichens may promote germination of birch seeds (Tømmervik et al., 2005). On the other hand, once the dwarf shrub-lichen cover has been destroyed and the organic layer has eroded, water-holding capacity, soil moisture and nutrient supply decrease. In the exposed upper mineral soil, temperature amplitudes increase (cf. Figures 20 and 22). High temperatures enhance moisture stress for potential birch seedlings. Extreme low temperatures may affect their root system (Holtmeier, 2003; Holtmeier et al., 2004; Anschlag, 2006).

Altogether, such conditions impede the establishment of birch seedlings on wind-swept eroded sites (Holtmeier et al., 2004; Broll et al., 2007) and will have a lasting effect (Sections 4.3.14.2 and 5.4). On more wind-protected sites, however, as well within the mountain birch forests open patches created by reindeer in the dense dwarf shrub-lichen cover may result in high birch seedling density because of reduced competition (Lehtonen, 1987; Sumoninen and Olofsson, 2000).

Ibex: Ibex (*Capra ibex*) and his subspecies are living in many high mountain areas of the northern hemisphere (Palaearctic region; Sedlag, 1995). However, serious damage caused by ibexes to the mountain forests and high-altitude afforestations up to the climatic treeline, on alpine vegetation and soils have been reported from limited areas in the European Alps only. The ibex reserve on the mountain slopes of 'Munt da la Bêscha' ('Sheep Mountain') and Piz Albris above Pontresina village (1.820 m; Photo 96) in the Upper Engadine probably is the most impressive example showing the complex interrelationships of the after-effects of grazing history (sheep) and the impact of a restored ibex population on the alpine and subalpine vegetation (Campell, 1958, Nievergelt, 1966; Bisaz, 1968; Holtmeier, 1968, 1969, 1987d, 2002). The present lower alpine zone had originally been covered with forests before the timberline became lowered by humans. Forest might recolonize this area in future, at least the lower parts.

The ibex habitat extends from the subalpine forest to the upper alpine zone (1850–2800 m). Ibex was re-introduced to this area in 1922 after it had been extirpated during the 17th century (for more details see Holtmeier, 2002). Prohibition of hunting, optimal habitat conditions and lack of any predator made the ibex population rapidly growing up to more than 700 individuals in the end of the 1960s already. Nowadays, the population is the largest in the European Alps (about 1.000 individuals). The great vertical range of the reserve, the relatively dry continental climate (precipitation at Pontresina just above 800 mm, winter snow cover about 6 months only), slope exposure to the southwest, large number of sunshine-hours, intense solar radiation, steep trough walls interspersed with bands of grass and often snow-free in the winter (exposure to the southwest), are the natural factors that favoured the growth of the ibex population.

Photo 96. View of the ibex reserve on the southwest-facing slope (*left side*) above Pontresina village. F.-K. Holtmeier, beginning of October 1973.

When the ibex population was growing larger, disturbances caused by ibex in both the subalpine forest and in the alpine zone increased rapidly to an intolerable extent. Large areas above the man-caused forest limit became afforested during the last century, in some places up to 2500 m, to reduce the danger of avalanches that are a permanent threat to the village of Pontresina located in the foot zone of the steep mountain slopes (inclination about 40°; cf. Photo 96). Ibexes frequent the afforestations mainly in the winter when the snow cover prevents access to ground vegetation. They damage many trees in the tree-line ecotone by browsing and rubbing their horns against the tree stems (Photo 97). Once damaged the trees become more vulnerable to the harsh high-altitude climate.

Photo 97. Swiss stone pine (*Pinus cembra*) severely damaged by ibex. High-altitude afforestations on the steep mountain slope above Pontresina at about 2.300 m. F.-K. Holtmeier, 5 October 1967.

Photo 98. Disintegration of the plant cover and soil erosion as a consequence of trampling by ibex, needle ice formation and subsequent deflation. South-west exposed slope of the Languard Valley above Pontresina village at about 2.500 m. F.-K. Holtmeier, 22 September 1985.

Moreover, ibex destroy the field layer by trampling thus enhancing soil erosion (Holtmeier, 1968, 1969, 1987d, 2002; Ten Houte de Lange, 1978).

Mainly on the southwest-facing slope, the mineral soil has become exposed over wide areas. In such places, frequent needle ice formation initiates downhill translocation and deflation of up-lifted fine particles. At the wind-exposed front of the remained vegetation patches wind-scarps develop. Overhanging sods break off and slide down the steep slopes (Photo 98).

Natural re-vegetation of the open patches is almost impossible because of high solar radiation loads (great altitude, southwest exposure), rapidly draining soil (lack of an organic layer) and needle ice that pushes seedlings out of the soil. The ibexes use the southwest exposure mainly in the winter, spring and autumn (Nievergelt, 1966; Holtmeier, 1987d, 2002). During the summer they move also to the north-facing slopes. On the sun-exposed slopes, however, daily freeze-thaw cycles and thus needle ice formation are most frequent and occur all year round with a maximum in spring and fall.

The severe soil erosion, however, cannot be attributed to the ibexes only. It was initiated by the large flocks of Bergamask sheep (Italian Alps) that had used the alpine pastures for centuries. However, since the beginning of the 20th century, the Italian shepherds were prohibited to graze their sheep in the Upper Engadine because of spreading mouth- and-foot disease. After sheep had already destroyed the plant cover and caused soil erosion, the ibexes carried on with these processes. During the 1950s already one tried to keep the ibexes away from the high-altitude afforestations by fence constructions, deterrents and also by dogs. However, the ibex population more than doubled since the late 1960s (about 700 individuals) and peaked in the 1990s with almost 1.700 (Holtmeier, 2002), despite reduction by shooting, decreasing reproduction (indicating natural control of population growth?) and natural accidents.

Ibex was re-introduced with the good intention to right the wrongs of the past and to restore the original environment. The rapid increase of the population was not foreseeable. Control, however, failed mainly because of the persistent opposition of the tourist managers and hotel owners who feared for the ibexes as an important tourist attraction. It was not before 1996 that in view of increasing soil erosion and serious damage to the plant cover the total number was reduced by culling and trapping to below 1000 ibexes (Holtmeier, 2002).

The magnitude of damage caused by the ibexes to the high-elevation forests and afforestations in the original timberline ecotone and resultant security risk for Pontresina village do not allow leaving the ibex population grow to the limit of the natural carrying capacity of the habitat. Effective control (shooting, trapping) combined with a profitable management (sale of the meet, sale of trapped individuals for re-introduction projects in other areas, etc.) will guarantee the existence of the ibex population at a tolerable level. The tolerable level can be defined as the ibex population size which would preclude serious damage to high-altitude afforestations and trees that have become established naturally above the old-growth forest. Also in other rapidly growing ibex reserves this aspect has to be considered.

4.3.13.2 Burrowing herbivorous mammals

Among the influences of burrowing animals in the timberline ecotone, the role of the northern pocket gopher (*Thomomys talpoides*; cf. Holtmeier, 1987b, 2002; Schütz, 1998, 2005) in the 'ribbon forest' on the Colorado Front Range (Section 4.3.12; Figure 77; Photos 93, 94) has been of particular interest to the present author.

Pocket gophers (Photo 99) are very common in the timberline ecotone and in the alpine zone at places where soil is deep enough for burrowing and an insulating snow cover accumulates in winter. Such conditions typically occur within 'ribbon forests', for example. Pocket gophers live within the 'snow glades' whereas they keep away from the 'ribbons'. In the 'snow glades', deep and late-lying snowpack not only provides shelter from predators but also from deep winter temperatures that would be lethal to the gophers. Moreover, soil moisture supply from the snowpack allows a lush herb-rich ground vegetation, which provides ample food to the pocket gophers and also to elk (*Cervus canadensis*) and mule deer (*Odocoileus hemionus*) (cf. Figure 77).

Photo 99. Northern pocket gopher (*Thomomys talpoides*) in front of its burrow. H.-U. Schütz, 9 August 1990.

Pocket gophers affect occasional tree seedlings and young growth in the 'snow glades' by girdling the stems and twigs buried under the deep snow-pack and by intense burrowing activities (Holtmeier, 1982, 1987b, 2002). Direct damage to seedlings by gnawing, for example, might be less important than damage caused by digging along the roots systems as was hypothesized by Schütz (2005). Below-ground damage, however, has not sufficiently been studied so far. Many seedlings become unearthed or covered with loose mineral material (mounds, casts, eskers) from below ground. Burrowing by gophers, in concert with trampling by elk and mule deer, promote soil erosion mainly by melt water runoff in the 'snow glades'. Wind erosion is less important because high surface roughness of the 'ribbon forest' reduces wind velocity. Moreover, the 'ribbons' themselves provide shelter from the strong winds Altogether, pocket gophers are one agent involved in a complex of factors keeping the 'snow glades' treeless thus contributing to the persistence of the spatial pattern of timberline 'ribbon forest' (Figures 76 and 77; see also Holtmeier, 1982, 2002).

On the other hand, loose mineral material brought to the surface by pocket gophers, ground squirrels, marmots and other burrowing mammals (e.g., Butler, 1995; Holtmeier, 2002) may locally facilitate tree seedling establishment (anemochoric tree species only) provided that sufficient soil moisture is available and other factors (e.g., too short growing season, snow fungus infection, frost action or trampling by cervids) do not impede the process. Hence, only in a few of these disturbed sites tree seedlings occur. In Glacier National Park, for example, Butler et al. (2004) found the number microsites caused by faunal pedoturbation (Butler, 1995) insufficient to explain the much higher number of tree seedlings that have become established in the alpine tundra.

4.3.13.3 Birds

Birds may influence the timberline ecotone by affecting the trees directly by consumption of seeds and buds and, what is more important, by dispersal of the seeds of some plant species, among them some subalpine tree species. There is no other bird, however, which is as important as are the nutcrackers (*Nucifraga caryocatactes*; *Nucifraga columbiana*). *Nucifraga caryocatactes*; Photos 100 and 101), several subspecies, races and varieties included, lives in boreal and mountain coniferous forests of Eurasia. A wide distribution gap, however, is apparent in the Lena drainage area (eastern Siberia), where both *Pinus sibirica* and *Pinus pumila* are missing (Figure 79). The Clark's nutcracker (*Nucifraga columbiana*) is native to the mountain conifer forest of western North America (Figure 80; Tomback and Linhart, 1990; Holtmeier, 1999c). The nutcrackers distribute the heavy wingless or almost wingless seeds of some subalpine pines such as *Pinus cembra*, *Pinus albicaulis*, and *Pinus flexilis* (Photo 102; Table 17).

The heavy stone pine nuts, being by far richer in nutrients than wind-borne seeds, are an important food for the nutcrackers and their nestlings, particularly in mid- to late winter when other energy-rich food is almost not

Photo 100. Thick-billed nutcracker (*Nucifraga carioca-tactes*) at a winter-feeding place in the Upper Engadine (Switzerland). F.-K. Holtmeier, March 1970.

Photo 101. Clark's nutcracker (*Nucifraga columbiana*) at a feeding place in Rocky Mountain National Park (Front Range, Colorado). F.-K. Holtmeier, September 1974.

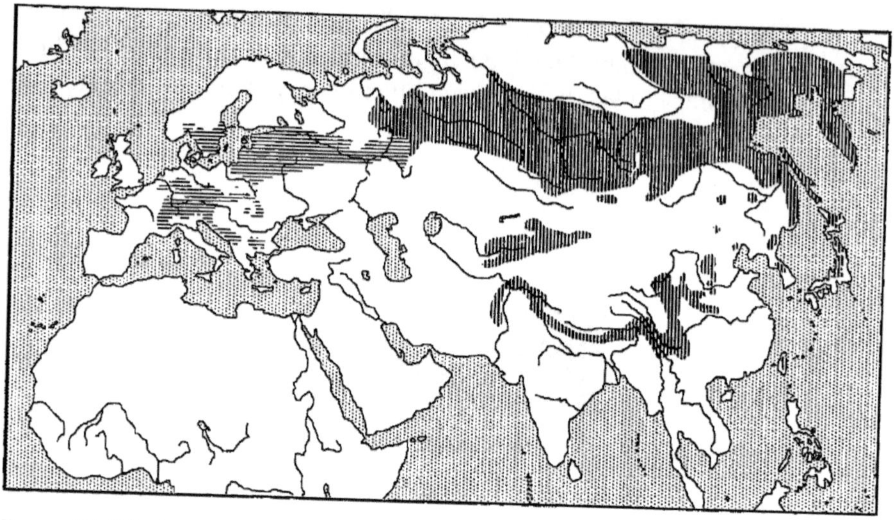

Figure 79. Distribution of the nutcracker in Eurasia (horizontal hatching = *Nucifraga c. caryocatactes*; vertical hatching = *Nucifraga c. macrorhynchos* and other subspecies). Modified from Mattes, in Glutz von Blotzheim and Bauer (1993).

available. The nestlings that already hatch in late winter depend completely on this diet (Campell, 1950, 1955; Oswald, 1956; Holtmeier, 1966; Tomback, 1977; Mattes, 1978, 1982; Hutchins and Lanner, 1982; Lanner, 1982, 1990; Schmidt and McCaughey, 1990). Furthermore, the seeds of whitebark pine are an important food source for granivorous mammals such as squirrels and also for grizzly (*Ursus arctos*) and black bears (*Ursus americanus*). The bears rely greatly on whitebark pine seeds for hibernation. The bears regularly plunder the middens of the red squirrel (*Tamasciurus hudsonicus*). More than 90% of the whitebark pine seeds consumed by grizzly bears come from the squirrel middens (Mattson et al., 1991; Mattson and Reinhart, 1994, 1997; Mattson et al., 2001). However, these mammals do not efficiently contribute to the expansion of whitebark pine within and beyond the timberline ecotone.

Nutcrackers establish food caches of stone pine nuts not only within the subalpine forest but also far beyond timberline and even above treeline. Usually, the birds hoard more seeds than they need for survival and rearing offspring until summer (Tomback, 1977; Vander Wall and Balda, 1977; Mattes, 1978, 1982; Tomback et al., 1993, 1994). Most seed caches above the forest remain unused and may give rise to seedling clusters (cf. Photos 56, 58, 84). Dispersal of the stone pine nuts beyond the closed forest and treeline depends on the nutcracker's activities whereas other seed-hoarding animals can be neglected. Distribution of stone pine seeds by the nutcrackers was very successful during evolution and thus a true mutualism has evolved between stone pines and nutcrackers (Mattes, 1978, 1982, 1985; Vander

Wall and Balda, 1981; Tomback, 1982, 1989; Linhart and Tomback, 1985; Tomback and Linhart, 1990).

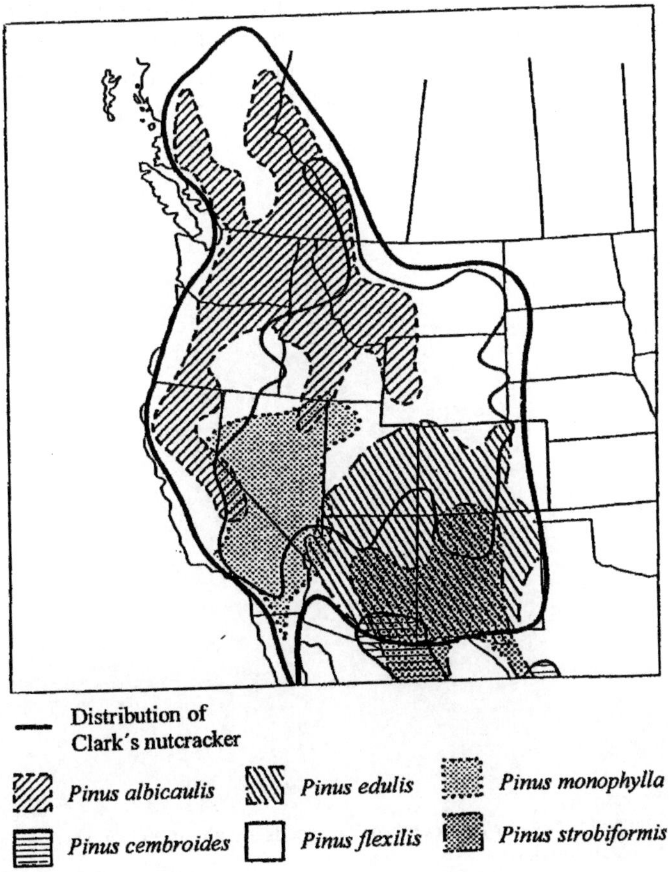

Figure 80. Distribution of the Clark's nutcracker (*Nucifraga columbiana*) and pine species producing windless seeds in western North America. Modified from Tomback and Linhart (1990).

The European nutcracker and the Clark's nutcracker both may cache several thousands of seeds per season (Kuznezov, 1959; Reijmers, 1959; Mezhennyi, 1964; Vander Wall and Balda, 1977; Mattes, 1978, 1982). Nutcrackers place food caches in the soil, in the organic layer, in rotten trunks and stumps, and in moss cushions at 2 to 4 cm depth. Seed caches in loose or soft ground usually contain more seeds than those established in compacted, hard substrate (Mattes, 1982, 1985).

Compared to wind-mediated seed dispersal, seed caching by nutcrackers is advantageous to the distribution of stone pines beyond the closed forests and the uppermost seed trees (Table 18). Seeds hoarded below the surface are in visible to seed predators and also to other nutcrackers, which might come across the seed caches accidentally, at the best. Moreover, the sub-surface caches usually provide sufficient moisture for germination, whereas too high temperatures and drought may impede germination of seeds lying on the surface, particularly at sun-exposed, wind-sheltered sites. Thus, high soil moisture in early summer will encourage germination of *Pinus pumila* seeds (Kajimoto et al., 1998). Precipitation in summer enhances also germination of *Pinus albicaulis* seeds (Tomback et al., 1993).

Photo 102. Cone and seeds of *Pinus cembra.* The seeds are about 90 heavier than larch seeds, for example. F.-K. Holtmeier.

Table 17. Subalpine pine species whose seeds are dispersed by nutcrackers (Modified after different sources, from Holtmeier, 1999c, 2002)

Pine species	Nutcracker species	Region
Pinus cembra	*Nucifraga caryocatactes*	Alps, Carpathian Mountains,
		Siberia, northern Mongolia
Pinus sibirica	*Nucifraga caryocatactes*	Siberia, northern Mongolia
Pinus pumila	*Nucifraga caryocatactes*	Northeastern Siberia, Korea, Kamchatka, Japan
Pinus koriaiensis	*Nucifraga caryocatactes*	Southeast Siberia, eastern Manchuria, Korea, Japan
Pinus albicaulis	*Nucifraga columbiana*	North America
Pinus flexilis	*Nucifraga columbiana*	North America

Table 18. Seed dispersal by nutcrackers or wind (Modified from Holtmeier, 2002)

Nutcrackers	Wind
Large energy-rich seeds	Small seeds, low energy content
Good seed quality (selective choice of the Seed-caching sites, seeds harvested mainly Below timberline)	Seed quality highly varying
Transport distance dependent on: > The nutcrackers' radius of action > Topographical structures, (convex, concave) > Plant cover and its structure (open, dense, low, high) > Attractiveness of seed-caching sites to the nutcrackers	Transport distance dependent on: > Height of the seed trees > Seed-wing size > Seed weight > Topography (convex, concave, gentle, steep) > Plant cover > Weather (dry, humid, calm, windy)
Selective choice of seed-caching sites	Landing of seeds accidental
Subterranean seed caches invisible for seed predators (also for other nutcrackers)	Seeds on the surface visible to seed predators
Seed-caching sites relatively favourable For germination and seedling growth	Because of accidental seed distribution prediction of potential conditions for germination and seedling growth at the seed landing place is impossible
Comparatively favourable moisture conditions for germination of seeds cached in the litter layer and topsoil	Seeds on the surface exposed to rapidly changing moisture conditions, high risk of desiccation

Additionally, seeds hoarded by the nutcrackers are usually of good quality because the birds collect the seeds mostly below timberline and also prove seed quality by 'bill-clicking' before filling their sublingual pouch (Mattes, 1978, 1982). After germination, the nutrient-rich stone pine nuts allow more rapid growth and establishment of seedlings (Tomback, 1978; Keane et al., 1990), which is advantageous to regeneration, particularly with respect to the short growing season at and above timberline.

Seedlings and young growth in the centre of a cluster are better protected from injurious climatic influences than those at the rim. About 10 years after seedlings became established, seedling clusters thin out gradually mainly because of competition for water, nutrients and light. Slow-growing individuals and in particular small seedlings that originated from the late-overlying cached seeds often fell victim to snow fungi (*Phacidium infestans*) as snow depth and length of the snow cover have increased due to the turbulences caused by the young growth cluster after projection above the snow (Holtmeier, 1974, 1990; see also Sections 4.3.7.1 and 4.3.11; Figure 62; Photo 85).

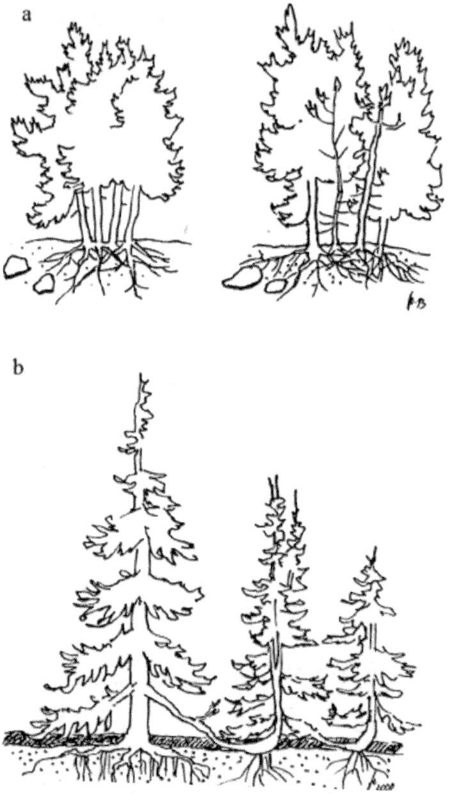

Figure 81. The influence of root competition.

a – Die-back of less competitive individuals in a *Pinus cembra* group that originated from a nutcracker's seed cache.

b – Clonal group acting as a nutrient-supply unit as long as the clonal members have not become independent from the mother tree.

Root competition (Figure 81) makes a substantial difference to clonal groups (spruce, fir, hemlock, and others; Section 4.3.10.2). As long as the clonal members have not become independent from the initial trees, clonal groups behave almost like one organism (nutrient supply unit; Kuoch and Amiet, 1970; Schönenberger, 1981). However, after several decades the root connections will decay and competition between the layers begin (Holtmeier, 1999c). On the other hand, root grafting was observed in tree groups that had originated from seed caches (Kuoch and Amiet, 1970; Holtmeier, 1986a, 1993a). We do not know, however, its effects on the further development of the individual trees. Obviously, root grafting occurs most frequently between tree individuals that are closely related to each other genetically (Tomback and Linhart, 1990; Schuster and Mitton, 1991).

Contrary to wind-mediated seed dispersal controlled by distance from the seed source, surface structure, wind velocity and direction and other physical factors (cf. Table 18), the distribution of stone pine nuts depends more on the nutcracker's home range and area of activity than on the distance from the seed trees. Nutcrackers carry stone pine seeds over horizontal distances of 15 or even more kilometres and easily cover 700 m in elevation (Sutter and Amann, 1953; Holtmeier, 1974; Mattes, 1978; Tomback, 1978). Thus, the whole timberline ecotone and also the alpine zone are within the nutcracker's range, and in case of improving climate a rapid advance of stone pines, following the favourable sites, can be expected.

While distribution of wind-borne seeds is usually rather accidental, nutcrackers carefully select the sites for placing seed caches. Obviously, they prefer convex topography not too long covered with snow. Small ridges, spurs, knolls, and rocky outcrops seem to be especially attractive to them, as is evidenced by direct observations of seed-caching nutcrackers and by great numbers of seedlings and young growth at such sites (Photo 103). Nevertheless, this distribution pattern might also be attributed to the site conditions on convex topography that are more favourable to seedling establishment compared to gullies and other snow-rich sites. On more gentle or almost flat topography lacking prominent convex sites, as is typical over wide areas in the timberline ecotone on the Colorado Front Range for example, seed dispersal appears to be more irregular compared to intensely sculptured terrain. Anyway, seedlings occur mainly in wind-exposed areas of the timberline ecotone, which are lacking snow in winter or are only occasionally snow-covered (Photo 104; Holtmeier, 1978, 1993a, 1996). Under the given climatic conditions, these sites are not very favourable to seedling establishment and seedling growth. If nutcrackers hoarded cached seeds, more seedlings and young growth could be expected as continuous snow cover in winter and high soil moisture provide conditions more suitable for germination and growth than exist on wind-exposed topography. In the broad timberline ecotone on Beartooth Plateau (Wyoming, Montana), for example, Mellmann-Brown (2002) found high concentrations of whitebark pine seedlings and seedling clusters leeward of tree groups and also in shallow depressions where snow usually lingers until midsummer. Obviously, the nutcracker has cached seeds at these sites which are comparatively favourable for regeneration under the given environmental conditions.

Although the nutcrackers have been intensively studied (Holtmeier, 1999c, 2002, further references there), the reasons for the specific site selection are still obscure. Undoubtedly, the nutcracker would save energy looking for seeds cached at sites with little or no snow. On intensely sculptured terrain, convex sites probably act as landmarks that help the nutcrackers retrieve their seed caches (Mattes, 1978, 1982). Also, nutcrackers might prefer prominent sites because they would allow surveying the surroundings more easily and thus would reduce the risk of falling victim of a predator (e.g., goshawk, great horned owl, peregrine falcon, fox; Holtmeier, 1974). This, however, could not be a plausible reason for site selection on flat topography. After all, shallow snow cover and early melt-out occur to make sites particularly attractive to seed-caching nutcrackers.

Photo 103. Swiss stone pines (*Pinus cembra*) that originated from a thick-billed nut-cracker's seed caches (at about 2.270 m) on the northwest-facing slope of the Upper Engadine main valley (Switzerland). F.-K. Holtmeier, 23 September 1968.

Photo 104. Limber pines (*Pinus flexilis*) that originated from a Clark's nutcracker's seed caches (at about 3.370 m) in the wind-swept forest-alpine tundra ecotone on Bald Mountain (Great Basin National Park, Nevada). F.-K. Holtmeier, 30 July 1994.

Altogether, seed dispersal by the nutcrackers contributes to an effective regeneration and thus to maintenance of timberline forests. I also encourages forest advance to higher elevation at improving climatic conditions. Clusters of young stone pines within or above the timberline ecotone may facilitate the establishment of other conifers such as spruce or fir in their immediate lee by providing shelter from the wind. At timberline east of the Continental Divide in Glacier National Park (Montana), for example, whitebark pine (*Pinus albicaulis*) apparently plays an important role in formation of tree groups (Resler and Tomback, 2008). A similar development of subalpine tree groups has been reported from the Cascade Mountains (Washington; Franklin and Dyrness, 1973). However, there and also in other areas of the Rocky Mountains as in the Colorado Front Range, for example, natural succession may run exactly the opposite direction as compact tree islands of spruce and fir that have persisted by layering for centuries facilitate recent establishment of pines at the margin of the groups (cf. Figure 67).

In contrast to synzoochoric distribution, endozoochoric seed dispersal (e.g., juniper, rowan) depends on where the animals defecate and this is more incidental. All rowans at treeline in the Alps and in the Scandinavian mountains (e.g. Holtmeier, 1965; Kullman, 1986b), for example, have originated from droppings that contained seeds usually from lower elevation. Occurrences of *Sorbus microphylla* far above (4,300 m) the uppermost birch forests in the central Himalayas result from endozoochoric seed dispersal, presumedly by birds. Regrettably, we lack precise information on the bird species involved. Also, juniper trees (*Juniperus indica*, *Juniperus turkestanica*) that are growing on rocky and other exposed sites on mountain slopes in the dry regions of the Himalayas and southern Tibet (Miehe and Miehe, 2000) probably originated from endozoochoric seed dispersal. Very likely seeds fell into rock crevices when crows (*Corvus* sp.) or jackdaws (*Pyrrhocorax* sp.) were feeding on ripe juniper berries (Schickhoff, 2005). Higher soil moisture resulting from accumulation of fine soil and reduced evaporation in the crevices (Schickhoff, 1993) may have facilitated juniper establishment in such places.

Some bird species such as black grouse (*Lyrurus tetrix*) and ptarmigan (*Lagopus mutus*, *Lagopus lagopus*, *Lagopus leucurus*) cause direct damage to the trees in the timberline ecotone, by clipping buds and fresh terminal shoots. However, clipping does usually not kill the trees but rather shapes the physiognomy mainly of young trees (e.g., forked or dwarfed growth forms). Such damage occurs only in some places and is by far less important than the effects of red deer, for example. In high-altitude afforestations, however, the damage caused to trees may matter a lot in the long-term (Schönenberger et al., 1990).

4.3.13.4 Defoliating insects (*Epirrita autumnata*, *Operophtera brumata*)

One of the probably most spectacular effects of leaf-consuming insects is defoliation of mountain birch (*Betula tortuosa*) in northernmost Europe. In northern Norway 16 mass-outbreaks of the autumnal moth (*Epirrita autumnata)* are reported since 1852 (Tenow et al., 2005). Also, in northernmost Finland (Utsjoki) mass-outbreaks repeatedly occurred (1844, 1905–1909, 1927, 1957, 1965–1966; Kalliola, 1941; Nuorteva, 1963; Kallio and Lehtonen, 1973; Holtmeier, 1974). A recent mass-outbreak occurred in 2004–2007. Birch forests on the mountain slopes up to the treeline became heavily damaged, except for the forests on lower slopes and on the valley floors where cold air accumulates and low temperatures kill the eggs of the autumnal moth. Another exception is the Finnmarksvidda (a highland plain in northern Norway) where the eggs of *Epirrita autumnata* usually do not survive the very low winter temperatures (Tenow et al., 2005). The rough bark and flakes as well as the lichens on the stems of older birches seem to be particularly suitable for egg deposition and survival as they provide protection from deep freezing temperatures and predators (Tammaru et al., 1995). In any case, more eggs can be found on older than on neighboured young trees (Bylund, 1997).

In northern Finnish Lapland, complete defoliation of the mountain birch forests during a mass outbreak of the autumnal moth during the 1960s combined with very cool summers destroyed extensive birch forests on the mountain slopes, partly up to timberline (Photo 105; see also Photo 86 (Kallio and Lehtonen, 1973, 1975; Holtmeier, 1974, 1999c, 2002; Heikkinen and Kalliola, 1989; Holtmeier, 1999c, 2002, further references there). About 5.000 km^2 of birch forest became defoliated (Nikula, 1992). In northern Utsjoki, the northern-most community of Finnish Lapland, 1.300 km^2 were affected. Cold weather during the growing season, lowest temperature sums (days >5°C, Mikola, 1971; Kärenlampi, 1972), prevented recovery of the birches and made them more susceptible to climatic stress and secondary parasites. About 50% of the birch trees died. Timberline stands were worst affected. (Kalliola, 1941; Nuorteva, 1963; Kallio and Lehtonen, 1973). Earlier mass outbreaks had the same effects (Figure 82; see also Kalliola, 1941; Palm, 1959; Nuorteva, 1963; Kallio and Lehtonen, 1973; Holtmeier, 1974). On the other hand, many mountain birch forests may recover from defoliation by reproducing from the root stock. Thus, short-term and long-term changes, system-internal and external changes are overlapping in a complex and often in-scrutable way. For example, birches which have survived mass outbreaks of the autumnal moth, normally show reduced radial growth for a couple of years after defoliation, as is often but not regularly reflected in the growth ring pattern. Reduced competition between

Photo 105. Mountain birch forest (*Betula tortuosa*) that was destroyed by a mass-outbreak of *Epirrita autumnata* during the 1960s. After dieback many birches have started to produce basal sprouts. Northwest slope of Jesnalvaara (northern Finnish Lapland) at about 300 m. F.-K. Holtmeier, 26 July 1994.

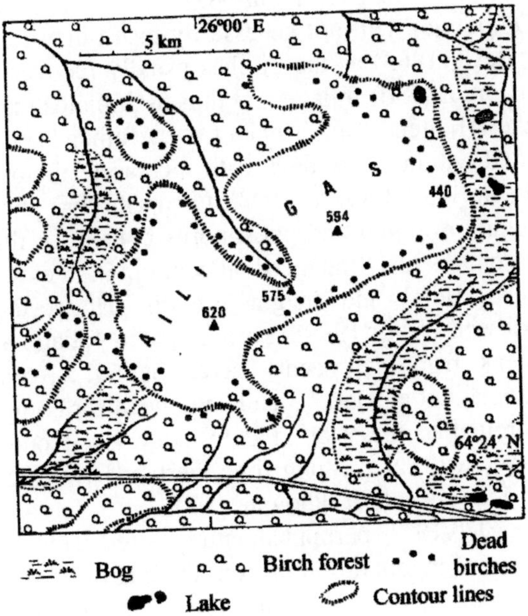

Figure 82. Depression of the upper birch forest limit (*Betula tortuosa*) following the defoliation by *Epirrita autumnata* on Ailigas (east of Karigasniemi, northern Finnish Lapland) in 1927. Modified from Nuorteva (1963).

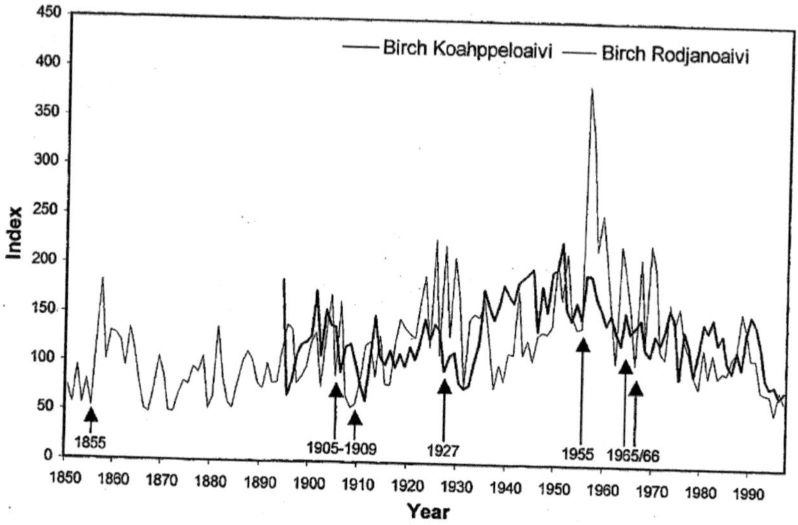

Figure 83. Annual growth increment of birch (*Betula tortuosa*) in the forest-alpine tundra ecotone on Koahppeloaivi and Rodjanoaivi (northern Finnish Lapland) and mass-outbreaks of *Epirrita autumnata*. From Holtmeier et al. (2003).

the remaining birch trees and also more favourable growing seasons following the outbreaks may compensate or even over-compensate for the negative effects of defoliation on radial growth thus disguising what really happened to the trees in the past (Figure 83).

Defoliation will not necessarily result in a depression of birch timberline (Nuorteva, 1963) as the birches usually recover from such impact (Section 4.3.10.2). Loss of foliage and thus assimilates may negatively affect nutrient uptake (Hjälten et al., 1993). On the other hand, the remaining leaves can respond within a few days with increased photosynthesis (Hoogesteger and Karlsson, 1992; Ovaska, 1993). Compensatory photosynthesis is accompanied by higher nitrogen and phosphorus content in the leaves (Hoogesteger and Karlsson, 1992). Enhanced allocation in the foliage may result again in reduced radial and height growth (Hoogesteger and Karlsson, 1992; Henriksson et al., 1999). Loss of green foliage also excludes resorption of nutrients from senescent leaves in fall (Nordell and Karlsson, 1995; Ruohomäki et al., 1997). Thus, the need for compensatory photosynthesis is not limited to the year of defoliation (Neuvonen et al., 2001). Recovery, however, will take a long time, particularly if delayed by grazing reindeer, voles and root rot (cf. Photo 62, 63).

In 1970/1971 permanent study sites were established for monitoring recovery of mountain birch from the mass-outbreak during the 1960s (Kallio and Lehtonen, 1973). Twenty-five years after defoliation, a fivefold number of viable seedlings had become established within the fenced sites (Helle and

Kajala, 1992; Lehtonen and Heikkinen, 1995; Helle, 2001). Obviously, reindeer did not significantly affect recovery from stump sprouts (see also Neuvonen et al., 2001), which is different from the present author's observations. Anyway, the mountain birch forests that fell victim to the mass-out-break in the 1960s have not been replaced by open tundra as was supposed in those days (Kallio and Lehtonen, 1973, 1975; Lehtonen and Yli-Rekola, 1979). Due to defoliation and decline of the timberline birch stands graminoids (e.g., *Deschampsia flexuosa* and *Festuca ovina*) and herbs may spread and temporarily improve grazing conditions for reindeer.

Re-establishment of birch may be extremely difficult on over-grazed and subsequently eroded topography, however (Holtmeier et al., 2003; Anschlag, 2006; Broll et al., 2007; Anschlag et al., 2008). Thus, in some fjeld areas birch timberline declined due to earlier mass-outbreaks of *Epirrita autumnata*. Very likely, the combined effects of defoliation and overgrazing by reindeer have impeded birch forest to comeback into these areas. Obviously, this also happened earlier during the Holocene (Holtmeier and Broll, 2006).

In northernmost Finnish Lapland and northern Norway, the winter moth (*Operophtera brumata*) is recently becoming an additional factor that seriously affects mountain birch. In Nuorgam-Pulmankijärvi area (Finnish Lapland), a mass outbreak of the winter moth started in 2004 and continued in 2007. It was preceded by an outbreak of the autumnal moth (*Epirrita autumnata*). The birches which had survived 1 to 2 years of defoliation by the autumnal moth seem to be dying now after several years of subsequent defoliation. Recovery from defoliation by second leafing was extremely low in 2008. Also in other areas as on Koahppeloaivi, for example, a mountain area west of Utsjoki village, treeline birch stands have been heavily damaged by the winter moth. Birch treeline is likely to decline on a horizontal distance of about 2 km. Tenow (1996) reported similar damage to mountain birch forests from the Torneträsk area (Swedish Lapland). It may be speculated whether this is already an effect of climate change or of a 'normal' gradual expansion of this species (see also Tenow et al., 2005). Transport of the larvae by strong winds is the most effective way of expansion of the winter moth as the females, having shortened wings, cannot fly. Climate warming may benefit egg survival because the eggs of the winter moth are less frost-tolerant than those of the autumnal moth and will not survive long-lasting frost periods with extremely low temperatures (<–33°C). However, other factors such as natural enemies and diseases, climate and weather variability, possible competition between the moth species, and temperature-controlled birch phenology, may impede population to grow to outbreak densities (cf. Bylund, 1999).

4.3.14 Anthropogenic impact on timberline

In many mountain regions, humans have, through history, played a role in lowering the upper timberline. Most of the mountains of Europe and Asia were already settled in prehistoric time (e.g., Geiger, 1901; Hess, 1923; Grabherr, 1934; Lampadius, 1937; Regel, 1938; Campell, 1944; Gams, 1951; Holtmeier, 1965, 1974, 1986b, 1989; Stern, 1966; Zoller, 1967a, b; Köstler and Mayer, 1970; Mayer, 1970; Kral, 1971, 1973; Bortenschlager, 1984; Messerli and Ives, 1984; Bortenschlager et al., 1992; Klimek, 1992, Ammann and Wick, 1993; Hüppe and Pott, 1993; Messerli and Winiger, 1994; Kosak et al., 1995; Aas and Faarlund, 1996; Tinner et al., 1996; Winkler, 1994, 1997). Thus, their situation is quite different from that of the high mountains of North America or New Zealand, where the natural environment has been comparatively little influenced by native people prior to the arrival of the Europeans 100 to 200 years ago.

In most cases, the depression of the upper timberline was caused not only by human impacts but also by the general deterioration of the climate after the postglacial optimum. In the Alps, for example, the effects of this general cooling have been compounded by the anthropogenic influences, at least since the Middle Ages (Holtmeier, 1986a, further references there). Human impact on high-elevation forests, in particular on the distribution, composition and structure of the present forest stands and on the position of the timberline, has been considerably underestimated if compared to the effects of climate (see also Chapter 5).

4.3.14.1 Lowering the timberline

In the European Alps, in the Carpathian Mountains and many other high mountains of Eurasia and in most tropical high mountains, almost no untouched nature was left (see also Miehe and Miehe, 2000; Bader, 2007). The kind and intensity of human impact were also different. Before Man started his continued influence on the mountain ecosystems, a more or less wide continuous forest covered most of the mountain slopes up to the climatic timberline, except for steep rocky valley sides, avalanche tracks, and similarly endangered locations where the forests occasionally happened to be destroyed (e.g., Holtmeier, 1987c). Due to the severe high-mountain climate and rugged topography, Man could use the mountains up to a certain altitude; there his influence was concentrated mainly in the forest belt. The extent of the timberline depression varied regionally. However, a depression of 150 to 300 m below the uppermost postglacial level of the climatic timberline can be accepted as an average value (e.g., Holtmeier, 1974, 1986b, 1994a; Burga, 1988; Tinner et al., 1996; Carcaillet et al., 1998; Thinon and Talon, 1998; Burga and Perret, 2001; Kaltenrieder et al., 2005).

According to Carnelli et al. (2004) maximum position of the tree limit was never higher than 350 m above the present potential treeline.

In many areas of the European Alps, the anthropogenic lowering of timberline by clearing the high-elevation forests, burning, and pastoral use began about 4500 years ago (e.g., Tinner and Theurillat, 2003). In the Engadine (Switzerland), for example, humans were influencing the mountain forests during the Bronze Age already (Conrad, 1940). The most extensive clearance of the high-elevation forests happened during the Middle Ages (e.g., Mayer, 1970; Damm, 1998). Emigrants from the Valais (Switzerland) who settled in the Avers (part of the upper drainage area of the Rhine river), in the interior parts of Vorarlberg (Austria), in the Paznaun Valley (Austria) and in western Tyrol caused extensive forest decline in these areas (cf. Photo 106; Krebs, 1961; Leidlmair, 1983). The Valaisians usually settled the upper section of high-elevation valleys close to timberline and removed the forest from all slopes suitable for agricultural or pastoral use. In particular, southern exposures became completely deforested. Alpine pasturing consumed huge masses of wood that was cut at close vicinity to the settlements, if possible. The need for firewood (heating, cheese production) surpassed the need for construction wood (houses, fences, furniture, artefacts) by far (Hess, 1923). In a report to the Swiss government the Swiss forester Landolt (1862) already emphasized the desolate condition of the high-elevation forests and blamed the mountain people for over-using the forests and neglecting any sustainable forest management.

Local climate and topography (accessibility) have controlled human influence on the mountain forests. Moderate slopes characterized by smooth topography and easily accessible were often completely deforested (Photo 106). On the relatively gentle southern slope of Mount Olympus (Greece), for example, forest was removed for the most part due to grazing, while on the steep and often inaccessibly eastern and northern slopes the forest and timberline were not significantly affected by humans (communication R. Brandes). In many temperate mountains, however, forest was removed from comparatively level slope sections suitable to grazing, even on northern exposures. In trough valleys, the upper timberline became located at the upper rim of the trough walls that were too steep to graze cattle. Due to agricultural use on the footslopes and grazing on the trough shoulders and similar almost level topography, the forest was restricted to the trough walls where it still forms a narrow band more or less dissected by avalanche tracks and rills (Photo 107).

A similar destruction of forest remains can be observed in the northern Norwegian fjordland. However, situation is somewhat different from the Alpine trough valleys in so far as the Norwegian farmers used the comparatively

Photo 106. On this relatively gentle south-facing slope above Cresta (1.962 m) in the Avers Valley (Switzerland) the forest was totally removed. Since that time the slope has been used for grazing and hay making. F.-K. Holtmeier, 28 September 1969.

Photo 107. On this southeast-facing slope above Samedan (1.795 m, Upper Engadine, Switzerland), forest was removed from all gentle topography to create alpine pastures (trough shoulders) and agricultural field (lower slope area). F.-K. Holtmeier, 9 October 1969.

gentle footslopes for agriculture and sheep grazing, while Laps from Norway, Sweden and also Finland grazed reindeer on the trough shoulders and similar gentle topography above the mountain birch forests during summer. The Laps cleared the uppermost forest belt (mainly firewood, fences) and reduced the timberline to the steep trough walls (Photo 108; Holtmeier, 1974). Obviously, the same holds true for the Fosen peninsula 300 km farther south (Lindemann, 1972).

In many regions of the European Alps, for example, in the High Tauern or on many mountain passes in eastern Switzerland (e.g., Bernina Pass, Ofen Pass, Julier Pass) extensive forests fell victim to medieval ore mining (Photo 109) Gaggi, 1990. Moreover, salt works consumed vast amounts of firewood (Grabherr, 1934, 1949; Boesch, 1936; Stoffel, 1938; Campell, 1944, 1949; Holtmeier, 1974, 1989, 1994a; Parolini, 1995). In the Avers (Grisons, Switzerland), many ore smelters were shut down because all forests had been cleared on the surrounding mountain slopes. The famous salt works of the town of Hall near Innsbruck (Austria) could only be kept running by firewood from far distant sources (e.g., from the Lower Engadine) after the local forests had been exploited (Calörtscher, 1969; see also Stern, 1966; Mayer, 1970). Also extraction of resin from the trees caused serious damage to the mountain forests. Not last, extensive high-elevation forests fell victim to pyromania (Hager, 1909; Grabherr, 1949; Holtmeier, 1974, and others). On many valley sides, the upper timberline became lower for about 150 to 300 m. Remnants of the former forests only survived on steep and almost inaccessible slopes. In the High Tatra Mountains, grazing and use of *Pinus cembra* and *Larix decidua* resulted in a depression of timberline for 120 to 150 m during the last 400 to 500 years (Rybnickova and Rybnicek, 1993).

In the Scottish Uplands, forest almost completely disappeared due to lumbering, grazing and in particular because of extremely high populations of red deer that considerably exceeded the carrying capacity of the plant cover (Watt and Jones, 1948). The present timberline is located at about 500 m (Pears, 1968). In the Cairngorm Mountains (north-eastern part of the Grampian Mountains) some pine stands (*Pinus sylvestris*) occur at 650 to 700 m, remnants of the previous continuous high-elevation conifer forest (Fielding and Haworth, 1999).

Grazing for hundreds of years might have caused also the treelessness of relatively level mountaintops in the upland of Middle Europe as for example on the Feldberg (Black Forest, Germany; Photo 110) or on the Vosges (France; Photo 111) (Aichinger, 1937; Liehl, 1958; Carbiener, 1969; Wilmanns, 1973; Wilmanns and Müller, 1977; Oberdorfer, 1982). Combined effects of ground fires, grazing of domestic herbivores, limited seed supply, competition of meadow vegetation preventing seedling establishment, moisture conditions, late-lying snow and other factors are the cause of comparatively persistent grassy meadows on montane ridges and peaks, such as in the Oregon Coast

Photo 108. Northeast-facing slope of Tromsdal (northern Norway). On gentle topography the mountain birch forest (*Betula tortuosa*) was destroyed by sheep and reindeer grazing. F.-K. Holtmeier, 4 July 1970.

Photo 109. Munt la Schera (Ofen Pass, Lower Engadine, Switzerland) at 2.200 m. The former forest fell victim to mining during the Middle Ages. Subsequently, the karstic substrate became widely exposed by erosion. Solitary prostrate mountain pines (*Pinus mugo*) are invading the former forest area. F.-K. Holtmeier, 20 September 1968.

Range (Magee and Antos, 1992), in the Medicine Bow Mountains of Wyoming, in the mountain ranges of Nevada or in the Appalachians (e.g., Billings and Mark, 1957; Gersmehl, 1973). The present author has observed such 'balds' in some high-mountain areas of Colorado west of the continental divide. Treeline bordering the 'balds' is usually abrupt and located below its true climatic limit that could be expected at the geographic latitudinal position. Therefore, Gersmehl (1973) called it 'pseudo-timberline'.

Human impact combined with mass-outbreaks of the autumnal moth (*Epirrita autumnata*) and over-grazing by reindeer influenced the birch forest and lowered the upper limit of the closed forests on the more gentle fjelds in the interior parts of northernmost Europe.

Throughout the Holocene timberlines in the Mediterranean high mountains have been strongly influenced by pastoral use and fire. However, the role of natural factors (e.g., temperature, moisture, snow, competition) in shaping their present location and patterns has not been sufficiently considered (e.g., Schreiber, 1998; Brandes, 2007). Furthermore, in comparison with reports in the older literature, regional variations are greater than might be expected (for references see Brandes, 2007).

In New Zealand, the Maori had already destroyed much of the mountain forests within a few centuries. It is after the arrival of the Europeans, however, that humans affected most heavily many high-elevation forests and

Photo 110. Timberline ecotone on Feldberg (Black Forest, Germany). Solitary spruces and spruce groups (*Picea abies*, many of them clonal groups) are growing in the summit area (about 1.400 m), which has been grazed for centuries. F.-K. Holtmeier, 29 September 1987.

Photo 111. Anthropogenic upper limit of beech forest (*Fagus sylvatica*) on the Hoheneck (Vosges, France) at about 1.300 m. R. Pott, Fall 1992.

timberline (e.g. Schweinfurth, 1966; Costin, 1967; Wardle, 1991). The use of fire and intensive grazing high-elevation areas with sheep and, not least, rapidly increasing populations of introduced cervids and other wild ungulates caused the most serious effects. New Zealand had no native mammalian herbivores before the Europeans introduced them during the late 1800s (e.g., Wodzicki, 1963; Kitching, 1986; King, 1990; Holtmeier, 1999c, 2002, further references there) among them fallow deer (*Dama dama*), European red deer (*Cervus elaphus scoticus)*, North American elk (Wapiti, *Cervus canadensis nelsoni*), Sikka deer (*Cervus nippon*), Himalayan Thar (*Hemitragus jemlahicus*) and chamois (*Rupicapra rupicapra*). Thus, we consider these browsers and grazers as a 'human impact'.

These animals could vigorously reproduce and spread at the given natural environmental conditions (not too strong winters, ample food supply, no native predators) and because hunting was not very effective on the rugged mountain terrain. Consequently, these herbivores have left their marks in and above the mountain forests (Section 4.3.14.2). Some remote ranges, however, remained relatively undisturbed.

European read deer was the most successful of all introduced cervids. To escape increasing hunting pressure red deer moved from the coastal lowland to the less accessible upper zone of the mountain forests and alpine tussock grassland. The undisturbed, dark and wet mountain forest itself was unsuitable for red deer (Schweinfurth, 1966). The high-elevation forest stands, however, provided shelter from the frequently severe weather conditions.

Red deer impeded tree regeneration and thereby increased the risk of over-aging and decay of high-elevation *Nothofagus* forests (Wardle, 1984, 2007).

However, red deer alone would not have seriously threatened the existence of the mountain forests. More probably, it was the red deer and brush-tailed opossum (*Trichosysrus vulpecula*) in combination, that lead to increased tree mortality and forest decline (Schweinfurth, 1966). The 'possum', as this animal is commonly known to New Zealanders, has already been introduced by the mid 1850s. Feeding mainly on the leaves of Mahoe (*Melyctus ramiflorus*), Rata (*Metrosideros umbellata*), Kamahi (*Weinmannia racemosa*) and other tree species, the possum opened the dense forest canopy. More light and higher temperatures on the forest floor stimulated the growth of herbs and shrubs in the forest understory. As food became more readily available red deer moved from the alpine tussock grassland (*Chionochloa pallescens*) into the open forest stands where they impeded tree regeneration by browsing and trampling. No longer protected by canopy interception the steep slopes eroded rapidly during heavy rains. The possum benefited from the now warmer conditions and better food supply in the thinning forest. In addition red deer caused soil erosion in the tussock grassland above the timberline by trampling and wallowing. One can safely assume that the effects of introduced cervids would not have been that detrimental to mountain vegetation and mountain sites if they were not at least locally exacerbated by intense sheep grazing and also careless use of fire by the sheep owners (Holtmeier, 1999c, 2002; further references there).

In the Rocky Mountains, sheep and cattle grazing as well as ore mining locally affected high altitude forests. In Colorado, for example, it was particularly during the 19th and early 20th that mining lead to total removal of the timberline forests in the vicinity of mines (Aldrich, 1990; Photo 112). However, also prehistoric hunters and gatherers were attracted by the broad timberline ecotones which provided shelter, fuel as well as plant and animal nourishment (Benedict, 1981, 1985). The primary attraction of the alpine region and the often broad timberline ecotone were not their plants but their large herbivores, in particular elk (*Cervus elaphus nelsoni*) and bighorn sheep (*Ovis canadensis*) (Benedict, 1992). Probably, the Indians had already affected the timberline forests before the arrival of the Europeans, though to a lesser extent. The role of human-caused fires should not be underestimated. We found charcoal almost regularly when digging soil pits in the timberline ecotone (Holtmeier and Broll, 1992; Schütz, 1998; see also Benedict, 2002). The oldest artifacts from sites above timberline date back more than 8.000 radiocarbon years (Benedict, 1974, 1981). Fires still frequently occur in the high-elevation forests up to timberline. The number of fires, however, has considerably decreased due to fire control, when compared to pioneer times.

Photo 112. Old mining area on Mosquito Pass (Colorado). The willow shrub has replaced the previous conifer forest that became a victim of the mining activities. F.-K. Holtmeier, 6 August 1989.

In many mountain regions of Central Asia, settled since early history, timberline declined due to human impact. In High Asia, undisturbed timberlines are unlikely. (e.g., Miehe and Miehe, 2000). In eastern Tibet, for example, the treelessness of sun-exposed slopes (first reported by Schäfer, 1938; Von Wissmann, 1960, 1961) results from extensive pastoral use throughout thousands of years and burning rather than from the direct influence of the dry continental climate (e.g., Winkler, 1994, 1997; Miehe et al., 1998) as was still assumed by Von Wissmann (1960, 1961). However, the climate plays its role as southern exposures are normally snow-free in winter which makes them suitable for grazing during the cold season. Nomads graze their herds mainly in the alpine belt but occasionally also in high-elevation forests (communication A. Bräuning). Altogether, in this region human impact on mountain vegetation and particularly on the timberline forests is by far more extensive than is generally supposed.

In the tropical high mountains, altitudinal timberlines that were not affected already by native people do very likely not exist, except for inaccessible steep topography. Thus, the upper timberlines in Africa were lowered by recurrent fires for several hundred metres. The groves occurring above the present forest limit are considered relics of former forests, which survived at sites protected from fires (Wesche et al., 2000; Wesche, 2002).

Meanwhile it has become apparent that abrupt timberlines, which Troll (e.g., 1973) considered being the true climatic timberline on tropical mountains, must be largely attributed to recurrent burning, particularly on slopes with smooth microtopography and almost even soil moisture conditions (Miehe and Miehe, 1996). Klötzli (1975) even held the view that natural timberlines in East Africa are never abrupt but rather occur as more or less wide ecotones characterized by varying mosaic of grassland alternating with forest stands. These mosaics are caused by many factors such as fire, extreme climatic conditions, locally varying soil moisture conditions, and animals.

At timberline on the comparatively humid high mountains of New Guinea, natural fires rarely occur. Instead, humans caused most fires (Paijmans and Löffler, 1972). Fire was used for hunting (e.g., Flenley, 1984), warming, signalling, easing travel through the tussock grassland and warfare. Most fires seem to have spread from easily walkable ridges into the high-elevation grassland (Smith, 1980). As was the same in temperate high mountains, the 'alpine' vegetation could extend into the former forest belt (e.g., Hamilton and Perrott, 1981; Corlett, 1984).

The Puna tussock grassland, which has been intensely grazed for thousands of years, must very likely be considered an anthropogenic-zoogenic substitute formation (Ellenberg, 1958, 1959, 1966, 1975, 1979, 1996; see also Figure 5).

Photo 113. Polylepis incana-stands on the west slope of the Eastern Cordillera of Quito (Ecuador) between 3.900 and 4.000 m. Very likely, the present distribution pattern must be attributed to anthropogenic influence, in particular to man-induced fires. M. D. Rafiqpoor, 31 April 1994.

Thus, the solitary stands of *Polylepis* high above the present upper forest limit in the Andes are supposed to be remains of a former continuous forest belt rather than highly tolerant 'outposts' that established themselves in climatically extreme sites (e.g., Sturm, 1998) as been assumed by many authors (cf. Section 4.1.3).The mosaic of solitary *Polylepis* groves alternating with open grassland results mainly from recurrent burning. The *Polylepis* stands may have survived because growing at relatively 'fireproof' sites (Photo 113) and/or access is difficult to people collecting wood (firewood, charcoal production, lime burning; Ruthsatz, 1983). Although the views are still different (e.g., Salgado-Labouriau, 1984; Goldstein et al., 1994), the forest-relic hypothesis has been supported by many studies (e.g., Hueck, 1966; Jordan, 1983; Ruthsatz, 1983; Fjeldså, 1992; Hensen, 1993, 1995; Laegaard, 1992; Kessler and Driesch, 1994; Kessler, 1995; Fjeldså and Kessler, 1996; Byers, 2000; Lauer, 2000; Lauer and Rafiqpoor, 2000).

However, regional differences have also become obvious. In the Ecuadorian Andes, for example, Di Pasquale et al. (2007) have found charcoal in soil profiles throughout the Páramo providing evidence that fires have occurred in this region since the beginning of the Holocene. However, the authors suppose that fires have not caused forest retreat but possibly have delayed Holocene forest advance. Accordingly, many forest patches above the present timberline are more likely to be foreposts of an expanding forest rather than being relic forests or extrazonal forest islands.

The lower quantity of *Polylepis* pollen in postglacial sediments in the Páramo of the Sierra Nevada de Merida (northern section of the eastern cordillera in Venezuela) also is inconsistent with the relic-hypothesis (Salgado-Labouriau, 1984). Very likely, no settlements existed in the highest part of the Venezuelan Andes prior to the arrival of the Spanish during the 16th century, and the Indians did not raise any cattle at his time. Cattle were first introduced from Europe in 1570.

However, in the Peruvian and also in the Ecuadorian Andes big llama and alpaca herds of the native people had already changed the natural vegetation considerably before the Europeans arrived (Flores Ochoa, 1979). At the time of the Spanish conquest, the herds of those camelides were so large that the available pasture area could not support them (De la Vega, 1602, mentioned by Crawford, 1989).

4.3.14.2 After-effects of timberline decline and present impact

The after-effects of human impact on high-elevation forests, particularly on the distribution, composition and structure of the present forest stands and on the position of the timberline, should not be underestimated if compared to the effects of climate. Humans not only lowered the upper timberline but also considerably changed the composition and age structure of the high-elevation forests. Most of the subalpine forests have become overmature due to the influence of prolonged grazing. In the central Alps, for example, larch (*Larix decidua*) which is relatively tolerant of grazing and trampling suffers to a lesser extent from pastoral use than Swiss stone pine (*Pinus cembra*). Moreover, forest fires favoured larch, which is fairly well protected by its thick cork-like bark from forest fires and also is able to replace burnt needles, while Swiss stone pine was usually killed by fire (Holtmeier, 1967a, b, 1974). Also, mineral soil often exposed by grazing cattle (trampling) enhanced seed-based reproduction of larch (cf. Section 4.3.10.1).

Thus, in the Upper Engadine and also in other central-alpine valleys, larch has through history spread at the cost of Swiss stone pine and often forms almost pure stands, particularly on sun-exposed and easily accessible foot-slopes (Auer, 1947; Holtmeier, 1967b). Nevertheless, also at timberline larch is more common than at undisturbed conditions. At present, however, gradual change is going on in consequence of decreasing use of alpine pastures and high-elevation forests. Larch and Swiss stone pine have rapidly invaded many former alpine pastures after cease of pastoral use. Larch profited still for a while from open patches of exposed mineral soil. After several decades, however, regeneration of larch considerably declined, as exposure of mineral soil by grazing cattle does not occur anymore. Now, closed dwarf-shrub vegetation, grassed and herbaceous vegetation widely prevents wind-mediated larch seeds from reaching a suitable seedbed. This is clearly reflected in the age of the trees. On many previous alpine pastures in the Upper Engadine (Switzerland), for example, 30 to 50 year old larches occur at great numbers, whereas Swiss stone pine of the same age are comparatively rare. Later, however, after the pastures were abandoned, Swiss stone pines increased rapidly and are by far more common now than young larches (Holtmeier, 1967a; Pohl, 1993; Müterthies, 2002). Locally only, earth slips and other kind of soil erosion promote establishment of larch seedlings.

In addition, dwarf-shrub vegetation producing thick raw humus layers have worsened moisture supply to larch seedlings (Auer, 1948). Raw humus dries more rapidly than fine mineral soil. Swiss stone pine, instead, which normally prevails at the climax stage of high-elevation forests in these regions, has become the dominant tree species at the given conditions on abandoned or only rarely used alpine pastures, mainly because of seed caching by the European nutcracker (Section 4.3.13.3). Thus, Swiss stone pine is at an advantage now.

After the removal and decay of subalpine forests vegetation considerably changed. Light-demanding dwarf shrubs, grasses and herbaceous vegetation spread downward into the former forested area. Dwarf juniper (*Juniperus nana*) and Alpine rose (*Rhododendron ferrugineum* on acidic substrate,

Rhododendron hirsutum on carbonatic substrate), for example, rapidly invaded abandoned or only rarely used alpine pastures where they became a real 'weed'. Fire (juniper) and excessive fertilization (Alpine rose) were used to remove these 'weeds' again. Often, fires got out of control and destroyed the neighbouring forest stands. Invasion of the 'weeds' into the existing often not well-maintained alpine pastures was also a reason to clear additional forest stands at timberline. Persistent intensive grazing often resulted in species-poor *Nardus stricta* grassland. On rarely grazed pastures, the 'weeds' Alpine rose and dwarf juniper protected occasional young growth, mainly of Swiss stone pine, from being grazed or trampled by cattle thus enhancing natural reforestation (Holtmeier, 1974).

In the northern limestone Alps, for example, and also in other limestone and dolomite mountains in the eastern Alps, in the High and Low Tatra Mountains, and in the Dinaric Alps, the range of prostrate mountain pine (*Pinus mugo*) and green alder (*Alnus viridis*) was considerably extended, whereas under undisturbed conditions those species had occurred mainly on avalanche tracks and similar sites unfavourable to erect tree growth (Sections 4.1.2 and 4.3.9.1). *Pinus mugo* colonized high-elevation clearings and burn areas (Photo 114, see also Photo 12; Grabherr, 1934; Mayer, 1965, 1966, 1970; Aichinger, 1967; Köstler and Mayer, 1970; Pitschmann et al., 1970; Wraber, 1970; Wilmanns, 1971; Kral, 1973; Wilmanns and Ebert, 1974; Hafenscherer and Mayer, 1986). After this, at some places they could

Photo 114. Prostrate mountain pine (*Pinus mugo*) with some flagged Norway spruce (*Picea abies*) on an old burn in the forest-alpine tundra ecotone, Low Tatra (Slovakia). Man-caused fires were very common in this mountain area. F.-K. Holtmeier, August 1970.

form a compact krummholz-belt above the man-made timberline. Thus, the krummholz-belt can be considered a substitute formation there (Holtmeier, 1981a).

Also in the central Alps (e.g., Upper Engadine; Rübel, 1912; Holtmeier, 1965, 1967b, 1974) prostrate mountain pine and green alder occasionally invaded deforested slopes. *Pinus mugo* colonized mainly sun-exposed dry valley sides, while *Alnus viridis* spread on shaded slopes offering better soil moisture conditions. Invading *Alnus viridis* reduces spatial diversity of habitats and original species richness in subalpine grasslands, mainly because of its dense foliage and high stem density (cf. Photo 18). Green alder full in leaf shades the ground almost completely and increases considerably air humidity. On the other hand, the effects of expanding green alder induce a new peculiar floristic composition (Anthelme et al., 2003).

Alnus viridis as well as *Pinus mugo* may impede or facilitate the establishment of spruce and other trees in the timberline ecotone (e.g., Anthelme et al., 2003). On alpine pasture (abandoned about 1900) in the Lechtaler Alps (Austria), for example, the most intense regeneration and highest vitality of spruce young growth (*Picea abies*, maximum age 40 years) can be observed in open stands of prostrate mountain pine and also in dwarf-shrub vegetation (*Rhododenron hirsutum*) that invaded the pastures simultaneously with spruce. Obviously, *Pinus mugo* stands and dwarf shrubs provide better

Photo 115. Green alder (*Alnus viridis*) invaded this area on the north-facing slope of the Albula Valley (Upper Engadine, Switzerland), where the forest had been removed by humans. F.-K. Holtmeier, October 1968.

protection from adverse external factors than grassland communities (Baader, 1995). However, Dullinger et al. (2005) found *Pinus mugo* having negative effects on recruitment and growth of both *Picea abies* and *Larix decidua* at timberline in the northern Limestone Alps. Mountain pine is able to invade rapidly former alpine pastures and alpine grassland above the current timberline when it has not to compete with other vegetation (Dullinger et al., 2003, 2004; Holtmeier and Broll, 2007).

Conifers will replace green alder that has colonized moist slopes after abandonment of alpine pastures (Photo 115). After 50 years or even longer, the green alder stands will gradually disaggregate thus letting the light pass to the ground and reducing root competition (see also Chapin III et al., 1994). Avalanches, landslides and also stagnant water may prevent this succession (Körner and Hilscher, 1978; Spatz et al., 1978). It remains open to question, however, to what extent the high amount of nitrogen accumulated by the alders will influence soil acidity (cf. Section 4.3.8). In Alaska, for example, soils covered with stands of Sitka alder (*Alnus sitchensis*) exhibit the highest nitrogen content in this region, and, with a pH-value of 3.3, they belong to the most acidified soils (Mitchell, 1968).

Extensive deforestation at timberline and over-utilization of the remaining forests resulted in serious consequences for the mountain people (cf. Figure 1). Avalanches, for example, became more frequent and destructive due to the extended avalanche-prone area above the forest and increased length of the avalanche tracks. About 60% of all avalanches that will cause damage to buildings, roads and railways, for example, are released within the previous forested area (Fromme, 1952). Also, soil erosion, land slides and earth slips, debris flows and flooding disasters have increased. They are a permanent threat to the people living in the mountain valleys. Over-aged high-elevation forests are not able to meet their protective functions. In the European Alps, for example, the safety of the people living in the steep-sided and densely settled valleys closely depends on the protection provided by the forest belt. Great efforts are made to maintain vigorous shelter forest and to improve its protective functions by reforestation up to the potential climatic timberline.

However, although the climatic limit of tree growth is located above the present timberline (e.g. Tessier et al., 1993), reforestation of the former forested area is very difficult (Holtmeier, 1965, 1967a, b, 1974). Site conditions deteriorated considerably in the previous forest area after the forest was destroyed hundreds of years ago. Thus, the man-caused timberline has become as pronounced an ecological boundary as the natural climatic timberline had been before (Holtmeier, 1965). In other words: the alpine zone became extended to lower elevation (Photo 116; Figure 84). Friedel (1967) called it therefore a 'pseudoalpine surrogate landscape'. Above the closed forests, tree growth is hampered more by unfavourable site conditions than should be expected at the present level of timberline. After the high-elevation forest stands were removed, site conditions have been controlled by strongly contrasting microclimates (exposure to solar radiation and wind) and resultant effects on site conditions (e.g., depth and length of the winter snow cover, distribution of soil temperature and soil moisture, etc.). The resulting site mosaics are reflected in the distribution of the plant communities (Section 4.3.7.2).

On top of Pru dal Vent (Grison, Switzerland; see Section 4.3.4), from which the original forest had been removed in consequence of pastoral use, Swiss stone pine (*Pinus cembra*) and European larch (*Larix decidua*) that were planted on a fully-wind exposed experimental site grew well when sufficiently protected by artificial constructions (cf. Photo 28). Natural reforestation, however, if not encouraged by artificial shelter, would fail under the present climatic conditions (Bednorz, 1993; Streule and Häsler, 2006) as was expected by the present author (Holtmeier, 1971b) before this experimental site became established.

Photo 116. The Bernina Pass (Switzerland) was formerly covered with forest (*Larix decidua*, *Pinus cembra*). After the forest became a victim of mining during the Middle Ages, site conditions changed, due mainly to the influence of microtopography on solar radiation and wind as is reflected in the distribution pattern of the winter snow cover. F.-K. Holtmeier, 15 March 1978.

Besides microclimates, unfavourable soil conditions may impede trees from invading the abandoned pasture areas. Loamy soils, for example, were often completely compacted by the trampling of grazing cattle, and also impoverished. Because of compaction, surface run-off and soil erosion have increased. On sandy soils, trampling by cattle and sheep enhanced soil

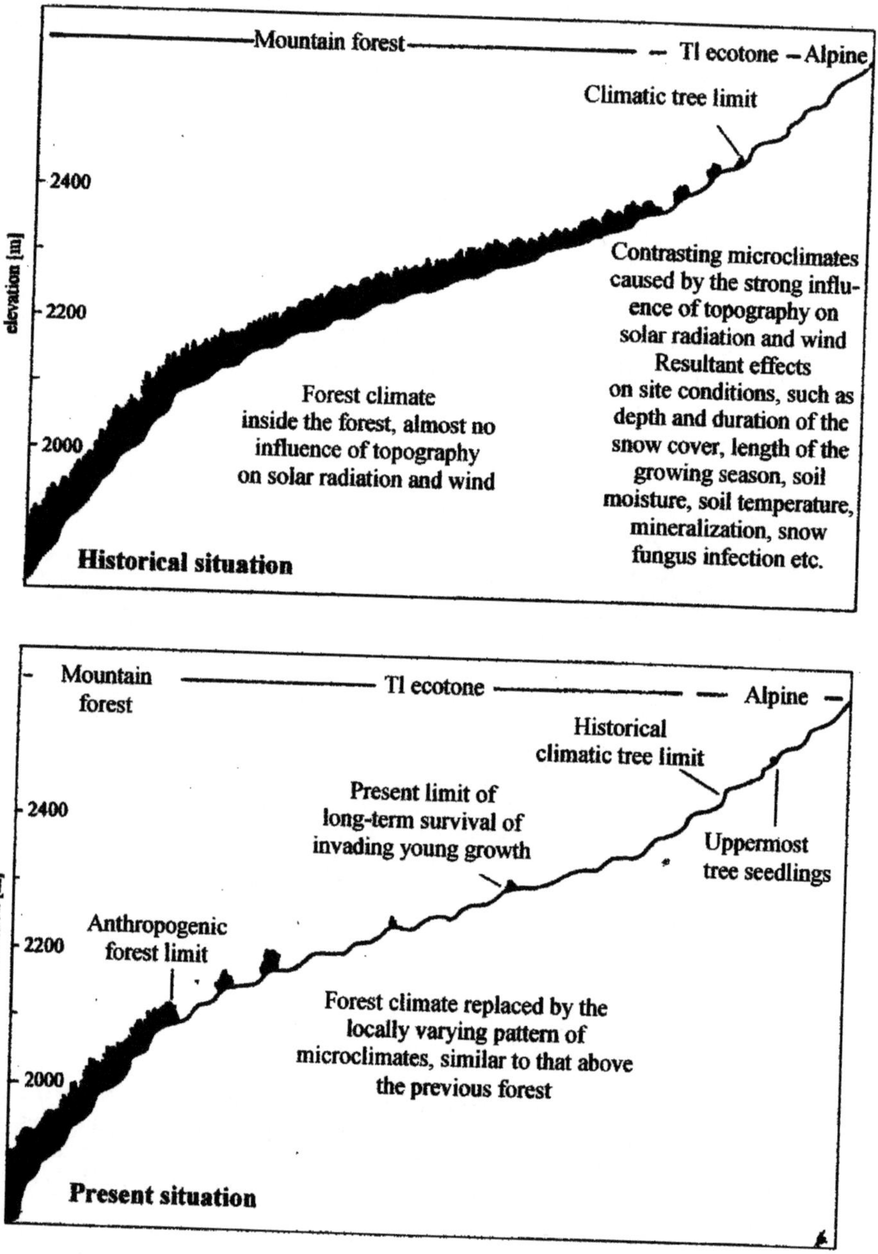

Figure 84. Change of the microclimatic conditions and some resultant effects after the uppermost forest was removed by humans. Modified from Holtmeier (1994a).

erosion by destroying the plant cover and loosening the soil (Holtmeier, 1999c). In addition, snow slides increased on many previous alpine pastures thus enhancing soil erosion. Slopes covered with relatively low and smooth dwarf-shrub vegetation are particularly affected (Newesely et al., 2000). More rigid woody dwarf shrubs, such as Alpine rose, and also young growth of trees, increased surface roughness and thus reduced snow slides. Especially in dry sites, which are exposed to the sun and strong winds or both, mycorrhizal flora has been destroyed, which hampers nutrient uptake. Thus, conifer seedlings that will be used for high-altitude afforestations are inoculated with mycorrhizal fungi in the tree nurseries to guarantee better growth after planting the seedling above the present forest limit.

The first efforts to restore the uppermost forests were made about 140 years ago. However, many of these early high-altitude afforestations failed completely because the very locally varying environmental conditions had not been sufficiently considered. Also, inappropriate conifer provenances had often been planted (Schlatter, 1935; Pitterle, 1985). Obviously, efforts were based more on good intention than on solid knowledge of the site-ecological conditions. Later it became evident that unsuitable sites can hardly be afforested within a foreseeable future (cf. Schönenberger, 1985), which means that afforestation should start with safe sites on which seedlings and young growth will very likely survive.

On many previous alpine pastures, also natural tree invasion can be observed and, in the long-term, a gradual natural forest advance to the potential climatic limit might be expected, if grazing and other kinds of use will be excluded (e.g., Spatz et al., 1978; Weis et al., 1982; Nola, 1994; Holtmeier and Broll, 2005). As to safety of the people living in the mountain valleys, however, high-altitude afforestation cannot be left to nature alone. Spontaneous establishment of trees follows favourable microsites and does not necessarily occur at those slope areas to which forest should be brought back as soon as possible to create a vigorous and effective shelter belt (Holtmeier, 1967a; Gunsch, 1972; Baader, 1995). Normally, young growth occurs spontaneously on convex topography (ridges, knolls, structural benches, etc.), while snow-rich sites (hollows, gullies and other depressions) will remain treeless for long periods of time. Consequently, reforestation on many sites above the anthropogenic timberline needs careful active management. High-altitude afforestations must often be combined with artificial constructions, such as avalanche walls and snow fences (Photo 117). These constructions will gradually be incorporated into the future forest and then lose their function. However, also spontaneous young growth should be encouraged in the same or a similar way in order to force reforestation.

In the Alps, high-elevation forests are now avalanche protection forests, almost without exception. In order to maintain these forests sustainable

Photo 117. Afforestation by Norway spruce (*Picea abies*), encouraged by artificial constructions on the extremely avalanche-prone slope of the Kirchberg near Andermatt (Switzerland). F.-K. Holtmeier, 8 September 1984.

Management objectives

Maintenance of remnant forests
- Natural forest structures
- Sufficient regeneration
- Different age classes
- Reduced pastoral use
- Reduced game population not exceeding the natural carrying capacity of the high-elevation forest

Restoration of climatic timberline
- Afforestation above the current, anthropogenic timberline with special respect to site conditions (microclimates, soils, plant competition, etc.)
- Support of natural restocking above the present timberline (abandoned alpine pastures)
- Control of snow fungi (mainly *Pinus cembra, Pinus mugo, Picea abies*)
- Reduction of damage by grazing (browsing) game animals and live-stock
- Fence constructions to protect young growth from moving snow

Restriction of skiing area and lifts

Restriction of mountain biking and off-road vehicles

Figure 85. Possible strategies to maintain high-elevation forest and to restore the climatic timberline. Modified from Holtmeier (1990).

management is required, trying to restore the climatic timberline and to produce natural structures and variety of successional stages to ensure their long-term vitality and protective function (Figure 85; Mayer, 1979), as was already postulated by the Swiss forester Landolt in 1862. Particularly during recent winters it has become apparent that at certain meteorological conditions avalanches may be released from starting zones only partly or not secured by avalanche constructions, as was also the case in the extreme winters in 1950/1951 and 1998/1999. In Switzerland, comparable avalanche disasters happened in 1720 and 1808. In the end, a continuous and compact unevenged forest belt reaching as high as possible is the best measure for preventing such catastrophes because prevention of natural hazards by technical means costs up to 20 times more than maintaining existing protective forests (Schönenberger, 2000).

It has become obvious that most of the actual problems of establishing and maintaining high-elevation forests have been caused by past human influences (cf. Figure 1). At present, however, these forests are being subjected to new impacts not less detrimental to existence. Preparation of new ski slopes, broadening of existent ski courses and skiing itself, particularly outside the official ski runs, cause severe damages to young growth and soils, mainly above the present upper forest limit (e.g., Cernusca, 1978).

In the high mountains of North America, on the other hand, the effects of winter sports on timberline are rather concentrated if compared to the total subalpine and alpine zone. Locally, damages caused by cutting firewood, campfires out of control, mountain biking and off-road vehicles have reached an extent critical to survival of trees, at least in the upper zone of the timberline ecotone, where trees are growing very slowly and almost do not regenerate from seeds at the given climatic conditions (Section 5.4).

In many high mountain regions, the high-elevation forests are suffering from too high densities of wild game. In the European Alps, for example, over-populations of red deer endanger the existence of the mountain forests, high-altitude forestations and trees that have spontaneously established themselves above the actual forest limit. Too high populations of wild ungulates (elk = Wapiti, *Cervus canadensis*; mule deer, *Odocoileus virgianus*) impede considerably natural regeneration also in the timberline ecotone on many mountain ranges in western North America (Holtmeier, 1999c, 2002; further references there). In New Zealand (Kuschel, 1975; King, 1990), the introduced wild ungulates seem to be no serious threat to the timberline forest these days. Hunting from helicopter since the 1960s has nearly eliminated red deer from the high country above the timberline (Challies, 1990). In places, however, feral goats (*Capra hircus*) being able to climb even very steep and rugged mountain terrain still cause damage to the timberline vegetation. Additionally, introduced brown hares (*Lepus europaeus*), widely browsing in the timberline ecotone, cause damage to *Nothofagus* seedlings and saplings by girdling (Wardle, 2007). Brown hare occupy alpine tussock grassland from timberline to the upper vegetation limit even in winter (Flux, 1990).

In northern Finnish Lapland, over-grazing by reindeer seems to be the factor most adverse to tree seedling establishment within the timberline ecotone and beyond the present tree limit (e.g., Holtmeier, 2002; Holtmeier et al., 2003; Anschlag, 2006; Broll et al., 2007; Anschlag et al., 2008). Reindeer were originally an integral component of the subarctic birch forest and fjeld tundra ecosystem. The number of reindeer has fluctuated in response to natural environmental factors (e.g., extreme winters, malnutrition, diseases, and predators such as bear, wolf, wolverine, and golden eagle; Väre et al., 1996). Thus, reindeer were a very important winter food source for lynx, for example, as can be concluded from studies on lynx predation on semi-domestic reindeer in northern Sweden (Pedersen et al., 1999). For centuries reindeer have been the main base of existence of the Laplanders without over-using natural vegetation. The situation changed when the state borders between Finland and its neighbouring countries were established (Norway 1853; Sweden 1889). They set an end to the traditional seasonal migrations of the reindeer herds between their winter pastures in the interior parts of Lapland to their distant summer grazing areas along the coast of northern Norway (Troms, Finnmarken). Overgrazing of winter pastures by reindeer is common in most reindeer herding areas of northernmost Eurasia.

In modern times, traditional reindeer herding has turned into a 'reindeer industry'. The mortality of semi-domestic reindeer has considerably declined due to additional winter-feeding with hay and imported lichens since 1974 (Helle and Kojola, 1993; Burgess, 1999). Winter-feeding has become necessary because the totally overgrazed lichen cover cannot support the great number of reindeer. Moreover, better supervising of the herds (using off-road vehicles), better veterinarian care, and the complete lack of natural predators are factors leading to over-grazing. Predators had been systematically removed mainly by the reindeer owners. Reindeer population more than doubled since the mid 1970s (Kumpula and Nieminen, 1992; Oksanen et al., 1995; Burgess, 1999) and peaked in the 1990s (5–10 reindeer/km^2, Oksanen et al., 1995; see also Danell et al., 1999; Sumoninen and Olofsson, 2000; Colpaert et al., 2003; Kashulina et al., 1997). In northernmost Norway (Finnmarksvidda) reindeer population reached its highest density in the same period of time (Thannheiser et al., 2003). The peak in reindeer population coincided with the Chernobyl nuclear disaster (26 April 1986) when animals could not be slaughtered because of low demand for reindeer meat (Colpaert et al., 2003). At this time, however, reindeer numbers had already grown beyond the natural carrying capacity. Besides too large populations of reindeer the lack of seasonal grazing practice has caused excessive summer grazing pressure (Käyhkö and Pellika, 1994). Over-grazing resulted in heavy disturbances of the ecological conditions (Holtmeier, 1974, 2002; Haapasaari, 1988; Heikkinen and Kalliola, 1989; Evans, 1995; Löffler, 2005, 2007).

The combination of excessive reindeer grazing and occasional extensive defoliation during mass-outbreaks of the autumnal moth (*Epirrita autumnata*) and the winter moth (*Operophtera brumata*) is the greatest threat to timberline birch forest in northern Lapland, (see also Chapter 5; Holtmeier, 1974, 1999c; Helle, 2001; Kallio and Lehtonen, 1975; Oksanen et al., 1995; Holtmeier et al., 2004; Lempa et al., 2005; Holtmeier and Broll, 2006).

In the high mountains of Central Asia, for instance, in the Karakoram and in the Himalayas, devastation of high-elevation forests is still going on or even increasing. Overgrazing and particularly burning are the most destructive factors. Jacobsen and Schickhoff (1995) and Schickhoff (1995a, b) report increasing over-utilization of the mountain forests and the timberline ecotone on the Karakoram and neighbouring mountains of northern Pakistan. In many regions over-utilization or removal of the forest from the mountain slopes has been followed by heavy soil erosion and general landscape degradation. Destruction of high-elevation forests is accelerating extremely (e.g., Schickhoff, 1995a, b, 2005). If this trend continues these forests will be gone within the next future. Reforestation of the cleared slopes is impossible due to lack of money. Moreover, the understanding of the mountain people for nature protection and sustainable use of such resources as mountain forests is still insufficient. Modern scientific exploration of the mountain forests was mainly initiated by Europeans and in current time is supported by economic aid programs. It is still an open question whether forest would invade these treeless slopes at the given climatic and edaphic conditions after human influence had ceased. The same is going on in the Himalayas (Schmidt-Vogt, 1990a, b). However, the intensity of recent human impacts on the mountain forests differs locally. Accessibility plays an important role. Thus, building forest-access roads has lead to increased exploitation of the forests. Near-natural forests can be found in remote thinly populated areas only (Schickhoff, 1995a, b, 2005). Over-utilization by cattle grazing and cutting firewood lowered the timberline that was originally located between 3.300 and 3.400 m for about 200 to 300 m, a few relic stands at higher elevation excepted (Photo 118; Schickhoff, 1993). In the Nanga-Parbat area on the other hand, birch timberline has been only locally depressed (Nüsser, 1998).

In the Jugal-Himalaya (Nepal), grazing and fire also are the agents most detrimental to the subalpine forest stands. In the high elevation forests, fire is largely a seasonal phenomenon. Most fires, however, are induced by humans at the early beginning of the rainy season to stimulate new growth of grass. Fires set by pastoralists often run out of control and spread into the adjacent forests. Pastoralists also induce fire to increase grazing area. Not last, hunters use fire to drive wild game. Junipers, usually growing on southern exposures, are particularly sensitive to burning because of their high resin content and the dry fire-prone environment. Obviously, man-caused fires (Schmidt-Vogt, 1990a, b) destroyed most previous juniper stands. The situation is similar in other regions of Nepal. Most sun-exposed slopes are nowadays devoid of forests, while forest covers the shaded slopes up to an elevation of 4.200 m (Haffner, 1972). Man-caused fires must be considered a permanent threat to the mountain forests in these regions.

Photo 118. On this northeast-facing slope (about 3.300 m) of the Saiful-Muluk Valley (Kaghan, Pakistan), the original *Betula utilis*-belt has been completely destroyed by intense pastoral use. The *Abies pindrow*-forest became open. Under undisturbed conditions the upper timberline is supposed to be located approximately 300 m higher than at present. U. Schickhoff, 31 April 1994.

The mountain forests in eastern Tibet (Min Chan, north-eastern Sichuan) are increasingly declining due to exploitation (Winkler, 1997). On the other hand, natural regeneration locally occurs on south-exposed slopes between 3.000 and 3.500 m. Birch has also invaded some remote grazing areas during the last decades. Remnants of the original birch forests can be found up to an elevation of 3.650 m. In Yunnan (China), the first Tibetan herders entered the region more than 2000 years ago. Since then humans have been continuously influencing the alpine zone and timberline, mainly by both pastoral use and fire. Since burning ceased (1988) trees (*Larix potaninii*) and other woody vegetation are invading the alpine meadows, probably encouraged by a warming climate (Baker and Moseley, 2007).

Also in East Africa and South America, fire was and still is used to prevent high-altitude pastures from being invaded by trees or to create new pastures. In the Andes, humans also used and still use fire to make the grazing area easy to survey and to destroy the habitats of predators (e.g., cougar). Moreover, religious rituals and the belief that smoke will bring rain or drive the clouds away (Kessler, 1995; Di Pasquale et al., 2007) and, not last, incendiarism (Seibert, 1983) played an important role. Burning also caused the abrupt upper limit of the Bolivian *Polylepis* forests, except for the western cordillera (Kessler, 1995). Also nowadays, grazing prevents natural reforestation of easily accessible slopes by *Polylepis* (Ruthsatz, 1983). This is indirectly supported by the recent development in protected areas, as for example in the upper Pisco-Valley (Huascarán National Park, Peru), where abundant regeneration occurs within protected *Polylepis* stands. Seedlings and young growth occur even 50 m above the uppermost groves within the grassland that is not grazed anymore. In contrast, in the Los Nevados National Park (west side of the Bolivian western cordillera) continued grazing and use of the forest remains (Verweij and Beukema, 1995) prevent any regeneration. Strict reduction of human use is necessary to prevent further decline of the upper treeline (Kok et al., 1995). Also in the eastern cordillera, young growth and coppice (basal sprouts) are so heavily grazed that any natural regeneration of the tree stands is impossible (Hensen, 1995).

In the Páramo grassland of Ecuador, regularly repeated burning has caused the decline of *Polylepis* groves. Forest stands growing in ravines and narrow valleys could survive probably because the fires, driven by upslope winds, spread normally faster and at greater intensity upslope than downslope (Laegaard, 1992). Grazing, on the other hand, seems to have been less important. Even on areas intensively grazed by cattle and horses *Polylepis* is never affected. Probably, secondary compounds keep off the grazers. Seedlings may grow up if not destroyed by trampling (Laegaard, 1992).

In New Guinea, many human-caused fires affect mountain timberline (Paijmans and Löffler, 1972). While fire was originally used for hunting (Flenley, 1984), warming, signalling, easing travel through the tussock grassland and warfare, nowadays pyromania plays an important role.

In East Africa, pastoralists set fires during the dry season to maintain the grazing area that would otherwise be rapidly invaded by *Erica* scrub (Miehe and Miehe, 1994, 1996). Outside the settled areas, fires are also caused by people collecting wild honey and by poachers (Hedberg, 1964). Also, many fires are caused by lightning and spontaneous ignition (Beck et al., 1986). More recently, Miehe and Miehe (1994, 1996) have reported impressive examples of the influence of fires regularly set to maintain or extend the grazing area on timberline in the Bale Mountains (Ethiopia). Fires usually spread from the easily burning (combustible) *Erica trimera* stands into the adjacent more humid primary forest (*Hypericum revolutum*) which is gradually reduced. Obviously, the upper timberlines in Africa were lowered by recurrent fires for several hundred metres. The groves occurring above the present forest limit are considered relics of former forests, which survived at sites protected from fires (Wesche et al., 2000; Wesche, 2002).

Altogether, tropical mountain timberlines have been and still are being influenced by human impact at least as much as the timberline in the temperate and boreal zones. Possible climatic limitation of forest is difficult to substantiate by direct evidence as natural tropical treelines that with certainty remained unaffected by humans are rare, if they really exist at all.

While the destruction of high-elevation forests had serious consequences to the mountain people and their environment, the effects on the mountain forelands have often been overestimated. In view of the high monsoonal precipitation, Messerli and Hofer (1992), for example, argue that the decline of the forests in the Himalayan valleys is of minor importance to runoff, erosion and sediment transport. On the footslope of the Himalayas about 5.000 m deep fluvial sediments have accumulated and tremendous sediment supply by the great Himalayan streams (e.g., Sapt Kosi, Nepal/Bihar) occurred long before deforestation and intensive agricultural use in the river head waters. Instead, tectonic events, substrates highly prone to erosion and heavy rainfalls exceed the effects of anthropogenic impacts by far.

5 TIMBERLINE FLUCTUATIONS

The previous chapters have provided a first insight into the heterogeneity, complexity and great physiognomic and ecological variety of the timberline phenomenon; a heterogeneity and variety that largely have faded out when scientists mainly focused on the one factor (mainly temperature) controlling the altitudinal position of timberline and on the physiological response of tree growth to the timberline environment. Spatial and temporal timberline structures reflect past and on-going changes caused by many external (e.g., macroclimate, effects of altitude) and internal factors specific to the ecotone (microclimates, soil conditions, succession, competition of trees and other vegetation, animals, ecological properties and requirements of the timberline forming tree species and others) interacting in a complex way. Topography is the only almost constant factor among them, if not affected by sudden changes due to landslides and similar events, for example, or when considered in geological terms. Mountain topography (convex, concave, gentle, rugged, smooth, etc.) determines the basic pattern of the spatial timberline structures (distribution of trees, soil moisture, soil temperatures, etc.) also under the influence of changing climate (Holtmeier and Broll, 2005).

5.1 General aspects

The desire to get a deeper insight into timberline dynamics confronts us with a couple of problems. The first problem is how to handle the often-ambiguous timberline-specific terminology (cf. Chapter 3). For example, minimum height used as criterion to define treeline, ranges from less than 1 to 8 m (cf. Table 1). Consequently, information on the actual or potential altitudinal position of timberline or treeline is often hard to compare. For example, Kullman (1986a) considers treeline of spruce (*Picea abies*) in the southern Swedish Scandes to be rising, after some stunted specimen that had already been established during the middle of the 19th century exceeded 2 m height between 1915 and 1975. Birches that existed as low-growing 'brushwood' in the beginning of the 20th century reached 'tree height' during the 1950s (Kullman, 1979). Thus, the rise of treeline is mainly the result of phenotypic response of the individual trees to more favourable conditions (e.g., Kullman, 1979, 1986a, b, 2000b, 2007b; Lescop-Sinclair and Payette, 1995) rather than a real advance of tree growth to higher elevation, as true seedlings are still very rare above the closed tree stands. On the other hand, there is no similar phenotypic

F.-K. Holtmeier, *Mountain Timberlines: Ecology, Patchiness, and Dynamics*,
Advances in Global Change Research 36, 293–333.

response of numerous bristlecone pines (*Pinus longaeva*) that locally established themselves in the timberline ecotone on the White Mountains (California) about 30 to 60 years ago. Most of them are straight grown like a Christmas tree, but many of them are still lower than 2 m. However, if we considered these pines not as trees, we would not meet the recent dynamics (development) or, in other words, the obvious positive effects of the present climate within and above the timberline ecotone would not have occurred yet or, in other words, the future position of the tree limit depends on what is defined a tree (Chapter 3).

Moreover, we must be aware that the present situation in the timberline ecotone has to be attributed for the most part to site history (cf. Figures 1 and 90), about which usually little information is available (Holtmeier, 1993b, 1994b, 1995a; Kullman, 1997). Quite possibly such a statement will seem trivial. If ignoring the regional and local differences, however, and reducing timberline ecotone to a 'line' more or less in balance with a factor considered being essential to tree growth (e.g., air temperature, soil temperature, carbon balance, etc.; see Chapter 1), the often mentioned dynamics of the timberline are confined to an upward or downward shift of this mostly ill-defined 'line'. Basically, this would throw us back to the beginning of timberline research when it became obvious that heat deficiency increasing by elevation and latitude controls the position of vegetation limits and thus timberline in one or another way (Holtmeier, 1974, 1994b, 1995a, 1999a; Lloyd and Graumlich, 1997; Luckman and Kavanagh, 1998). This, however, will not be enough to assess the potential effects of global warming on timberline. When investigating changes in timberline position and spatial patterns we have to consider both regionally varying timberline history and current processes. Last but not least, comparatively little information is available on timberline response to climate change in the southern hemisphere and particularly in the tropical high mountains.

The altitudinal position of timberline and the width of the timberline ecotone should not be readily ascribed to the present climate, even though it actually shapes tree physiognomy. If climate deteriorates, for example, seed-based regeneration will fail and timberline stands become overaged. The mature trees, however, may survive for hundreds of years, then forming a relic timberline the altitudinal position of which is not in accordance with the cooler climate. The position of the upper timberline on Tierra del Fuego (*Nothofagus pumilio*), for example, has not changed for at least 150 years (Cuevas, 2002). Likewise, the 'Little Ice Age' (approx. 1150 to 1870) did not essentially influence the position of the upper timberline (*Pinus balfouriana*) on the southern Sierra Nevada in California (Scuderi, 1987), for instance. Also, the ancient bristlecone pines (*Pinus longaeva*, Photo 119) at timberline on the White Mountains (California), on the Snake Range (Nevada) and on the southern Rocky Mountains (*Pinus aristata*; Colorado,

New Mexico) are further examples of the great inertia even of seed-produced trees (LaMarche and Mooney, 1967, 1972; LaMarche, 1969, 1973; Krebs, 1972, 1973; Brunstein and Yamaguchi, 1992).

Photo 119. Several-thousand-years-old bristlecone pine (*Pinus longaeva*) on Sheep Mountain (White-Inyo Mountains, California) at about 3.470 m. This tree has survived many changes of climate. F.-K. Holtmeier, 26 July 1994.

Even more persistent are clonal tree islands (Section 4.3.10.2). Many of them originated from seeds far beyond the closed forest during relatively favourable warm periods and survived for hundreds or thousands of years although the climate became cooler (cf. Ives, 1973b, 1978; Hansen-Bristow, 1981; Ives and Hansen-Bristow, 1983; Holtmeier, 1985b, 1986b; Kullman, 2000a). The same holds true for many tree islands at the northern timberline (e.g., Larsen, 1965, 1980, 1989; Tolmachev, 1970; Nichols, 1974, 1975a, b, 1976; Elliott, 1979; Payette and Gagnon, 1979; Légère and Payette, 1981; Payette and Morneau, 1993).

Thus, timberlines do not respond to climatic changes as automatically and promptly as snow line does, for example. For this reason, when using the position of former timberlines reconstructed by palynological evidence, charcoal, wood remains and fossil soils as climatic indicators, fluctuations of the paleoclimate may be easily disguised by the inertia of long-lived individual trees and clonal tree islands in particular (e.g., Faegri, 1972; Idso, 1989; Holtmeier, 1993b). Consequently, former timberlines are less appropriate as indicators of the paleoclimatic situation than is usually assumed. The same is true for extrapolation of the rough coincidences of actual thermal conditions (e.g., 10°C-isotherm of the warmest month, tritherm, tetratherm, etc.; Section 4.3.1) and existing timberlines that have been documented by many studies. Therefore, maps and graphs projecting the future position of vegetation zones and altitudinal belts (e.g., Kauppi and Posch, 1988; Environment Canada, 1989; Solomon, 1989; Ozenda and Borel, 1991; Gates, 1993; Neilson and Chaney, 1997; Batchelet and Neilson, 2000) should be considered with reservation.

Caution is also required in automatically interpreting high seedling density, as reported from existing timberlines to be an effect of the warming climate. In many cases increased seedling density might result from a more thorough and direct searching than before timberline response to climate change became popular in research (Crawford, 2008).

Changes in timberline position and spatial pattern take place at different speeds. Some changes are going on comparatively slowly, such as aging of trees and forest stands in the ecotone. Successful seedling establishment, changes in tree physiognomy and coverage, effects of competition, positive and negative feedbacks of growing tree population on their environment become visible after a decade at the earliest. In contrast, phenology (e.g. bud burst, needle flush, shoot extension, maturation of newly formed tissue, ripening of seeds) varies from year to year. Climatically-driven trends may become apparent after a decade or more.

Feedbacks of expanding forest on the regional climate would take many decades or hundreds of years (Figure 86). On the other hand, fire, storms, plant diseases, severe droughts, avalanches, extremely snow-rich winters or winters with nearly no snow, mass outbreaks of pathogenic or leaf-eating insects, volcanic eruptions, landslides, etc. cause abrupt changes that, however, may effectively influence the further development of tree growth, regeneration, and also site conditions. Prediction of extreme events is impossible, particularly as to their potential effects on tree growth at timberline under future climatic conditions.

Medium-term response (Several years to a few decades)	⟹	**Long-term response (Several decades to hundred or more years)**
• Increasing supply of viable seeds • Facilitation of seedling establishment and survival by the increasing tree population (e.g., reduced sky exposure, reduced wind velocity) • Icreasing survival rate in young trees (survival > mortality) • Increasing height and diameter growth in trees • Increasing tree population in the timberline ecotone • Establishment of trees beyond the current tree limit • Increasing competitive dominance of the arborescent vegetation in the timberline ecotone • Changing tree physiognomy (e.g., 'recovery' of suppressed mat and table trees by release of vertical leaders) • Changing phenology (e.g., period of shoot extension, date of bud burst, date of needle flush or leafing of broad-leaved trees)		• Adaptation of the tree species to the warmer climate • Upward shift of the timberline ecotone • Development of uneven-aged forest stands • 'Stabilization' of timberline at its higher position, where heat deficiency and climatic injury will limit tree growth • Higher mean position of the upper limit of production of viable seeds • Reduced biodiversity and landscape diversity by forest expansion into the alpine zone (fragmentation of alpine plant cover)

Figure 86. Timberline response to the warming climate in the medium- and long-term. The medium-term change can be measured and documented by field data while the long-term response is speculative. It may be assessed by modelling.

5.2 Timberline fluctuations in the past

Climatically caused fluctuations of timberline repeatedly happened during the Holocene. Literature on this subject is extensive. Most studies refer to northern and upper timberlines in Northern Europe, North America, and to upper timberline in the European Alps. MacDonald et al. (2000) have recently presented a Holocene treeline history in relation to climate change across northern Eurasia based on new radio carbon dates for tree macrofossils from the northern Russian treeline (Kola Peninsula, Petchora, Taymyr and Lena) combined with previously published Russian data. Esper and Schweingruber (2004) analyzed broad-scale treeline change in Siberia (see also Moiseev and

Shiyatov, 2003; Moiseev et al., 2004; Van der Meer et al., 2004; Mazepa, 2005). The authors suggest treeline advance in this region to be part of a long-term reforestation process of the tundra from which the forest had receded earlier in the last millennium probably because of a general cooling. Bräuning (1994, 1999) provides local chronologies of the last 1300 years from Tibet. Aas and Faarlund (2001) give an updated review of the Holocene history of the Nordic mountain birch belt and treeline in Fennoscandia, Greenland, Faroer Islands and Scotland. Comparatively little information exists on timberline fluctuations in the Karakoram, in the Himalayas (e.g., Esper et al., 1995), in northern Asia (Shiyatov, 1993, 2000; Earle et al., 1994; MacDonald et al., 1998a; Kremenetski et al., 1998) and in the high mountains of the southern hemisphere (Villalba et al., 1997).

While the postglacial history of climate and timberline can be reconstructed to a certain extent by means of pollen analysis, dating charcoal and other wood remains, tree-ring chronology and fossil soils, timberline history since the 'Little Ice Age' is well documented by direct observation and studies. Assuming that the ecological requirements of timberline tree species have not significantly changed during the Holocene, studying carefully the present situation will help to better understand the history of the postglacial climate and its effect on timberline (Holtmeier, 1979b, 1993b). Conversely, the advanced knowledge about the Holocene history of climate and timberlines should prevent us from projecting the actual causal relationships between tree growth and site conditions to the far future (see also Scott et al., 1997). For example, increased regeneration that has occurred at timberline on the eastern dry slope of the southern Sierra Nevada (California) during the comparatively humid 20th century, might suggest a rise of timberline at continued warming of the climate. Palynological studies, however, covering the last 3.500 years (Lloyd and Graumlich, 1997) provide evidence that at least two periods of extreme drought, lasting several decades each, occurred during a warm phase of climate between 950 and 550 years BP, resulting in a break-down of the high-elevation forests and subsequent decline of timberline. In the same way, the future dynamics of the present forest stands at timberline might be controlled by moisture supply rather than by higher temperatures. Also, any extrapolation into the future of oscillation cycles in climate and tree growth, such as did Siren (1963) based on his studies on tree growth in the European Subarctic since the 12th century, must remain pure speculation. He predicted a general decrease of summer temperature in northern Europe and expected summer temperature to remain low in the future.

5.3 Driving processes and adverse factors controlling present timberline dynamics

As to the potential effects of changing global climate, research has focused mainly on the influence of increased temperature and CO_2 on the growth at timberline. The timberline-raising effect of increasing temperatures is proven. However, seedling establishment and survival depends also on precipitation, particularly in regions with summer droughts (e.g., Brandes, 2007; Grunewald and Scheithauer, 2008). The fertilising effect of CO_2 has been evidenced by many experimental studies. However, most of these experiments were done in the laboratory with herbaceous plants (e.g., Kramer, 1981; Kimball, 1983; Gates, 1985; Strain and Cure, 1985; Conroy et al., 1986; Luxmoore et al., 1986; Drake, 1992). Since CO_2-partial pressure is less at timberline than at lower elevation, enriched CO_2 has been expected to stimulate tree growth and trigger altitudinal advance of timberline (e.g., Walsh et al., 1992). However, there is no evidence of significant correlation of enriched CO_2 and increase of tree growth at timberlines, thus far.

Increased diameter growth that has occurred in *Pinus longaeva* at timberline on the White Mountains (California) and in *Pinus flexilis* on Mt. Washington (Nevada; LaMarche and Mooney, 1972; LaMarche et al., 1984) since the middle of the 19th century must be attributed to increased precipitation rather than to enriched CO_2 (Stockton, 1984). As to the future development in the timberline ecotone, the locally abundant regeneration surely is more important than increased diameter growth in the ancient bristlecone pines. Likewise, subalpine trees in the Cascade Mountains were unaffected by increased CO_2. Instead, growth has declined since the favourable 1940s, following the regional trend of temperature (Graumlich and Brubaker, 1986; Graumlich et al., 1989). In the southern Sierra Nevada, last year's precipitation and actual summer temperature override completely the effects of CO_2-fertilisation (Graumlich, 1991). Also, analyses of increment cores samples at 34 sites in the northern hemisphere do not provide any correlation between enriched CO_2 and tree growth (Kienast and Luxmoore, 1988; see also Innes, 1991). On the other hand, Hari and Arovaara (1988) found growth of Scots pine (*Pinus sylvestris*) at the subarctic timberline in northern Finland 15.5% to 14.3% greater than expected when related to the predicted climatic data. These differences might be ascribed to enriched CO_2. However, the authors themselves hesitate to consider their results to be unambiguous, because they are highly sensitive to an autocorrelation parameter that predicts current growth on the basis of past growth. Also, enhanced diameter growth in subalpine *Pinus cembra* on the central Alps (Nicolussi et al., 1995) was ascribed to elevated CO_2. However, the results were confounded either by changes in soil moisture supply (*Pinus longaeva*) or by increased nitrogen input (*Pinus cembra*), as also was admitted by the authors. Light-saturated photosynthesis in *Larix decidua* and

Pinus uncinata at timberline (2.180 m) on Stillberg (Dischma Valley, Swiss central Alps) significantly increased at elevated CO_2 (Hättenschwiler et al., 2002; Handa et al., 2005). This effect gradually declined, however, during the 3 years of treatment with enriched CO_2. In the long-term, increasing competition among the trees will override the effects of elevated CO_2 (Hättenschwiler et al., 1997). Tree response to enriched CO_2 showed also delayed initiation of growth (Murray et al., 1994; Körner, 1995), which might counterbalance increased growth due to enhanced photosynthesis or longer periods with higher soil temperature (see also Sveinbjörnsson, 2000).

Recently, warm growing seasons that allow efficient carbon use (for formation of new tissue) have been put forward to be more important in timberline dynamics than enriched CO_2 (e.g., Körner, 1999, 2002, 2003a; Paulsen et al., 2000). Körner (2007a) has emphasized the role of soil being the source of nutrients and water which are needed to match the carbon demand of meristems that controls photosynthesis. In the view of soil ecology it is just natural. However, both carbon cycle and nutrient cycle (mineralization) must be considered as one functional system when analyzing the possible response of tree growth at timberline to enriched CO_2. Enriched CO_2 and rising temperatures would increase netto primary production and respiration only as long as the soil provides nutrients and water. Under generally warmer conditions the soil organic matter may decompose more rapidly and become a growth limiting factor. Thus, as to explaining tree growth at timberline it would be very promising if tree physiologists paid more attention to soil ecological conditions and processes, particularly in the rooting horizons. Altogether, the discussion on the role of enriched CO_2 as a possible driving factor in timberline advance has not come to an end yet.

Both hemispheres have warmed significantly during the 20th century. The amounts of warming have been greatest from 1920 to 1944 and from 1977 to 2001; Jones and Moberg, 2003), and temperature is still rising (see also Figure 87). For about the last 1.000 years it has never been warmer than at present. For the end of the 21st century a rise of the mean global temperature by 1.8°C to 4.0°C (best estimate, 1.1°C to 6.4°C, likely range) has been predicted (Solomon et al., 2007). Regionally, the increase of temperature might be even higher (Wohlgemuth et al., 2006).

Theoretically, an increase of mean temperature by 6°C would mean an upward shift of timberline by 1.200 m (Zimmermann et al., 2006). This schematic linear extrapolation should not be taken too literally, however. Sensitivity of the timberline tree species and their response to the annual, interannual and inderdecadal climate variability and particularly to extreme climatic events (e.g., drought) are likely to play an important role (see below). In addition, competition between tree species and between tree seedlings and ground vegetation as well as the adaptability of the tree species will modulate the process.

If warming continues for a long time (e.g., Førland et al., 2004), high-elevation forests will certainly respond to higher temperatures. Young growth will probably advance beyond the present treeline, provided that other factors such as drought (mainly in dry regions), mass outbreaks of pathogenous or leaf-eating insects will not adversely affect regeneration. Changes that have occurred after the 'Little Ice Age', particularly during the previous century, may provide some insight into ecological dynamics of timberline.

5.4 Regional variation in timberline response after the 'Little Ice Age'

At the northern and upper timberline, waves of increased seed-based regeneration have repeatedly occurred since the 1920s. This obvious climatically driven change has attracted the attention of many researchers in North America, Europe and in the former Soviet Union. Anyway, most of the observations and studies conducted in the North American mountains (e.g., Brink, 1959; Franklin and Mitchell, 1967; Fonda and Bliss, 1969; Arno, 1970; Franklin et al., 1971; Douglas, 1972; Vale, 1981; Rochefort and Peterson, 1991, 1996; Rochefort et al., 1994; Hessl and Baker, 1997; Miller and Halpern, 1998; Moir et al., 1999) are concerned mainly with the invasion of previously treeless meadows in the so-called 'subalpine parkland' by conifer seedlings rather than on the very upper timberline ecotone. As it has become obvious from investigations on timberline dynamics during the 20th century, fluctuations of climates and timberline were not synchronous and occurred also at different regional and local intensity.

At treeline (*Picea glauca*) in Alaska, warming since the 'Little Ice Age' resulted in widespread increases in tree growth, as evidenced by tree-ring data presented by Lloyd and Fastie (2002). Since the 1950s, however, tree growth began to decline at two of the three treeline sites investigated, despite continued warming. As growth decrease is more pronounced in warmer and drier sites, the authors suppose drought to be a possible local factor impeding tree growth. On Seward Peninsula and in the Tanana-Yukon uplands (near Fairbanks, Alaska) as well as in many places on the Alaska Range, white spruce (*Picea glauca*) advanced far beyond the outmost forest stands (Hopkins, 1972; Viereck, 1979), while in the eastern part of the mountain range no comparable changes happened (Denton and Karlen, 1977). In the northern section of the Kluane Ranges (Saint Elias Mountains, south-west Yukon), *Picea glauca* treeline moved upslope and tree stand density increased during the early to mid 20th century (Danby and Hik, 2007).

About 80% of all white spruce in the timberline ecotone became established after 1900. However, local differences became apparent as white spruce rapidly advanced up to 85 m on southern exposures while treeline did not shift on the north-facing slopes where permafrost is present. Nevertheless, stand density increased. Reduced recruitment after the 1950s has been attributed to the comparatively cool period from 1950 to 1980. Regrettably, Danby and Hik (2007) do not provide information on recent young growth because they did not sample trees lower in height than 50 cm.

In the Ennadai Lake area (central Canada), any young growth was missing at the northern treeline until the end of the 1970s (Elliott, 1979), whereas seedlings were abundant at the treeline on the Labrador Peninsula (Elliott and Short, 1979; Elliott-Fisk, 1983). In northern Quebec, on the other hand, young growth has not occurred until the present. Very likely, the amplitude and magnitude of recent warming is not large enough to compensate at such locations for what is left of the negative effects of the 'Little Ice Age', which again emphasizes the importance of site history to the present timberline dynamics (Payette et al., 1989; Lescop-Sinclair and Payette, 1995).

Gamache and Payette (2005) report recent changes at treeline east of Hudson Bay. In the southern forest-tunra ecotone, treeline has slightly risen by the establishment of black spruce seedlings (*Picea mariana*) during the 20th century, while seedlings do not occur in the northern part. The abundance of spruce trees (<2.5 m) originated from seeds decreases exponentially with latitude. In the northern section of the forest-tundra ecotone, treeline advance results from increasing growth of vertical leaders released from stunted spruces already growing on the hilltops. This is considered to be an effect of 20th century climate change. However, it has also become apparent that spruce regeneration was not synchronous with short-term fluctuations of temperature and precipitation what has been supposed to result from the irregular release of seed from the cones. In black spruce seed release may be delayed for up to 25 years (Haavisto, 1975).

Scott et al. (1987) reported that climatic warming resulted in an increased tree population within the timberline ecotone, whereas treeline did not change. Kullman (1990) observed a similar situation in the timberline ecotone on the Swedish Scandes, where no regeneration had occurred since the early 1970s. In contrast, on the northern Urals, where the upper timberline (*Larix sibirica*) had declined to about 280 m since the 13th century, larch young growth (20 to 30 years old) advanced for about 100 to 500 m beyond the ancient forest, which corresponds to an altitudinal rise of treeline of 20 to 30 m (Shiyatov, 1993, 2000). However, trees have not yet reached the potential tree limit, probably because of insufficient seed supply from the distant seed sources at lower elevation to the upper part of the timberline ecotone (Mazepa, 2005). Wind-mediated seed dispersal of larch is not very effective (Shiyatov, 1966).

In the southern Urals, timberline (*Picea obovata*) has intermittently moved upslope about 100 m during the last 300 years, with a maximum during the 20th century (Moiseev et al., 2004; Van der Meer et al., 2004). Regeneration was most intense between 1973 and 1990. The upper limit of closed forest advanced 60 to 80 m compared to 70 years ago (Moiseev and Shiyatov, 2003). The upward shift has been triggered by increased temperature during the growing season. Siberian spruce is the pioneer tree invading the mountain tundra first. Mountain birch (*Betula tortuosa*) and rowan (*Sorbus aucuparia*) follow, facilitated by shelter providing groups of Siberian spruce, The deciduous tree species will be overruled by spruce in the course of time until gaps are created by the natural decay of the spruce stands. In western Siberia (Esper and Schweingruber, 2004), pulses of regeneration occurred in the middle and in the end of the previous century. Regeneration varied regionally, however. Trees that became established at shelter-providing micro-sites are likely to grow several metres high after about 50 years, whilst others growing in windy places will survive as suppressed growth forms.

In the Tianshan Mountains (northwestern China), timberline (*Picea schrenkiana*) did not noticeably move upslope during the last decades. A few trees already became established at treeline 200 years ago. Tree recruitment increased until the early 20th century and then decreased during the period 1950–2000. Several subsequent years of high minimum summer temperatures and high precipitation in spring had likely favoured seedling establishment at timberline (Wang et al., 2006).

Occasionally, a couple of subsequent favourable growing seasons have accelerated height growth in previously low-growing (table or mat growth) conifers. Changing tree physiognomy has been considered an advance of treeline, if trees exceeded 2 m height (Kullman, 1979, 1986a, b, 2000b; Lescop-Sinclair and Payette, 1995; cf. Chapter 3). Along the eastern coast of Hudson Bay, for example, only crippled scrub-like black spruces (*Picea mariana*) were represented at timberline in the end of the 19th century. They had persisted in unfavourable periods of climate by layering. Seedlings, although abundant in the forest, are still rare at treeline. After the 'Little Ice Age', however, when the winters became comparatively mild, many terminal leaders have grown up to 2 m height and taller. On the other hand, in view of many dead stems projecting beyond the average winter snowpack, we might assume that more unfavourable future conditions (e.g., a couple of extreme winters or too cold summers) might interrupt this trend resulting in a dieback of the prominent leaders. Such dieback happened at least twice to the above-mentioned about 500 years old clonal tree islands, namely in the middle of the 17th century and at the end of the 19th century (Payette et al., 1994). The physiognomy of clonal tree islands studied by the present author at timberline on the Rocky Mountains and in Finnish Lapland reflects a comparable history (cf. Photos 125 and 126).

Partly, regeneration occurred only within the timberline ecotone thus resulting in a higher tree density. At the northern treeline near Churchill (Hudson Bay, Manitoba), for example, no young growth could be found until the end of the 1980s, whereas many trees had established themselves in the ecotone itself (Scott et al., 1987a, b). In principle, the same holds true for many high mountains in North America. In the Northwest Territories, for example, tree population increased considerably in the timberline ecotone during the last 100 to 150 years. At treeline, however, no appreciable regeneration had occurred, very likely because of insufficient supply of viable seeds and lack of suitable seedbeds (Szeicz and MacDonald, 1995). Also, in the mountains farther south, the number of trees conspicuously increased at favourable sites in the middle and lower zone of the timberline ecotone. However, almost no seedlings were established near the uppermost stunted trees and clonal tree islands that must have originated under a more favourable climate (see Holtmeier, 1995a).

In most high mountain ranges of North America, relatively favourable climatic conditions during the previous century released advance of conifers to until then almost treeless subalpine sites. In the northern Cascade Mountains (Washington) invasion of subalpine meadows by conifers set in about 1920 at above-average temperatures. Simultaneously, existing tree groups expanded rapidly (Lowery, 1972). *Abies lasiocarpa* colonised mainly dry convex topography, while *Tsuga mertensiana* largely invaded moist and cooler locations. At timberline in the Mount Baker area, trees established themselves mainly on ridges, spurs and similar topography with little snow cover in winter, whereas almost no regeneration occurred in snow-rich and waterlogged small valleys and other depressions (Heikkinen, 1984). In the 1950s, when the climate became more humid and cooler again, the expansion of tree growth came to an end. On Mt. Rainier, invasion of subalpine meadows started about 1910 (Photo 120) with several peaks in the 1990s (cf. Franklin et al., 1971; Rochefort and Peterson, 1996). The establishment patterns on the subalpine zone vary considerably at different locations. Thus, the recruitment has been almost continuous on the west side of Mt. Rainier since about 1930, but has occurred in short, discrete periods on the east side. Obviously, warm, dry summers facilitated tree establishment on the extremely snow-rich west side, whereas cool, wet summers enhanced establishment of trees on the east side where winter snowpack is lower (Rochefort and Peterson, 1996). However, microsites play an important role. On the south-facing slope, mature tree stands as well as young growth are restricted usually to convex topography (Photo 121). The lower slopes and the valley bottom remain treeless as snow may linger here far into the summer as a consequence of extremely snow-rich winters. This causal connection is clearly reflected in Photo 121 which shows the lower distribution limit of the younger tree generations coinciding with the rim of the late-lying winter snowpack.

Also, in the central Cascade Mountains (Three sisters Wilderness Area, Oregon), trees (*Tsuga mertensiana*) could establish in the meadows of the 'subalpine parkland' during the 20th century. This process, however, varied considerably in different locations. *Tsuga mertensiana* colonised mainly northern exposures during the relatively dry period from 1920 to 1945. On southern exposures, young growth rarely occurred at that time but considerably increased when the summers became more humid in the following years (Miller and Halpern, 1998). Thus, at higher temperatures precipitation was the factor controlling the advance of tree growth into until then almost treeless locations in the 'subalpine parkland'.

Similar regional and local differences occurred also on the Olympic Mountains, where conifers invaded subalpine meadows between the early 1920s and 1930s and during the 1940s and 1950s. In normally snow-rich areas, young growth established during years with little snowpack, whereas above-average wet summers favoured tree establishment in areas normally receiving little snow in winter (Fonda and Bliss, 1969; Kuramoto and Bliss, 1970; Woodward et al., 1995; see also Section 4.3.7.2). In Jasper National Park (Alberta, Canada), *Abies lasiocarpa* and *Picea engelmannii* invaded treeless locations in the subalpine between the 1940s and 1960s, and in particular by great numbers between 1965 and 1975 (Kearney, 1982).

From repeat photography used to detect landscape change in the timberline ecotone in Glacier National Park (Montana) during the last 7 to 9 decades, it became obvious that tree population has grown in most sites by about 60% on average. This has partly been attributed to increased snow-pack during the period 1950–1975 (Roush et al., 2007). On Lee Ridge (Glacier National Park) 90% of the trees at treeline became established from the 1930s to the 1980s. In the following, however, tree establishment abruptly decreased. According to Klasner and Fagre (2002; Butler et al., 1994) treedensity has mainly increased within open patches between existing 'krummholz', while a general advance of treeline to higher elevation has not yet occurred. Nevertheless, numerous seedlings have become established in the alpine tundra at microsites protected from harsh climatic influences (Butler et al., 2004). Alftine et al. (2003) suppose altitudinal forest advance in the Pacific Northwest to be related to warmer and drier periods. Surviving trees facilitated establishment of young growth mainly by influencing snow distribution pattern. Future timberline advance is likely to occur by degrees in response to climatic oscillations (Alftine et al., 2003).

As a rule, seed-based regeneration at timberline on the strongly maritime high mountains in western North America depends on the length of the snow-free season (cf. Photos 120 and 121), whereas moisture conditions are the controlling factor in the drier regions (Peterson, 1998). In the White-Inyo Mountains (California), for example, comparatively much young growth

Photo 120. Young growth of *Abies lasiocarpa* that became established in the forest-alpine tundra ecotone during the favourable 1920s to 1940s. South-facing slope of Mt. Rainier (Washington) at about 2.050 m. F.-K. Holtmeier, 3 August 1985.

Photo 121. Same location as above, 12 years later. Obviously, snow-rich winters with extremely late-laying snow prevent tree establishment on the lower slope, probably by shortening the growing season and/or increasing snow fungus infection (e.g., *Herpotrichia juniperi*). F.-K. Holtmeier, 25 July 1997.

established itself during the previous century. Occasionally, *Pinus flexilis* also occurred that originated very likely from the nutcracker's seed caching. Local differences, however, are conspicuous also in this area. While regeneration has been relatively abundant on Sheep Mountain (Photo 122), young growth is almost missing on Campito Mountain (Photo 123), southwest and a few kilometres distant from Sheep Mountain. The differences in regeneration are caused mainly by substrates; loamy soil covered by stone shards on dolomite on Sheep Mountain, permeable sandy soils on sandstone on Campito Mountain. In the Sierra Nevada (California), the invasion of subalpine meadows, the shrinking of persistent snowfields by trees as well as increased growth of horizontal branches and release of vertical leaders above the canopy of *Pinus albicaulis* 'krummholz' reflect the influence of changing climate during the 20th century (Millar et al., 2004). The release of terminal leaders occurred in one pulse in the mid- to late 20th century probably having been primarily caused by increased minimum temperature and reduced numbers of freezing events. Stem release terminated in 1980 thus pointing to less favourable growing conditions than before. Such effects running counter to the century-long climatic trend reflect the role of interdecadal climate variability. The same holds true for the dieback of terminal leaders of conifers in the timberline ecotone in the Rocky Mountain (cf. Photo 24) and in northernmost Finland (cf. Photos 125 and 126).

Photo 122. Young growth of bristlecone pine (*Pinus longaeva*) in the forest-alpine tun-dra ecotone on Sheep Mountain (White-Inyo Mountains, California) at about 3.470 m. F.-K. Holtmeier, 31 July 1985.

Photo 123. At the upper edge of the forest-alpine tundra ecotone (at about 3.420 m) on Campito Mountain (White-Inyo Mountains, California) almost no successful regeneration occurred. F.-K. Holtmeier, 26 July 1994.

Altogether, tree establishment within and above the timberline ecotone during the 20th century varies considerably in the different mountain ranges, and even at different slopes within a single mountain range (cf. Photos 1, 2). The further development is hard to predict. Weisberg and Baker (1995), for example, reported intense recruitment of young growth at timberline (2.900 to 3.500 m) in the Rocky Mountain National Park (Colorado). Young conifers have colonized open patches between the clonal tree islands for several decades. The leaders of the tree islands and also of young individual trees grow relatively vigorously at the given climatic conditions, particularly at wet sites. Pine leaders (*Pinus flexilis*) grow faster than those of Engelmann spruce and subalpine fir. The authors suppose that in case the conifer-scrub patches grew taller the conditions for establishment of seedlings in the upper part of the timberline ecotone would improve. At the present rate of height growth, the apical shoots projecting above the conifer mats will have escaped the critical zone (abrasion, etc.; see Section 4.3.11) just above the snowpack within 13 to 14 years, if no adverse events cause dieback to the apical shoots.

Intense regeneration of conifers occurred at timberline also in our study areas on the Rocky Mountains and other mountain ranges in western North America as well as in northern Europe during the favourable decades of the previous century. Most of the trees, however, have already died or become crippled by climatic influences (Holtmeier, 1970, 1971a, 1974, 1979b, 1979c, 1993a, 1994b, 1995a, 2003; Kullman, 1983a, 1989b, 1990, 1993).

The present authors also studied regeneration in the timberline ecotone in the Rocky Mountain National Park. The sample sites were located on Tombstone Ridge and on a northwest-exposed slope (about 3.500 m) above Cache la Poudre Valley. Our results, however, do not support the findings of Weisberg and Baker (1995). Young growth was established mainly during the middle of the 20th century. Younger trees (10 to 20 years old) are rare, except for moist locations. Most regeneration occurs in the middle and lower zone of the timberline ecotone where the density of trees and tree islands is relatively high. Locally, we found up to eight specimens (≥40 years) per hundred square metres, mainly *Picea engelmannii* growing usually between willow-scrub patches and other wind-protected snow-rich sites, such as leeward slopes of slightly convex microtopography and in shallow depressions. Climatic influences have left their marks on almost all the young growth which, for the most part, exhibit stunted wind-shaped growth forms. At relatively wind-protected sites, the individual trees have lost about 30% of their needles and twigs, while the damages ranged from 75% to total loss on exposed locations. On the average, these trees grow only 2 cm in height per year, may be 4 cm during favourable years. Thus, rapid growth through the critical zone above the snowpack appears very unlikely and long-term distorted growth will also be normal. There are exceptions, however. In some places, wind-depressed 'krummholz' has spontaneously released vertical stems during the last years. The stems have grown fast enough within a couple of favourable year reaching beyond the zone of permanent abrasion by wind-driven ice crystals (cf. Figure 56). Locally, however, trees are colonizing the Alpine, without being promoted by microtopographical structures or alpine vegetation (Photo 124) as it often happened in the past (e.g., Photo 87).

The observation in the Rocky Mountain National Park is in accordance with those in other timberline ecotones on the Colorado Front Range. Probably, they reflect the cooling at high elevation that has occurred since the middle of the 20th century as is evidenced by the temperature data from Niwot Ridge (about 25 km south of Rocky Mountain National Park). Annual mean temperature has decreased by 1°C since 1953. The cooling has run parallel to an increase in precipitation. At the meteorological station D-1, which is located on Niwot Ridge at 3.750 m, the amount of precipitation has increased by 300 mm, while global radiation in summer considerably declined because of the higher cloudiness (Williams et al., 1996). At lower elevation, temperature has risen during the same period (Pepin, 2000). A similar trend has been observed in western Europe (Garnett et al., 1997) and in the northern part of Fennoscandia (see below).

Photo 124. Young Engelmann spruce (*Picea engelmannii*) invading the alpine tundra on a south-facing slope on Ida Ridge (Rocky Mountain National Park, Colorado). The highest new colonist is at an altitude of 3.612 m). J.B. Benedict, 28 July 2006.

The response of the timberline tree species (pine, larch, fir, birch) to the improving climate was different according to their specific ecological properties and requirements and to the site conditions (e.g., Griggs, 1934; Zinserling, 1934; Bathen, 1935; Blüthgen, 1937a, b, 1942, 1943, 1952, 1970; Hustich, 1937, 1942, 1948, 1949, 1958; Söyrinki, 1939; Aario, 1940; Marr, 1948; Regel, 1949, 1950; Mikola, 1952; Kujala, 1952; Erkamo, 1956; Grigorjew, 1956; Kallio and Mäkinen, 1957; Siren, 1958, 1961; Haugen, 1965; Holtmeier, 1965, 1970, 1971a, 1974; Franklin et al., 1971; Nichols, 1975a, b; MacDonald and Gajewski, 1992; Earle et al., 1994; Holtmeier and Broll, 2007). On Beartooth Plateau (Montana/Wyoming), for example, great numbers of whitebark pine seedlings (*Pinus albicaulis*) invaded the treeline ecotone while young Engelmann spruce (*Picea engelmanni*) and subalpine fir (*Abies lasiocarpa*) are nearly absent (Mellmann-Brown, 2002, 2005). Also, the tree species at timberline in our study areas in the Rocky Mountains respond to changing climate in a different way. Thus, young growth of *Picea engelmannii* is usually more frequent in the timberline ecotone than young growth of *Abies lasiocarpa*. For example, along a transect (length 910 m, width 5 m) crossing the timberline ecotone on the south-facing slope of Niwot Ridge (Colorado Front Range) 73% of 'young' conifers (<2 m) were spruce, 9% subalpine fir and 18% limberpine (Holtmeier, 1999b). Usually, zoochoric pines (*Pinus albicaulis*, *Pinus flexilis*) regenerate more successfully at timberline than anemochorous Engelmann spruce and sub-alpine fir (Holtmeier, 1995a).

In northern Europe, differences in timberline response to climate change also are apparent. In the Swedish Scandes, for example, recently and locally increasing numbers of fast-growing seedlings of Norway spruce (*Picea abies*), Scots pine (*Pinus sylvestris*) and mountain birch (*Betula tortuosa*) have become established 400 to 500 m or even 700 m above the present tree limit (Kullman, 2003, 2004a, 2005a, b, c; Kullman and Kjällgren, 2006). This may support the hypothesis of Tinner and Kaltenrieder (2005) of prompt timberline response to climate change. The authors suggested a rapid upslope shift of treeline for about 800 m within a few decades in the Swiss Alps. In the Handölan Valley (southern Swedish Scandes), a consistent expansion of Scots pine has occurred since the late 1980s. Very likely, the exceptionally warm summers since 1997 and low mortality due to milder winters have been the driving factors (Kullman, 2007a). Advance of birch and spruce treeline in this area has resulted almost exclusively from the release of upright stems from old-growth 'krummholz' (Kullman, 2007b). In many places, trees seem to be growing more vigorously than just a few decades before and tree crowns also appear to be denser (see also Wielgolaski et al., 2005). In other places, however, height growth of climatically-stunted trees growing on wind-swept topography with little or no snow in winter is still impeded, despite the generally warmer climate (cf. Holtmeier, 2003, 2005a, b).

In northern Norway, stable or advancing treelines are common in the southernmost and probably in the middle regions whereas treeline is declining in the north (Dalen and Hofgaard, 2005).

On the mountains in western central Finnish Lapland (Pallastunturi area), Scots pine (*Pinus sylvestris*) became established at great numbers in the timberline ecotone during the favourable period from the 1920s to the 1940s (cf. Hustich, 1937, 1942, 1958; Blüthgen, 1942, Holtmeier, 1974; Holtmeier et al., 1996, 2003; Autio and Colpaert, 2005) while comparatively few new pines occurred within and beyond the birch-treeline ecotone on the mountains in the northernmost part of Finland during the same period of time (Holtmeier et al., 2003; Holtmeier, 2005a). It is possible that climatic conditions were less favourable to seedling establishment when compared to the more southern mountains. The present situation is similar. Scots pine has been invading the birch-treeline ecotone and locally even the alpine zone during the last 3 to 4 decades. Isolated mature pine stands (outliers of the boreal coniferous forest) on the valley floors and valley sides are acting as a seed source (Holtmeier et al., 2003). Again, however, also recent pine recruitment within and beyond the treeline ecotone is sparse, compared to southwestern Finnish Lapland where pine has considerably increased since the 1970s, particularly since the mid-1980s, until the end of the previous century (Holtmeier et al., 1996, 2003; Tasanen et al., 1998; Holtmeier, 2005a; Holtmeier and Broll, 2007).

The annual mean temperature (1962–2007) in northernmost Finnish Lapland does not reflect an exceptional warming (see also Juntunen et al., 2002; Timonen, 2002). Mean temperature of the growing season (June–September, Figure 87), however, better characterize the thermal conditions that are influencing tree regeneration and tree growth. After the growing season temperatures had ranged mostly below the average during the 1960s (falling to at minimum in 1968) and they culminated in the beginning of the 1970s. In the following years they fluctuated closely around the average. Since the end of the 1990s mean temperature of the growing season have continuously stayed above the long-term mean (9.65°C) though not reaching the maximum of the early 1970s. This may be taken for an episode or an effect of a gradually warming climate.

Although, regeneration of pine does not require exceptionally warm summers, some above-average summers (June, July and August in 1972–1974, during the 1990s and in the beginning of the present century) are likely to have favoured production of viable seeds and seedling establishment. In general, production of great quantities of good quality seeds in occasional warm summers may be so important for pine regeneration that intermediate years will have only a limited effect (Renvall, 1912; Ågren and Zackrisson, 1990; see also Harju et al., 1996).

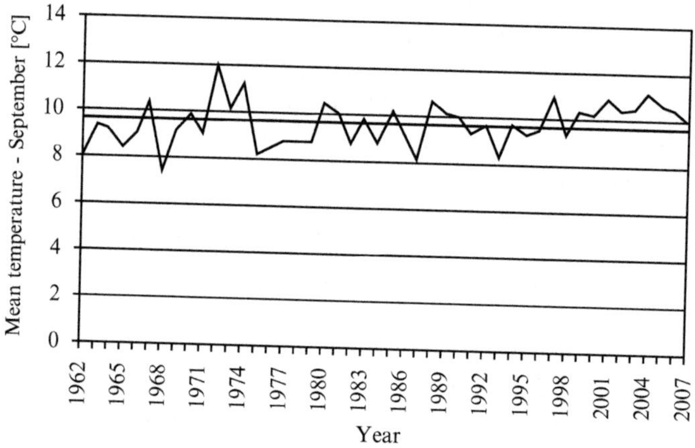

Figure 87. Mean temperature of the growing season at Kevo Subarctic Research Station (about 90 m, northern Finnish Lapland). The horizontal line shows the long-term average. Temperatures at the upper timberline are generally lower than at the relatively wind-protected Kevo site. – Data provided by Kevo Subarctic Research Station.

However, climatic conditions have not been favourable enough for undisturbed growth. Many of the young pines have lost their terminal leaders and show reduced height growth, clipped surface corresponding to the winter snow surface and other growth disturbances. They may survive for decades as suppressed growth forms 'waiting' for a couple of more favourable years that would allow undisturbed height growth and attaining 'tree size'. If applying the conventional minimum tree height to the new pine generations on the northernmost fjelds tree-limit advance as has been reported from the Swedish Scandes (Kullman, 2000b, 2001, 2004a, c, 2007b) would not yet have occurred.

Scots pine (*Pinus sylvestris*) and, to a lesser extent, Norway spruce (*Picea abies*) that became established at comparatively favourable climatic conditions in the timberline ecotone (400 to 500 m) in Pallas-Ounastunturi National Park (Finnish Lapland) from the 1920s to the 1940s (Blüthgen, 1942; Hustich, 1937, 1942, 1958; Holtmeier, 1974; Holtmeier et al., 1996) showed climatically-shaped growth of 30 years ago (Photo 125). Many had already died. A great number of younger specimens, which were still protected from injurious climatic influences by the snowpack at that time, have died since. Most of the survivors exhibit crippled, shrubby growth caused mainly by loss of the apical shoots. A few spruces had become established 80 to 100 years before the present have continuously reproduced by layering. Thus, compact tree islands developed, partly exceeding 2 m height (Photo 126). During the last 30 years, most of the shoots that were projecting above the snowpack have died or lost their foliage, particularly at their windward sides.

Also, Scots pines that have vigorously grown up at the leeward edge of the clonal spruce islands in this area during the last 15 to 25 years exhibit damaged apical shoots now and will not be able any more to develop 'normal' growth. Thirty years ago, almost all age classes were represented in the lower part of the timberline ecotone, while seedlings and young growth were comparatively rare in the more open upper ecotone. In the following years, regeneration considerably increased until the 1980s. Spruce in particular has intensively reproduced by seeds. The proportion of spruce seedlings is three to five times higher if compared to pine seedlings. Nevertheless, individuals younger than 10 years rarely occur. Young growth that originated from the second third of the previous century has become damaged by external factors (climate, reindeer), though not exceeding 20 to 40 cm in height and still being protected in winter by the snowpack for the most part. Obviously, these individuals could not recover in the following years. This corresponds to the negative trend of diameter growth that has occurred in the older pines in the lower part of the ecotone after the 'thermal optimum' of the 20th century. Radial growth is declining since the late 1970s, as has been evidenced by age-detrended increment cores (Holtmeier et al., 1996). Also, other tree-ring data of Scots pine in Finnish Lapland and on Kola Peninsula reflect sharp decrease in radial growth (Raspopov, 2002).

As to the future development in the middle and upper zone of the timberline ecotone one should not be too optimistic, as was for example Blüthgen (1942). He wrote that in view of the intensive regeneration during the favourable period of the previous century pine forests were advancing to their uppermost and northernmost Holocene position. Also, response of the tree species to changing environment is different. Young growth and seedlings of Scots pine and mountain birch are rare compared to spruce and pine (Table 19). Being less sceptical, Autio and Colpaert (2005) consider recent regeneration of Scots pine and Norway spruce in the timberline ecotone to be successful. Studies on some fjelds located south and southeast of Pallas-tunturi (Aakenustunturi, Yllästunturi, Pyhätunturi) during the last 3 decades showed this successful regeneration to be an indicator of climatically-driven forest advance.

Photo 125. Pinus sylvestris that became established at about 500 m on Sammaltunturi (Pallastunturi area, Finnish Lapland) during the favourable decades of the 20th century. Most of the pines dating from that period are more or less heavily damaged by climatic injuries. F.-K. Holtmeier, 19 October 1996.

In our study areas in northernmost Finnish Lapland (Utsjoki commune), the forest-alpine tundra ecotone is characterized by a mosaic of birch groves alternating with open almost treeless areas covered by fjeld vegetation. It is an open question whether birch forests will invade the treeless location in the foreseeable future and which factors might impede forest advance.

Photo 126. Clonal group of *Picea abies* on the southwest-facing slope of Pallaskero (Pallastunturi area) at about 500 m. The terminal leaders, which are no longer protected by the winter snow cover, have lost almost all needles, and some have died. F.-K. Holtmeier, 26 July 1970.

Table 19. Percentage of tree seedlings in transects of Pallastunturi [1]

Tree species [%]	Transect 1 (Palkaskero) n = 52	Transect 2 (Palkaskero) n = 84	Transect 3 (Palkaskero) n = 83	Transect 4 up to 550 m n = 37	(Pyhäkero) 550 m – 745 m n = 15
Spruce	77	70	70	30	67
Pine	17	25	21	20	27
Birch	6	5	9	50	6

[1] From Holtmeier et al. (2003).

Our studies were conducted on several mountains (420 to 500 m) south of the Tana River during the last 5 years. Closed forests end at about 300 m, giving way to low alpine (subarctic) dwarf shrub-lichen-heath. The crystalline bedrock is covered by 2 to 4 m of sandy-skeletal glacial till. The sandy substrate drains rapidly and thus is highly susceptible to deflation if not covered by vegetation (see also Laine and Nurmi, 1971; Ukkola, 1995). Commonly, convex topography is heavily eroded by strong winds and bare mineral soil is exposed (Holtmeier, 1979b, 1974). Only small patches of low alpine (subarctic) health are left. Intensive reindeer grazing and trampling have enhanced deflation. Within the valleys, alongside little streams at moderately wet sites, and relatively wind-protected snow-rich locations, 2 to 4 m high solitary birches and birch groves can be found up to 350 m. The average age of these trees ranges from 80 to 90 years. The oldest living mountain birch we found was 225 years old, however.

Large treeless areas extend between valleys and gullies. Previously, birch forest had covered these areas (Holtmeier and Broll, 2006), before they were destroyed, very likely by mass-outbreaks of the autumnal moth (*Epirrita autumnata*; cf. Photo 105; Figure 82). Such 'catastrophes' are peculiar to the mountain birch forest in this area. However, also humans might have contributed to the birch forest decline by firewood cutting and grazing reindeer. The comparatively large sporadic birches that often occur far above the present closed tree stands as well as eroded relic Podzols and wood remains which are found mainly in eroded peaty layers and in eroded peaty hummocks very likely indicate the former extent of the mountain birch forest (cf. Photo 16; Holtmeier and Broll, 2006). Thus, our view is somewhat different from Mattson (1995), who supposes that birch timberline on the valley slopes has not declined.

Seedling density, which is relatively high within the open birch forests, rapidly declines by altitude. Locations covered by dwarf-shrub heath are comparatively favourable to seedling establishment and growth, if no dense vegetation prevents wind-borne birch seeds from reaching a seedbed. Also, allelopathic effects of dwarf shrubs (e.g., *Empetrum hermaphroditum*) may be a critical factor to birch seedlings. Thus, grazing may enhance seedling establishment by reducing competition and exposing bare mineral soil.

On typical deflation areas (cf. Photo 36 and Figure 31), sparsely covered with vegetation, seedlings are usually rare. There are exceptions, however. Occasionally clusters of up to ten, only a few centimetres high birches per square meter have been established, mainly in open patches between dwarf shrub-patches and also on open mineral soil in the deflation areas. At close sight, however, supposed 'seedlings' often reveal themselves as sprouts thriving from old buried rootstocks.

As almost all recent birches are more or less damaged by reindeer and also by snow hares and voles, one might assume that grazing and trampling had also destroyed the older generations. Browsing by snow hare seems to be most intense during the winter on wind-swept topography with little or without snow cover and does not occur in comparatively high dwarf shrub vegetation. The same has been observed in Scottish mountains (Rao et al., 2003).

In contrast to wind-exposed dry sites, birch seedlings are abundant at wet and moderately wet places close to the little streams, where dense willow and dwarf-birch thickets give way to open sedge, grass and herb vegetation (cf. Photo 45; Figure 88). At such places (at about 380 m) we found up to 25 only a few years old seedlings per square meter. Also, in willow shrubs birch seedlings occur at relatively great numbers (Anschlag, 2006; Broll et al., 2007). On the other hand, almost no seedlings occur within compact stands

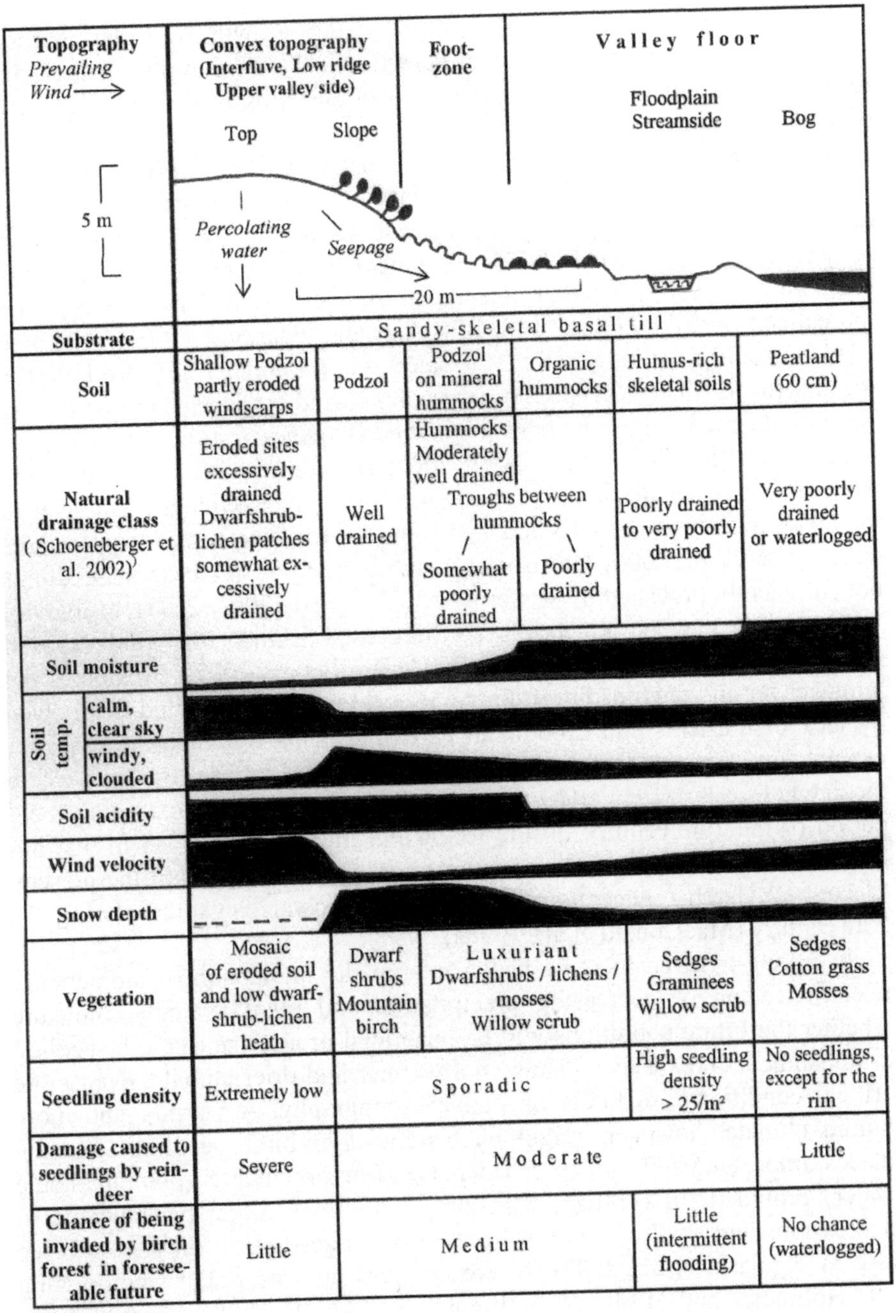

Figure 88. Topography dependent site conditions and their effects on regeneration of mountain birch (*Betula tortuosa*) in the forest-alpine tundra ecotone (schematic) and chance of present treeless sites to be invaded by mountain birch in foreseeable future. The model is based on field data from Mt. Rodjanoaivi (northern Finnish Lapland). Modified from Broll et al. (2007).

of dwarf-birches. Older seed-produced young growth of birch is rare. Damages caused by grazing are not very frequent, and trampling effects are restricted to reindeer paths crossing the wet and mosquito-rich valleys and other depressions.

With continued warming, a climatically driven advance of the birch forest to higher elevation could be expected in the shallow valleys and in similar debirch stands (*Betula tortuosa*) that had locally become established in snowrich wet sites above the Scots pine forest (*Pinus sylvestris*) in the southern and central Swedish Scandes during the 'warm' Holocene period (boreal-sub-boreal, 9.000–2.500 BP; Aas and Faarlund, 2001; Kullman, 1983b, 2004b) seemingly reflect the same rule (Holtmeier and Broll, 2005). Birch will not likely invade almost treeless wind-eroded locations, mainly because of extreme site conditions (high wind velocities, insufficient snowpack, reindeer; cf. Holtmeier et al., 2003, 2004; Broll et al., 2007). Thereby low winter soil temperatures in the rooting zone of the birch seedlings might play an important role (Sections 4.3.5 and 4.3.6). In case of rising winter temperatures (e.g., Jylhä et al., 2004) late frost damage is likely to increase in seedlings not sufficiently protected by winter snow (Skre et al., 2003, 2008). Moreover, wind erosion may be adverse to seedling establishment on wind exposed topography (Holtmeier et al., 2003, 2004; Broll et al., 2007). Erosion of the humus layer and the topsoil (cf. Photo 36 and Figure 31) by wind is followed by lack of nutrients and drought in particular. This was also the case, for example, on the uplands along the lower Lena River (northern Siberia), where the larch forests (*Larix sibirica*) had collapsed during the extremely cool period of the 19th century. In the following, intense erosion set in that has prevented recolonisation of the former forest area by larch until the present. In contrast, larch regenerated prosperously at lower elevations during the 20th century (MacDonald et al., 1998a).

In a long-term perspective it depends on the interaction of temperature rise, increasing or decreasing precipitation and plant available moisture whether the future conditions will be beneficial or adverse to birch seedling establishment and survival. Thus, in a warmer and drier climate, even more critical conditions are likely on convex topography. A warmer and more humid climate, however, might be beneficial to birch seedlings in such places, improving soil properties (increase of organic matter, plant available water) provided. In contrast, worsening conditions might be expected in depressions and shallow valleys, because of longer-lasting snow cover (see also Wielgolaski et al., 2003), shortened growing season and waterlogging (cf. Holtmeier and Broll, 2005; Figure 89). Abiotic soil properties are a relative constant factor which does not spontaneously respond to the changing climate (see also Woodward, 1998).

In all our study areas in Finnish Lapland, over-grazing by reindeer is an important if not the main factor preventing the advance of birch in the present treeless areas within the timberline ecotone (see also Lehtonen and Heikkinen, 1995; Neuvonen et al., 1996, 2001; Anschlag, 2006). Thus, the

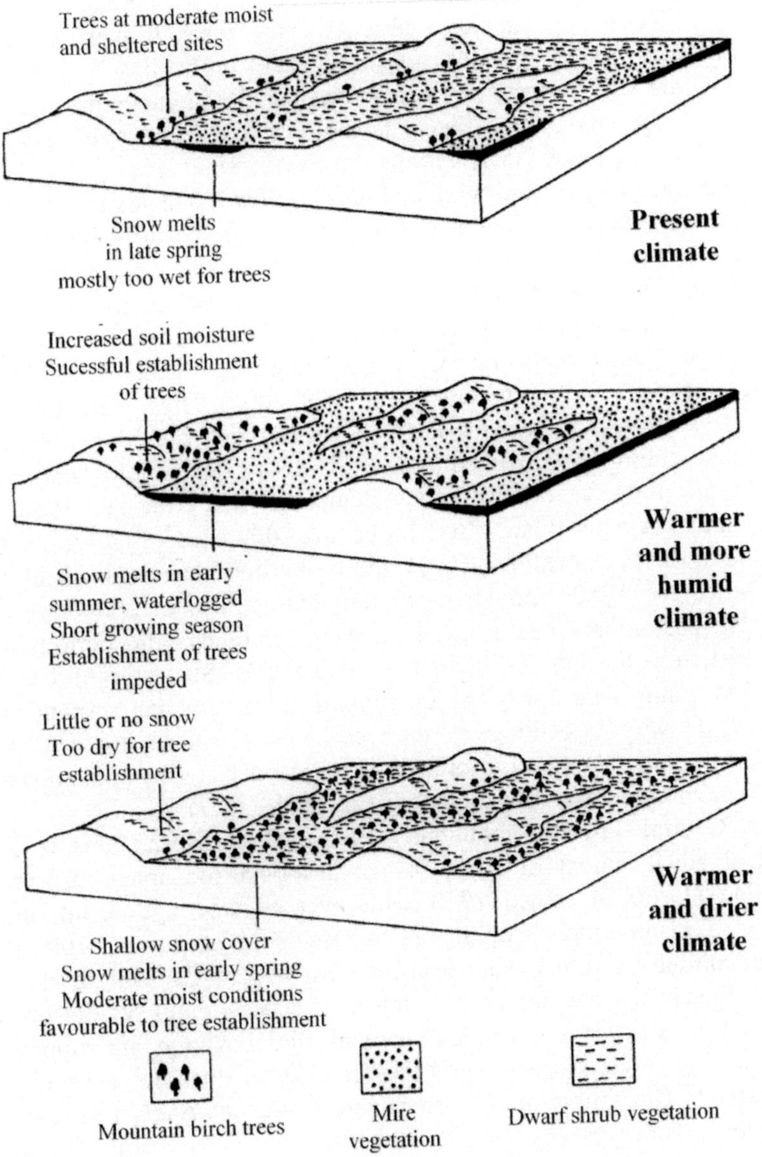

Figure 89. Scenario showing possible changes of a topographically-controlled vegetation pattern in the mountain birch (*Betula tortuosa*) timberline ecotone in northern Finnish Lapland. Modified from Holtmeier and Broll (2005).

situation is somewhat different from the southern Swedish Scandes, where the effects of reindeer on present vegetation change and long-term timberline dynamics seem to be less important (Kullman, 2007b). In addition, recurrent defoliation by the autumnal moth (*Epirrita autumnata*) and the winter moth (*Operophtera brumata*) may impede regeneration and spread of birch into the present treeless patches within the birch-timberline ecotone. Locally the mass-outbreaks may even cause retreat of the birch timberline (Section 4.3.13.4). If warming continues, defoliation will probably affect also birch forests that are still protected by extremely low frost temperatures regularly occurring in the cold air on the valleys floors and foot-slopes (Neuvonen et al., 1996, 1998, 1999, 2001; Virtanen et al., 1998).

In the Spanish Pyrenees, treeline (*Pinus uncinata*) had advanced since the middle of the 18th century and reached its maximum elevation between 1900 and 1950. Between 1955 and 1975, warm springs and wet summers again enhanced regeneration in the timberline ecotone. Treeline, however, did not advance further (Camarero and Gutiérrez, 2000). Notably, *Pinus uncinata* expanded parallel to increasing May temperatures, before any major decrease in pastoral use (Camarero and Gutiérrez, 1999, 2004). Thus, warming climate is likely to have been the driving factor (Camarero and Gutiérrez, 2007). Pine recruitment temporarily declined in the early 1970s due to cold summers and failed in the northeast of the Iberian Peninsula. Since the late 1980s pine encroachment decreased (see also Ninot et al., 2008).

In the northern Appenines (Italy), the timberline has remained relatively stable since the 1950s though mean temperature has increased by about 1.3°C and pastoralism declined. There were relatively cold periods in the 1960s and from the late 1970s to the early 1980s. Since the mid-1980s a sudden warming occurred to which timberline has not yet responded (cf. Photo 19). However, seedlings might become established in the near future, very likely close to the present forest limit, especially in *Vaccinium myrtillus-Vaccinium gaultherioides* shrublands (Pezzi et al., 2007).

In the Central Bulgarian mountains (Vezhen Peak; Meshinev et al., 2000) a gradual advancement of *Pinus peuce* has occurred above the present timberline (1.760 m) since 1970 with average tree age decreasing by elevation. The uppermost seedlings (1 to 2 years old) occur at 2.100 m. The original coniferous forest was destroyed by fire in the end of the 19th century. Survived trees are now acting as a seed source. As the growing season temperature have not much changed milder winters are supposed to be the driving force. Norway spruce (*Picea abies*) does not expand to the same extend. The timberline (*Pinus peuce, Pinus heldreichii*) on the Pirin Mountains (south-west Bulgaria) seems to be caused mainly by aridity during the growing season rather than by heat deficiency. Particularly regeneration is affected by longer dry periods and may fail for decades.

Although drier summers have been predicted for the 21st century *Pinus heldreichii* is likely to profit from rising temperatures as long as the amount of winter snow will not decrease (Grunewald and Scheithauer, 2008).

In many areas, recent establishment of trees in the timberline ecotone and, locally, above treeline results from declined pastoral use or other human impact and thus should not be confused with a climatically triggered advance of natural timberline. This holds particularly true for the Alps (Holtmeier, 1965, 1967a, 1974, 1986b; see also Stützer, 2000, 2002, 2004, 2005). However, also the rise of birch forest limit in Norway is commonly a consequence of decreased grazing and wood-cutting (Holtmeier, 1974; Aas and Faarlund, 1996; Bryn and Dangstad, 2001); see also Photo 108). Hofgaard (1997a, b, 1999), referring to timberline in central Norway, points out that vegetation response caused by changes in grazing pressure might be confused with a response to climate change. In eastern Norway (Vågå Upland), recovery of fragmented mountain birch forests (*Betula tortuosa*) on former alpine pastures and advance of single trees to greater altitude obviously result from decrease in land use rather than from direct response to climate change (Rössler and Löffler, 2007). However, regional differences exist. Thus, Kullman (2007b), for example, does not share this opinion as the magnitude of the treeline rise during the 20th century is not statistically related to past or present land use in his study area in the Swedish Scandes (Kjällgren and Kullman, 1998).

Warming climate might have favoured tree invasion into the previous high-elevation pastures. It has become obvious from our field studies however, that its positive effects are overridden by microclimates and resulting effects that became exacerbated after removal of the natural high-elevation forests by humans (cf. Figure 84; Section 4.3.14.2). The now locally varying mosaic of strongly contrasting site conditions controls the advance of tree growth to higher elevation, also at a slightly warmer climate.

At high elevation in the Northern Corries (northern Cairngorm Mountains, Scotland), Scots pine has been expanding since the end of the 1950s when burning was stopped and grazing by sheep and red deer ceased (French et al., 1997). Most of the trees, however, display crippled growth as is typical of a natural climatic timberline (see also Watt and Jones, 1948). Above 700 m elevation, young growth was established only in protected microsites. Apparently, the breaking of new ground for winter sports and construction of an access road and a ski lift in 1960 had a positive effect on advance of pine, as red deer retreated because of increasing disturbance by the growing number of visitors in this area. At the same time, fences were erected in some places to keep red deer off the area. Although a few warm summers and mild winters occurred between 1950 and 1971, a causal relationship to the expansion of pine was not evident. Also, decrease of regeneration after 1971 could not be attributed to deteriorating climatic conditions. Probably, pine had colonized the most favourable sites at that time. If grazing remains low and mortality of young pines does not increase due to any possible catastrophe, pine will gradually advance to its upper potential limit (French et al., 1997). Locally, however, the above-mentioned positive effect of increasing number of visitors on the expansion of pine is countered by increasing damage caused to the vegetation and soil (Watson, 1991; Gordon et al., 1998).

Just as intensive grazing causes long-lasting effects to timberline so do forest fires. They have caused timberline decline on many temperate and tropical mountains. Humans set most of the fires. In some regions of the Himalayas, for example, and at many tropical timberlines, fire still is one of the most important factors controlling vegetation and site conditions (Section 4.3.14.2). In the Alps, forest fires have become rare in modern time, particularly at timberline. The situation is different at mountain timberline in North America. Fires frequently occur in the high-elevation forests up to timberline though the number of fires has considerably decreased due to fire control, when compared to pioneer times.

Often, fire prevents any reforestation and timberline advance to its present climatic upper limit. The direct and indirect effects of forest fires are normally not sufficiently considered when discussing the potential rise of timberline. They influence population dynamics and spatial structure in the timberline ecotone more effectively than the warming trend (see also Suffling, 1989; Sirois and Payette, 1991).

Fires have shaped the spatial distribution pattern of vegetation and age structure of the forest stands, particularly at high elevations where the climate seriously impedes natural regeneration. While burns at lower elevation will promptly be colonized by aspen (*Populus tremuloides*) and/or lodgepole pine (*Pinus contorta*) as a rule (see also Stahelin, 1947; Daubenmire, 1953; Oosting, 1956; Hoff, 1957), post-fire succession is usually considerably delayed or even missing for long times at climatically extreme high-elevation sites, where sexual reproduction is almost an exception (Section 4.3.10.1). Regeneration on burn areas will proceed extremely slowly if the thick organic layer and even the humus-rich top soil have been destroyed. Wind, heavy rainfall, and meltwater-runoff remove the charred soil remains and expose the mineral soil (Photos 127 and 128).

Grasses, herbs and shrubs invading the burns are favourable to pedogenesis. At increasing coverage, however, they may prevent seed germination and establishment of tree seedlings. Intense solar radiation will normally kill sporadic seedlings after 2 to 3 years. Conditions are most favourable to seedling establishment at sites where shrubs (e.g., willows) provide shade to the seedlings and enhance accumulation of snow and litter. Thus, they improve moisture conditions (Bollinger, 1973). Altogether, extreme conditions

Photo 127. Old burn with *Pinus flexilis* on Trail Ridge (Front Range, Colorado) at about 3.350 m. Seedlings were still rare in the late 1970s. F.-K. Holtmeier, 1 July 1979.

Photo 128. One year old burn near Coney Lake (Front Range, Colorado) at about 3.250 m. The area was covered by clonal groups of *Picea engelmannii* and *Abies lasiocarpa* that had reproduced mainly by layering. After the fire, strong winds and runoff have eroded the organic layer, except for some burned remains (in the middle of the photo). F.-K. Holtmeier, 6 August 1989.

on burn areas in the ectone may prevent the forest from coming back for a hundred or more years, despite seed supply from the nearby seed trees. Thus, the situation is somewhat different from burns at the northern treeline where sexual regeneration mostly depends on long-distance transport of viable seeds.

Temperature has been increasing also in High Asia during the last decades (Böhmer, 1996). The warming climate has caused glacier retreat (e.g. Qin et al., 2000; Karma et al., 2003) and is likely to trigger an altitudinal shift of timberline. Upslope expansion of *Pinus wallichiana* on both south-facing and north-facing slopes has been reported from the western Himalayas. Pine advance was ascribed to the changing climate (Dubey et al., 2003).

Information on timberline response to the changing climate in the southern hemisphere is comparatively scarce, except for New Zealand. On the South Island of New Zealand, young growth of silver beech (*Nothofagus menziesii*) and mountain beech (*Nothofagus solandri*) has advanced beyond timberline. However, most young growth occurs within 9 m of the outer canopy edge. A few individuals can be found up to 30 m distance from the closed forest stands. Thus, young growth has advanced less than would be expected at the 0.6–1.0°C rise of the mean temperature during the last 100 years (Wardle and Coleman, 1992; Wardle, 2007). Very likely, seed rain rapidly decreasing by distance from the seed source and extreme micro-climates (intense solar radiation; frequent frosts, etc.) above the forest stands have impeded establishment of seedlings. Also, competition of the dense snow tussock-grass vegetation might have suppressed tree seedlings. On the other hand, the roots, which usually extend for several metres from the marginal upper tree stands into the grassland, very likely made inoculation of the seedling roots with mycorrhiza easier. Comparatively sparse regeneration above the forest line must be attributed to the relic character of the present upper forest limit that became established long ago under warmer climatic conditions and persisted during the most recent cool interval due to the longevity (360 years for mountain beech and 600 years for silver beech, Wardle, 1984) and vegetative reproduction of *Nothofagus*. On spurs and northern (sun-exposed) slopes, young growth of silver beech advanced highest above the high-stemmed forest. At similar locations, east of the main divide, where mountain beech displays only severely stunted growth forms, no young growth can be found. Likely, strong downslope winds and frequent clear sky (intense solar radiation) cause severe stress to the plants that increased regional mean temperature cannot overcome (Wardle and Coleman, 1992). Cullen et al. (2001) emphasize that there has been no recent climatically-driven upward movement of the tree line or increase in seedling establishment. Instead, natural disturbances creating open patches with reduced competition from herbs and shrubs obviously play an important role for recruitment in the closed-canopy tree line stands of southern beech.

At the treeline in Patagonia, *Nothofagus pumilio* has not successfully regenerated although temperature has increased since 1970 (Daniels, 2000). Regeneration failed presumedly because of moisture deficiency. On Tierra del Fuego, spatial restriction of recruitment, episodic regeneration and also herbivory by the native guanaco (*Lama guanicoe*) seem to prevent timberline advance (Cuevas, 2002).

The consequences of global and regional climatic change for timberlines and mountain forests in tropical mountains have almost not been considered so far. More frequent and persistent droughts resulting from climate change might affect timberlines more than would the rise of average temperature (Rundel et al., 1994). However, this is pure speculation, in particular as geoecological paradigms developed from studies on temperate treelines are unlikely to be of any value for understanding timberline on tropical mountains (Biondi, 2001).

From recent studies in the Andes of northern Ecuador (Bader et al., 2008b) it becomes apparent that a spontaneous response of tropical timberline to a warming climate is rather unlikely. Timberline has not advanced for the last 40 years although it appears to be located below its natural climatic limit (e.g., Lauer et al., 2001). Obviously, regular Páramo fires override possible timberline-rising effects of climate change. The Páramo vegetation is burned every 3 to 6 years these days (Di Pasquale et al., 2007), locally even every year (Ellenberg, 1979; Balslev and Luteyn, 1992). Hence, a cessation of burning would very likely have a much greater effect on timberline spatial patterns and dynamics than a rise of temperature, for example (cf. Photos 22 and 113). Recurrent Páramo fires are suggested to prevent forest advance in a closed front, possibly combined with other factors such as excessive solar radiation and extreme daily amplitudes of temperature that impede tree establishment in the Páramo. Also in East Africa, fire still is used to prevent high-altitude pastures from being invaded by trees or to create new pastures.

Bader et al. (2008b) present some scenarios of possible timberline response to a changing (warming) climate. Cessation of recurrent Páramo fires might result in an upward shift o timberline if radiation-tolerant tree seedlings became established. Otherwise timberline would not advance. If burning ceased and radiation-tolerant tree seedling were present timberline would advance very slowly. In this case, however, timberline would shift upslope anyway without being driven by a warming climate because it is commonly located below its climatic altitudinal limit. Besides human impact delayed response of the timberline to postglacial warming is supposed to be a reason for the relatively low position of the present timberline. The comparatively high stability of the present abrupt timberlines seems to be at least partly attributed to the shade-dependence of most of the forest tree species. Shading by the trees as well as lower and less fluctuating temperatures facilitate seedling establishment and thus self-maintenance of the forests (Bader et al., 2008b). This hypothesis should be tested also at other tropical timberlines.

5.5 Conclusions and perspectives

Response of altitudinal and polar timberlines to changing climate is ubiquitous. The general trend, however is being modified by regional, local and temporal variations and thus is different as to its extent, intensity and process of change. Consequently, it seems difficult to make generally acceptable statements without restricting them at the same time in view of the many regional and local peculiarities. However, when passing the previous chapters in review, some aspects become apparent, that are of general relevance for timberline fluctuations (Table 20).

- The present advance of timberline to greater elevation and a more northern position often is a reforestation of sites that had been under tree cover earlier in the last millennium. The after-effects of landscape and site history often have a long-lasting impact and probably are the most important agents shaping the present timberline (Figure 90, see also Figure 1).
- In case the thermal level were distinctly higher compared to the present conditions and if no other factors adversely affected regeneration and tree growth, the tree population in the timberline ecotone would increase and timberline advance to higher elevation. However, establishment of trees beyond the current tree limit would be relatively moderate compared to the increase of tree density in the timberline ecotone itself.
- Higher tree density and real upward shift of the treeline would result from effective sexual regeneration rather than from increasing growth of mature trees and release of erect stems from suppressed growth forms ('krummholz'). Regeneration would be effective if the trees were able to survive and mature at their utmost potential climatic limit. At timberline on temperate mountains and in the Subarctic, the situation would be exacerbated when the young trees start growing beyond the average winter snowpack. From then, height growth of the leaders would depend on their physiological and mechanical tolerance to climatic influences.
- Occasional extreme events, such as strong frosts during the growing season, particularly late and early frost, extremely snow-rich winters or winters with nearly no snow, forest fires, heavy wind storms, droughts, insect infestations, pathogens, diseases, volcanic eruptions, etc., might cause dieback and control timberline dynamics into the far future, in a similar way as site history has influenced the present situation at timberline (cf. Figures 1 and 90; see also Holtmeier and Broll, 2005, 2007). However, extreme events and their relative importance for timberline change cannot

be predicted. It seems likely that changes in mean climate will be associated with changes in climatic extremes as well (Jylhä et al., 2004).

- While climate is changing, local topography and microtopography are the only relatively constant factors in the timberline environment. Thus, topography and substrate will play an important role as they did during the Holocene vegetation history (e.g., Pregitzer et al., 2000). The effects of microtopography on site conditions and timberline dynamics would be different under continental climates compared to maritime climates. As topography is a key factor determining the site pattern, high-elevation forests would not advance as a closed front to higher elevations, parallel to the upward shift of any isotherm considered to be essential to tree-growth. Trees would first establish themselves at the favourable sites, while other locations would remain tree-less for the long term (Schönenberger, 1985; Holtmeier, 1989). The effects of microtopography on solar radiation and wind and resulting consequences to the microenvironment would primarily control site conditions and the distribution pattern of the surviving young growth (Sections 4.3.6 and 4.3.7.2). Microtopography also would control relocation of solid and suspended substances by surface runoff and seepage (Section 4.3.9.1).

Table 20. Factors promoting timberline advance to higher elevation

Climatic factors	Other factors
• Warm an frost-free growing seasons • No extreme climatic events fatal to seedling establishment and tree growth (e.g., summer drought, extremely snow-rich winters, winters with nearly no snow, strong late and early frost) • Winter snow cover (outside the tropics) providing seedlings and saplings with shelter from climatic injuries and herbivores • Low to moderate wind velocities	• Sufficient viable seed supply • Effective seed dispersal • Suitable seed beds • No or little competition of tree seedlings and saplings with non-arborescent vegetation • Seedling and sapling survival > mortality • Unlimited nutrient supply • Balanced soil moisture conditions • Facilitation of seedling establishment and survival by microsites and increasing tree population (e.g., reduced sky exposure and wind velocities) • Recovery of hitherto suppressed trees (e.g., by releasing vertical leaders) • No lasting disturbances by pathogens, diseases and insects • Absence of wildfires and burning by humans • Absence of destructive effects of snow movements • Stable substrates • Adaptation of the tree species to the new environmental conditions

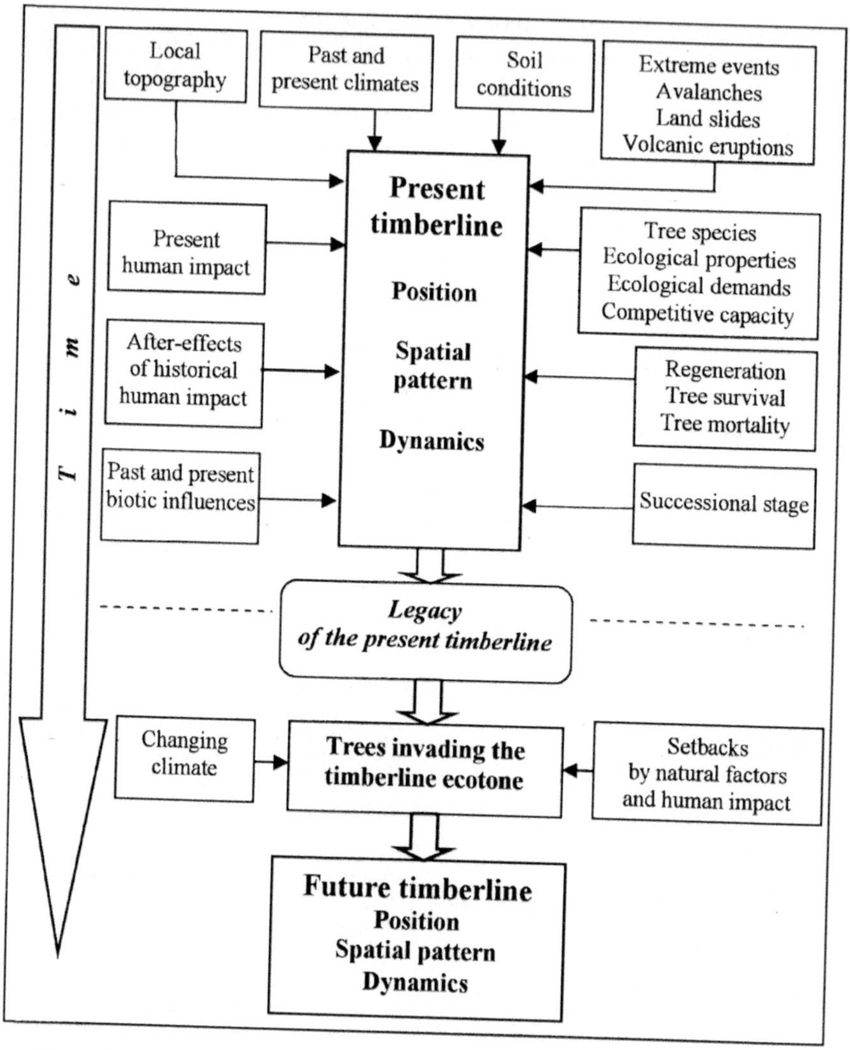

Figure 90. Altitudinal position, spatial pattern and dynamics of the timberline ecotone as a result of historical timberline legacy and environmental change. Modified from Holtmeier and Broll (2007).

- In addition to the effects of microtopography, trees, and tree groups (compact, open), in particular, would influence their sites and close environment (Section 4.3.12; Figure 62). In the beginning of tree invasion into and beyond the timberline ecotone snow depth and its side effects may vary abruptly and widely (Holtmeier, 1978, 2005b; Daly, 1984; Holtmeier and Broll, 1992; Broll and Holtmeier, 1994; Kullman, 2005d). Growing tree population will reduce wind velocity and gradually enhance deposition of snow in the timberline ecotone. Consequently, snow will last longer than in the wind-swept Alpine and in the open tundra (Hare, 1971; Holtmeier, 1978, 1993a, 1996, 2005b; Hiltunen, 1980; Walsh et al., 1994; Hiemstra et al., 2002; Geddes et al., 2005). In the timberline ecotone, snow accumulation will also be greater than under dense forest canopies. The deeper snowpack providing shelter from injurious climatic influences will facilitate seedling establishment (e.g., Germino and Smith, 1999; Germino et al., 2002; Smith et al., 2003; Johnson et al., 2004). Moreover, seedlings and saplings may benefit from increased meltwater-runoff, particular in dry years and dry regions.

However, there may also be negative effects. Thus, late-lying wet snow would probably cause higher losses in evergreen conifer young growth by snow fungus infection. Moreover, late-lying snow will shorten the growing season and result in lower soil temperatures during early summer. Mechanical damage caused to seedlings and saplings by settling snow will very likely increase. Thus, it will depend on the local situation whether increased snowpack will facilitate or impede further tree establishment and timberline advance. In the summer, growing tree population may lower the risk of summer frost damage by reducing the exposure of seedlings to high solar radiation loads. Such changes in the ecotone spatial and temporal structures would completely overlap the effects of slightly higher average air temperatures. Thus, predictions of the future position of timberline based only on expected change of the thermal conditions may be questionable because they disregard the complexity of the ongoing process.

- Melting of local permafrost at high elevation will destabilize steep mountain slopes thus increasing soil erosion of potential forest sites (Burga and Perret, 2001; see also Harris, 2005). In subarctic lowland such as the Hudson Bay area and western Siberia increasing paludification is likely to cause a southward retreat of the boreal forest (e.g., Crawford, 1978, 2005, 2008; Crawford et al., 2003).
- Changing climate will alter snow fall conditions (Beniston, 1997, 2001) and may thus influence frequency, extent of action and destructive forces of avalanches what might have lasting effects on timberline pattern and dynamics, particularly on steep slopes in temperate and northern mountains (e.g., Reardon et al., 2008).
- Timberline shift to greater elevation will bring about exposure to a much windier climate (see also Kullman, 2007b). Consequently, damage caused to the trees by direct and indirect effects of strong winds is likely to increase as long as tree population is too small to enhance snow accumulation. The relative importance of shelter provided by geomorphic features, such as leeward sides of convex topography, solifluction lobes or steplike

solifluction terrace risers, to seedlings and saplings will increase (see also Kullman, 2005d; Resler et al., 2005; Resler, 2006).

- As a rule, timberlines are highly complex biological boundaries that developed over long periods of time controlled by the interactions of vegetation with repeated changes of environmental conditions. In case of a sudden rise of temperature as has been predicted by many models, mountain and polar timberlines would not respond spontaneously, but rather with a time lag of several decades or even centuries (e.g., Agee and Smith, 1984; Holtmeier, 1985b, 1986a; Lassoie et al., 1985; Davis, 1986; Noble, 1993; Little and Peterson, 1994; Woodward, 1998; Hofgaard and Wilman, 2002; Lloyd, 2005; Holtmeier and Broll, 2007), contrary to the assumptions made by many authors (e.g., Peters, 1990; Franklin et al., 1991; Roos, 1996; Kellomäki et al., 1997). Thus, in many mountain regions treeline advance has been less than could be expected at the degree of climatic warming (e.g., Wardle and Coleman, 1992; Holmgren and Tjus, 1996; Crawford, 1997; Lloyd and Graumlich, 1997; MacDonald et al., 1998b; Peterson, 1998; Tasanen et al., 1998).
- Timberline outside of the tropics as well as tropical timberline would not linearly respond to changing climatic conditions, as the snow line would.
- Moreover, global warming cannot be expected to cause a synchronous adjustment of environmental conditions and timberline to a complex climatic change. More likely, the adjustment would be asynchronous as can be concluded from climatically driven timberline fluctuations during the Holocene (cf. LaMarche and Mooney, 1967, 1972; LaMarche, 1973, 1977; Henderson, 1973; Andrews et al., 1978; Elliott, 1979; Elliott and Short, 1979; Viereck, 1979; Kullman, 1988, 1989a, b; Moser and MacDonald, 1990; Kullman and Engelmark, 1991; Graumlich, 1991; Koster, 1991). Thus, advance of seedlings to higher elevation in widely separated temperate mountain regions in Europe and North America may suggest a more general pattern. It may be speculated whether the 'new' treeline will stabilize for longer (Kullman, 2000b). In the same regions, however, we also find evidence that seed-produced regeneration is still scarce or even missing in the uppermost zone of the timberline ecotone and particularly above treeline as was demonstrated in the foregoing chapter.
- In case of rapidly rising temperature trees would not have adequate time to adapt to the new thermal conditions (Skre, 1993).
- Timberline response to climatic change would be different in arid or semiarid regions, when compared to humid environments as lack of available moisture might become the controlling factor, particularly on permeable soils and on convex topography.
- Additionally, global warming would not affect tree species at timberline in the same way, as can be concluded, for example, from timberline

history in the Swedish Scandes. Growth conditions for Scots pine deteriorated during 6000 to 5300 years before the present, while conditions for mountain birch improved (Kullman, 1983b, 1987, 1988).

- Orographically caused altitudinal position and spatial timberline structures would hardly change, not even in the long term, as long as mass wasting, debris slides and avalanches occur (Holtmeier and Broll, 2005). On steep valley sides these factors will prevent the forest from reaching its possible thermal limit while these factors do not occur on flat or gently sloping topography. Tree colonization of unvegetated block fields by trees may be impeded because mycorrhiza are lacking, at least temporarily (e.g., Cázares et al., 2005).
- In general, anthropogenic timberlines will show the greatest altitudinal shifts after the cessation of pastoral use and other human disturbances (Holtmeier and Broll, 2005). This process should not be misunderstood as a result of climate change.
- Grazing, browsing and trampling by wild-living herbivorous mammals (e.g., red deer, reindeer, ibex, guanaco) at high population density may locally delay timberline advance. A positive effect on regeneration resulting from reduced ground vegetation is often outweighted by lack of moisture and nutrients. The role of herbivorous mammal as effective dispersers of seeds of timberline-forming tree species is usually overestimated. On the other hand, dispersal of tree seeds (e.g., stone pines, juniper, and rowan) by birds encourages forest advance to higher elevations at improving climatic conditions.
- Almost certainly, the warming climate would increase insect harassment of reindeer in the Subarctic. Forest advance to higher elevation would reduce open areas usually frequented by reindeer to escape from molestation by nose bot flies and warble flies. Thus, the impact of reindeer on the remained open sites is likely to increase. The same would apply to ibex habitats.
- The possible consequences of present and future changes at timberline and treeline will be different at the altitudinal timberline compared to the northern timberline. In the mountains, only a relatively narrow zone will be affected, a few hundred metres wide at the maximum. An altitudinal advance of timberline would improve the protective function of high-elevation forest (avalanches, erosion, debris flows, etc.) and thus increase the safety for the people living in the mountain valleys. The effects of an advance of the northern timberline would be somewhat different. As a consequence of northward shift of polar treelines considerable changes in primary production, for example, might be expected (e.g., Alcamo et al., 2007). However, commercial timber production is unlikely to take place in and beyond the present timberline ecotone in the foreseeable future as predicted by ACIA (2004). Increase of tree population within and beyond

the present timberline ecotone will lag behind climatic change (e.g. Chapin III et al., 2005). Moreover, the trees that will become established beyond the current tree limit are likely to grow at too slow a rate for the production of commercial timber also in a warmer climate.

- Increasing tree population within the 30 to 50 km (e.g., west of Hudson Bay) or several hundred kilometres wide forest-tundra ecotone (e.g., east of Hudson Bay) would not only cause structural changes to the living space of the polar people but would also lead to far-reaching supraregional consequences, partly by affecting the interactions of forest cover and thaw depth of the permafrost (e.g., Brown and Péwé, 1973; Viereck, 1973; Viereck and Van Cleve, 1984; Koster and Nieuwenhuijszen, 1992), partly by lowering the albedo compared to the present tundra vegetation and by a northward shift of the Arctic front, and thus affect global atmospheric circulation (e.g., Hare and Ritchie, 1972; Brown and Péwé, 1973; Viereck, 1973; Monteith, 1975; Viereck and Van Cleve, 1984; Bonan et al., 1992, 1995; Koster and Nieuwenhuijzen, 1992; Lafleur et al., 1992; Foley et al., 1994; Pielke and Vidale, 1995; Hobbie and Chapin III, 1998). Studies on the effects of shrub and tree expansion in western Alaska (Chapin III et al., 2005) show however, that the recent vegetation change has not yet much contributed to regional summer heating. This may change if the current expansion process continues. In contrast, possible changes at mountain timberline would not cause any effects of comparable order of magnitude.
- Scientists as well as the general public are concerned about the possible effects of continued global warming. Indeed, many of these effects such as increasing climatic extremes (e.g., severe storms, flooding, drought, diseases and loss of biological diversity, etc.) may be fatal for humans. However, what is about timberline advance? It could be considered to be a 'positive' consequence because it may compensate for the worldwide destruction of original subalpine forests. On the other hand, advance of timberline to greater elevation and resulting fragmentation of alpine vegetation are likely to increase the risk of alpine species extinction and to reduce biological diversity (e.g., Theurillat et al., 1998; Theurillat and Guisan, 2001; Moiseev and Shiyatov, 2003; Carnelli et al., 2004; Moen et al., 2004; Tinner and Kaltenrieder, 2005; Baker and Moseley, 2007; Malanson et al., 2007; Sundqvist et al., 2008). An advance of timberline of more than 200 m up to nearly 700 m as suggested by a simulation study for the Swedish mountains (Moen et al., 2004) would result in a complete landscape change. A timberline advance of only 100 m would already reduce alpine heath area by 41% and much of the remaining area would be boulder fields and slope debris. In the Ural highlands, for example, the increased tree density within the forest stands and upward shift of the timberlines has reduced the area of non-woody vegetation by

about 10–30% (Moiseev and Shiyatov, 2003). However, many alpine species would survive on cliffs and slopes too steep for the establishment of closed forests and on disturbed sites. This has happened in the past at least (Bruun and Moen, 2003). Thus, the risk of species extinction would be reduced. Kullman (2007b) even explicitly objects to the widespread opinion of alpine species extinction in response to climate warming and timberline advance as existing field data do not support this hypothesis.

- Upward shift of timberline would also reduce landscape diversity. This might affect the esthetic values and would thus have negative consequences for the economy. In the European Alps, for example, an advance of the sub-alpine forest to higher elevation, probably up to its postglacial maximum or even higher, would make the landscape less attractive for tourists, who are the main base of the local economy these days.

6 TIMBERLINE PROSPECTS AND RESEARCH NEEDS

Future timberline research will focus mainly on the response of timberline to the changing environment and possible feedbacks. In the foregoing chapter, some aspects of past and present timberline fluctuations have been highlighted, which must be considered when predicting timberline response to climatic change. Apparently, great heterogeneity as well as great regional and local variety are the main problems.

The many factors that have influenced timberline during history and those controlling timberline at present are interrelated to each other in a manifold, complex way that is hard to understand. When projecting these 'mechanisms' into a 'warmer future' we cannot be sure whether they would work in the same way as at present or as they worked in the past (Holtmeier, 2000; Giorgi and Hewitson, 2001; Holtmeier and Broll, 2005, 2007). It is still an open question which effects will result from a delayed response of high-elevation forests to a sudden climatic change (Wardle and Coleman, 1992; Lloyd and Graumlich, 1997; Peterson, 1998).

Research on carbon use of trees and its dependence from the site conditions (temperature, moisture, nutrients, etc.) in different climates needs to be continued. In this context, soil biological, soil physical and soil chemical conditions, in particular temperature and moisture regimes, decomposition and availability of nutrients, are factors that need to be more intensively studied in the timberline ecotones, in particular with respect to the many negative and positive feedbacks between the factors (see also Bekker et al., 2001). The role of soil organisms in carbon and nutrient cycles in the timberline environment has not been sufficiently considered. As the changing climate is also likely to sustainably influence also soil ecological conditions research on this issue should be intensified. In particular, the role of mycorrhiza (Section 4.3.8) within and beyond the present timberline ecotone, its interaction with tree and dwarf shrub vegetation (symbiontic and negative effects) and its dependence on physical and chemical properties of soil must be studied in detail in different timberline environments.

The present knowledge on response of the roots systems is also insufficient. We urgently need more information on cold tolerance, acclimation, and deacclimation (Bigras et al., 2001) of seedling-roots, in particular, to better understand successes or failure of seedling establishment in the harsh timberline environment (cf. Section 4.3.5).

F.-K. Holtmeier, *Mountain Timberlines: Ecology, Patchiness, and Dynamics*,
Advances in Global Change Research 36, 335–341.

Systematic studies on the change of soil temperatures during the last decades (e.g., Kullman, 2007b) should be carried out in different timberline environments.

Moreover, investigations on the role of phenotypic plasticity of tree species under the influence of the harsh timberline environment (Section 4.3.11) are needed. In this context the interactions of soil temperatures and apical meristem temperatures in shaping tree life-form should be more extensively studied in different timberline environments. Moreover, investigation on hardiness of timberline trees under different climatic scenarios and nutrient supply (e.g., increased nitrogen input) would be welcome. The same holds true for photooxidative stress and photoinhibition.

At the tropical timberline, in particular, research on these issues is behind the times with a few exceptions (e.g., Beck et al., 2008) and needs to be intensified to create a more reliable basis for understanding the current situation and to speculate about the future.

Most studies have focused hitherto on the possible response of mature trees to environmental change. However, successful regeneration seems to be a bottleneck in timberline advance (cf. Section 4.3.10). Thus, there is a great need of research on the response of tree seedlings and saplings to a warming climate and its side effects. The role of facilitation for seedling establishment and survival by microtopographical structures and increasing tree population within and beyond the present timberline ecotone needs more consideration as a factor which may strongly influence the timberline spatial pattern and timberline advance.

More attention should be paid to the influence of animals, pathogens and diseases, in particular on the ecosystem level (e.g., Holdenrieder et al., 2004). The quality of mass-outbreaks of defoliation insects, for example, and their effects on timberline may change. Pathogens and diseases may increase and have stronger effects on timberline than at present. At timberline in some areas of the high-mountains in the American West, for example, white pine blister rust (*Cronartium ribicola*) is seriously affecting whitebark pines (*Pinus albicaulis*), in particular as pines stressed by blister rust are highly susceptible to attack by mountain pine beetle (*Dendroctonus ponderosae*; Keane and Arno, 1993). Stands at high altitude are usually less infected than stands at lower elevation, probably because of fewer frost-free days (Campbell and Antos, 2000). Blister rust was introduced by chance from Europe to North America in 1910 (Hoff and Hagle, 1990). Some species of *Ribes* and, more recently of *Pedicularis* and *Castilleja* have been found (McDonald et al., 2006) functioning as natural hosts of blister rust. Blister rust now has infected 25% of the whitebark pines in the Greater Yellowstone and over 80% farther north in Glacier National Park, for example (communication D. F. Tomback). Locally it destroys pine young growth also in the timberline ecotone thus acting opposite to climatically-driven timberline advance. Blister rust will probably have a lasting effect on the whole subalpine ecosystem (e.g., Arno and Hoff, 1989; Kendall and Arno, 1990; Mattson and Reinhart, 1994; Resler and Tomback, 2008; Smith et al., 2008; Tomback and Resler, 2008). In the long-term, many timberline and dry sites, currently dominated by white-bark pine, are likely to turn into hebaceous and shrub-dominated communities (Keane and Morgan, 1994, Campbell and Antos, 2000).

Forest-ungulate interactions must be studied with special regard to timberline dynamics to lay the foundations for game management in a changing environment (see also Reimoser, 2003). This holds true particularly for densely populated mountain areas where high elevation forests have a protective function.

As has been emphasized in the introduction to this book, it is differentiating the timberline phenomenon in a global perspective being a main objective of timberline research. Completing our knowledge about the present local and regional knowledge gaps on spatial and temporal structures of timberline and exploring the functional relationships behind them is an indispensable step in future timberline research rather than further focusing on 'better' coincidences between the position of timberline and certain temperatures considered to be essential to tree growth.

Thus, thorough studies are required, especially on reproduction (production of viable seeds, germination success, survival rate of seedlings, vegetative reproduction and success), on distribution pattern of tree species and ground vegetation as related to site conditions (microclimates, soil biological, soil physical and soil chemical conditions), and on site history.

These studies should be supported by physiological research in the field and in the laboratory (Figure 91), in particular on seedlings and young growth rather than on mature trees. The actual processes in and beyond the timberline ecotone can only be assessed in view of the regional climatic conditions. It is obvious from the available data that local and regional fluctuations may run counter to the long-term warming trend, as has been the case in the northern hemisphere in the period between 1940 and 1970. In this context, the interannual and annual variability of climatic factors (temperature, hygric conditions, snowmelt patterns, etc.) and their effects on regeneration and tree growth should be studied in different timberline regions. Long-term monitoring is needed of seedling recruitment as well as of disturbances (e.g., heavy storms, insect infestations, diseases, summer droughts) and resulting vegetation changes within and beyond the present timberline ecotone. Such studies should go along with manipulation experiments such as transplantation of tree seedlings to different controlled conditions or experiments on the effects of different grazing practices (e.g., reindeer). Experimental studies are essential to improve our knowledge on fundamental processes controlling tree growth and timberline dynamics.

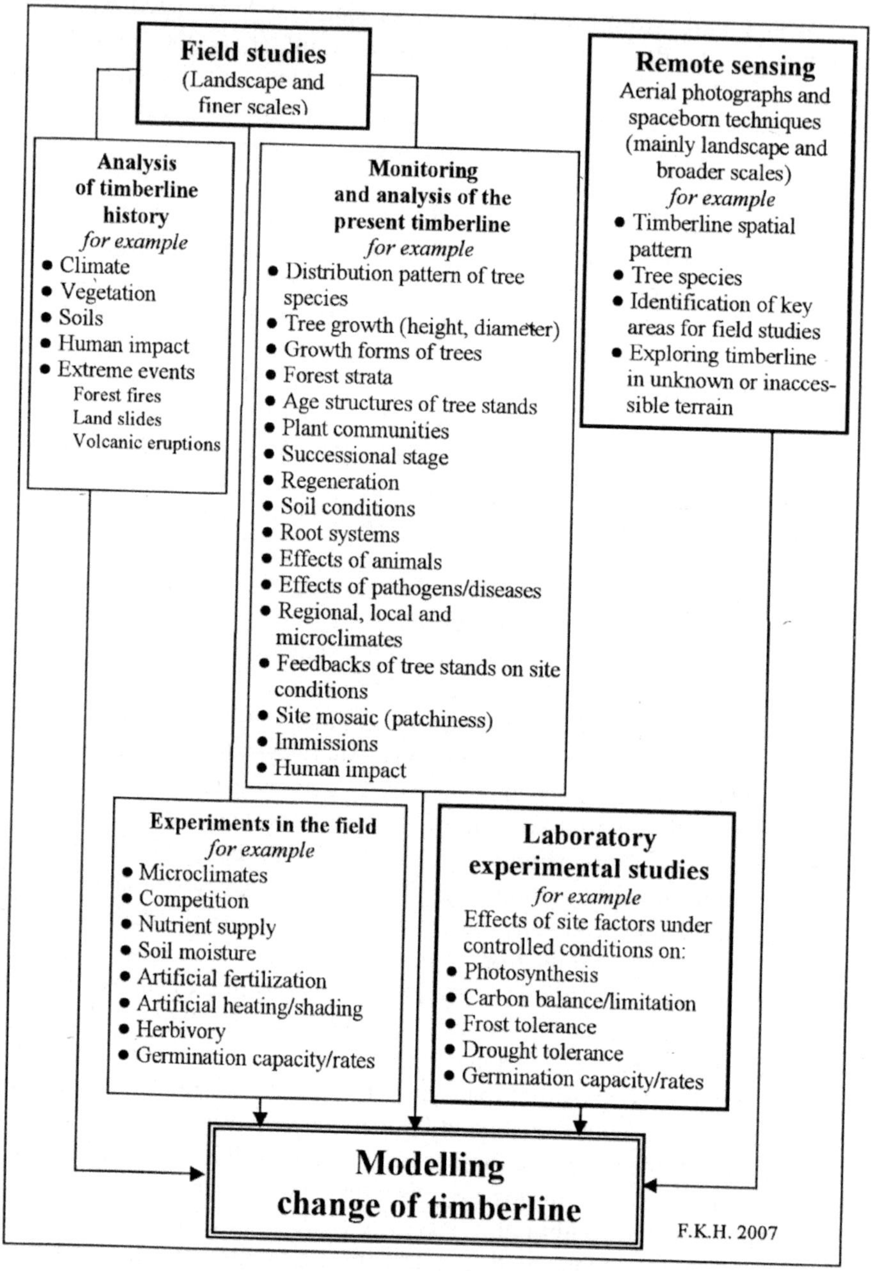

Figure 91. Basic research needed for modelling potential changes in timberline due to changing climate. Modified from Holtmeier (1993b).

Not least, the worldwide phenomenon 'timberline' should be studied with comparable methods. For long-term monitoring of the on-going changes in the timberline ecotones, the site mosaics must also be comparable as far as land-forms, parent material and soils are concerned.

During the last 2 decades, modelling the relationships between tree growth (normally mature trees) and site conditions, especially temperature, enriched CO_2 and increased nitrogen input, under changed climatic conditions has increasingly become popular in ecology (e.g., Kauppi and Posch, 1988; Ozenda and Borel, 1991; Gates, 1993; Foley et al., 1994; Bonan et al., 1995; Neilson and Chaney, 1997; Batchelet and Neilson, 2000; Rupp et al., 2001; Bekker et al., 2001). Model-based scenarios supported by geographical information systems might help to get an idea of the potential regional and local changes. This holds particularly true for the comparatively broad forest-tundra ecotone in the north, where topography and its effects on regional climates might be an essential factor controlling northward advance of vegetation boundaries. West–east oriented mountain ranges, for example, can be expected to exert a barrier effect on timberline advance to the north. This is also demonstrated by simulations on the response of subarctic vegetation to climatic warming in northern Alaska. Regardless of the degree of warming, the forest limit needed at least 1000 years to advance from its present position to the northern side of the Brooks Range (Rupp et al., 2001).

In contrast to first-generation 'equilibrium models' (e.g., Franklin, 1995; Cramer, 1997; Skre et al., 2002), dynamic vegetation models (e.g., Wolf et al., 2008) are able to represent continuous changes by including processes such as establishment, growth, reproduction and mortality, physiological adaptation and competition. Every assessment of timberline response to future climate must consider the effects of local site conditions and feedbacks of growing tree population in modulating this change (e.g., Holtmeier, 1985b, 1989, 1995a; Luckman and Kavanagh, 1998). So-called 'ground truths' are imperative as ever. Models, however, must be as simple as possible in order to work, and thus they cannot meet the complex reality, and great regional and local variety. Adding all the complexities to the existing models would increase uncertainties in the predictions (Batchelet and Neilson, 2000). In the end it will be the regionally and locally different response of timberline, which is the main characteristic of the effects of changing macroclimate on timberline (see also Luckman and Kavanagh, 1998). Models of future timberline position considering only climate as the driving factor will necessarily fail on a regional scale (Holtmeier and Broll, 2005, 2007; Rössler and Löffler, 2007).

Regional scenarios cannot be better than the data they are based on, and these data are still insufficient in most timberline areas. Thus, modern timberline research might run the risk of increasingly happening in 'cyber space'. Thus, from the present author's opinion, there is a great need for careful local and regional field studies based on a complex landscape-ecological approach that also considers the historical aspect. This might appear a little antiquated; it is however an indispensable condition for differentiating the ecological, spatial and temporal structures in the mountain timberline ecotone.

Up- and downscaling of statistically significant relationships between environmental factors and ecosystem spatial patterns and temporal structures have become popular in ecology (e.g., Curran et al., 1997). The relative importance of the factors varies by the scale of consideration (cf. Figure 2). Nevertheless, downscaling of statistical relationships existing between the timberline position and one or two obviously controlling factors (e.g., heat deficiency, lack of moisture) found that the global or zonal scale would produce simplistic scenarios that disguise the complexity and heterogeneity of the timberline phenomenon (Holtmeier and Broll, 2007) rather than contribute to a better causal understanding. Conversely, the possibility of upscaling local relationships existing, for example, between soil conditions, seedling establishment and survival, tree growth and timberline patchiness is also limited as they depend on the topographical context and thus vary locally. Timberline-specific soils, for example, do not exist (Holtmeier, 2000). Instead, mosaics of different soil types closely related to the geological substrate, varying microtopography and plant cover (e.g., dwarf shrub vegetation, grassland) are typical of the timberline ecotone (e.g., Burns, 1980; Broll, 1994, 1998, 2000; Holtmeier, 2000; Broll et al., 2007). It depends on the specific question whether and to which extent the results of studies at both broad and landscape scales can be combined.

It would be a promising strategy to compare responses of different types of timberline to environmental change: for example timberline on steep valley sides, on trough walls, on glacial moulded trough shoulders or on cirque floors, on gentle slopes with smooth microtopography and on uplifted old land surfaces (Section 4.3.9; see also Holtmeier and Broll, 2005). Response of timberline located on a gentle slope or on almost level terrain, for example, would be completely different from timberline on a steep valley side that is dissected by avalanche chutes down to the valley floor. In the latter case timberline advance would follow the slope ribs (rib and groove topography) and similar convex topography not affected by avalanches, while on gentle fjeld slopes, for example, the timberline would rise first within the shallow, wind-protected and moist valleys. On the other hand, orographic timberlines caused by steep rock walls or slope debris would probably not respond to any climatic warming, which seems important to the future forest cover, as orographic timberlines are more common in high mountains than true climatic timberlines. Remote sensing techniques such as oblique air photos and satellite imagery provide excellent tools to explore and map timberline spatial structures, timberline types (Section 4.3.9, photos and figures there; see also Allen and Walsh, 1996; Klasner and Fagre, 2002) and temporal variation even in almost unknown and inaccessible areas thus providing an unprecedented broad data base for circumpolar monitoring of timberline changes (e.g., Rees et al., 2002, there further specific literature). Interpretation, however, must be supported by field studies. Repeat photography has proven to be an excellent tool in assessing historical changes in the timberline exotone (e.g., Moiseev and Shiyatov, 2003; Roush et al., 2007).

Timberline research always is multidisciplinary. However, it will only be successful if researchers from different sciences would really cooperate (interdisciplinary research in its original sense), because the knowledge of the results of studies on single components does not automatically allow understanding the response of the forest transitions ecosystems. This is and will probably remain the most difficult problem on the way to better understanding of timberline dynamics. On the occasion of an international workshop on 'Dynamics of the Taiga-Tundra Interface' that was held in late winter 2000 in Abisko Biological station (Sweden), Robert Crawford (University of St. Andrews, Scotland) illustrated this to the timberline specialists accordingly by a remark of the playwright G. B. Shaw that was also used by W. Churchill: '*Just as America and England are two countries divided by a common language* (George Bernard Shaw), *so timberline researchers find themselves separated by the nature of a common problem – the timberline*'.

REFERENCES

Aario, L. (1940): Waldgrenzen und subrezente Pollenspektren in Petsamo-Lappland. Annales Academiae Scientarum Fennicae, Serie A, Bd. 54 (8): 1–120

Aas, B. (1964): Björke – og barskoggrenser in Norge: En undersökelse av de övre klimatiske skoggrenser. Hovedfagsoppgave i naturgeografi, University of Oslo

Aas, B. (1969): Climatically raised birch lines in southeastern Norway. Norsk Geogr. Tidsskr. 23: 119–130

Aas, B. and Faarlund, T. (1996): The present and the holocene birch belt in Norway. Paläoklimaforschung 20: 18–24

Aas, B. and Faarlund, T. (2001): The holocene history of the nordic mountain birch belt. In: Nordic mountain birch ecosystems (Wielgolaski, F., ed.), Man and Biosphere Series 27: 5–22

ACIA (Arctic Climate Impact Assessment) (2004): Impacts of a warming arctic. Cambridge University Press, Cambridge

Agee, J. K. and Smith, I. (1984): Subalpine tree establishment after fire in the Olympic Mountains, Washington. Ecology 65: 810–819

Agnew, A. D. Q. and Hedberg, O. (1969): Geocarpy as an adaptation to Afroalpine solifluction soils. J. East Africa Nat. Hist. Soc. 27 (33): 215–216

Ågren, J. and Zackrisson, O. (1990): Age and size structure of Pinus sylvestris populations on mires in central and northern Sweden. J. Ecol. 78: 1049–1062

Aichinger, E. (1937): Die Waldverhältnisse Südbadens. Karlsruhe

Aichinger, E. (1967): Pflanzen als forstliche Standortsanzeiger. Österr.- Agrarverlag, Wien

Akhaltkatsi, M., Adaladze, O., Nakhutsrishvili, G. and Smith, W. K. (2006): Facilitation of seedling microsites by Rhododendron caucasicum extends the Betula litwinowii alpine treeline, Caucasus Mountains, Georgia. Arct. Antarct. Alp. Res. 38 (4): 481–488

Alberdi, M., Romero, M. Rios, D. and Wenzel, H. (1985): Altitudinal gradients of seasonal frost resistance in Nothofagus communities of southern Chile. Acta Oecol. Oec. Plant. 6 (20): 21–30

Alcamo, J., Moreno, J. M., Nováky, B., Bindi, M., Corobov, R., Devoy, R. J. N. Giannkopoulos, C., Martin, E., Olesen, J. E. and Shvidenko, A. (2007): Europe. Climate Change 2007: Impacts, adaptation and vulnerability. Contribution of working group II to the fourth assessment report of the Intergovernmental Panel on Climate Change (Pary, M. L., Canziani, O. F., Palutikof, J. P., van der Linden, P. J. and Hanson, C. E., eds.): 541–580, Cambridge University Press, Cambridge

Aldrich, J. K. (1990): Ghosts of Boulder County. A guide to the ghost towns and mining camps of Boulder County, Colorado. Centennial Graphics, Lakewood, CO

Alftine, K. J. and Malanson, G. P. (2004): Directional positive feedback and pattern at an alpine tree line. J. Veg. Sci. 15: 3–12

Alftine, K. J., Malanson, G. P. and Fagre, D. B. (2003): Feedback-driven response to multidecadal climatic variability at an alpine treeline. Phys. Geogr. 24 (6): 520–535

Allen, M. F. (1991): The ecology of mycorrhiza. Cambridge University Press, Cambridge/New York/Port Chester/Melbourne/Sidney

Allen, T. R. and Walsh, S. J. (1996): Spatial and compositional pattern of alpine treeline, Glacier National Park, Montana. Photogramm. Eng. Rem. Sens. 62: 1261–1268

Allison, I. and Bennet, J. (1976): Climate and microclimate. In: The equatorial glaciers of New Guinea (Hope, G. S. et al., eds.): 61–80, A.A. Balkema, Rotterdam
Ammann, B. and Wick, L. (1993): Analysis of fossil stomata of conifers as indicators of the alpine tree line fluctuations during the Holocene. Paläoklimaforschung 9: 175–184
Andersson, G. (1902): Zur Pflanzengeogrphie der Arktis. Geographische Zeitschr. 8: 1–23
Andrews, J. T., Carrara, P. E., King, F. B. and Stuckenrath, R. (1975): Holocene environmental changes in the alpine zone, northern San Juan Mountains, Colorado: Evidence from bog stratigraphy and palynology. Quat. Res. 5: 173–197
Andrews, J. T., Carrara, P. E., Bartos, F. and Stuckenrath, R. (1978): Holocene stratigraphy and geochronology of four bogs (3700 a.s.l.), San Juan Mountains, SW Colorado, and implications to the neoglacial record. Geol. Soc. Am. Abst. Prog. 5 (6): 460–461
Anschlag, K. (2006): Regeneration der Fjellbirke (Betula pubescens ssp. czerepanovii) und Wurzelsysteme ihres Jungwuchses im Waldgrenzökon, Finnische Subarktis. Dissertation Fachbereich Geowissenschaften, Mathematisch-Naturwissenschaftliche Fakultät der Westfälischen Wilhelms-Universität, Münster
Anschlag, K., Broll, G. and Holtmeier, F.-K. (2008): Mountain birch seedlings in the treeline ecotone, Subarctic Finland. Variation in above- and below-ground growth in relation to microtopography. Arct. Antarct. Alp. Res. 40 (4): 609–616
Anthelme, F., Michalet, R., Barbaro, L. and Brun, J.-L. (2003): Environmental and spatial influences of shrub cover (Alnus viridis DC.) on vegetation diversity at the upper treeline in the inner Western Alps. Arct. Antarct. Alp. Res. 35 (1): 48–55
Armand, A. D. (1992): Sharp and gradual mountain timberlines as a result of species interaction. Ecol. Stud. 92: 360–378
Arno, S. F. (1966): Interpreting the timberline: An aid to help park naturalists to acquaint with the subalpine-alpine ecotone of Western North America. Master thesis, University of Montana, Missoula, MT
Arno, S. F. (1970): Ecology of alpine larch (Larix lyallii Parl.) in the Pacific Northwest. Ph.D. thesis, University of Montana, Missoula, MT
Arno, S. F. (1984): Timberline, mountain and arctic frontiers. The Mountaineers, Seattle, WA
Arno, S. F. (1990): Larix lyallii. Agricultural handbook 654, USDA For. Serv., Washington, DC, 152–159
Arno, S. F. and Habeck, J. R. (1972): Ecology of alpine larch (Larix lyallii Parl.) in the Pacific Northwest. Ecol. Monogr. 42: 417–450
Arno, S. F. and Hoff, R. J. (1989): Silvics of whitebark pine (Pinus albicaulis). USDA For. Serv., Intermtn. Res. Stat., Gen. Tech. Rep. INT 253
Arno, S. F., Worall, J. and Carlson, C. E. (1995): Larix lyallii: Colonist of tree line and talus sites. Proc. int. symp. ecology and management of Larix forests: A look ahead, Whitefish, MT, October 5–9, 1992. USDA For. Serv., Intermtn. Res. Stn., Gen. Tech. Rep. GTR – INT 310: 72–78
Askawa, S. (1957): Studies on hastening of seeds of five-leaved pines (Japanese, English summary). Bull. Govern. Forest Exp. Stat. Tokyo 10: 41–54
Askew, G. P. (1964): The mountain soils of the east ridge of Mt. Kinabalu. Roy. Soc. Proc. Ser. B Biol. Ser. 161: 65–74
Atkinson, K. (1981): Vegetation zones in the Canadian Subarctic. Area 13: 13–17
Auer, C. (1947): Untersuchungen über die natürliche Verjüngung der Lärche im Arven-Lärchenwald des Oberengadins. Schweizerische Anstalt für das forstliche Versuchswesen, Mitteilungen 25: 3–140
Auer, C. (1948): Untersuchungen über Dürreresistenz und Wachstum von Lärchenkeimlingen auf verschiedenen Horizonten des Eisenpodsols subalpiner Arven-Lärchenwaldungen. Vegetatio 1: 79–92

Aulitzky, H. (1961): Die Bodentemperaturen in der Kampfzone oberhalb der Waldgrenze und im subalpinen Zirben-Lärchenwald. Mitteilungen der Forstlichen Bundes-Versuchsanstalt Mariabrunn 59: 155–208
Aulitzky, H. (1963a): Grundlagen und Anwendung des vorläufigen Wind-Schnee-Ökogrammes. Mitteilungen der Forstlichen Bundes-Versuchsanstalt Mariabrunn 60: 763–834
Aulitzky, H. (1963b): Bioklima und Hochlagenaufforstung in der subalpinen Stufe der Inneralpen. Schweizerische Zeitschrift für Forstwesen 114 (1/2): 1–25
Aulitzky, H. (1968): Die Lufttemperaturverhältnisse einer zentralalpinen Hanglage. Archiv für Meteorologie, Geophysik und Bioklimatologie. Ser. B 16: 18–69
Aulitzky, H., Turner, H. and Mayer, H. (1982): Bioklimatische Grundlagen einer standortsgemäßen Bewirtschaftung des subalpinen Lärchen-Arvenwaldes. Eidgenössische Anstalt für das forstliche Versuchswesen, Mitteilungen 58 (4): 327–577
Autio, J. and Colpaert, A. (2005): The impact of elevation, topography and snow load damage of trees in the position of the actual timberline on the fells in central Finnish Lapland. Fennia 183: 15–36
Autio, J. and Heikkinen, O. (2002): The climate of northern Finland. Fennia 180 (1–2): 61–66
Azócar, A. and Monasterio, M. (1979): Variabilidad ambiental en el Páramo de Mucubají. In: El Medio Ambiente 'Páramo' (Saldago-Labouriau, ed.), Ediciones CIET-IVIC/MAB-UNESCO, Caracas, 149–159
Baader, U. (1995): Zur natürlichen Wiederbewaldung aufgelassener Almweideflächen durch die Fichte im Bereich der Waldgrenze der nördlichen Kalkalpen (Lechtaler Alpen). Diploma thesis, Institut für Geographie, Westfälische Wilhelms-Universität, Münster
Bader, F. (1960): Die Verbreitung borealer und subantarktischer Holzgewächse in den Gebirgen des Tropengürtels. Eine arealgeographische Studie in dreidimensionaler Sicht. Nova Acta Leopoldina, Neue Folge 148 (23)
Bader, M. Y. (2007): Tropical alpine treelines. How ecological processes control patterning and dynamics. Ph.D. thesis, Wageningen University, Wageningen, The Netherlands
Bader, M. Y. and Ruijten, J. J. A. (2008): A topography-based model of forest cover at the alpine treeline in the tropical Andes. J. Biogeogr. 35: 711–723
Bader, M. Y., Rietkerk, M. and Bregt, A. K. (2007): Vegetation structures and temperature regimes of tropical alpine timberlines. Arct. Antarct. Alp. Res. 39 (3): 353–364
Bader, M. Y., Rietkerk, M. and Bregt, A. K. (2008a): Simulated climate change in a simple spatial model of treeline dynamics influenced by excess solar radiation and fire. Arct. Antarct. Alp. Res. 40 (2): 269–278
Bader, M. Y., van Geloof, I. and Rietkerk, M. (2008b): High solar radiation hinders tree regeneration above the alpine treeline in northern Ecuador. Plant Ecol. 191: 33–45
Baig, M. N. and Tranquillini, W. (1976): Studies on upper timberline: Morphology and anatomy of Norway Spruce (Picea abies) and stone pine (Pinus cembra) needles from various habitat conditions. Can. J. Bot. 54: 1622–1632
Baig, M. N. and Tranquillini, W. (1980): The effects of wind and temperature on cuticular transpiration of Picea abies and Pinus cembra and their significance in desiccation damage at the alpine timberline. Oecologia 47: 252–256
Baig, M. N., Tranquillini, W. and Havranek, W. (1974): Cutikuläre Transpiration von Picea abies- und Pinus cembra-Zweigen aus verschiedener Seehöhe und ihre Bedeutung für die winterliche Austrocknung der Bäume an der alpinen Waldgrenze. Centralblatt für das gesamte Forstwesen 91: 195–211
Bailey, S. I. (1908): Peruvian meteorology: Observations made at the auxiliary stations 1892–1895. Ann. Astron. Obs. Harvard Coll. 49 (2): 104–232
Bailey, V. (1936): The mammals and life zones of Oregon. USDA North American Fauna 55
Baker, B. B. and Moseley, R. K. (2007): Advancing treeline and retreating glaciers: Implications for conservation in Yunnan, P. R. China. Arct. Antarct. Alp. Res. 39 (2): 200–209

Ball, M. C., Hodges, V. S. and Laughlin, G. P. (1991): Cold-induced photoinhibition limits regeneration of snow gum at tree-line. Ecology 5: 663–668
Balslev, H. and Luteyn, J. L. (1992): Páramo – An Andean ecosystem under human influence. Academic Press, London
Banks, J. C. B. and Paton, D. M. (1993): Low temperatures as an ecological factor in the cool-climate eucalypt forest. In: Management of structure and productivity of boreal and sub-alpine forests (Linders, S. and Kellomäki, S., eds.), Studia Forestalia Suecica 191: 25–32
Bannan, M. W. and Bindra, M. (1970): The influence of wind on ring width and cell length in conifer stems. Can. J. Bot. 48: 255–259
Barclay, A. M. and Crawford, R. M. M. (1982): Winter desiccation stress and resting bud viability in relation to high altitude survival. Flora 72: 21–34
Barclay, A. M. and Crawford, R. M. M. (1984): Seedling emergence in the rowan (Sorbus aucuparia) from an altitudinal gradient. J. Ecol. 72: 627–636
Barrera, M. D., Frangi, J. L., Richter, L., Perdomo, M. H. and Pinedo, L. B. (2000): Structural and functional changes in the Nothofagus pumilio forest along an altitudinal gradient in Tierra del Fuego, Argentina. J. Veg. Sci. 11: 179–188
Barry, R. G. (1978): Diurnal effects on topoclimate on an equatorial mountain. Arb. Zentralanst. Met. Geodynam. 32 (72): 1–8
Barry, R. G. (1992): Mountain weather and climate. Routledge, London/New York
Barsch, D. (1963): Wind, Baumform und Landschaft. Eine Untersuchung des Windeinflusses auf Baumform und Kulturlandschaft unter besonderer Berücksichtigung des Mistralgebietes im französischen Rhonetal. Freiburger Geographische Hefte 1
Batchelet, D. and Neilson, R. P. (2000): Biome redistribution under climate change. USDA For. Serv., Gen. Tech. Rep. TMRS-GTR-59: 18–33
Bathen, R. (1935): Vara polare furuskoger. Tidskrift for Skogbruk: 75–83
Baumann, F. (1988): Geographische Verbreitung und Ökologie südamerikanischer Hochgebirgspflanzen. Beitrag zur Rekonstruktion der quartären Vegetationsgeschichte der Anden. Physische Geographie 28
Bazzigher, G. (1956): Pilzkrankheiten in Aufforstungen. Eidgenössische Anstalt für das forstliche Versuchswesen., Kurzmitteilungen 12
Bazzigher, G. (1976): Reservespritzung mit hochkonzentrierten Fungiziden für die Bekämpfung des Schwarzen Schneeschimmels im Herbst. Eidgenössische Anstalt für das forstliche Versuchswesen, Berichte 166
Bazzigher, G. (1978): Die Bekämpfung des Arven-Schneepilzes Phacidium infestans Karst. Eidgenössische Anstalt für das forstliche Versuchswesen, Berichte. 182
Beaman, J. H. (1962): The timberlines of Iztaccihuatl and Popocatepetl, Mexico. Ecology 43 (3): 377–385
Beaudoin, A. B. (1989): The response of the treeline zone to climatic change in the Canadian Rocky Mountains. Can. Geogr. 33 (1): 83–86
Beck, E. (1990): Jede Nacht Winter: Jeden Tag Sommer Tropenpflanzen müssen frostfrei sein. forschung Mitteilungen DFG 90 (4): 4–7
Beck, E., Schulze, E.-D., Sensor, J. and Scheibe, R. (1984): Equilibrium freezing of leaf water and extracellular ice formation in afroalpine "giant rosette" plants. Planta 162: 276–282
Beck, E., Scheibe, R. and Schulze, E.-D. (1986): Recovery from fire: Observations in the alpine vegetation of western Mt. Kilimanjaro (Tanzania). Phytocoenologia 14 (1): 55–77
Beck, E., Bendix, J., Kottke, I., Makeshin, F. and Mosandl, R. (eds.) (2008): Gradients in a tropical mountain ecosystem of Ecuador. Ecol. Stud. 198: 291–303
Bednorz, F. (1993): Vergleichende ökologische Untersuchungen von Kleinstandorten am Pru dal Vent (Puschlav/Schweiz). Diploma thesis, Fachbereich Geowissenschaften, Westfälische Wilhelms-Universität, Münster

Bednorz, F. (2000): Der Abbau der organischen Substanz im Waldgrenzökoton am Stillberg (Dischmatal/Schweiz): Arbeiten aus dem Institut für Landschaftsökologie 7, Westfälische Wilhelms-Universität, Münster
Bednorz, F., Reichstein, M. and Broll, G. (1999): Die mikrobielle Biomasse der Böden im Waldgrenzökoton am Stillberg (Dischmatal bei Davos/Schweiz). Mitteilungen der Deutschen Bodenkundlichen Gesellschaft 89: 205–208
Bekker, M. F. (2005): Positive feedback between tree establishment and patterns of subalpine forest advancement, Glacier National Park, Montana, U.S.A. Arct. Antarct. Alp. Res. 37 (1): 97–107
Bekker, M. F., Malanson, G. P., Alftine, K. J. and Cairns, D. M. (2001): Feedback and pattern in computer simulations of the alpine treeline ecotone. Kluwer Int. Ser. Eng. Comp. Sci. 626: 123–138
Bendix, J. and Rafiqpoor, M. D. (2001): Studies in the thermal conditions of soils at the upper treeline in the Páramo of Papallacta (Eastern Cordillera of Ecuador). Erdkunde 55: 257–276
Benecke, U. (1972): Wachstum, CO_2-Gaswechsel and Pigmentgehalt einiger Baumarten nach Ausbringung in verschiedenen Höhenlagen. Angewandte Botanik 46: 117–135
Benecke, U. and Havranek, W. M. (1980): Phenological growth characteristics of trees with increasing altitude, Craigieburn Range, New Zealand. In: Mountain environments and subalpine tree growth (Benecke, U. and Davis, M. R., eds.), New Zeal. For. Serv. Tech. Paper 70: 165–173
Benecke, U. and Nordmeyer, A. (1982): Carbon uptake and allocation by Nothofagus solandri var. cliffortioides (Hook. f.) Poole and Pinus contorta Douglas ex Loudon ssp. contorta at montane and subalpine altitudes. In: Carbon uptake and allocation in subalpine ecosystems as a key to management (Waring, R. H., ed.): 9–12, Forest Research Laboratory, Oregon State University, Corvallis, OR
Benedict, J. B. (1973): Chronology of cirque glaciation, Colorado Front Range. Quatern. Res. 3 (4): 585–599
Benedict, J. B. (1974): Early occupation of the Caribou Lake site, Colorado Front Range. Plains Anthropol. 19 (63): 1–4
Benedict, J. B. (1981): The Fourth of July Valley, glacial geology and archeology of the timberline ecotone. Research Report 2, Center for Mountain Archeology Ward, Colorado
Benedict, J. B. (1984): Rates of tree-island migration, Colorado Rocky Mountains, USA. Ecology 65: 820–823
Benedict, J. B. (1985): Arapaho Pass. Glacial geology and archeology at the crest of the Colorado Front Range. Research Report 3, Center for Mountain Archeology Ward, Colorado
Benedict, J. B. (1992): Footprints in the snow: High altitude cultural ecology of the Colorado Front Range, U.S.A. Arct. Alp. Res. 24 (1): 1–16
Benedict, J. B. (2002): Eolian deposition of forest-fire charcoal above tree limit, Colorado Front Range, U.S.A.: Potential contamination of AMS radiocarbon samples. Arct. Antarct. Alp. Res. 34 (1): 33–37
Beniston, M. (1997): Variations of snow depth and duration in the Swiss Alps over the last 50 years: Links to changes in large-scale climatic forcings. In: Climatic change at high elevation sites (Diaz, M. F., Beniston, M. and Bradkey, R. S., eds.): 49–68, Dordrecht
Beniston, M. (2001): The effects of Global Warming on mountain regions: a summary of the 1995 report of the Intergovernmental Panel on Climate Change. Advances in Global Change Research 9 (Visconti, G., Beniston, M., Iannorelli, E. D. and Barba, D., eds.): 155–185, Dordrecht, Boston, London
Berg, N. H. (1986): Blowing snow at a Colorado alpine site: Measurements and implications. Arct. Alp. Res. 18 (2): 147–161

Bernadzki, E. (1976): Der ökologische Waldbegriff und das Problem der Abgrenzung von Wald und Nicht-Wald. Eidgenössische Anstalt für das forstliche Versuchswesen, Berichte 167: 17–28

Bernbeck, O. (1907): Der Wind als pflanzenpathologischer Faktor. Diss. Bonn, Stuttgart

Bernbeck, O. (1911): Wind und Pflanzenwachstum. Forstwissenschaftliches Centralblatt: 210–211

Bichsel, M. (1995): Ökologie und Verbreitung der Arve. Bündnerwald 6: 19–25

Bigras, F. J., Ryyppö, A., Lindström, A. and Stattin, E. (2001): Cold acclimation and deacclimation of shoots and roots of conifer seedlings, In: Conifer cold hardiness (Bigras, F. and Colombo, S. J., eds.), Tree Physiology 1: 57–88

Billings, W. D. (1969): Vegetational pattern near alpine timberline as affected by fire-snow drift interactions. Vegetatio 19: 192–207

Billings, W. D. (1973): Arctic and alpine vegetations: Similarities, differences, and susceptibility to disturbance. BioScience 23: 679–704

Billings, W. D. (1974): Arctic and alpine vegetation: Plant adaptations to cold summer climates. In: Arctic and alpine environments (Ives, J. D. and Barry, R. G., eds.): 403–443, Methuen, London

Billings, W. D. (1979): Alpine ecosystems of western North America. In: Special management needs of alpine ecosystems (Johnson, D. A., ed.), Range Sci. Series 5: 6–21, Denver

Billings, W. D. (1990): The mountain forests of North America and their environment. In: Plant Biology of the Basin and Range (Osmond, C. B., Pitelka, L. F. and Hidy, G. M., eds.), Ecol. Stud. 80: 47–86

Billings, W. D. and Bliss, L. C. (1959): An alpine snowbank environment and its effect on vegetation, plant development and productivity. Ecology 40: 388–397

Billings, W. D. and Mark, A. F. (1957): Factors involved in the persistence of montane treeless balds. Ecology 38: 140–142

Biondi, F. (2001): A 400-year tree-ring chronology from the tropical treeline of North America. Ambio 30 (3): 162–166

Biondi, F., Hartsough, P. C. and Estrada, I. G. (2005): Daily weather and tree growth at the tropical treeline of North America. Arct. Antarct. Alp. Res. 37 (1): 16–24

Bisaz, O. (1968): Das Lawinenverbauungs- und Aufforstungsprojekt Munt da la Bês-cha (Schafberg) der Gemeinde Pontresina. Bündnerwald 21 (5): 201–207

Bischoff, N. (1995): Tamangur. Bündnerwald 6: 9–13

Björkman, E. (1948): Studier över snöskyttetssvampens (Phacidium infestans Karst.) biologi samt metoder för snöskyttets bekämpande. Meddelanden från statens skogsforskingsinstitut. 37 (2): 43–128

Björkman, E. (1962): The phacidium blight of pine fungus: Phacidium infestans Karst. 13. IUFRO-Kongreß (Wien 1961), Berichte 2 (1): 24–26

Björkman, E. (1963): Resistance to snow blight (Phacidium infestans Karst.) in different provenances of Pinus silvestris L. Studia Forestalia Suecica 5

Black, R. A. and Bliss, L. C. (1980): Reproductive ecology of Picea mariana (MILL.) BSP., at tree line near Inuvik, Northwest Territories, Canada. Ecol. Monogr. 50 (3): 331–354

Blaser, P. (1980): Der Boden als Standortsfaktor bei Aufforstungen in der subalpinen Stufe (Stillberg, Davos). Schweizerische Anstalt für das forstliche Versuchswesen, Berichte 56 (3)

Bliss, L. C. (1963): Alpine zone of the Presidential Range. Edmonton

Bliss, L. C. and Woodwell, G. M. (1965): An alpine podzol on Mt. Katahdin, Maine. Soil Sci. 100 (4): 274–279

Block, J. and Treter, U. (2003): Growth conditions at the upper and lower forest limits in the mountain-forest steppe of Northwest Mongolia. In: TRACE-tree rings in archaeology, climatology and ecology, Vol. 1, Proceedings of the Dendrosymposium 2002, April 11–13, Bonn (Schleser, G., Winiger, M., Bräuning, A., Gärtner, H., Helle, G., Jansma, E.,

Neuwirth, B. and Treydte, K., eds.): 33: 101–107, Schriftenreihe des Forschungszentrums Jülich, Reihe Umwelt
Bloom, A. J., Chapin, F. S. and Mooney, H. A. (1985): Resource limitation in plants. An economic analogy. Ann. Rev. Ecol. Syst. 16: 363–392
Blüthgen, J. (1937a): Vegetations-Beobachtungen in Skandinavien. Natur und Volk 67 (2): 102–208
Blüthgen, J. (1937b): Baumgrenzen in Lappland. Geographischer Anzeiger: 532–533
Blüthgen, J. (1942): Die polare Baumgrenze. Veröffentlichungen des Deutschen Wissenschaftlichen. Instituts zu Kopenhagen, Reihe I, Arktis 10
Blüthgen, J. (1943): Zur Dynamik der polaren Baumgrenze in Lappland. Forschung und Fortschritte 19: 158–160
Blüthgen, J. (1952): Baumgrenze und Klimacharakter in Lappland. Ein Beitrag zur Frage der Veränderlichkeit von Klimawerten und deren geographischer Auswirkung. Berichte des Deutschen Wetterdienstes in der U. S. Zone 42: 362–371
Blüthgen, J. (1960): Der skandinavische Fjällbirkenwald als Landschaftsformation. Petermanns Geographische Mitteilungen 2/3: 119–144
Blüthgen, J. (1970): Problems of definition and geographical differentiation of the subarctic with special regard to northern Europe. Ecol. Conserv. 1, UNESCO: 11–33
Blüthgen, J. (1980): Allgemeine Klimageographie (3rd edition, reworked by W. Weischet). Springer, Berlin/New York
Böcher, T. W. (1979): Birch woodlands and tree growth in southern Greenland. Holarct. Ecology 2: 218–221
Bochter, R., Neuerburg, W. and Zech, W. (1983): Humus und Humusschwund im Gebirge. Nationalpark Berchtesgaden, Forschungsberichte 2 (2nd edition)
Bockheim, J. G. (1968): Vegetational transition and soil morphogenesis on Saddleback Mountain, western Maine. Unpubl. M.S. thesis, Department of Plant and Soil Sciences, University of Maine, Maine
Bockheim, J. G. (1972): Effects of alpine and subalpine vegetation on soil development, Mount Baker, Washington. Ph.D. thesis, University of Washington, Washington, DC
Boesch, H. (1936): Der Bergbau am Ofenpaß nördlich der Paßhöhe. Jahresbericht der Naturforschenden Gesellschaft Graubündens 74, Neue Folge: 109–122
Böhmer, J. (1996): Säkulare Klimaschwankungen und rezente Klimatrends Zentral- und Hochasiens. Göttinger Geograpische Abhandlungen 101
Bollinger, W. H. (1973): The vegetation patterns after fire at the alpine forest-tundra ecotone in the Colorado Front Range. Ph.D. thesis, University of Colorado, Boulder, CO
Bonan, G. B., Pollard, D. and Thompson, S. L. (1992): Effects of boreal forest vegetation on global climate. Nature 359: 716–718
Bonan, G. B., Chapin, F. S. and Thompson, S. L. (1995): Boreal forest and tundra ecosystems as part of the climate system. Clim. Change 29: 145–167
Bortenschlager, S. (1984): Beiträge zur Vegetationsgeschichte Tirols I: Inneres Ötztal und unteres Inntal. Berichte des Naturwissenschaftlich-Medizinischen Vereins in Innsbruck 71: 19–56
Bortenschlager, S., Kofler, W., Oeggl, K. and Schock, W. (1992): Erste Ergebnisse der Auswertung der vegetabilischen Reste vom Hauslabjochfund. In: Der Mann im Eis (Höpfel, F., Platzer, W. and Spindler, K., eds.), Symp. Innsbruck 1: 307–312
Boysen-Jensen, P. (1932): Die Stoffproduktion der Pflanze. Jena
Boysen-Jensen, P. (1949): Causal plant geography. Dansk. Vedensk. Selsk., Biol. Medd. 21 (3)
Braathe, P. (1977): Biological and technological aspects of forest management in Norway. In: North American forest lands at latitudes north of 60 degrees. Proceedings of a symposium, September 19–22, 1977 at the University of Alaska, School of Agriculture and Land Resources Management and Agriculture Experiment Station, Fairbanks, AK: 35–54

Bramer, H. (1985): Geographische Zonen der Erde. Lehrbuch der Physischen Geographie (Hendl, M. and Bramer, H., eds.): 436–522

Brandes, R. (2007): Waldgrenzen griechischer Hochgebirge. Unter besonderer Berücksichtigung des Taygetos (Waldgrenzdynamik, dendrochronologische Untersuchungen). Erlanger Geographische Arbeiten, Sonderband 36

Brandes, R. and Ise, M. (2007): Fingerprints of climate change in Mediterranean mountain forests? – Observations in Mediterranean fir-species threatened by climate change. Geoöko 28: 1–26

Brass, L. J. (1964): Results of the Archbold Expeditions No. 86. Results of the sixth Archbold Expedition to New Guinea (159). Bull. Am. Mus. Nat. Hist. 127: 145–216

Braun, G. (1988): Verbreitung und Ökologie der Gattung Polylepis in den Anden – unter besonderer Berücksichtigung des Vulkans Sajama in der Westkordillere Boliviens. Diploma thesis, Geographisches Institut der Rheinischen Friedrich-Wilhelms-Universität, Bonn

Braun-Blanquet, J. (1964): Pflanzensoziologie (3rd edition). Wien/New York

Bräuning, A. (1994): Dendrochronologische Untersuchungen an osttibetischen Waldgrenzstandorten. Göttinger Geographische Abhandlungen 95: 185–192

Bräuning, A. (1999): Zur Dendroklimatologie Hochtibets während des letzten Jahrtausends. Dissertationes Botanicae 312: 127–164

Brink, V. C. (1959): A directional change in the subalpine forest-heath ecotone in Garibaldi-Park, British Columbia. Ecology 40 (1): 10–16

Brockmann-Jerosch, H. (1919): Baumgrenze und Klimacharakter. Beiträge zur geobotanischen Landesaufnahme 6

Brockmann-Jerosch, H. (1928): Die südpolare Baumgrenze. Vierteljahresschrift der Naturforschenden Gesellschaft Zürich 15: 705–718

Brodersen, C. R., Germino, M. J. and Smith, W. K. (2006): Photosynthesis during an episodic drought in Abies lasiocarpa and Picea engelmannii across an alpine treeline. Arct. Antarct. Alp. Res. 38: 34–41

Broll, G. (1994): Influence of the soil mosaic at heath sites in the European Subarctic. Transactions 15th World Congress of Soil Science (Mexico) 4a: 220–231

Broll, G. (1998): Diversity of soil organisms in alpine and arctic soils in Europe. Review and research needs. Pireneos 151–152: 43–72

Broll, G. (2000): Influence of overgrazing by reindeer on soil organic matter and soil microclimate of well-drained soils in the Finnish Subarctic. In: Global climate change and cold regions ecosystems (Lal, R., Kimble, J. M. and Stewart, B. A., eds.), Boca Raton Adv. Soil Sci. 2000: 163–172

Broll, G. and Holtmeier, F.-K. (1994): Die Entwicklung von Kleinreliefstrukturen im Waldgrenzökoton der Front Range (Colorado, USA) unter dem Einfluß leewärts wandernder Ablegergruppen (Picea engelmannii and Abies lasiocarpa). Erdkunde 48: 48–59

Broll, G. and Keplin, B. (eds.) (2005): Mountain ecosystems. Studies in treeline ecology. Springer, Berlin/Heidelberg/New York

Broll, G., Holtmeier, F.-K., Anschlag, K., Brauckmann, H.-J., Wald, S. and Drees, B. (2007): Landscape mosaic in the treeline ecotone on Mt. Rodjanoaivi, subarctic Finland. Fennia 185 (2): 89–105

Brooks, P. D., Williams, M. W. and Schmidt, S. K. (1996): Microbial activity under alpine snowpacks, Niwot Ridge, Colorado. Biogeochemistry 32: 93–113

Brown, C. L. (1974): Growth and form. In: Tree structure and function (Zimmermann, M. H., Brown, C. L. and Tyree, M. T., eds.): 125–167, Berlin/Heidelberg/New York

Brown, R. J. E. and Péwé, T. L. (1973): Distribution of permafrost in North America and its relationship to the environment review 1963–1973: In North American contribution, Permafrost International Conference Yakutsk, July 13–28, 1973, USSR (Nat. Acad. Sci., Washington DC, ed.): 71–100

Brown, R. T. and Mikola, P. (1974): The influence of fruticose soil lichens upon the mycorrhizae and seedling growth in forest trees. Acta Forestalia Fennica 141: 5–22

Bruijnzeel, L. A. and Proctor, J. (1995): Hydrology and biogeochemistry of tropical montane cloud forests: What do we really know? In: Tropical montane cloud forest (Lange, O. L., Mooney, H. A. and Remmert, H., eds.), Ecol. Stud. 110: 38–78

Brunstein, F. C. and Yamaguchi, D. K. (1992): The oldest known Rocky Mountains bristlecone pines (Pinus aristata Engelm.). Arct. Alp. Res. 24 (3): 253–256

Bruun, H. H. and Moen, J. (2003): Nested communities of alpine plants on isolated mountains; relative importance of colonization and extinction. J. Biogeogr. 30: 297–303

Bryn, A. and Dangstad, K. (2001): Summer farming in the subalpine birch forest. In: Nordic mountain birch ecosystems (Wielgolaski, F. E., ed.), Mand and Biosphere Series 27: 307–315

Bryson, R. A., Irwin, W. N. and Larsen, J. A. (1965): Radiocarbon and soil evidence of former forest in the southern Canadian tundra. Science 147: 46–48

Buckner, D. L. (1977): Ribbon forest development and maintenance in the central Rocky Mountains of Colorado. Ph.D. thesis, University of Colorado, Boulder, CO

Bugmann, H. and Weisberg, P. J. (2003): Forest ungulate interactions: Monitoring, modeling and management. J. Nat. Conserv. 10: 193–201

Burga, C. (1990): Vegetationsgeschichte und Paläoklimatologie. Vierteljahresschrift der Naturforschenden Gesellschaft Zürich 135 (1): 17–30

Burga, C. A. (1988): Swiss vegetation history during the last 18000 years. New Phytologis 110: 581–602

Burga, C. A. and Perret, R. (2001): Monitoring of eastern and southern Swiss alpine timberline ecotones. In: Biomonitoring: General and applied aspects on regional and global scale (Burga, C. and Kratochwil, A., eds.): 179–194, Kluwer, Dordrecht

Burgess, P. (1999): Human environmental interactions in Upper Lapland, Finland. Arctic Centre Reports 27

Burnham, C. P. (1974): The role of the soil forming factors in controlling altitudinal zonation on granite in Malaysia. In: Altitudinal zonation in Malaysia (Flenley, R., ed.). Jr. Misc. Ser. 16: 59–74, University of Hull, Dept. Geogr., Hull, England

Burns, S. (1980): Alpine soil distribution and development, Indian Peaks, Colorado Front Range. Ph.D. thesis, University of Colorado, Boulder, CO

Burns, S. F. and Tonkin, P. J. (1982): Soil-geomorphic models and the spatial distribution and development of alpine soils. In: Space and time in geomorphology (Thorn, C. G., ed.), Internat. Ser. 12: 25–43

Bussmann, R. W. (1993): The forest of Mt. Kenya (Kenya). A phytosociological approach with special reference to ecological problems. Diploma thesis, Fakultät für Biologie, Universität Tübingen

Bussmann, R. W. (1994): The forest of Mt. Kenya (Kenya), vegetation, ecology, destruction and management of a tropical mountain forest ecosystem. Diss. Fakultät für Biologie, Chemie und Geowissenschaften, Universität Bayreuth

Butin, H. (1996): Krankheiten der Wald- und Parkbäume (3rd edition). Thieme, Stuttgart/New York

Butler, D. R. (1995): Zoogeomorphology. Animals as geomorphic agents. Cambridge University Press, Cambridge/New York

Butler, D. R., Malanson, G. P. and Cairns, D. M. (1994): Stability of alpine treeline in Glacier National Park, Montana, U.S.A. Phytocoenologia 22: 485–500

Butler, D. R., Malanson, G. P. and Resler, L. M. (2004): Turf-banked terrace treads and risers, turf exfoliation and possible relationships with advancing treeline. Catena 58: 259–274

Butler, D. R., Malanson, G. P., Walsh, S. J. and Fagre, D. B. (2007): Influences of geomorphology and geology on alpine treeline in the American west – More important than climatic influences? Phys. Geogr. 28: 434–450

Byers, A. C. (2000): Contemporary landscape change in the Huascarán National Park and buffer zone, Cordillera Blanca, Peru. Mt. Res. Develop. 20 (1): 52–63
Bylund, H. (1997): Stand age-structure influence in a low population peak of Epirrita autumnata. Ecography 20: 319–326
Bylund, H. (1999): Climate and the population dynamics of two insect outbreak species in the north. In: Animal responses to global change in the north (Hofgaard, A., Ball, J. P., Danell, K. and Callaghan, T., eds.), Ecol. Bull. 47: 54–62
Cairns, B. B. and Fonda, R. W. (1974): The influence of subalpine snowbanks on vegetation pattern, production and phenology. Bull. Torrey Bot. Club 101 (6): 340–350
Cairns, D. M. (1998): Modeling controls on pattern at alpine treeline. Geogr. Environ. Model. 2 (1): 43–63
Cairns, D. M. (2001): Patterns of winter desiccation in Krummholz forms of Abies lasiocarpa at treeline sites in Glacier National Park, Montana, USA. Geografiska Annaler 83A (3): 157–168
Cairns, D. M. and Malanson, G. P. (1998): Environmental variables influencing the carbon balance at the alpine treeline: A model approach. J. Veg. Sci. 9: 679–692
Cairns, D. and Moen, J. (2004): Herbivory influences tree lines. J. Ecol. 92: 1019–1024
Cairns, D., Moen, J. and Young, A. (2007): Influences of animal activity on treeline position and pattern: Implications for treeline responses to climate change. Phys. Geogr. 28: 419–433
Caldas, E. F. and Delgado, A. G. (1971): Condiciones de formacion y evolucion de los suelos de Tenerife. Anales de Edafologia y Agrobiologia 30: 565–610
Caldwell, M. M. (1970): The wind regime at the surface of the vegetation layer above timber-line in the Central Alps. Centralblatt für das gesamte Forstwesen 87 (2): 65–74
Callaghan, T. V., Crawford, R. M. M., Eronen, M., Hofgaard, A., Payette, S., Rees, W. G., Skre, O., Sveinbjörnsson, B., Vlassova, T. K. and Werkman, B. R. (2002a): The dynamics of the tundra-taiga boundary: An overview and suggested coordination and integrated appproach to research. Ambio Sp. Rep. 12: 3–5
Callaghan, T. V., Werkman, B. R. and Crawford, R. M. M. (2002b): The tundra-taiga interface and its dynamics: Concepts and applications. Ambio Sp. Rep. 12: 6–14
Calörtscher, H. (1969): Die Schutz- und Wohlfahrtswirkungen des Waldes. Bündner Wald, Beiheft 7: 73–107
Camarero, J. J. and Gutiérrez, E. (1999): Structure and recent recruitment at alpine forest ecotone in the Spanish Central Pyrenees. Écosciene 6: 451–464
Camarero, J. J. and Gutiérrez, E. (2000): Recent spatiotemproal variability of Pinus uncinata regeneration in altitudinal treeline ecotones (Spanish Central Pyrenees). Int. Conf. Dendrochron. for the Third Millenium, 2–7 April, Mendoza, Argentina: 222
Camarero, J. J. and Gutiérrez, E. (2004): Pace and pattern of recent treeline dynamics: Response of ecotones to climate variability in the Spanish Pyrenees. Climatic Change 63: 181–200
Camarero, J. J. and Gutiérrez, E. (2007): Response of Pinus uncinata recruitment to climate warming and changes in grazing pressure in an isolated population of the Iberian system (NE Spain). Arct. Antarct. Alp. Res. 39 (2): 210–217
Campbell, E. M. and Antos, J. A. (2000): Distribution and severity of white bark pine rust and mountain pine beetle on whitebark pine in British Columbia. Can. J. For. Res. 30: 1051–1059
Campell, E. (1944): Der Wald des Oberengadins im Wandel der Zeiten. Festschrift. Jahresversammlung der Schweizerischen Naturforschenden Gesellschaft 124: 93–102
Campell, E. (1949): Raubbau am Il Fuorn. Der Schweizer Nationalpark, Zürich
Campell, E. (1950): Der Tannen-oder Nusshäher und die Arvenverbreitung. Bündner Wald, Beiheft 4 (1): 3–7
Campell, E. (1955): Der Lärchen-Arvenwald. Bündner Wald, Beiheft 5: 14–26

Campell, E. (1958): Steinwild im Engadin. Bündnerwald 11: 287–294
Carbiener, R. (1969): Subalpine primäre Hochgrasprärien im hercynischen Gebirgsraum Europas mit besonderer Berücksichtigung der Vogesen und des Massif Central. Mitteilungen der Floristisch-soziologischen Arbeitsgemeinschaft. Neue Folge 14: 322–345
Carcaillet, C., Talon, B. and Barbero, M. (1998): Pinus cembra et incendies pendant l'Holocène, 300 m au-dessus de la limite actuelle des arbres dans le massif de la Vanoise (Alpes du nord-ouest). Ecologie 29: 277–282
Carnelli, A. L., Theurillat, J.-P., Thinon, M., Vadi, G. and Talon, B. (2004): Past uppermost tree limit in the central European Alps (Switzerland) based on soil and soil charcoal. The Holocene 14 (3): 393–405
Carrara, P. E., Mode, W. N., Rubin, M. and Robinson, S. W. (1984): Deglaciation and postglacial timberline in the San Juan Mountains, Colorado. Quatern. Res. 21: 42–55
Cavieres, L. A., Rada, F., Azocar, A., Núñez, C. G. and Cabrera, H. M. (2000): Gas exchange and low temperature resistance in two tropical high mountain tree species from the Venezuelan Andes. Acta Oecol. 21 (3): 203–211
Cázares, E., Trappe, J. M. and Jumpponen, A. (2005): Mycorrhiza-plant colonization patterns on subalpine glacier forefront as model system of primary succession. Mycorrhiza 15: 405–416
Cernusca, A. (1978): Ökologische Veränderungen im Bereich aufgelassener Almen. Zusammenfassung der Ergebnisse einer interdisziplinären Ökosystemstudie im Gasteiner Tal. In: Ökologische Analysen von Almflächen im Gasteiner Tal, Veröffentlichungen des Österreichischen Zentralblatts 2: 7–27
Challies, C. N. (1990): Red deer. In: The handbook of New Zealand mammals (King, C. M., ed.): 436–458, Auckland Oxford University Press, Melbourne/Oxford/New York
Chapin, D. M. and Bliss, L. C. (1989): Seedling growth, physiology, and survivorship in a subalpine volcanic environment. Ecology 70: 1325–1334
Chapin III, F. S., Walker, L. R., Fastie, C. L. and Sharman, L. C. (1994): Mechanisms of primary succession following deglaciation at Glacier Bay, Alaska. Ecol. Monogr. 64: 149–175
Chapin III, F. S., Sturm, M., Serreze, M. C., McFadden, J. P., Key, J. R., Lloyd, A., H., McGuire, A. D., Rupp, T. S., Lynch, A. H., Schimel, J.P., Beringer, J., Chapman, W. L., Epstein, H. E., Euskirchen, E. S., Hinzman, L. D., Jia, G., Ping, C. L., Tape, K. D., Thompson, C. D. C., Walker, D. A. and Welker, J. M. (2005): Role of land-surface changes in arctic summer warming. Science Express Report/www.scienceexpress.org/22 September 2005/Page 1/10.1126/science.11173368
Chapman, J. A., Romer, J. I. and Stark, J. (1955): Ladybird beetles and army cutworms adults as food of grizzly bears in Montana. Ecology 36: 156–158
Chew, R. M. (1994): Consumers as regulators of ecosystems: An alternative to energetic. Ohio J. Sci. 74: 359–370, Ohio Academy of Sciences, Columbus
Christersson, L. (1978): The influence of photoperiod and temperature on the development of frost hardiness in seedlings of Pinus silvestris and Picea abies. Physiol. Plant. 44: 288–294
Christersson, L. and Sandstedt, R. (1978): Short-term temperature variation in needles of Pinus sylvestris L. Canadian Journal of Forest Research 8: 480–482
Cierjacks, A., Rühr, N. K., Wesche, K. and Hensen, I. (2008): Effects of altitude and livestock on the regeneration of two tree line forming Polylepis species in Ecuador. Plant Ecol. 194: 207–221
Clapham, A. R., Tutin, T. G. and Moore, D. M. (1989): Flora of the British Isles. Cambridge University Press, Cambridge
Clausen, J. (1962/1963): Studies on the distribution of tree species. Carnegie Institution Year Book 62: Ann. Rep. Dir. Dept. Plt. Biol.
Clausen, J. (1963): Tree lines and germ plasm – a study in evolutionary limitations. Proc. Nat. Acad. Sci. 50: 860–868

Clements, F. E. (1936): Nature and structure of the climax. J. Ecol. 24: 255–284
Cochrane, P. M. and Slatyer, R. O. (1988): Water relations of Eucalyptus pauciflora near the alpine tree line in winter. Tree Physiol. 4: 45–52
Coe, M. J. (1967): The ecology of the alpine zone of Mt. Kenya. Monographiae Biologicae 17, The Hague
Cogbill, C. V. and White, P. S. (1991): The latitude-elevation relationship for spruce-fir forest and treeline along the Appalachian mountain chain. Vegetatio 94: 153–175
Coleman, M. D., Hinckley, T. M., McNaughton, G. and Smit, B. A. (1992): Root cold hardiness and native distribution of subalpine conifers. Can. J. For. Res. 22: 932–938
Colpaert, A., Kumpula, J. and Nieminen, M. (2003): Reindeer pasture biomass assessment using satellite remote sensing. Arctic 56 (2): 147–158
Conrad, H. (1940): Beitrag zur urgeschichtlichen Besiedlung des Engadins. Jahresbericht der Historisch-Antiquarischen Gesellschaft Graubündens 70: 1–40
Conroy, J. P., Smillie, R. M., Küppers, M., Bevege, D. I. and Barlow, E. W. (1986): Chlorophyll A fluorescence and photosynthetic and growth responses of Pinus radiata to phosphorus deficiency, drought stress, and high CO_2. Plant Physiol. 81: 423–429
Cooper, W. S. (1911): Reproduction by layering among conifers. Contributions from the Hull Botanical Laboratory 149. Botanical Gazette 52: 369–379
Cooper, W. S. (1931): The layering habit in Sitka spruce and the two western hemlocks. Bot. Gaz. 91: 441–451
Cordell, S., Goldstein, G., Melcher, P. J. and Meinzer, F. C. (2000): Photosynthesis and freezing avoidance in Ohia (Metrosideros polymorpha) at treeline in Hawaii. Arct. Antarct. Alp. Res. 32 (4): 381–387
Corlett, R. T. (1984): Human impact on the subalpine vegetation of Mt. Wilhelm, Papua New Guninea. J. Ecol. 72: 841–854
Corlett, R. T. (1987): Post-fire succession on Mt. Wilhelm, Papua New Guinea. Biotropica 19: 157–160
Costin, A. B. (1967): Alpine ecosystems of the Australasian region. In: Arctic and alpine environments (Wright, H. E. and Osburn, W. H., eds.): 55–87, Indiana University Press, Bloomington, IN
Costin, A. B. (1981): Alpine and subalpine vegetation. In: Australian vegetation (Groves, R. H., ed.), Cambridge University Press, Cambridge
Cox, C. F. (1933): Alpine plant succession on James Peak, Colorado. Ecol. Monogr. 3: 299–372
Cramer, T. (1997a): Climatic gradients in the Karakorum and their effects on the natural vegetation. In: Perspectives on history and change in the Karakorum, Hindukush and Himalaya (Stellrecht, I. and Winiger, M., eds.), Culture Area Karakorum Scientific Studies 3: 265–276
Cramer, W. (1997b): Modeling the possible impact of climate change on broad-scale vegetation structure: Example from northern Europe. In: Global change and arctic terrestrial ecosystems (Oechel, W. C., Callaghan, T., Gilmanov, T., Holten, J. I., Marwell, B., Molau, U. and Sveinbjörnsson, B., eds.): 312–329, Springer, New York
Crawford, R. M. M. (1978): Ecological hazards of ocenanic environments. New Phytol. 128: 257–281
Crawford, R. M. M. (1989): Studies in plant survival. Ecological case histories of plant adaptation to adversity (reprinted 1990, 1994). London/Edinburgh/Boston, MA/Melbourne/Paris/Berlin/Wien
Crawford, R. M. M. (1997): Consequences of climatic warming for plants of the northern and polar regions of Europe. Flora Colonia 5/6: 65–78
Crawford, R. M. M. (2005): Peripheral plant populations survival in polar regions. In: Mountain ecosystems. Studies in treeline ecology (Broll, G. And Keplin, B., eds.): 44–76, Springer, Berlin/Heidelberg

Crawford, R. M. M. (2008): Plants at the margin. Ecological limits and climate change. Cambridge University Press, Cambridge
Crawford, R. M. M., Jeffree, C. E. and Rees, W. G. (2003): Paludification and forest retreat in northern oceanic environments. Ann. Bot. 91: 213–226
Critchfield, W. B. and Allenbaugh, G. L. (1969): The distribution of Pinaceae in and near northern Nevada. Madrono 20 (1): 12–25
Cuevas, J. G. (2002): Episodic regeneration at the Nothofagus punilio alpine timberline in Tierra del Fuego, Chile. J. Ecol. 90: 52–60
Cui, M. and Smith, W. K. (1991): Photosynthesis, water relations and mortality of Abies lasiocarpa seedlings during natural establishment. Tree Physiol. 8: 37–46
Cullen, L. E., Stewart, G. H., Duncan, R. P. and Palmer, J. G. (2001): Disturbance and climate warming influences on New Zealand Nothofagus tree-line population dynamics. J. Ecol. 89: 1061–1071
Curran, P. J., Foody, G. M. and Van Gardeningen, P. R. (1997): Scaling-up. In: Scaling up: From cells to landscape (Van Gardeningen, P. R., Foody, G. M. and Curran, P. J., eds.): 1–5, Cambridge
Currey, D. R. (1968): An ancient bristlecone pine stand. Ecology 44: 564–566
Curry, R. R. (1962): Geobotanical correlation in the alpine and subalpine regions of the Tenmile Range, Summit County, Colorado. Masters thesis, University of Colorado, Boulder, CO (cited in Arno, 1984)
Czell, A. (1967): Beitrag zum Wasserhaushalt subalpiner Böden. Mitteilungen der forstlichen Bundesversuchsanstalt Wien 75: 305–332
Czell, A. (1972): Wasserhaushaltsmessungen in subalpinen Böden. Mitteilungen der forstlichen Bundesversuchsanstalt Wien 98
Dahl, E. (1986): Zonation in arctic and alpine tundra and fellfield ecobiomes. In: Ecosystem theory and application (Polunin, N., ed.): 35–62, Wiley, Chichester
Dahl, E. and Mork, E. (1959): Om sambandet mellom temperatur, anding og vekst hos Gran (Picea abies (L.) Karst.). Meddelanden fran det Norske Skogforsöksvesen 53: 82–93
Dahms, A. (1984): Die natürliche Vermehrung verschiedener Baumarten im oberen Waldgrenzbereich der Colorado Front Range in ökologischer Sicht. Staatsarbeit, Institut für Geographie, Westfälische Wilhelms-Universität. Münster
Dahms, A. (1992): Wachstumsbedingungen bei Picea engelmannii (Parry) Engelm. und Abies lasiocarpa (Hook) Nutt. an unterschiedlich windexponierten Standorten im Waldgrenzbereich der Colorado Front Range, U.S.A. Diss. Mathematisch-Naturwissenschaftliche Fakultät, Westfälische Wilhelms-Universität, Münster
Dalen, L. and Hofgaard, A. (2000): Differential regional treeline dynamics in the Scandes Mountains. Arct. Antarct. Alp. Res. 37: 284–296
Daly, C. (1984): Snow distribution patterns in the alpine krummholzzone. Prog. Phys. Geogr. 8: 157–175
Damm, B. (1998): Waldgrenze, Baumgrenze, Waldentwicklung und Siedlungstätigkeit in den Tauferer Bergen (Südtirol) während des Postglazials. Geoökodynamik XIX: 19–42
Danby, R. K. and Hik, D. S. (2007): Variability, contingency and rapid change in recent subarctic alpine tree line dynamics. J. Ecol. 95: 352–363
Danell, Ö., Staatland, H. and Nieminen, M. (1999): Rennäring i Nordvest Europa. In: Reindrift i Nordvest Europa i 1998 – biologiske muligether of begrensinger (Dahle, H. K., Danell, Ö., Gaare, E. and Nieminen, M., eds.): 19–30, Nordisk ministerråd, København
Daniels, L. D. (2000): The dynamics of altitudinal treelines in northern Patagonia: Spatio-temporal influences. Ph.D. thesis, University of Colorado, Boulder, CO
Däniker, A. (1923): Biologische Studien über Wald- und Baumgrenze, insbesondere über die klimatischen Ursachen und deren Zusammenhänge. Vierteljahresschrift der Naturforschenden Gesellschaft Zürich 63

Danilkin, A. (1996): Behavioural ecology of Siberian and European deer. Chapman & Hall, London/Glasgow/Weinheim/New York/Tokyo/Melbourne/Madras
Daubenmire, R. (1953): Vegetational zonation in the Rocky Mountains. Bot. Rev. 19 (6): 325–393
Daubenmire, R. (1954): Alpine timberlines in the Americas and their interpretation. Butler Univ. Bot. Stud. 11: 119–136
Davidson, N. J. and Reid, J. B. (1985): Frost as a factor influencing the growth and distribution of subalpine eucalypts. Austr. J. Bot. 33: 657–667
Davidson, N. J. and Reid, J. B. (1987): The influence of hardening and waterlogging on the frostresistance of subalpine eucalypts. Austr. J. Bot. 35: 91–101
Davis, J., Schober, A., Bahn, M. and Sveinbjörnsson, B. (1991): Soil carbon and nutrient turnover at and below the elevational treeline in northern Fennoscandia. Arct. Alp. Res. 23: 279–286
Davis, M. B. (1986): Climate instability, time lags, and community disequilibrium. In: Community ecology (Diamond, J. and Case, T., eds.): 269–284, Harper & Row, New York
De Quervain, A. (1904): Die Hebung der atmosphärischen Isothermen in den Schweizer Alpen und ihre Beziehung zu den Höhengrenzen. Gerlands Beiträge zur Geophysik 6: 481–533
Del Moral, R. and Grishin, S. Y. (1999): Volcanic disturbances and ecosystem recovery. In: Ecosystems of disturbed ground (Walker, L. R., Hrsg.), Ecosystems of the world 16: 137–160, Amsterdam/Lausanne/New York/Oxford/Shannon/Singapore/Tokyo
Delucia, E. V. and Berlyn, G. P. (1984): The effect of increasing elevation on leaf cuticle thickness and cuticular transpiration in balsam fir. Can. J. Bot. 62: 2423–2431
Den Herder, M., Virtanen, R. and Roininen, H. (2008): Reindeer-herbivory reduces willow growth and grouse forage in a forest-tundra ecotone. Basic Appl. Ecol. 9: 324–331
Denton, G. H. and Karlen, W. (1977): Holocene glacial and treeline variation in the White River valley and Skolai pass, Alaska and Yukon Territory. Quatern. Res. 7: 63–71
Di Pasquale, G., Marziano, M., Impagliazzo, S., Lubritto, C., de Natale, A. and Bader, M. Y. (2007): The Holocene treeline in the northern Andes (Ecuador): First evidence from soil charcoal. Palaeogeography, Palaeoclimatology, Palaeoecology, accepted
Dolukhanov, A. G. (1978): The timberline and the subalpine belt in the Caucasus Mountains, USSR. Arct. Alp. Res. 10 (2): 409–422
Donaubaur, E. (1963): Über die Schneeschütte-Krankheit (Phacidium infestans KARST.) der Zirbe (Pinus cembra L.) und einige Begleitpilze. Mitteilungen der forstlichen Bundesversuchsanstalt Mariabrunn 60: 575–600
Douglas, G. W. (1972): Subalpine plant communities of the Western North Cascades, Washington. Arct. Antarct. Alp. Res. 4: 147–166
Drake, B. G. (1992): The impact of rising CO_2 on ecosystem production. Water Air Soil Pollut. 64: 25–44
Drude, O. (1890): Handbuch der Pflanzengeographie. Engelhorn, Stuttgart
Dubey, B., Yadav, R. R., Singh, J. and Chaturvedi, R. (2003): Upward shift of Himalayan pine in Western Himalaya, India. Curr. Sci. 85: 1135–1136
Dullinger, S., Dirnböck, T. and Grabherr, G. (2003): Patterns of shrub invasion into high mountain grassland of the northern Calcareous Alps, Austria. Arct. Antarct. Alp. Res. 35 (4): 434–441
Dullinger, S., Dirnböck, T. and Grabherr, G. (2004): Modelling climate change-driven treeline shifts: Relative effects of temperature increase, dispersal and invisibility. J. Ecol. 92: 241–252

Dullinger, S., Dirnböck, T., Köck, R., Hochbichler, F., Englisch, T., Sauberer, N. and Grabherr, G. (2005): Interactions among tree-line conifers: Differential effects of pine on spruce and larch. J. Ecol. 93: 948–957
Dyunin, A. K. (1967): Fundamentals of the mechanics of snow storms. In: Physics of snow and ice 1 (2) (Oura, H., ed.): 1 (2): 1065–1073, Institute of Low Temperature Science, Sapporo, Japan
Earle, C. J., Brubaker, L. B., Lozhkin, A. V. and Anderson, P. M. (1994): Summer temperature since 1600 for the upper Kolyma region, northeastern Russia, reconstructed from tree rings. Arct. Alp. Res. 26 (1): 60–65
Eggler, W. A. (1971): Quantitative studies on vegetation on sixteen young lava flows on the island of Hawaii. Trop. Ecol. 12: 66–100
Egorov, O. V. (1967): Wild ungulates in Yakutia (Translated from Russian). Jerusalem
Elkington, T. T. (1968): Introgressive hybridization between Betula nana L. and B. pubescens Erh. in north-west Iceland. New Phytol. 67: 109–118
Ellenberg, H. (1958): Wald oder Steppe? Die natürliche Pflanzendecke Perus. Umschau 21: 645–648; Umschau 22: 679–681
Ellenberg, H. (1959): Typen tropischer Urwälder in Peru. Schweizerische Zeitschrift für Forstwesen 110: 169–187
Ellenberg, H. (1963): Vegetation Mitteleuropas mit den Alpen in ökologischer Sicht (1st edition). Verlag Eugen Ulmer, Stuttgart
Ellenberg, H. (1966): Leben und Kampf an den Baumgrenzen der Erde. Naturwissenschaftliche Rundschau 19 (4): 133–139
Ellenberg, H. (1975): Vegetationsstufen in perhumiden bis perariden Bereichen der tropischen Anden. Phytocoenlogia 2 (3/4): 368–387
Ellenberg, H. (1978): Vegetation Mitteleuropas mit den Alpen in ökologischer Sicht (2nd edition). Verlag Eugen Ulmer, Stuttgart
Ellenberg, H. (1979): Man's influence on tropical mountain ecosystems in South America. J. Ecol. 67: 401–416
Ellenberg, H. (1996): Páramos und Punas der Hochanden Südamerikas, heute größtenteils als potentielle Wälder anerkannt. Verhandlungen der Gesellschaft für Ökololgie 25: 17–23
Ellenberg, H. and Mueller-Dombois, D. (1967): Tentative physiognomic-ecological classification of plant formations of the earth. Berichte des Geobotanischen Instituts ETH Stiftung Rübel 37: 21–55
Elliott, D. (1979): The current regenerative aspect of the northern Canadian trees, Keewatin, N.W.T., Canada: Some preliminary observations. Arct. Alp. Res. 11 (2): 243–251
Elliott, D. and Short, S. (1979): The northern limit of trees in Labrador: A discussion. Arctic 32 (3): 201–206
Elliott-Fisk, D. (1983): The stability of the northern Canadian tree-limit. Ann. Am. Geogr. 73: 560–576
Elliott-Fisk, D. and Peterson, A. M. (1991): Trees. In: Natural history of the White-Inyo Range, Eastern California (Hall, C. A., ed.): 87–107
Em, H. (1970): Höhengürtel in der Vegetation von Hochgebirgen Mazedoniens. Mitt. Ostalpin-dinarische Gesellschaft für Vegetationskunde. 11: 33–44
Enquist, F. (1929): Studier över samtidiga växlingar i klimat och växlighet. Svensk Geografisk Årsb.: 7–48
Enquist, F. (1933): Trädgränsundersökingar. Svensk Skogsvardsför. Tidskr. 2: 145–214
Environment Canada (1989): Atmospheric Environment Service, Changing Atmosphere Fact Sheet, The greenhouse effect: Impacts on the Arctic
Erkamo, V. (1956): Untersuchungen über die pflanzensoziologischen und einige andere Folgeerscheinungen der neuzeitlichen Klimaschwankung in Finnland. Annales Botanici Societatis Zoologicae Botanicae Fennicae "Vanamo" 28 (3)

Ern, H. (1966): Die dreidimensionale Anordnung der Gebirgsvegetation auf der Iberischen Halbinsel. Bonner Geographische Abhandlungen. 37

Eskuche, U. (1973): Estudios fitosociológicos en el norte de Patagonia. I. Investigacion de algunos factores de ambiente en comunidades de bosque y de chapparal. Phyotocoenologia 1: 64–113

Esper, J. and Schweingruber, F. H. (2004): Large-scale treeline changes recorded in Siberia. Geophys. Res. Lett. 31, L06202, doi: 10.1029/2003GL019178

Esper, J., Bosshard, A., Schweingruber, F. H. and Winiger, M. (1995): Tree-rings from the upper timberline in the Karakoram as climatic indicators from the last 1000 years. Dendrochronologia 13: 79–88

Evans, R. (1995): Impacts of reindeer grazing on soils and vegetation in Finnmark. NORUT informasjonsteknologi

Eyre, S. R. (1968): Vegetation and soil: A world picture (2nd edition). Edward Arnold, London

Faegri, K. (1966): A botanical excursion to the Steen's Mountain, SE Oregon, USA. Blyttia 24: 173–181

Faegri, K. (1972): Geoökologische Probleme der Gebirge Skandinaviens. Erdwissenschaftliche Forschung IV: 98–105, Wiesbaden

Fanta, J. (1973): Die vegetative Vermehrung der Fichte in der Lüneburger Heide. Mitteilungen der Deutschen Dendrologischen Gesellschaft 66: 39–48

Fanta, J. (1981): Fagus silvatica L. und das Aceri-Fagetum an der alpinen Waldgrenze in mitteleuropäischen Gebirgen. Vegetatio 44: 13–24

Farjon, A., Miehe, G. and Miehe, S. (2000): The taxonomy, distribution and ecology of Juniperus in High Asia. In: Proceedings – International Symposium of Juniper Forests, 6–11 August, 2000 (Forest and Walnut Research Institute of National Academy of Sciences/Kyrgyz-Swiss Forestry Support Programme, eds.): 70–90, Osh, Kyrgyzstan, Bishek

Farmer, R. E. Jr. (1997): Seed ecophysiology of temperate and boreal zone forest trees. St. Lucie Press, Delray Beach, FL

Fielding, A. H. and Haworth, P. F. (1999): Upland habitats. Routledge, London/New York

Fink, S. (1980): Anatomische Untersuchungen über das Vorkommen von Sproß- und Wurzelanlagen im Stammbreich von Laub- und Nadelbäumen. Diss. Univ. Freiburg/Br

Fischer, E. (1996): Die Vegetation des Parc National de Kahuzi-Biéga, Süd-Kivu, Zaire. Erdwissenschaftliche Forschung 35, Wiesbaden

Fjeldså, J. (1992): Biogeography of the birds of the Polylepis woodlands of the Andes. In: Páramo, an Andean ecosystem under human influence (Balslev, H. and Luteyn, J. L., eds.): 30–44, Academic Press, London

Fjeldså, J. and Kessler, M. (1996): Conserving the biological diversity of Polylepis woodlands of the Highlands of Peru and Bolivia: A contribution to sustainable natural resource management in the Andes. Nordic Foundation for Development and Ecology (NORDECO), Kopenhagen

Flemming, G. (1969): Zur Umlagerung der Schneedecke durch den Wind. Archiv für Naturschutz und Landschaftsforschung 9 (2): 175–194

Flenley, J. R. (1984): Late quaternary changes of vegetation and climate in the Malesian mountains. Erdwissenschaftliche Forschung XVII: 261–267, Wiesbaden

Flenley, J. R. (1992): UV-B insolation and the altitudinal forest limit. In: Nature and dynamics of forest-savanna boundaries (Furley, P. A., Proctor, J. and Ratter. J. A., eds.): 273–283, Chapman & Hall, London

Flenley, J. R. (1995): Cloud forest, the Massenerhebung effect, and ultraviolet insolation. Ecol. Stud. 110: 150–155

Fliri, F. (1975): Das Klima der Alpen im Raum von Tirol. Monographien zur Landeskunde Tirols, Folge 1 (Leidlmair, A. and Huter, F., eds.), Innsbruck, München
Flores Ochoa, J. A. (1979): Desarollo de las culturas humanas en las altas montanas tropicales: Estrategias adaptivas. In:El Medio Ambiente Páramo (Saldago-Labouriau, M. L., ed.), Ediciones Centro de Estudios Avanzados IVIC: 225–234, Venezuela
Flückiger, O. (1906): Die obere Grenze der menschlichen Siedlungen in der Schweiz. Bern
Flux, J. E. C. (1990): Brown hare. In: The handbook of New Zealand mammals (King, C. M., ed.): 161–172, Auckland University Press, Melbourne/Oxford/New York
Foley, J. A., Kutzbach, J. E., Coe, M. T. and Levis, S. (1994): Feedbacks between climate and boreal forests during the Holocene. Nature 371: 52–54
Fonda, R. W. and Bliss, L. (1969): Forest vegetation of the montane and subalpine zones, Olympic Mountains, Washington. Ecol. Monogr. 39: 271–296
Førland, E. J., Skaugen, T. E., Benestad, R. E., Hanssen-Bauer, I. and Tveito, O. (2004): Variations in thermal growing, heating, and freezing indices in the Nordic Arctic, 1900–2050. Arct. Antarct. Alp. Res. 36 (3): 347–356
Formozov, A. N. (1969): Snow cover as an integral factor of the environment and its importance in the ecology of mammals and birds. Boreal Institute, The University of Alberta, Occ. Paper 1 (2nd edition)
Franklin, J. (1995): Predictive vegetation mapping: Geographic modelling of biospatial pattern in relation to environmental gradients. Prog. Phys. Geogr. 19 (4): 474–499
Franklin, J. F. and Dyrness, C. T. (1973): Natural vegetation of Oregon and Washington. USDA For. Serv., Gen. Tech. Rep. PNW-8
Franklin, J. F. and Mitchell, R. G. (1967): Successional status of subalpine fir in the Cascade Range. USDA For. Serv. Res. Paper, PNW-46
Franklin, J. F., Moir, W. H., Douglas, G. W. and Wiberg, C. (1971): Invasion of subalpine meadows by trees in the Cascade Range, Washington and Oregon. Arct. Alp. Res. 3: 215–224
Franklin, J. F., Swanson, F. J., Harmon, M. E., Pery, D. A., Spies, T. A., Dale, V. H., McKee, A., Ferell, W. K., Means, J. E., Gregory, V., Lattin, J. D., Schowalter, T. D. and Larsen, D. (1991): Effects of global climatic change on forests in northwestern North America. In: Global warming and biologica (Peters, R. L. und Lovejoy, T. E., eds.): 244–257, New Haven, CT
Franz, H. (1979): Ökologie der Hochgebirge. Ulmer, Stuttgart
Frei, E. (1958): Eine Studie über den Zusammenhang zwischen Bodentyp, Klima und Vegetation in Ecuador. Plant Soil IX (3): 215–236
Freiray, S. and Schweingruber, F. H. (1994): A dendrochronological study on the dying off of young conifers in an afforestation at the alpine timberline. Dendrochronologia 11: 159–164
Freitag, H. (1971): Die natürliche Vegetation Afghanistans. Beiträge zur Flora und Vegetation Afghanistans I. Vegetatio 22: 285–344
French, D. D., Miller, G. R. and Cummins, R. P. (1997): Recent development of high-altitude Pinus sylvestris scrub in the northern Cairngorm Mountains, Scotland. Biol. Conserv. 79: 133–144
Frenzel, B. (1994): Über Probleme der holozänen Vegetationsgeschichte Osttibets. Göttinger Geographische Abhandlungen. 95: 143–166
Frey, W. (1977): Wechselseitige Beziehungen zwischen Schnee und Pflanze – eine Zusammenstellung anhand der Literatur. Eidgenössisches Institut für Schnee- und Lawinenforschung, Mitteilungen 34
Frey, W. (1983): The influence of snow on growth and survival of planted trees. Arct. Alp. Res. 15 (2): 241–251

Friedel, H. (1961): Schneedecken-Andauer und Vegetations-Verteilung im Gelände. Mitteilungen der forstlichen Bundes-Versuchsanstalt Mariabrunn 59: 317–368

Friedel, H. (1965): Kleinklima-Kartographie. Mitteilungen der forstlichen Bundes-Versuchsanstalt Mariabrunn 66: 13–32

Friedel, H. (1966): Die Waldgrenze in ihrem Bezug zu anderen höhenzonalen Erscheinungen. Allgemeine Forstzeitung 77 (6): 114–116

Friedel, H. (1967): Verlauf der alpinen Waldgrenze im Rahmen anliegender Gebirgsgelände. Mitteilungen der forstlichen Bundes-Versuchsanstalt Mariabrunn 75: 81–172

Fries, T. C. E. (1913): Botanische Untersuchungen im nördlichen Schweden. Ein Beitrag zur Kenntnis der alpinen und subalpinen Vegetation in Torne Lappmark. Flora och Fauna 2

Fries, R. E. and Fries, Th. E. (1948): Phytogeographical researches on Mt. Kenya and Mt. Aberdare, British East Africa. Kungl. Svenska vetenskapsakademiens handligar, Tredje serien 25 (5), Stockholm

Fries, R. E. and Fries, Th. E. (1952): Über Riesensenecionen der afrikanischen Hochgebirge. Svensk Botanisk Tidskrift

Fröhlich, W. and Willer, M. (1977): Bodentemperature und obere Waldgrenze. Vorläufiger Bericht über Untersuchungen an den Hochbergen Ostafrikas. Die Erde 4: 347–353

Fromme, G. (1952): Schach der Waldverwüstung. Österreichisches Produktivitätszentrum, Wien: 1–74

Fukarek, P. (1970): Beitrag zur Kenntnis der oberen Waldgrenze in einigen Gebirgszügen der südlichen Dinariden. Mitteilungen der ostalpin-dinarischen Gesellschaft für Vegetationskunde 11: 45–54

Furrer, E. (1966): Kümmerfichten und Kaltluftströme in den Alpen der Ost- und Innerschweiz. Schweizerische Zeitschrift für Forstwesen: 720–733

Furrer, G. and Fitze, P. (1970): Beitrag zum Permafrostproblem in den Alpen. Vierteljahresschrift der Naturforschenden Gesellschaft Zürich 115 (3): 355–368

Furrer, G. and Graf, K. (1978): Die subnivale Höhenstufe am Kilimandjaro und in den Anden Boliviens und Ecuadors. In Geoecological relations between the southern temperate zone and the tropical mountains. Erdwissenschaftliche Forschung 11: 441–457, Wiesbaden

Gaggi, S. (1990): I forni die fusione del minerale di ferro in Valmalenco. Rassegna Economica della Provincia die Sondrio 2: 23–31

Gallenmüller, F., Bogenrieder, A. and Speck, T. (1999): Biomechanische und ökologische Untersuchungen an Alnus viridis (Cahaix) DC. in verschiedenen Höhenlagen der Schweizer Alpen. Eidgenössische Forschungsanstalt für Wald, Schnee und Landschaft, Berichte 347

Gamache, I. and Payette, S. (2005): Latitudinal response of subarctic tree lines to recent climate change in eastern Canada. J. Biogeogr. 32: 849–862

Gams, H. (1931): Die klimatische Begrenzung von Pflanzenarealen und die Verteilung der hygrischen Kontinentalität in den Alpen. I. Teil. Zeitschrift der Gesellschaft für Erdkunde Berlin 19/10: 321–346

Gams, H. (1951): Aus der Pflanzenwelt des Schlern und der Seiseralm. Der Schlern 25: 301–306, Bozen

Gansert, D. (2002): Betula ermanii, a dominant subalpine and subarctic treeline tree species in Japan: Ecological traits of deciduous tree life in winter. Arct. Antarct. Alp. Res. 34 (1): 57–64

Gansert, D., Baches, K. and Kakubari, Y. (1999): Altitudinal and seasonal variation of frost resistance of Fagus crenata and Betula ermanii along the Pacific slope of Mt. Fuji, Japan. J. Ecol. 87: 382–390

Garnett, M., Ineson, P. and Adamson, J. K. (1997): A long-term upland temperature record: No evidence for recent warming. Weather 52: 342–351

Gates, D. M. (1985): Global biospheric response to increasing atmospheric carbon dioxide concentration. In: Direct effects of increasing carbon dioxide (Stain, B. R. and Cure, J. D., eds.): 171–184, US Department of Energy, Washington, DC

Gates, D. M. (1993): Climate change and its biological consequences. Sinauer, Sunderland, MA

Gäumann, R., Roth, C. and Anliker, J. (1934): Über die Biologie der Herpotrichia nigra HARTIG. Zeitschrift für Pflanzenkrankheiten und Pflanzenschutz 44: 97–116

Geddes, C. A., Brown, D. G. and Fagre, D. B. (2005): Topography and vegetation as predictors of snow water equivalent across the alpine treeline ecotone at Lee Ridge, Glacier National Park, Montana, U.S.A. Arct. Antarct. Alp. Res. 37 (2): 197–205

Geiger, E. (1901): Das Bergell. Forstbotanische Monographie. Diss. Philosophische Fakultät der Universität Zürich, Chur

Geiger, R. (1961): Das Klima der bodennahen Luftschicht. Ein Lehrbuch der Mikroklimatologie. Die Wissenschaft 78, Braunschweig

Germino, M. J. and Smith, W. K. (1995): Structural and physiological factors influencing young seedling establishment in conifers of the upper timberline in the Medicine Bow Mountains of southeastern Wyoming. Bull. Ecol. Soc. America 76 (3 Suppl.): 331

Germino, M. J. and Smith, W. K. (1999): Sky exposure, crown architecture, and low-temperature photoinhibition in conifer seedlings at alpine treeline. Plant Cell Environ. 22: 407–415

Germino, M. J. and Smith, W. K. (2000): Differences in microsite, plant form, and low temperature photoinhibition of photosynthesis in alpine plants. – Arctic, Antarctic, and Alpine Research 32: 388–396

Germino, M. J., Smith, W. K. and Resor, A. C. (2002): Conifer seedling distribution and survival in an alpine tree ecotone. Plant Ecol. 162: 157–168

Gersmehl. P. (1973): Pseudotimberline: The southern Appalachian grassy balds. Arct. Alp. Res. 5 (3, Pt. 2): 137–138

Gilfedder, L. (1988): Factors influencing the maintenance of an inverted Eucalyptus coccifera tree-line on the Mt. Wellington Plateau, Tasmania. Aust. J. Ecol. 13: 495–503

Gillison, A. N. (1970): Structure and floristics of a montane grassland-forest transition, Doma Peaks Region, Papua. Blumea 18: 71–89

Giorgi, F. and Hewitson, B. (2001): Regional climate information – evaluation and projections. Climate change 2001. The scientific basis. Contributions of working group 1 to the third assessment report of the intergovernmental panel on climate change (Houghtons, J. T., Ding, Y., Griggs, D. J., Noguer, P., van der Linden. S., Dai, X., Maskell, K. and Johnsons, C. A., eds.): 583–638, Cambridge University Press, Cambridge

Gleeson, S. K. and Tilman, D. (1992): Plant allocation and the multiple limitation hypothesis. Am. Nat. 139: 1322–1343

Glerum, C. (1973): Annual trends in frost hardiness and electrical impedance for seven coniferous species. Can. J. Plant Sci. 53: 881–889

Glutz von Blotzheim, U. N. and Bauer, K. M. (1993): Handbuch der Vögel Mitteleuropas 13 (III). Wiesbaden

Godley, E. J. (1960): The botany of southern Chile in relation to New Zealand and the Subantarctic. Proc. Roy. Soc. B 152: 457–475

Goldstein, G., Meinzer, F. C. and Rada, F. (1994): Environmental biology of a tropical tree-line species, Polylepis sericea. In: Tropical alpine environments – plant form and function (Rundel, P. W., Smith, A. P. and Meinzer, F. P., eds.), Cambridge University Press, Cambridge

Gorbunov, A. P. (1978): Permafrost investigations in high-mountain regions. Arct. Alp. Res. 10 (2): 283–294

Gorchakovsky, P. L. and Shiyatov, S. G. (1978): The upper forest limit in the mountains of the boreal zone of the USSR. Arct. Alp. Res. 10 (2): 349–363

Gordon, J. E., Thompson, D. B. A., Haynes, V. M., Brazier, V. and MacDonald, R. (1998): Environmental sensitivity and conservation management in the Cairngorm Mountains, Scotland. Ambio 27 (4): 335–344
Grabherr, W. (1934): Der Einfluß des Feuers auf die Wälder Tirols in Vergangenheit und Gegenwart. Centralblatt für das gesamte Forstwesen 60: 260–273; 289–302
Grabherr, W. (1949): Die Waldbrandfrage in den früheren Haller Salinenwäldern im Unterengadin. Schweizerische Zeitschrift für Forstwesen 100: 366–374
Grace, J. (1977): Plant response to wind. Academic Press, London
Grace, J. (1980): The functional significance of short stature in montane vegetation. Plant form and vegetation structure, SPH Academie publishing bv, The Hague: 201–209
Grace, J. (1988): The functional significance of short stature in montane vegetation. Plant form and vegetation structure: 201–209, The Hague
Grace, J. (1989): Tree lines. Philosophical Trans. Roy. Soc. Lond. B 234: 233–245
Grace, J. (1990): Cuticular water loss unlikely to explain treeline in Scotland. Oecologia 84: 64–68
Grace, J. and Morton, A. (1990): Climate and growth of Pinus sylvestris at its upper altitudinal limit in Scotland. Evidence from tree growth-ring. J. Ecol. 78: 601–610
Granström, A. (1987): Seed viability of fourteen species during five years of storage in a forest soil. J. Ecol. 75: 321–331
Grant, M. C. and Mitton, J. B. (1977): Genetic differentiation among growth forms of Engelmann spruce and subalpine fir at tree line. Arct. Alp. Res. 9 (3): 259–263
Graumlich, L. J. (1991): Subalpine tree growth, climate and increasing CO_2: An assessment of recent growth trends. Ecology 72 (1): 1–11
Graumlich, L. J. and Brubaker, L. B. (1986): Reconstruction of annual temperature (1590–1979) for Longmire, Washington, derived from tree rings. Quatern. Res. 25: 223–234
Graumlich, L. J., Brubaker, L. B. and Grier, C. C. (1989): Long-term trends in forest net primary productivity: Cascade Mountains, Washington. Ecology 70 (2): 405–410
Grebenchikov, O. S. (1978): Vegetation structure in the high mountains of the Balkan peninsula and the Caucasus, USSR. Arct. Alp. Res. 10 (2): 441–447
Gregory, S. C., Risbeth, J. and Shaw, C. G. III (1991): Pathogenicity and virulence. In Armillaria root disease (Shaw, C. G. III and Kile, G. A., eds.): 76–84, USDA Forest Service Agricultural Handbook 691
Griggs, R. F. (1934): The edge of the forest in Alaska and reasons for its position. Ecology 15 (2): 80–96
Griggs, R. F. (1938): Timberline in the northern Rocky Mountains. Ecology 19: 548–564
Griggs, R. F. (1946): The timberlines of northern America and their interpretation. Ecology 27 (4): 275–289
Grigorjew, A. (1956): Subarctica. Moscow
Grishin, S. Y. (1988): The upper timberline at Kluchevskoy volcano group (Kamchatka) (in Russian, typewritten English abstract enclosed by the author). In: Plant cover of high mountain ecosystems of the USSR. Far East Branch of the USSR Academy of Sciences (Kharkevich, S. S., ed.): 193–201
Grishin, S. Y. (1995): The boreal forests of north-eastern Eurasia. Vegetatio 121: 11–21
Grishin, S. Y. and Del Moral, R. (1996): Dynamics of forests after catastrophic eruptions of Kamchatkas's volcanoes. In: Biodiversity and the dynamics of ecosystems (Turner, I. M., Diong, C. H., Lim, S. S. and Ng, P. K. L., eds.), DIWPA Series 1: 133–146
Grishin, S. Y., Krestov, P. and Okitsu, S. (1996a): The subalpine vegetation of Mt. Vysokaya, Central Sikhote-Alin. Vegetatio 127: 155–172
Grishin, S. Y., Del Moral, R., Krestov, P. V. and Verkholat, V. P. (1996b): Succession following the catastrophic eruption of Ksudach volcano (Kamtchatka, 1907). Vegetatio 127: 129–153

Gross, M. (1989): Untersuchungen an Fichten der alpinen Waldgrenze. Dissertationes Botanicae 139, Berlin, Stuttgart

Gross, M., Rainer, I. and Tranquillini, W. (1991): Über die Frostresistenz der Fichte mit besonderer Berücksichtigung der Zahl der Gefrierzyklen und der Geschwindigkeit der Temperaturänderung beim Frieren und Auftauen. Forstwissenschaftliches Centralblatt 110: 207–217

Grubb, P. J. (1971): Interpretation of the "Massenerhebung" effect on tropical mountains. Nature 229: 44–45

Grunewald, K. and Scheithauer, J. (2008): Untersuchungen an der alpinen Waldgrenze im Piringebirge (Bulgarien). Geoöko 29: 1–32

Guillaumin, J. J., Mohammed, C., Anselmi, N., Courtecuisse, R., Gregor, S. C., Holdenrieder, O., Intini, M., Lung, B., Marxmüller, H., Morrison, D., Risbeth, J., Termorshuizen, A. J., Tirró, A. and Van Dam, B. (1993): Geographical distribution and ecology of the Armillaria species in western Europe. Eur. J. For. Pathol. 23: 321–341

Gunsch, J. (1972): Vergleichende ökologische Untersuchung von Kleinstandorten im Bereich der subalpinen Zirben-Waldgrenze. Diss. Univ. Innsbruck

Haapasaari, M. (1988): The oligotrophic heath vegetation of northern Fennoscandia and its zonation. Acta Botanica Fennica 135: 1–219

Haase, R. (1983): Verbreitungsmuster und Standorte der alpinen Vegetation in der Indian Peaks Area, Colorado Front Range, U. S. A. Diss. Mathematisch-Naturwissenschaftliche Fakultät, Westfälische Wilhelms-Universität., Münster

Haase, R. (1987): An alpine vegetation map of Caribou Lake Valley and Fourth of July Valley, Front Range, Colorado, U. S. A. Arct. Alp. Res. 19 (1): 1–10

Haavisto, V. F. (1975): Peatland black spruce seed production and dispersal in northeastern Ontario. In: Black Spruce Symposium, Symposium Proceedings O-P-4: 250–264, Great Lake Forest Research Centre, Sault Ste-Marie, Canada

Habeck, F. and Reif, A. (1994): Die Waldgesellschaften der montanen und subalpinen Stufe des Ostabfalls des Olymp, Griechenland. Phytocoenologia 22 (4): 501–536

Habeck, J. R. (1969): A gradient analysis of a timberline zone at Logan Pass, Glacier Park, Montana. Northwest Sci. 43 (2): 65–73

Habeck, J. R. and Hartley, E. (1968): A glossary of alpine terminology. University of Montana, Dept. Bot., Missoula, MT

Hacquet, B. (1779): Mineralogisch-botanische Lustreise von dem Berge Terglou in Krain zu dem Berge Glockner in Tirol 1779 (cited in Marek, 1910)

Hader, F. (1954): Nordostalpine Seehöhenmittel der Niederschlagsmenge (1851–1950). Archiv für Meteorologie, Geophysik und Bioklimatologie, B. 5

Hadley, J. L. and Smith, W. K. (1983): Influence of wind exposure on needle desiccation and mortality for timberline conifers in Wyoming, U.S.A. Arct. Alp. Res. 15: 127–135

Hadley, J. L. and Smith, W. K. (1986): Wind effects on needles of timberline conifers: Seasonal influences on mortality. Ecology 67: 12–19

Hadley, J. L. and Smith, W. K. (1987): Influence of krummholz mat microclimate on needle physiology and survival. Oecologia (Berlin) 73: 82–90

Hadley, J. L. and Smith, W. K. (1989): Wind erosion of leaf surface wax in timberline conifers. Arct. Antarct. Alp. Res. 21 (4): 392–398

Hadley, J. L. and Smith, W. K. (1990): Influence of leaf surface wax and leaf area to water content ratio on cuticular transpiration in western conifers, U. S. A. Can. J. For. Res. 20: 1306–1311

Hafenscherer, J. and Mayer, H. (1986): Standort, Aufbau, Entwicklungsdynamik und Verjüngung von Latschenbeständen im Karwendeltal, Tirol. Schweizerische Zeitschrift für Forstwesen 137 (3): 177–203

Haffner, W. (1972): Khumbu Himalaya – Landschaftsökologische Untersuchungen in den Hochtälern des Mt. Everest-Gebietes. Erdwissenschaftliche Forschung IV: 244–263

Haffner, W. (1982): Die Gebirge und Hochländer der Tropen und Subtropen. Gießener Beiträge zur Entwicklungsforschung, Reihe I (8): 1–33
Hagemoen, R. I. M. and Reimers, E. (2004): Reindeer summer activity pattern in relation to weather and insect harassment. J. Ecol. 71: 883–892
Hager, K. (1909): Streifzüge zwischen den Arven und Bergföhren am Lukmanier. Jahrbuch des Schweizerischen Alpenclubs 44
Hämet-Ahti, L. (1963): Zonation of the mountain birch forest in northernmost Fennsoscandia. Annales Botanici Societas Zoologicae-Botanicae Fennicae "Vanamo" 34 (4)
Hämet-Ahti, L. (1979): The dangers of using the timberline as the "zero line" in comparative studies on altitudinal vegetation zones. Phytocoenologica 6: 49–54
Hämet-Ahti, L. (1987): Mountain birch and mountain birch woodland in NW Europe. Phytocoenologia 15 (4): 449–453
Hämet-Ahti, L. and Ahti, T. (1969): The homologies of the fennoscandian mountain and coastal birch forests in Eurasia and North America. Vegetatio 19 (1–6): 208–219
Hamilton, A. C. and Perrott, R. A. (1981): A study of altitudinal zonation in the montane forest belt of Mt. Elgon, Kenya/Uganda. Vegetatio 45: 107–125
Handa, T., Körner, C. and Hättenschwiler, S. (2005): A test of treeline carbon limitation hypothesis by in situ CO_2 enrichment and defoliation. Ecology 86: 1288–1300
Hansen-Bristow, K. J. (1981): Environmental controls influencing the altitude and form of the forest-alpine tundra ecotone, Colorado Front Range. Ph.D. thesis, University of Colorado, Boulder, CT
Hansen-Bristow, K. J. (1986): Influence of increasing elevation on growth characteristics at timberline. Can. J. Bot. 64: 2517–2523
Hardon, H. J. (1936): Podzol-profiles in the tropics. Naturrwetenschappelijk Tijdschrift voor Nederlandsch. Indie 96: 25–41
Hare, F. K. (1971): Snow cover problems near the arctic treeline of North America. Rep. Kevo Subarct. Res. Stat. 8: 31–40
Hare, F. K. and Ritchie, J. C. (1972): The boreal bioclimates. Geogr. Rev. 62: 333–365
Hari, P. and Arovaara, H. (1988): Detection of CO_2 induced enhancement in the radial increment of trees: Evidence from northern timberline. Scand. J. For. Res. 3: 64–74
Harju, A. M., Kärkkäinen, K. and Ruotsalainen, S. (1996): Phenotypic and genetic variation in the seed maturity of Scots pine. Silvae Genet. 45 (4): 205–211
Harries, H. (1966): Soils and vegetation in the alpine and subalpine belts of the Presidential Range. Unpubl. Ph.D. thesis., Dept. of Crops and Soils, Rutgers University, New Brunswick, NJ
Harris, C. (2005): Climate change, mountain permafrost degradation and geotechnical hazard. In: Global change and mountain regions. An overview of current knowledge (Huber, U. M., Bugmann, H. K. M. and Reasoner, M. A., eds.), Adv. Glob. Change Res. 23: 2151–524
Harris, S. A. (1971): Podzol development on volcanic ash deposits in the Talamanca Range, Costa Rica. In: Paleopedology: Origin, nature and dating of paleosols (Yaalon, D. H., ed.): 191–209, Jerusalem
Harris, S. A. and Brown, R. J. E. (1978): Plateau Mountain: A case study of alpine permafrost in the Canadian Rocky Mountains. Proc. Can. Conf. Permafrost 1: 385–391, Nat. Res. Council of Canada, Ottawa
Härtel, O. and Eisenzopf, R. (1953): Zur Physiologie und Ökologie der kutikulären Wasseraufnahme durch Koniferennadeln. Zentralblatt für die gesamte Forst- und Holzwirtschaft 72: 47–59

Haselwandter, K. (2007): Mycorrhiza in the alpine timberline ecotone: Nutritional implications. In: Trees at their upper limit. Treelife limitation at the alpine timberline (Wieser, G. and Tausz, M., eds.), Plant Ecophysiol. 5: 57–66

Häsler, R., Streule, A. and Turner, H. (1999): Shoot and root growth of young Larix decidua in contrasting microenvironments near the alpine timberline. Phyton 39: 47–52

Hastenrath, S. (1968): Certain aspects of the three-dimensional distribution of climate and vegetation belts in the mountains of central America and southern Mexico. Colloquium Geographicum 9: 122–130

Hättenschwiler, S. and Körner, C. (1998): Biomass allocation and canopy development in spruce model ecosystems under elevated CO_2 and increased N deposition. Oecologia 113: 104–114

Hättenschwiler, S., Miglietta, F., Raschi, A. and Körner, C. (1997): Thirty years in situ tree growth under elevated CO_2: a model for future forest responses? Glob. Change Biol. 3: 463–471

Hättenschwiler, S., Handa, T., Egli, L., Asshof, R., Ammann, W. and Körner, C. (2002): Atmospheric CO_2 enrichment of alpine conifers. New Phytol. 156: 363–375

Haugen, R. K. (1965): Climatic implications of timberlines in interior Alaska (Abstract). Abstracts Int. Ass. Quat. Res., VII Int. Congr., Boulder, CO: 197

Haukioja, E. and Heino, J. (1974): Birch consumption by reindeer (Rangifer tarandus) in Finnish Lapland. Rep. Kevo Subarct. Res. Stat. 11: 22–25

Hauman, L. (1933): Esquisse de la végétation de la haute altitudes sur la Ruwenzori. Bull. Acad. Belg. Cl. Sci. 5 (19): 602–616; 702–717; 900–917

Havis, J. R. (1970): Water movement in woody stems during freezing. Cryobiology 8: 581–585

Havranek, W. H. and Tranquillini, W. (1995): Physiological processes during winter dormancy and their ecological significance. Ecophysiology of forest (Smith, W. K. and Hinckley, T. M., eds.): 95–124, Academic Press, New York

Havranek, W. M. (1972): Über die Bedeutung der Bodentemperatur für die Photosynthese und Transpiration junger Forstpflanzen und für die Stoffproduktion an der Waldgrenze. Angewandte Botanik 46: 101–116

Havranek, W. M. (1993): The significance of frost and frost-drought for the alpine timberline. In: Ecologia delle foreste d'alta quota (Anfodillo, T. and Urbinati, C., eds.), Atti del XXX Corso di Cultura in Ecologia XXX: 115–127

Hedberg, I. and Hedberg, O. (1979): Tropical-alpine life-forms of vascular plants. Oikos 33: 297–307

Hedberg, O. (1951): Vegetation belts of the east African mountains. Svensk Botanisk Tidskrift 45: 140–202

Hedberg, O. (1964): Features of afroalpine plant ecology. Acta Phytogeographica Suecica 49, Uppsala

Hegi, G. (1958): Alpenflora. München

Heikkinen, O. (1984): Forest expansion in the subalpine zone during the past hundred years. Erdkunde 38: 194–202

Heikkinen, O., Obrebska-Starkel, B. and Tuhkanen, S. (1995): Introduction: The timberline – a changing battle front. Prace Geograficzne 98: 7–16

Heikkinen, R. K. and Kalliola, R. (1989): Vegetation types and maps of the Kevo nature reserve, northernmost Finland. Kevo notes 8

Heiniger, M. and Kanzler, E. (1988): Shoot dieback in subalpine reforestations in Switzerland. Mitteilungen der forstlichen Bundesversuchsanstalt Wien 162: 67–71

Helland, A. (1912): Traedgrenser og sommervarmen. Tidskrift für Skogbruk: 131–146; 169–175; 303–313

Helle, T. (2001): Mountain birch and reindeer husbandry. In: Nordic mountain birch ecosystems (Wielgolaski, F. E., ed.), Man and the Biosphere Series 27: 279–291

Helle, T. and Kajala, L. (1992): Browsing of subarctic birch forests by reindeer. In: Disturbance related dynamics of birch and birch dominated ecosystems. Nordic Symp. Iceland (Iceland Forest Research Institute, Mogilsa), Extended Abstracts: 25–27

Helle, T. and Kojola, I. (1993): Reproduction and mortality of Finnish semi-domesticated reindeer in relation to density and management strategies. Arctic 46: 72–77

Henderson, J. A. (1973): Composition, distribution and succession of subalpine meadows in Mt. Rainier National Park, Washington. Unpubl. Ph.D. thesis, State University, Corvallis, OR

Henning, I. (1974): Geoökologie der Hawaii-Inseln. Erdwissenschaftliche Forschung IX, Wiesbaden

Henriksson, J., Haukioja, E. and Ruohomäki, K. (1999): Impact of leaf damage on growth of mountain birch shoots. New Phytol. 142: 469–474

Hensen, I. (1993): Vegetationsökologische Untersuchungen in Polylepis-Wäldern der Ostkordillere Boliviens. Diss. Universität Göttingen

Hensen, I. (1995): Die vegetation von Polylepis-Wäldern in der Ostkordillere Boliviens. Phytocoenologia 25 (2): 235–277

Henttonen, H., Kanninen, M., Nygren, M. and Ojansuu, R. (1986): The maturation of Pinus sylvestris seeds in relation to temperature climate in Northern Finland. Scand. J. For. Res. 1: 243–249

Hermes, K. (1955): Die Lage der oberen Waldgrenze in den Gebirgen der Erde und ihr Abstand zur Schneegrenze. Kölner Geographische Arbeiten 5

Herrick, G. T. and Friedland, A. J. (1991): Winter desiccation and injury of subalpine red spruce. Tree Physiol. 8: 23–26

Hertel, D., Therburg, A. and Villalba, R. (2008): Above- and below-ground response by Nothofagus pumilio to climatic conditions at the transition from the steppe-forest boundary to the alpine treeline in southern Patagonia, Argentina. Plant Ecol.d Diver. 1 (1): 21–33

Hertel, D. and Wesche, K. (2008): Tropical moist polylepis stands at the treeline in East Bolivia: The effect of elevation on stand microclimate, above- and below-ground structure, and regeneration. Trees 22: 303–315

Hess, E. (1923): Waldstudien im Oberhasli. Beiträge zur Geobotanischen Landesaufnahme 13, Zürich

Hesselman, H. (1934): Nagra studier över fröspridning samt kalhyggets besaning. Meddelanden fran Statens Skogsförsöksanstalt 27: 145–182

Hesselman, H. (1938): Fortsatta studier over tallens och granens fröspridning samt kalhyggets besaning. Meddelanden från Statens Skogsförsöksanstalt 31: 1–64

Hessl, A. E. and Baker, W. L. (1997): Spruce and fir regeneration in the forest-tundra ecotone of Rocky Mountain National Park, Colorado, U. S. A. Arct. Alp. Res. 29: 173–183

Hibsch-Jetter, C. (1994): Birken in den Alpen. Taxonomisch-ökologische Untersuchungen an Betula pubescens Ehrh. und Betula pendula Roth. Contributiones biologiae arborum 6, Landsberg am Lech

Hiemstra, C. A., Liston, G. E. and Reiners, W. A. (2002): Snow redistribution by wind and interactions with vegetation at upper treeline in the Medicine Bow Mountains, Wyoming, U.S.A. Arct. Antarct. Alp. Res. 34 (3): 262–273

Higgins, P. D. and Spomer, G. G. (1976): Soil temperature effects on root respiration and the ecology of alpine and subalpine plants. Bot. Gaz. 137 (2): 110–120

Hilbig, W., Helmecke, K. and Bumzaa, D. (1989): Mikroklima-Untersuchungen in Pflanzengesellschaften verschiedener Höhenstufen in den Hochgebirgen der Nordwest- und Südmongolei. Erforschung biologischer Ressourcen in der MVR 7: 5–97, Halle (Saale)

Hildebrand-Vogel, R., Godoy, R. and Vogel, A. (1990): Subantarctic-Andean Nothofagus pumilio forests. Vegetatio 89: 55–68

Hiller, B. (1996): Landschaftsökologische Untersuchungen zu Böden und Vegetation von Schneetälchen auf Silikatgestein im Oberengadin. Diploma thesis, Institut für Landschaftsökologie., Westfälische Wilhelms-Universität, Münster

Hiller, B. (2001): Humusformen imWaldgrenzökoton (Oberengadin, Schweiz). Arbeiten aus dem Institut für Landschaftsökologie 9, Westfälische Wilhelms-Universität, Münster

Hiller, B., Müterthies, A., Holtmeier, F.-K. and Broll, G. (2002): Investigations on spatial heterogeneity of humus forms and natural regeneration of larch (Larix decidua Mill.) and Swiss stone pine (Pinus cembra L.) in an alpine timberline ecotone (Upper Engadine, Central Alps, Switzerland). Geographica Helvetica 2: 81–90

Hiltunen, R. (1980): Tuulenpieksemän ja lumenviipymän lämpötila ja lumioloista Pikku Mallala (Temperature and snow conditions on snow beds and wind-exposed places on Pikku Malla). Luonnon Tukija 84: 11–14

Hinckley, Th. M., Goldstein, G. H., Meinzer, F. and Teskey, R. O. (1985): Environmental constraints at arctic, temperate-maritime and tropical treelines. Establishment and tending of subalpine forest: Research and management. In: Proc. 3rd IUFRO workshop P. 107-00 1984 (Turner, H. and Tranquillini, W., eds.): 270:21–30, Eidgenössische Anstalt für das forstliche Versuchswesen, Berichte

Hinrichsen, D. (1986): Multiple pollutants and forest decline. Ambio 15: 258–265

Hjälten, J., Danell, K. and Ericson, L. (1993): Effects of simulated herbivory and intraspecific competition on the compensatory ability of birches. Ecology 74: 1136–1142

Hnatiuk, R. J. (1994): Plant form and function in alpine New Guinea. In: Tropical alpine environment – plant form and function (Rundel, P. W., Smith, A. P. and Meinzer, F. C., eds.): 307–318, Cambridge University Press

Hobbie, S. E. and Chapin III, F. S. (1998): An experimental test of limits to tree establishment in Arctic tundra. J. Ecol. 86: 449–461

Hoch, G. and Körner, C. (2003): The carbon charging of pines at the climatic treeline: A global comparison. Oecologia 135: 10–21

Hoch, G. and Körner, C. (2005): Growth, demography and carbon relations of Polylepis trees at the world's highest treeline. Funct. Ecol. 19: 941–951

Hoch, G and Körner, C. (2006): A test of treeline theory on a montane permafrost island. Arct. Antarct. Alp. Res. 38 (1): 113–119

Hoch, G., Popp, M. and Körner, C. (2002): Altitudinal increase of mobile carbon pools in Pinus cembra suggests sink limitation of growth at the Swiss treeline. Oikos 98: 361–374

Hoff, C. C. (1957): A comparison of soil, climate and biota of conifer and aspen communities in the central Rocky Mountains. Amer. Midl. Naturalist 58

Hoff, R. and Hagle, S. (1990): Diseases of whitebark pine with special emphasis on white blister rust. In: Proceedings – Symposium on whitebark pine ecosystems: Ecology and management of a high-mountain resource, Bozeman, MT, March 29–31, 1989 (Schmidt, W. C. and McDonald, K. J., compilers): 270: 179–190, USDA For. Serv., Intermtn. Res. Stat., Gen. Techn Rep. INT 270

Hofgaard, A. (1997a): Structural changes in the forest-tundra ecotone: A dynamic process. In: Past and future rapid environmental changes: The spatial and evolutionary responses of terrestrial biota (Huntley, B., Cramer, W., Morgan, A. V., Prentice, H. C. and Allen, J. R. M., eds.): 255–263, Springer, Berlin

Hofgaard, A. (1997b): Inter-relationships between treeline position, species diversity, land use and climate change in the central Scandes Mountains of Norway. Glob. Ecol. Biogeogr. Lett. 6: 419–429

Hofgaard, A. (1999): The role of 'natural' landscapes influenced by man in predicting responses to climate change. Ecol. Bull. 47: 160–167

Hofgaard, A. and Wilmann, B. (2002): Plant distribution across the forest-tundra ecotone: The importance of treeline position. Ecoscience 9: 375–385

Hofstede, R. G. M., Chilito, E. J. and Sandovals, E. M. (1995): Vegetative structure, microclimate, and leaf growth on a páramo tussock grass species in undisturbed, burned and grazed conditions. Vegetatio 119: 53–65

Holdenrieder, O., Pautasso, M., Weisberg, P. J. and Lonsdale, D. (2004): Tree diseases and landscape processes: The challenge of landscape pathology. Trend. Ecol. Evol. 19 (8): 446–452

Höllermann, P. and Poser, H. (1977): Grundzüge der räumlichen Ordnung in der heutigen periglazialen Höhenstufe der Gebirge Europas und Afrikas. Rückblick und Ausblick. Abhandlungen Akademie der Wissenschaften Göttingen, Mathematisch-Physikalische Klasse, 3. Folge, Nr. 31: 333–354

Höllermann, P. W. (1972): Zur natürlichen Höhenstufung der Pyrenäen. Erdwissenschaftliche Forschung IV: 36–60, Wiesbaden

Höllermann, P. W. (1978): Geoecological aspects of the upper timberline in Tenerife, Canary Islands. Arct. Alp. Res. 10 (2): 365–382

Höllermann, P. W. (1980): Naturräumliche Höhengrenzen und die Hochgebirgsstufe in Gebirgen des westlichen Nordamerika. In: Höhengrenzen in Hochgebirgen (Jentsch, Ch. and Liedtke, H., eds.): 29:75–117, Arbeiten aus dem Geographischen Institut der Universität des Saarlandes

Höllermann, P. W. (1982): Studien zur aktuellen Morphodynamik und Geoökologie der Kanareninseln Teneriffa und Furteventura. Arbeiten Akademie der Wissenschaften Göttingen, Mathematisch-Physikalische Klasse, 3. Folge, Nr. 34

Höllermann, P. W. (1995): Wald- und Buschbrände auf den westlichen Kanarischen Inseln. Abhandlungen Akademie der Wissenschaften Göttingen, Mathematisch-Physikalische Klasse, 3. Folge, Nr. 46

Holmgren, B. and Tjus, M. (1996): Summer air temperatures and tree line dynamics at Abisko. Ecol. Bull. 45: 159–169

Holroyd, E. W. (1970): Prevailing winds on Whiteface Mountains as indicated by flag trees. For. Sci. 16: 222–292

Holtmeier, F.-K. (1965): Die Waldgrenze im Oberengadin in ihrer physiognomischen und ökologischen Differenzierung. Diss. Mathematisch-Naturwissenschaftliche Fakultät, Rheinische Friedrich-Wilhelms-Universität Bonn

Holtmeier, F.-K. (1966): Die ökologische Funktion des Tannenhähers im Zirben-Lärchenwald und an der Waldgrenze im Oberengadin. J. für Ornithologie 4: 337–345

Holtmeier, F.-K. (1967a): Zur natürlichen Wiederbewaldung aufgelassener Alpen im Oberengadin. Wetter und Leben 19: 195–202

Holtmeier, F.-K. (1967b): Die Verbreitung der Holzarten im Oberengadin unter dem Einfluß des Menschen und des Lokalklimas. Erdkunde 21 (4): 249–258

Holtmeier, F.-K. (1968): Ergänzenden Beobachtungen in der Steinwildkolonie am Schafberg und Piz Albris bei Pontresina. Bündnerwald 21: 244–249

Holtmeier, F.-K. (1969): Zur Waldgrenze im Oberengadin. Bündner Wald 23 (3)

Holtmeier, F.-K. (1970): Zur vergleichenden Geographie und Ökologie der alpinen und polaren Waldgrenze. Deutscher Geographentag Kiel 1969, Tagungsbericht und wissenschaftliche Abhandlungen: 519–528, Wiesbaden

Holtmeier, F.-K. (1971a): Waldgrenzstudien im nördlichen Finnisch-Lappland und angrenzenden Nordnorwegen. Rep. Kevo Subarct. Res. Stat. 8: 53–62

Holtmeier, F.-K. (1971b): Der Einfluß der orographischen Situation auf die Windverhältnisse im Spiegel der Vegetation dargestellt an Beispielen aus dem Val Maroz (Bergell), aus dem Oberengadin und vom Pru del Vent (Puschlav). Erdkunde 25 (3): 178–195

Holtmeier, F.-K. (1973): Geoecological aspects of timberline in northern and central Europe. Arct. Alp. Res. 5 (3, Pt. 2): 45–54

Holtmeier, F.-K. (1974): Geoökologische Beobachtungen und Studien an der subarktischen und alpinen Waldgrenze in vergleichender Sicht (nördliches Fennoskandien/Zentralalpen). Erdwissenschaftliche Forschgung VIII, Wiesbaden

Holtmeier, F.-K. (1978): Die bodennahen Winde in den Hochlagen der Indian Peaks Section, Colorado Front Range. Münstersche Geographische Arbeiten 3: 3–47

Holtmeier, F.-K. (1979a): Luftbild Arapaho-Paß, Colorado Front Range, USA. Die Erde 110: 97–106

Holtmeier, F.-K. (1979b): Remarks on oscillations of polar timberline. Acta Universitas Ouluensis, A 82, Geol. 3: 165–171

Holtmeier, F.-K. (1979c): Die polare Waldgrenze (forest-tundra ecotone) in geoökologischer Sicht. Trierer Geographische Studien, Sonderheft 2: 230–246

Holtmeier, F.-K. (1980): Influence of wind on tree physiognomy at the upper timberline in the Colorado Front Range In: Mountain environments and subalpine tree growth, Proc. IUFRO workshop, Nov. 1979, Christchurch, New Zealand (Benecke, U. and Davis, M. R., eds.), New Zeal. For. Serv. Tech. Paper 70: 247–261

Holtmeier, F.-K. (1981a): What does the term "krummholz" really mean? Observations with special reference to the Alps and the Colorado Front Range. Mt. Res. Develop. 1/3–4: 253–260

Holtmeier, F.-K. (1981b): Einige Besonderheiten des Krummholzgürtels in der Colorado Front Range. Wetter und Leben 33: 150–160

Holtmeier, F.-K. (1982): "Ribbon-forest" und "Hecken". Streifenförmige Verbreitungsmuster des Baumwuchses an der oberen Waldgrenze in den Rocky Mountains. Erdkunde 36: 142–153

Holtmeier, F.-K. (1985a): Climatic stress influencing the physiognomy of trees at the polar and mountain timberline. In: Establishment and tending of subalpine forest: Research and management (Turner, H. and Tranquillini, W., eds.): 270: 31–40, Proceedings of the 3rd IUFRO workshop P 1.07-00 1984, Eidgenössische Anstalt für das forstliche Versuchswesen, Berichte

Holtmeier, F.-K. (1985b): Die klimatische Waldgrenze – Linie oder Übergangssaum (Ökoton)? – Ein Diskussionsbeitrag unter besonderer Berücksichtigung der Waldgrenzen in den mittleren und hohen Breiten der Nordhalbkugel. Erdkunde 39 (4): 271–285

Holtmeier, F.-K. (1986a): Über Bauminseln (Kollektive) an der klimatischen Waldgrenze – unter besonderer Berücksichtigung von Beobachtungen in verschiedenen Hochgebirgen Nordamerikas. Wetter und Leben 38: 121–139

Holtmeier, F.-K. (1986b): Die obere Waldgrenze unter dem Einfluß von Klima und Mensch. Abhandlungen aus dem Landesmuseum für Naturkunde 48 (2): 395–412

Holtmeier, F.-K. (1987a): Der Baumwuchs als klimaökologischer Faktor an der oberen Waldgrenze. Münstersche Geographische Arbeiten 27: 145–151

Holtmeier, F.-K. (1987b): Beobachtungen und Untersuchungen über den Ausaperungsverlauf und einige Folgewirkungen in "ribbon-forests" an der oberen Waldgrenze in der Front Range, Colorado. Phytocoenologia 15 (3): 373–396

Holtmeier, F.-K. (1987c): Human impacts on high altitude forests and upper timberline with special reference to middle latitudes. In: Human impacts and management of mountain forests. Proceedings of an International Workshop held in Japan, 5–13 September 1987 (Fujimori, T. and Kimura, M., eds.): 2–20, Japan
Holtmeier, F.-K. (1987d): Biber und Steinwild. Ökologische Aspekte der Wiederansiedlung von Tieren in ihren ehemaligen Lebensräumen. Münstersche Geographische Arbeiten 26: 99–117
Holtmeier, F.-K. (1989): Ökologie und Geographie der oberen Waldgrenze. Berichte der Reinhold-Tüxen-Gesellschaft 1: 15–45
Holtmeier, F.-K. (1990): Disturbance and management problems in larch-cembra pine forests in Europe. In: Proceedings – Symposium on whitebark pine ecosystems: Ecology and management of a high-mountain resource, Bozeman, MT, March 29–31, 1989 (Schmidt, W. C. and McDonald, K. J., compilers): 270: 25–36, USDA For. Serv., Intermtn. Res. Stat., Gen Tech. Rep. INT 270
Holtmeier, F.-K. (1993a): Der Einfluß der generativen und vegetativen Verjüngung auf das Verbreitungsmuster der Bäume und die ökologische Dynamik im Waldgrenzbereich. Beobachtungen und Untersuchungen in Hochgebirgen Nordamerikas und den Alpen. Geoökodynamik XIV: 153–182
Holtmeier, F.-K. (1993b): Timberlines as indicators of climatic changes: Problems and research needs. Paläoklimaforschung 9: 211–222
Holtmeier, F.-K. (1994a): Introduction to the Upper Engadine and its forest. In: Proc. int. workshop on subalpine stone pines and their environment: The status of our knowledge, St. Moritz, Switzerland, September 5–11, 1992 (Schmidt, W. C. und Holtmeier, F.-K., compilers): 9–17, USDA For. Serv., Intermtn. Res. Stat., Gen. Tech. Rep. INT-GTR-309
Holtmeier, F.-K. (1994b): Ecological aspects of climatically-caused timberlines fluctuations – review and outlook. In: Mountain environment in changing climates (Beniston, M., ed.): 220–232, London/New York
Holtmeier, F.-K. (1995a): Waldgrenze und Klimaschwankungen – Ökologische Aspekte eines vieldiskutierten Phänomens. Geoökodynamik XVI: 1–24
Holtmeier, F.-K. (1995b): European Larch in Middle Europe with special reference to the Central Alps. In: Ecology and management of Larix forests: A look ahead. In: Proc. int. symp., Whitefish, MT, October 5–9, 1992 (Schmidt, W. C. and McDonald, K. J., eds.): 41–49, USDA For. Serv., Intermtn. Res. Stat., Gen. Techn. Rep. GTR-INT-319
Holtmeier, F.-K. (1996): Der Wind als landschaftsökologischer Faktor in der subalpinen und alpinen Stufe der Front Range, Colorado. Arbeiten aus dem Institut für Landschaftsökologie, Westfälische Wilhelms-Universität, Münster 1: 19–45
Holtmeier, F.-K. (1999a): Timberline dynamics in view of climate change. In: A terrestrial transect for Scandinavia/Northern Europe: Proc. int. Scantran conference (Turunen, M., Hakkinen, J., Heal, O. W., Saelthun, R. and Holten, J., eds.), Eur. Comm. Ecosystems research report 31: 53–55
Holtmeier, F.-K. (1999b): Ablegerbildung im Hochlagenwald und an der oberen Waldgrenze in der Front Range, Colorado. Mitteilungen der Deutschen Dendrologischen Gesellschaft 84: 39–61
Holtmeier, F.-K. (1999c): Tiere als ökologische Faktoren in der Landschaft. Arbeiten aus den Institut für Landschaftsökologie, Westfälische Wilhelms-Universität, Münster 6
Holtmeier, F.-K. (2000): Die Höhengrenze der Gebirgswälder. Arbeiten aus dem Institut für Landschaftsökologie, Westfälische Wilhelms-Universität, Münster 11
Holtmeier, F.-K. (2002): Tiere in der Landschaft – Einfluß und ökologische Bedeutung. Stuttgart
Holtmeier, F.-K. (2003): Mountain timberlines - ecology, patchines, and dynamics (1st English edition). - Advances in Global Change Research 14, Dordrecht, Boston, London
Holtmeier, F.-K. (2005a): Change in the timberline ecotone in northern Finnish Lapland during the last thirty years. Rep. Kevo Subarct. Res. Stat. 23: 97–113

Holtmeier, F.-K. (2005b): Relocation of snow and its effects in the treeline ecotone – with special regard to the Rocky Mountains, the Alps and Northern Europe. Die Erde 136 (4): 343–373

Holtmeier, F.-K. and Broll, G. (1992): The influence of tree islands on microtopography and pedoecological conditions in the forest-alpine tundra ecotone on Niwot Ridge, Colorado Front Range, U.S.A. Arct. Alp. Res. 24 (3): 216–228

Holtmeier, F.-K. and Broll, G. (2005): Sensitivity and response of northern hemisphere altitudinal and polar treelines to environmental changes at landscape and local scales. Glob. Ecol. Biogeogr. 14: 395–410

Holtmeier, F.-K. and Broll, G. (2006): Radiocarbon-dated peat and wood remains from the Finnish Subarctic: Evidence of treeline and landscape history. Holocene 16 (5): 743–751

Holtmeier, F.-K. and Broll, G. (2007): Treeline advance driving processes and adverse factors. Landscape Online 1: 1–33, doi: 10.3097/L0.200701

Holtmeier, F.-K., Müterthies, A. and Stevens, G. E. (1996): Effektive Verjüngung und Zuwachs der Kiefer (Pinus sylvestris) und Fichte (Picea abies) an ihrer Höhengrenze in Finnisch-Lappland während der letzten 100 Jahre. Arbeiten aus dem Institut für Landschaftsökologie. Westfälische Wilhelms-Universität, Münster 1: 85–99

Holtmeier, F.-K., Broll, G., Anschlag, K. and Müterthies, A. (2003): Regeneration of trees in the treeline ecotone: Northern Finnish Lapland. Fennia 181 (2): 103–128

Holtmeier, F.-K., Broll, G. and Anschlag, K. (2004): Winderosion und ihre Folgen im Waldgrenzbereich und in der alpinen Stufe einiger nordfinnischer Fjelle. Geoöko 25: 203–224

Holway, J. G. and Ward, R. T. (1963): Snow and meltwater effects in an area of Colorado alpine. Am. Midl. Nat. 69: 189–197

Holzer, K. (1958): Die winterlichen Veränderungen der Assimilationszellen von Zirbe (Pinus cembra L.) und Fichte (Picea excelsa LINK) an der alpinen Waldgrenze. Österreichische Botanische Zeitschrift 105 (4): 324–346

Holzer, K. (1959): Winterliche Schäden an Zirben nahe der alpinen Baumgrenze. Centralblatt für das gesamte Forstwesen 76: 232–244

Holzer, K. (1967): Das Wachstum der Bäume in seiner Anpassung an zunehmende Seehöhe. Mitteilungen der forstlichen Bundes-Versuchsanstalt Wien 75: 427–456

Hoogesteger, J. and Karlsson, P. S. (1992): Effects of defoliation on radial stem growth and photosynthesis in the mountain birch (Betula pubescens ssp. tortuosa). Funct. Ecol. 6: 317–323

Hope, G. S. (1976): The vegetation history of Mt. Wilhem, Papua New Guinea. J. Ecol. 64: 627–664

Hopkins, D. M. (1959): Some characteristics of the climate in forest and tundra region in Alaska. Arctic 12: 214–220

Hopkins, M. (1972): The paleogeography and climatic history of Beringia during Cenozoic time. Inernord 12: 121–150

Hornstedt, R. and Venn, K. (1980): Frost injury to Norway spruce during the winter 1976–1977. Eur. J. For. Pathol. 10: 71–76

Horvat, I., Glavac, V. and Ellenberg, H. (1974): Vegetation Südosteuropas. Gebotanica selecta 4

Hueck, K. (1953): Urlandschaft, Raublandschaft und Kulturlandschaft in der Provinz Tucuman im nordwestlichen Argentinien. Bonner Geographische Abhandlungen 10

Hueck, K. (1962): Der Polylepiswald in den venezolanischen Anden, eine Parallele zum mitteleuropäischen Latschenwald. Angewandte Pflanzensoziologie 17: 57–76

Hueck, K. (1966): Die Wälder Südamerikas. Stuttgart

Hueck, K. and Seibert, P. (1981): Vegetationskarte von Südamerika. Vegetationsmonographien der einzelnen Großräume IIa (Walter, H., ed.) Stuttgart

Hultèn, E. (1974): The plant cover of southern Kamchatka. Arkiv för Botanik 7 (3): 181–257

Hüppe, J. and Pott, R. (1993): Man-induced changes of the alpine timberline of the Val Fenga (Silvretta, Switzerland) and their reflections in pollen diagrams (Preliminary report). Paläoklimaforschung 9: 137–161
Hustich, I. (1937): Pflanzengeographische Studien im Gebiet der Niederen Fjelde im westlichen Finnischen Lappland. Teil I. Acta Bot. Fenn. 19
Hustich, I. (1942): Pflanzengeographische Forschungen und Probleme in Finnisch-Lappland. Botanikai Közlemények 36 (5): 233–246
Hustich, I. (1948): The scotch pine in northernmost Finland and its dependence on the climate in the last decades. Acta Botanica Fennica 42
Hustich, I. (1949): On the correlation between growth and the recent climatic fluctuation. Geografiska Annaler 31: 99–105
Hustich, I. (1958): On the recent expansion of the scotch pine in northern Europe. Fennia 82 (3)
Hustich, I. (1966): On the forest-tundra and northern tree-lines. Rep. Kevo Subarct. Res. Stat. (3): 7–47
Hutchins, H. E. and Lanner, R. M. (1982): The central role of Clark's Nutcracker in the dispersal and establishment of whitebark pine. Oecologia 55: 195–201
Hygen, G. (1965): Water stress in conifers during winter. In: Water stress in plants (Slavik, B., ed.): 89–98, Junk, The Hague
Idso, B. (1989): A problem for paleoclimatology? Quatern. Res. 31: 433–434
Imhof, E. (1900): Die Waldgrenze in der Schweiz. Gerlands Beiträge zur Geophysik 4: 241–330
Innes, J. L. (1991): High-altitude and high-latitude tree growth in relation to past, present and future global climate change. The Holocene 1 (2): 168–173
International Society of Soil Science Working Group RB (1998): World reference base for soil sciences, an introduction. Leuven
Isard, S. A. and Belding, M. J. (1986): Evapotranspiration from the alpine tundra of Colorado, USA. Arct. Alp. Res. 21: 71–82
Ishizuka, K. (1974): Mountain vegetation In: The flora and vegetation of Japan (Numata, M., ed.) (in Japanese with German summary): 151–172, Kodansha, Tokyo
IUSS Working Group WRB (2006): World reference base for soil resources 2006. World Soil Resources Reports 103, FA, Rome
Ives, J. D. (1973a): Permafrost and its relationship to other environmental parameters in a midlatitude, high-altitude setting, Front Range, Colorado Rocky Mountains. In: Permafrost: The North American Contribution to the Second International Conference: 121–125, National Academy of Science, Washington, DC
Ives, J. D. (1973b): Studies in high altitude geoecology of the Colorado Front Range. A review of the research program of the Institute of Arctic and Alpine Research. Arct. Alp. Res. 5 (3, Pt. 2): 65–66
Ives, J. D. (1978): Remarks on the stability of timberline. Erdwissenschaftliche Forschung 11: 313–317
Ives, J. D. and Fahey, B. D. (1971): Permafrost occurrence in the Front Range, Colorado Rocky Mountains, U. S. A. J. Glaciol. 10 (58): 105–111
Ives, J. D. and Hansen-Bristow, K. J. (1983): Stability and instability of natural and modified upper timberline landscapes in the Colorado Rocky Mountains, U. S. A. Mt. Res. Develop. 3 (2): 149–155
Jacobsen, J. P. and Schickhoff, U. (1995): Untersuchungen zur Besiedlung und gegenwärtigen Waldnutzung im Hindukush/Karakorum. Erdkunde 49 (1): 49–59
Jalkanen, R. (1996): Repeated red belts at Lusmavaara, western Lapland. Pallas-Symposium 1996, Finnish Forest Research Institute, Research Papers 623: 91–95

Jalkanen, R. and Konôpka, B. (1998): Snow-packing as a potential harmful factor on Picea abies, Pinus sylvestris and Betula pubescens at high altitudes in northern Finland. Eur. J. For. Pathol. 28: 373–382
Jegerlehner, J. (1903): Die Schneegrenze in der Schweiz. Gerlands Beiträge zur Geophysik 5
Jenic, J . and Locvenc, T. (1962): Die alpine Waldgrenze im Krkonose Gebirge. Prag
Jenny, H. (1948): Great soil groups in the equatorial regions of Colombia, South America. Soil Sci. 66: 5–28
Johnson, D. D. and Cline, A. J. (1965): Colorado mountain soils. Adv. Agron. 17: 233–281
Johnson, D. M., Germino, M. J. and Smith, W. K. (2004): Abiotic factors limiting photosynthesis in Abies lasiocarpa and Picea engelmannii seedlings below and above the alpine timberline. Tree Physiol. 24: 377–386
Johnson, E. A. (1987): The relative importance of snow avalanche disturbance and thinning on canopy plant populations. Ecology 68 (1): 43–53
Johnson, P. L. and Billings, W. D. (1962): The alpine vegetation of the Beartooth Plateau in relation to cryopedogenic processes and patterns. Ecol. Monogr. 32: 105–135
Johnston, W. F. (1990): Larix laricina. Agricultural handbook 654, USDA For. Serv., Washington, DC, 141–151
Jones, P. D. and Moberg, A. (2003): Hemispheric and large-scale surface air temperature variations: An extensive revision and update to 2001. J. Climate 16: 206–223
Jordan, D. N. and Smith, W. K. (1995): Microclimate factors influencing the frequency and duration of growth season frost for subalpine plants. Agric. For. Meteorol. 77: 17–30
Jordan, E. (1983): Die Verbreitung von Polylepis-Beständen in der Westkordillere Boliviens. Tuexenia, Neue Serie 3: 101–116
Jordan, E. (1996): Umweltrisiken in den tropischen Anden. HGG-Journal 1: 68–96
Junghuhn, F. (1852): Java, seine Gestalt, Pflanzendecke und innere Bauart. Ins Deutsche übertragen von J. K. Hasskarl, Bd. I-III
Juntunen, V., Neuvonen, S., Norokorpi, Y. and Tasanen, T. (2002): Potential for timberline advance in northern Finland, as revealed by monitoring during 1983–1999. Arctic 55: 348–361
Jylhä, K., Tuomenvirta, H. and Ruosteenoja, K. (2004): Climate change projection for Finland during the 21st century. Boreal Environ. Res. 9: 127–152
Kaitaniemi, O., Neuvonen, S. and Nyyssönen, T. (1995): Effects of cumulative defoliation on growth, reproduction, and insect resistance in the mountain birch. Ecology 80: 524–532
Kajimoto, T. (1992): Dynamics and dry matter production of belowground woody organs of Pinus pumila trees growing on the Kiso Mountains Range in central Japan. Ecol. Res. 7: 333–339
Kajimoto, T., Onodera, H., Ikeda, S., Daunarz, H. and Seki, T. (1998): Seedling establishment of subalpine stone pine (Pinus pumila) by nutcracker (Nucifraga) seed dispersal on Mt. Yumori, Northern Japan. Arct. Alp. Res. 30 (4): 408–417
Kajimoto, T., Seki, T., Shigeto, I., Daimaru, H., Okamoto, T. and Onodera, H. (2002): Effects of snow fall fluctuations on the growth and establishment of subalpine Abies mariesii upper forest-limit on Mt. Yumori, Northern Japan. Arct. Antarct. Alp. Res. 34 (2): 191–200
Kajimoto, T., Daimaru, H., Okamoto, T., Otani, T. and Onodera, H. (2004): Effects of snow avalanche disturbance on regeneration of subalpine Abies mariesii forest, northern Japan. Arct. Antarct. Alp. Res. 36 (4): 436–445
Kalela, E. K. (1941a): Über die Holzarten und die durch die klimatischen Verhältnisse verursachten Holzartenwechsel in den Wäldern Ostpatagoniens. Annales Academiae Scientarum Fennicae Ser. A IV (2): 1–151
Kalela, E. K. (1941b): Über die Entwicklung der herrschenden Bäume in den Beständen verschiedener Waldtypen Ost-Patagoniens. Suomalainen Tiedeaktemiean Toimituksia, Seris A. 4, Biologica 3: 5–67

Kallio, P. and Lehtonen, J. (1973): Birch forest damage caused by Oporinia autumnata (Bkh.) in 1965–1966 in Utsjoki, N- Finland. Rep. Kevo Subarct. Res. Stat. 10: 55–69
Kallio, P. and Lehtonen, J. (1975): On the ecocatastrophe of birch forest caused by Oporinia autumnata (Bkh.) and the problem of reforestation. In: Fennoscandian tundra ecosystems (Wielgolaski, F. E., ed.): Part 2, New Ecol. Stud. 17: 174–180, Berlin/Heidelberg/New York
Kallio, P. and Mäkinen, Y. (1957): Untersuchungen über die Flora von Utsjoki in Nordfinnland II. Arch. Societas Zoologicae-Botanicae Fennicae Vanamo 12 (1): 12–29
Kalliola, R. (1941): Tunturimittari (Oporinia autumnata) subalpiniisten koivikoiden tuholainen (cited by Nuorteva, 1993). Luonnon Ystäva 45: 45–60
Kaltenrieder, P., Tinner, W. and Ammann, B. (2005): Zur Langzeitökologie des Lärchen-Arvengürtels in den südlichen Walliser Alpen. Botanica Helvetica 115: 137–154
Kankaanpää, S. (1999): Northern timberline forests – a review. In: Sustainable development in the northern timberline forests. Proceedings of the timberline workshop, May 10–11, 1998, Whitehorse, Canada (Kankaanpää, S., Tapanen, T. and Sutinen, M.-L., eds.) – The Finnish Forest Research Institute, Research Papers 743: 101–116
Kärenlampi, L. (1972): On the relationships of Scots pine annual ring width and some climatic variables at the Kevo Subarctic Research Station. Rep. Kevo Subarct. Res. Stat. 9: 78–81
Karlsson, P. S. and Nordell, K. O. (1996): Effects of soil temperature on nitrogen economy and growth of mountain birch seedlings near its presumed low temperature distribution limit. Ecoscience 3: 183–189
Karlsson, P. S. and Weih, M. (2001): Soil temperatures near the distribution limit of the mountain birch (Betula pubescens ssp. czerepanovii): Implications for seedling nitrogen economy and survival. Arct. Antarct. Alp. Res. 33 (1): 88–92
Karlsson, P. S. and Wielgolaski, F.-E. (2005): Synthesis Section 1: A dynamic forest in a changing environment. In: Plant ecology, herbivory, and human impact in nordic mountain birch forests (Wielgolaski, F.-E., ed.), Ecol. Stud. 180: 117–122
Karma, K., Ageta, Y., Naito, N., Iwata, S. and Yabuki, H. (2003): Glacier distribution in the Himalayas and glacier shrinkage from 1963 to 1993 in the Bhutan Himalayas. Bull. Glaciol. Res. 20: 29–40
Kashulina, G., Reimann, C., Finne, T. E., Halleraker, J. H., Äyräs, M. and Chekushin, V. A. (1997): The ecosystems of the central Barents Region: Scale, factors and mechanisms in disturbance. Sci. Total Environ. 206: 203–255
Kasthofer, K. (1818): Bemerkungen über die Wälder und Alpen des bernischen Hochgebirges. (2nd edition) Aarau
Kasthofer, K. (1822): Bemerkungen auf einer Alpenreise über den Susten, Gotthard, Bernhardin, und über die Oberalp, Furka und Grimsel. Aarau
Katz, C., Oren, R., Schulze, E.-D. and Milburn, J. A. (1989): Uptake of water and solutes through twigs of Picea abies (L.) Karst. Trees 3: 33–37
Kaufmann, M. R. (1975): Leaf water stress in Engelmann spruce. Influence of the root and shoot environments. Plant Physiol. 56: 841–844
Kauhanen, H. (1987): On growth problems of mountain birch near its distributional limits. UNGI Rep. 65: 183–190
Kauppi, P. and Posch, M. (1988): A case study of the effect of CO_2-induced climatic warming on forest growth and the forest sector: A. Productivity reactions of northern boreal forest. In: The impact of climatic variations on agriculture, Vol. I (Paary, M. L., Carter, T. R. and Konijin, N. T., eds.): 183–195, Kluwer, Dordrecht
Käyhkö, J. and Pellika, P. (1994): Remote sensing of the impact of reindeer grazing on vegetation in northern Fennoscandia using SPOT XS data. Polar Res. 13: 115–124
Keane, R. E. and Arno, S. F. (1993): Rapid decline of whitebark pine in western Montana: Evidence from 20-year measurements. Western Journal of Applied Forestry 8: 44–47
Keane, R. E. and Morgan, P. (1994): Decline of whitebark pine in the Bob Marshall Wilderness Complex of Motana, U.S.A. In: Proc. int. workshop on subalpine stone pines

and their environment: The status of our knowledge, St. Moritz, Switzerland, September 5–11, 1992 (Schmidt, W. C. and Holtmeier, F.-K., compilers): 245–253, USDA For. Serv., Intermtn. Res. Stat., Gen. Tech. Rep. INT-GTR-309

Keane, R. E., Arno, S. F., Brown, J. K. and Tomback, D. F. (1990): Simulating disturbances and conifer succession in whitebark pine forests. Proc. symp. Whitebark pine ecosystems: Ecology and management of a high-mountain resource, USDA For. Serv., Intermtn. Res. Stat., Gen. Tech. Rep. INT-270: 274–288

Kearney, M. S. (1982): Recent seedling establishment at timberline in Jasper National Park, Alta. Can. J. Bot. 60: 2283–2287

Kelletat, D. (1972): Zum Problem der Abgrenzung und ökologischen Differenzierung des Hochgebirges in Schottland. Erdwissenschaftliche Forschung 4: 110–130

Kellomäki, S., Väisänen, H. and Kolström, T. (1997): Model computations on the effects of elevating temperature and atmosphere CO_2 on the regeneration of scots pine at the timberline in Finland. Climate Change 37: 683–708

Kendall, K. C. and Arno, S. F. (1990): Whitebark pine – an important but endangered wildlife resource. In: Proc. symp. on whitebark pine ecosystems: Ecology and management of a high-mountain resource, Bozeman, MT, March 29–31, 1989 (Schmidt, W. C. and McDonald, K. J., compilers): 264–273, USDA For. Serv., Intermtn. Res. Stat., Gen. Techn. Rep. INT 270

Kernaghan, G. and Harper, K. A. (2001): Community structure of ectomycorrhizal fungi across an alpine/subalpine ecotone. Ecography 24: 181–188

Kerner, A. (1864/1865): Studien über die obere Waldgrenze in den österreichischen Alpen. Österreichische Revue

Kessler, M. (1995): Polylepis-Wälder Boliviens: Taxa, Ökologie, Verbreitung und Geschichte. Berlin, Stuttgart

Kessler, M. and Driesch, P. (1994): Causa e historia de la destrucción de bosques Altoandinos en Bolivia. Ecología en Bolivia 21: 1–18

Kessler, M. and Hohnwald, S. (1998): Bodentemperaturen innerhalb und außerhalb bewaldeter und unbewaldeter Blockhalden in den bolivianischen Hochanden. Ein Test der Hypothese von Walter und Medina 1969. Erdkunde 52: 54–62

Khutornoy, O. V., Velisevich, S. N. and Vorob'ev, V. N. (2001): Ecological variation in the morphological structure of the crown in Siberian stone pine at the timberline. Russ. J. Ecol. 32 (6): 393–399

Kienast, F. (1985): Dendroökologische Untersuchungen an Höhenprofilen aus verschiedenen Klimabereichen. Diss. Philosphische Fakultät II, Universität Zürich

Kienast, F. and Luxmoore, R. J. (1988): Tree-ring analysis and conifer growth responses to increased atmospheric CO_2 levels. Oecologia 76: 487–495

Kienast, F. and Schweingruber, F. H. (1986): Dendroecological studies in the Front Range, Colorado, U. S. A. Arct. Alp. Res. 18 (3): 277–288

Kihlman, A. O. (1890): Pflanzenbiologische Studien aus Russisch Lappland. Acta Societas Fauna et Flora Fennica VI (3)

Kilander, S. (1965): Alpine zonation in the southern part of the Swedish Scandes. In: The plant cover of Sweden, Acta Phytogeographica Suecica 50: 78–84

Kimball, B. A. (1983): Carbon dioxide and agricultural yield: An assemblage and analysis of 430 prior observations. Agron. J. 75: 779–788

Kimble, J. M., Ping, C. L., Summer, M. E. and Wilding, L. P. (1999): Andisols. In: Handbook of soil science (Summer, M. E., ed.): 209–224, Boca Raton, FL/London/New York/Washington, DC

Kincaid, D. T. and Lyons, E. E. (1981): Winter water relations of red spruce on Mount Monadnock, New Hampshire. Ecology 62: 1155–1161

Kind, R. J. (1981): Snow drifting. In: Handbook of snow – principles, processes, management and use (Gray, D. M. and Male, D. H., eds.): 338–359, Toronto

King, C. M. (ed.) (1990): The handbook of New Zealand mammals (paperback edition 1995). Auckland/Melbourne/Oxford/New York
Kitching, R. I. (1986): Exotics in Australia and elsewhere. In: The ecology of exotic animals and plants. Some Australian case histories (Kitching, R. L., ed.): 1–5, Brisbane/New York/Chichester/Toronto/Singapore
Kjällgren, L. and Kullman, L. (1998): Spatial pattern and structure of the mountain birch-treeline in the southern Swedish Scandes – a regional perspective. Geografiska Annaler 80 A: 1–16
Klasner, F. L. and Fagre, D. B. (2002): A half century of change in alpine treeline pattern at Glacier National Park, Montana, U. S. A.: 49–56
Klikoff, G. (1965): Microenvironmental influence on vegetational pattern near timberline in the Central Sierra Nevada. Ecol. Monogr. 35: 187–211
Klimek, K. (1992): Historic slope degradation above timberline in the Balkan Mts., Bulgaria. Geographia Polonica 60: 45–50
Klink, H. J. (1973): Die natürliche Vegetation und ihre räumliche Ordnung im Puebla-Tlaxcala-Gebiet (Mexiko). Erdkunde XXVII (3): 213–225
Klink, H.-J. and Lauer, W. (1978): Die räumliche Anordnung der Vegetation im östlichen Hochland von Zentralmexico. Pflanzengeographie: 472–506, Darmstadt
Klink, H.-J. and Mayer, E. (1983): Vegetationsgeographie. Das Geographische Seminar, Braunschweig
Klink, H.-J., Lauer, W. and Ern, H. (1973): Erläuterungen zur Vegetationskarte 1: 200 000 des Puebla Tlaxcala-Gebietes. Erdkunde 27 (3): 225–229
Klötzli, F. (1958): Zur Pflanzensoziologie des Südhanges der alpinen Stufe des Kilimandscharo. Berichte des Geobotanischen Institutes ETH, Stiftung Rübel 1957: 33–59
Klötzli, F. (1975): Zur Waldfähigkeit der Gebirgssteppen Hoch-Semiens (Nordäthiopien). Berichte zur naturkundlichen Forschung in Südwest-Deutschland 34: 131–147
Klötzli, F. (1977): Wild und Vieh im Gebirgsgrasland Äthiopiens. Berichte über das Internationale Symposium der Internationalen Vereinigung für Vegetationskunde, Vegetation und Fauna (Rinteln 12–15.4.1976): 498–512, Vaduz
Klötzli, F. (1984): Die Nische der Pinus-Vikarianzen in der südlichen Hemisphäre. Schweizerische Zeitschrift für Forstwesen, Beiheft 72: 89–101
Klötzli, F. (1991): Niches of longevity and stress. In: Modern ecology, basis and applied aspects (Esser, G. and Overdieck, D., eds.): 97–110, Amsterdam/London/New York/Tokyo
Kneisel, C., Hauck, C. and Vondermühlen, D. (2000): Permafrost below the timberline confirmed and characterized by geoelectrical resistivity measurements, Bever Valley, Eastern Swiss Alps. Permafrost Periglac. Process. 11: 295–304
Kobayashi, K. (1971): Phytosociological studies on the scrub of dwarf pine (Pinus pumila) in Japan. J. Sci. Hiroshima Univ. Ser. B. Bot. 14: 1–52
Koch, M. (1909): Beiträge zur Kenntnis der Höhengrenzen der Vegetation im Mittelmeergebiet. Diss. Univ. Halle-Wittenberg, Halle
Koepke, H.-W. (1961): Synökologische Studien an der Westseite der peruanischen Anden. Bonner Geographische Abhandlungen 29
Kok, K., Verweij, P. A. and Beukema, H. (1995): Effects of cutting and grazing on Andean treeline vegetation. In: Biodiversity and conservation of neotropical montane forests (Churchill, S. P. et al., eds.): 527–539, New York
Kokelj, S. V. and Burns, C. R. (2004): Tilt of spruce trees near ice wedges, Mackenzie Delta, Northwest Territories, Canada. Arct. Antarct. Alp. Res. 36 (4): 615–623
Komarkova, V. (1976): Alpine vegetation of the Indian Peaks area, Front Range, Colorado Rocky Mountains. Ph.D. Dissertation, University of Colorado, Boulder, CT
Komarkova, V. (1979): Alpine vegetation of the Indian Peaks area, Front Range, Colorado Rocky Mountains. Flora et Vegetatio Mundi 7, Vaduz

Kontic, R., Niederer, M., Nippel, C.-A. and Winkler-Seifert, A. (1986): Jahrringanalysen an Nadelbäumen zur Darstellung und Interpretation von Waldschäden (Wallis, Schweiz). Eidgenössische Anstalt für das forstliche Versuchswesen, Berichte 283

Köppen, W. (1919): Baumgrenze und Lufttemperatur. Petermanns Geographische Mitteilungen : 201–203

Köppen, W. (1920): Verhältnis der Baumgrenze zur Lufttemperatur. Meteorologische Zeitschrift 37: 39–42

Körner, Ch. (1994): Impact of atmospheric changes on high mountain vegetation. In: Mountain environments in changing climates (Beniston, M., ed.): 155–166, London/New York

Körner, C. (1995): Towards a better experimental basis for upscaling plant responses to elevated CO_2 and climate warming. Plant Cell Environ. 18: 1101–1110

Körner. C. (1998a): A re-assessment of high elevation treeline positions and their explanation. Oecologia 115: 445–459

Körner, C. (1998b): Worldwide positions of alpine treelines and their causes. In: The impacts of climatic variability on forests (Bensiton, M. and Innes, J. L., eds.): 221–229, Springer Heidelberg

Körner, C. (1999): Alpine plant life. Functional plant ecology of high mountain ecosystems. Springer, Berlin/Heidelberg

Körner, C. (2002): Treelines in a changing world. Centralblatt für das gesamte Forstwesen 119 (3/4): 307–308

Körner, C. (2003a): Carbon limitation in trees. J. Ecol. 91: 4–7

Körner, C. (2003b): Ein morphologiebedingter Wärmemangel bestimmt die Waldgrenze. Colloquium Geographicum 26: 114–119

Körner, C. (2007a): Drivers and the driven in a warming, high CO_2 world. The BES Annual Meeting and AGM 2007 Report, British Ecological Society, The Bulletin December 2007: 19–20

Körner, C. (2007b): Climatic treelines: Conventions, global patterns, causes. Erdkunde 61: 316–324

Körner, C. and Hilscher, H. (1978): Wachstumsdynamik von Grünerlen auf ehemaligen Almflächen an der zentralalpinen Waldgrenze. In: Ökologische Analysen von Almflächen im Gasteiner Tal (Cernusca, A., ed.): 187–193, Veröffentlichungen des Österreichischen MaB-Hochgebirgsprogramms Hohe Tauern, Innsbruck

Körner, C. and Hoch. G. (2006): A test of treeline theory on a montane permafrost island. Arctic, Antarctic and Alpine Research 38: 113–119

Körner, C. and Paulsen, J. (2004): A world-wide study of high altitude treeline temperatures. J. Biogeogr. 31: 713–732

Korotaev, A. A. (1994): Untersuchungen zur Frostresistenz von Baumwurzeln. Forstarchiv 65: 93–95

Kosak, J., Troll, M. and Widacki, W. (1995): The anthropogenic upper timberline in the Silesian Beskid Mts. Zeszyty Naukowe Uniwersytetu Jagellonskiego, Prace Geograficzne 98: 199–207

Koster, E. A. (1991): Assessment of climate change impact in high-altitude regions. Terra 103 (1): 3–13

Koster, E. A. and Nieuwenhuijzen, M. E. (1992): Permafrost response to climatic change. Catena Suppl. 22: 37–58

Köstler, N. and Mayer, H. (1970): Waldgrenzen im Berchtesgadener Land. Jahrbuch des Vereins zum Schutze der Alpenpflanzen und -Tiere 35: 1–33

Kral, F. (1971): Pollenanalytische Untersuchungen der Waldgeschichte des Dachsteinmassives. Wien

Kral, F. (1973): Zur Waldgrenzdynamik im Dachsteingebiet. Jahrbuch des Vereins zum Schutze der Alpenpflanzen und -Tiere 38: 71–79

Kramer, P. J. (1981): Carbon dioxide concentration, photosynthesis and dry matter production. BioScience 31: 29–33
Krebs, N. (1961): Die Ostalpen und das heutige Österreich. Bd. I (3rd edition). Stuttgart
Krebs, P. V. (1972): Dendrochronology and the distribution of bristlecone pine (Pinus aristata Engelm.) in Colorado. Ph.D. thesis, University of Colorado, Boulder, CT
Krebs, P. V. (1973): Dendrochronology of bristlecone pine (Pinus aristata Engelm.) in Colorado. Arct. Alp. Res. 5 (2): 149–150
Kremenetski, C. V., Sulerzhitsky, L. D. and Hantemirov, R. (1998): Holocene history of the northern range limits of some trees and shrubs in Russia. Arct. Alp. Res. 30 (4): 317–333
Krivsky, L. (1958): Bestimmung der vorherrschenden Windrichtung aus Windfahnenbäumen. Meteorologische Rundschau 11: 86–90
Kronfuss, H. (1967): Schneelage und Ausaperung an der Waldgrenze. Mitteilungen der Forstlichen Bundes-Versuchsanstalt Mariabrunn 75: 207–241
Kronfuss, H. (1970): Räumliche Korrelation zwischen der Windstärke in Bodennähe und der Schneedeckendauer. Centralblatt für das gesamte Forstwesen 82 (2): 96–116
Kronfuss, H. (1972): Kleinklimatische Vergleichsmessungen an zwei subalpinen Standorten. Mitteilungen der Forstlichen Bundes-Versuchsanstalt Wien 96: 159–176
Kronfuss, H. and Havranek, W. M. (1999): Effects of elevation and wind on the growth of Pinus cembra L. in a subalpine afforestation. Phyton (Austria) Special Issue: "Eurosilva" 39 (4): 99–106
Kryuchkov, V. V. (1973): The effect of permafrost on northern tree line. Proc. Internat. Conf. On Permafrost, USSR Contribution, Washington, DC. Proc. Nat. Acad. Sci 1973: 136–138
Kujala, V. (1929): Untersuchungen über Waldtypen in Petsamo und angrenzenden Teilen von Inari-Lappland. Communicationes ex Instituto Quaestionum Forestalium Finlandiae Editae 13, Helsinki
Kujala, V. (1952): Diskussionsbeitrag Ilmari Hustich: The recent climatic fluctuation in Finland and its consequences. A Symposium. Fennia: 108–109
Kullman, L. (1979): Change and stability in the altitude of the birch tree-limit in the southern Swedish Scandes 1915–1975. Acta Phytogeographica Suecica 65: 1–121
Kullman, L. (1983a): Short-term population trends of isolated tree-limit stands of Pinus sylvestris L. in central Sweden. Arct. Antarct. Alp. Res. 15 (3): 369–382
Kullman, L. (1983b): Past and present treelines of different species in the Handölan valley, Central Sweden. Collection Nordicana 47: 25–45
Kullman, L. (1984): Transplantation experiments with saplings of Betula pubescens ssp. tortuosa near the tree-limit in central Sweden. Holarctic Ecol. 7 (3): 289–293
Kullman, L. (1986a): Recent tree-limit history of Picea abies in the southern Swedish Scandes. Can. J. For. Res. 16 (4): 761–771
Kullman, L. (1986b): Temporal and spatial aspects of subalpine populations of Sorbus aucuparia in Sweden. Annales Botanici Fennici 23: 215–275
Kullman, L. (1987): A decade of tree-line monitoring in the southern Swedish Scandes. UNGI Rep. 65: 191–202
Kullman, L. (1988): Holocene history of the forest-alpine tundra ecotone in the Scandes Mountains (Central Sweden). New Physiol. 108: 101–110
Kullman, L. (1989a): Cold-induced dieback of montane spruce forest in the Swedish Scandes – a modern analogue of paleoenvironmental processes. New Phytol. 113: 377–389
Kullman, L. (1989b): Rapid decline of upper montane forests in the Swedish Scandes. Arctic 42 (3): 217–226
Kullman, L. (1990): Dynamics of altitudinal tree-limits in Sweden: A review. Norsk Geografisk Tidsskrift 44: 103–116
Kullman, L. (1991): Cataclysmic response to recent cooling of a natural boreal pine (Pinus sylvestris L.) forest in northern Sweden. New Phytol. 117: 351–360

Kullman, L. (1993): Pine (Pinus sylvestris L.) tree-limit surveillance during recent decades, Central Sweden. Arct. Alp. Res. 25 (1): 24–31

Kullman, L. (1997): Tree-limit stress and disturbance. A 25-year survey of geoecological change in the Scandes Mountains of Sweden. Geografiska Annaler 79 A: 139–165

Kullman, L. (2000a): The geoecological history of Picea abies in northern Sweden and adjacent parts of Norway. A contrarian hypothesis of postglacial tree immigration patterns. Geoöko XXI: 141–172

Kullman, L. (2000b): Tree limit rise and recent global warming: A geoecological case study from the Swedish Scandes. Norsk Geografisk Tidsskrift 54: 49–59

Kullman, L. (2001): 20th century warming and tree-limit rise in the southern Scandes of Sweden. Ambio 30: 72–80

Kullman, L. (2002): Rapid recent range-margin rise of tree and shrub species in the Swedish Scandes. J. Ecol. 90: 68–77

Kullman, L. (2003): Changes in alpine plant cover: Effects of climate warming. Svensk Botanisk Tidskrift 97: 210–221

Kullman, L. (2004a): The changing face of the alpine world. Glob. Change Newsletter 57: 12–14

Kullman, L. (2004b): Early Holocene appearance of mountain birch (Betula pubescens ssp. tortuosa) at unprecedented high elevations in the Swedish Scandes: Megafossil evidence exposed by recent snow and ice recession. Arct. Antarct. Alp. Res. 36 (2): 172–180

Kullman, L. (2004c): Tree-limit landscape evolution at the southern fringe of the Swedish Scandes (Dalarna Province). Holocene and the 20th century. Fennia 182 (2): 73–94

Kullman, L. (2005a): Trädgränsen i Dalafjällen. Del. 1. Gamla och nya träd på Fulufjället vegetationshistoria på hög nivå. Miljövärdsenheten Rapport 2005, 10

Kullman, L. (2005b): Mountain taiga of Sweden. In: The physical geography of Fennoscandia (Seppälä, M., ed.): 453–497, Oxford University Press, Oxford

Kullman, L. (2005c): Gamla og nya träd på hög nivå. Svensk Botanis Tidskrift 99 (6): 315–329

Kullman, L. (2005d): Wind-conditioned 20th century decline of birch treeline vegetation in the Swedish Scandes. Arctic 58: 286–294

Kullman, L. (2005e): Pine (Pinus sylvestris) tree dynamics during the past millenium – a population study in west-central Sweden. Annales Botanici Fennici 42: 95–106

Kullman, L. (2006): Transformation of alpine and subalpine vegetation in a potentially warmer future, the Anthropocene era. Tentative projections based on long-term observations and paleovegetation records. Curr. Trend. Ecol. 1: 1–16

Kullman, L. (2007a): Tree line population monitoring of Pinus sylvestris in the Swedish Scandes, 1973–2005: Implications for tree line theory and climate change ecology. J. Ecol. 95: 41–52

Kullman, L. (2007b): Modern climate change and shifting ecological states of the subalpine/alpine landscape in the Swedish Scandes. Geoöko 28: 187–221

Kullman, L. and Engelmark, O. (1991): Historical biogeography of Picea abies (L.) Karst. at its subarctic limit in northern Sweden. J. Biogeogr. 18: 63–70

Kullman, L. and Högberg, N. (1989): Rapid natural decline of upper montane forests in the Swedish Scandes. Arctic 42 (3): 217–226

Kullman, L. and Kjällgren, L. (2006): Holocene pine tree-line evolution in the Swedish Scandes: Recent treelin-rise and climate change in a long-term perspective. Boreas 35: 159–168

Kumpula, J. and Nieminen, M. (1992): The inventory of reindeer pastures in Finland by satellite imagery. In: Extended abstracts. Symposium on the state of the environment and environmental monitoring in Northern Fennoscandia and Kola Peninsula, October 6–8, 1992, Rovaniemi, Finland (Tikkanen, E. , Varmola, M. and Katermaa, T., eds.): 294–296

Kuoch, R. (1965): Der Samenanflug 1963/1964 an der oberen Fichtenwaldgrenze im Sertigtal. Schweizerische Anstalt für das forstliche Versuchswesen, Mitteilungen 41 (3): 63–85
Kuoch, R. and Amiet, R. (1970): Die Verjüngung im Bereich der oberen Waldgrenze der Alpen. Schweizerische Anstalt für das forstliche Versuchswesen, Mitteilungen 46: 159–328
Kuoch, R. and Schweingruber, F. H. (1975): Baumarten an der alpinen Waldgrenze in der Schweiz. Schweizerische Zeitschrift für Forstwesen 126 (1): 13–40
Kuramoto, R. T. and Bliss, L. C. (1970): Ecology of subalpine meadows in the Olympic Mountains, Washington. Ecol. Monogr. 40: 317–345
Kuschel, G. (1975): Biogeography and ecology in New Zealand. Monographiae Biologicae 27
Kuuluvainen, T., Sprugel, D. G. and Brooks, J. R. (1996): Hydraulic architecture and structure of Abies lasiocarpa seedlings in three subalpine meadows of different moisture status in the Eastern Olympic Mountains, Washington. Arct. Alp. Res. 28 (1): 60–64
Kuznezov, N. I. (1959): On the ecology of the nutcracker Nucifraga caryocatactes in the Mid Urals. Bjull. Moip. SSSR 64: 132–133
Laberge, M.-J., Payette, S. and Bousquet, J. (2000): Life span and biomass allocation of stunted black spruce clones in the subarctic environment. J. Ecol. 88: 584–593
Laegaard, S. (1992): Influence of fire in the grass páramo vegetation of Ecuador. Páramo: An Andean ecosystem under human influence: 151–170, London
Lafleur, P. M., Rouse, W. R. and Carlson, D. W. (1992): Energy balance differences and hydrologic impacts across the northern treeline. Int. J. Climatol. 12: 193–205
Laine, U. and Nurmi, J. (1971): Factors affecting vegetation and flora of anorthosite and granulite areas in Western Inari, Finnish Lapland. Rep. Kevo Subarct. Res. Stat. 8: 104–115
Lakusic, R. (1970): Natürliche und anthropogene Höhengrenzen in den südöstlichen Dinariden. Mitteilungen der ostalpin-dinarischen Gesellschaft für Vegetationskunde 11: 89–94
LaMarche, V. C. (1969): Environment in relation to age of bristlecone pines. Ecol. 50: 53–59
LaMarche, V. C. (1973): Holocene climatic variations, inferred from tree line fluctuations in the White Mountains, California. Quatern. Res. 3: 632–660
LaMarche, V. C. (1977): Dendrochronological and paleoecological evidence for holocene climatic fluctuations in the White Mountains. Erdwissenschaftliche Forschung 13: 151–155
LaMarche, V. C. and Mooney, H. A. (1967): Altithermal timberline advance in western United States. Nature 213: 980–982
LaMarche, V. C. and Mooney, H. A. (1972): Recent climatic change and development of the bristlecone pine (Pinus longaeva Bailey) krummholz zone, Mt. Washington, Nevada. Arct. Alp. Res. 4: 61–72
LaMarche, V. C., Graybill, D. A., Fritts, H. C. and Rose, M. R. (1984): Increasing atmospheric carbon dioxide: Treering evidence for growth enhancement in natural vegetation. Science 225: 1019–1021
Lämmermayr, L. (1932): Die Legföhre in den Alpen. Jahrbuch des Deutschen Alpenvereins 63: 52–61
Lampadius, G. (1937): Die Höhengrenzen der Cima d'Asta und des Lagorai-Gebirges. Berliner Geographische Arbeiten 15
Landolt, E. (1862): Bericht an den hohen schweizerischen Bundesrath über die Untersuchung der schweizerischen Hochgebirgswaldungen, vorgenommen in den Jahren 1858, 1859 und 1860. Bern
Lange, S., Bussmann, R. W. and Beck, E. (1997): Stand structure and regeneration of the subalpine Hagenia abyssinica forests of Mt. Kenya. Bot. Acta 110: 473–480
Langlet, O. (1929): Nagra egendomliga frosthärjingar pa tallskog jämte ett försök att klarläga deres orsak. Svenska Skogsvardsföreningens Tidskrift. 27: 423–459
Lanner, R. M. (1982): Adaptations of whitebark pine for seed dispersal by Clark's Nutcracker. Can. J. For. Res. 12: 391–402

Lanner, R. M. (1990): Biology, taxonomy, evolution and geography of stone pines of the world. Proc. Symp. whitebark pine ecosystems: Ecology and management of a high-mountain resource, USDA For. Serv., Intermtn. Res. Stat., Gen. Tech. Rep. INT-270: 14–24

Larcher, W. (1957): Frosttrocknis an der Waldgrenze und in der alpinen Zwergstraucheide. Veröffentlichungen Museum Ferdinandeum 37: 49–81

Larcher, W. (1963): Zur spätwinterlichen Erschwerung der Wasserbilanz von Holzpflanzen an der Waldgrenze. Berichte des Naturwissenschatlich-Medizinischen Vereins in Innsbruck 53: 125–137

Larcher, W. (1972): Der Wasserhaushalt immergrüner Pflanzen im Winter. Berichte der Deutschen Botanischen Gesellschaft 85: 315–327

Larcher, W. (1975): Pflanzenökologische Beobachtungen in der Páramostufe der venezolanischen Anden. Österreichische Akademie der Wissenschaften, Sitzungen der mathematisch-naturwissenschaftlichen Klasse, 23. Okt. 1975: 194–213

Larcher, W. (1980a): Ökologie der Pflanzen auf physiologischer Grundlage (3rd edition). Stuttgart

Larcher, W. (1980b): Klimastreß im Gebirge – Adaptionstraining und Selektionsfilter für Pflanzen. Rheinische Akademie der Wissenschaften, Vorträge 291: 49–88

Larcher, W. (1985): Winter stress in high mountains. In: Establishment and tending of sub-alpine forest: Research and management, Proc. 3rd IUFRO workshop P. 1.07-00, 1984 (Turner, H. and Tranquillini, W., eds.), Eidgenössische Anstalt für das forstliche Versuchswesen, Berichte 270: 11–19

Larsen, J. A. (1965): The vegetation of Ennedai Lake Area, NWT: Studies in arctic and sub-arctic bioclimatology. Ecol. Monogr. 35 (1): 37–59

Larsen, J. A. (1980): The boreal ecosystem. Academic Press, New York

Larsen, J. A. (1989): The northern forest border in Canada and Alaska – biotic communities and ecological relationships. Ecol. Stud. 70

Larsen, J. B. (1978): Untersuchungen über die Bedeutung der Kalium- und Stickstoffversorgung für die Austrocknungsresistenz der Douglasie (Pseudotsuga menziesii) in winter. Flora 167: 197–207

Larsen, J. B. (1993): Testing winter desiccation resistance for species and provenance selection at timberlines. In: Forest development in cold climates (Alden, J., Mastrantonio, J. L. and Ødum, S., eds.), NATO ASI Series 244: 59–64

Lassoie, J. P., Hinckley, T. M. and Grier, C. C. (1985): Coniferous forests of the Pacific Northwest. In: Physiological ecology of North American plant communities (Chabot, B. F. and Mooney, H. A., eds.): 127–161, Chapman & Hall, New York

Lauer, W. (1973): The altitudinal belts of vegetation in the central Mexican Highlands and their climatic conditions. Arct. Alp. Res. 5 (3, Pt. 2): 99–113

Lauer, W. (1975): Klimatische Grundzüge der Höhenstufung tropischer Gebirge. 40. Deutscher Geographentag Innsbruck, Tagungsbericht und wissenschaftliche Abhandlungen: 76–90

Lauer, W. (1978): Timberline studies in central Mexico. Arct. Alp. Res. 10: 383–396

Lauer, W. (1979a): Die hypsometrische Asymmetrie der Páramo-Höhenstufe in den nördlichen Anden. Innsbrucker Geographische Studien 5: 115–130

Lauer, W. (1979b): Im Vorland der Apolobamba-Kordillere. Physisch-geogaphische Beobachtungen auf einer kurzen Studienreise nach Bolivien. In: Estudios Americanistas II (Hartmann, E. and Oberem, U., eds.), Coll. Inst. Anthropol. 21: 9–15

Lauer, W. (1986): Die Vegetationszonierung der Neotropis und ihr Wandel seit der Eiszeit. Berichte der Deutschen Botanischen Gesellschaft 99: 211–235

Lauer, W. (1999): Forschungen zur Landschaftsökologie der Erde. 1. Geoökologische Studien in Südamerika. Akademie der Wissenschaften und der Literatur, Mainz, Jahrbuch für 1998: 178–183

Lauer, W. (2000): Forschungen zur Landschaftsökologie der Erde. 1. Geoökologische Studien in Südamerika. Kommission für Erdwissenschaftliche Forschung, Bericht Lauer. Akademie der Wissenschaften und der Literatur, Mainz, Jahrbuch für 1999: 178–183
Lauer, W. and Erlenbach, W. (1987): Die tropischen Anden. Geökologische Raumgliederung und ihre Bedeutung für den Menschen. Geographische Runddschau. 39 (2): 86–95
Lauer, W. and Klaus, D. (1975a): Geoecological investigations on the timberline of Pico de Orizaba, Mexico. Arct. Alp. Res. 7 (4): 315–330
Lauer, W. and Klink, H.-J. (1978): Die räumliche Anordnung der Vegetation im östlichen Hochland von Zentralmexiko. Pflanzengeographie: 472–506, Darmstadt
Lauer, W. and Rafiqpoor, M. D. (2000): Páramo de Papallacta. Erdkunde 54: 20–33
Lauer, W. and Rafiqpoor (2002): Die Klimate der Erde. Eine Klassifikation auf der Grundlage der ökophysiologischen Merkmale der realen Vegetation. Erdwissenschaftliche Forschung XI, Wiesbaden
Lauer, W., Rafigpoor, M. D. and Theisen, I. (2001): Physiogeographie, Vegetation und Syntaxonomie der Flora des Páramo de Papallacta. Erdwissenschaftliche Forschung 29, Stuttgart
Lavender, D. P. and Silim, S. N. (1987): The role of plant growth regulators in dormancy in forest trees. Plant Growth Regul. 6: 171–191
Lavoie, C. and Payette, S. (1992): Black spruce growth forms as a record of a changing environment at treeline, Quebec, Canada. Arct. Alp. Res. 24 (1): 40–49
Lawrence, D. B. (1939): Some features of the vegetation of the Columbian River Gorge with special reference to asymmetry of forest trees. Ecol. Monogr. 9: 217–257
Ledgard, N. J. (1980): Specific and intraspecific growth differences in tree trials up to timberline in the Craigieburn Range, New Zealand. In: Mountain environments and subalpine tree growth (Benecke, U. and Davis, M. R., eds.), New Zeal. For. Serv. Tech. Paper 70: 137–153
LeDrew, E. F. (1975): The energy balance of a mid-latitude alpine site during the growing season 1973. Arct. Alp. Res. 7: 301–314
Légère, A. and Payette, S. (1981): Ecology of black spruce (Picea mariana) clonal population in the hemiarctic zone, northern Québec: Population dynamics and spatial development. Arct. Alp. Res. 13: 261–276
Lehmkuhl, F. (1997): Der Naturraum Hoch- und Zentralasien. Geographische Rundschau 49 (5): 300–306
Lehto, J. (1957): Tutkimuksia männyn luontaissesta undistumisesta eteläsuom, en kangasmailla (Ref.: Studies on natural reproduction of scotspine on the upland soil in southern Finland). Acta Botanica Fennica 66 (2)
Lehtonen, J. (1987): Recovery and development of birch forests damaged by Epirrita autumnata in Utsjoki area, North Finland. Rep. Kevo Subarct. Res. Stat. 20: 35–39
Lehtonen, J. and Heikkinen, R. K. (1995): On the recovery of mountain birch after Epirrita damage in Finnish Lapland with a particular emphasis on reindeer grazing. Ecoscience 2: 349–356
Lehtonen, J. and Yli-Rekola, M. (1979): Field and ground layer vegetation in birch forests after Oporinia damage. Rep. Kevo Subarct. Res. Stat. 15: 27–32
Leibundgut, H. (1938): Wald- und Wirtschaftstudien im Lötschental. Beiheft zur Zeitschrift des Schweizerischen Forstvereins 18
Leidlmair, A. (1983): Länderkunde Österreich. München
Lempa, K., Neuvonen, S. and Tømmervik, H. (2005): Effects of reindeer grazing on pastures – A necessary basis for sustainable reindeer herding. In: Plant ecology, herbivory, and human impact in Nordic mountain birch forests (Wielgolaski, F. E., Karlsson, P. S., Neuvonen, S. and Thannheiser, D., eds.), Ecol. Stud. 180: 157–164

Lenz, O. (1967): Action de la neige et du gel sur les arbres de montagne, en particulier sur les formes et l'anatomie de la tige. Schweizerische Anstalt für das forstliche Versuchswesen, Berichte 43 (3): 289–316

Leonardis, R. F. J. (1976): Libro del árbol. 2. Esencias Forestales Indigenas de la Argentina de Aplicacción Industrial (Celulosa Argentina, S. A., eds.). Buenos Aires

Lescop-Sinclair, K. and Payette, S. (1995): Recent advance of the arctic treeline along the eastern coast of Hudson Bay. J. Ecol. 83: 929–936

Leser, H., Haas, H.-D., Mosimann, T. and Paesler, R. (1991): Wörterbuch der Allgemeinen Geographie, Band 2 N-Z (5th edition). Braunschweig, München

Leuschner, C. (1996): Timberline and alpine vegetation on the tropical and warm-temperate oceanic islands of the world: Elevation, structure and floristics. Vegetatio 123: 193–206

Leuschner, C. (1998): Vegetation an der Waldgrenze auf tropischen und subtropischen Inseln. Geographische Rundschau 50 (12): 690–697

Leuschner, C. and Schulte, M. (1991): Microclimatical investigations in the tropical alpine scrub of Maui, Hawaii: Evidence for a drought-induced alpine timberline. Pac. Sci. 45 (2): 152–168

Li, B. S. (1993): The alpine timberline of Tibet. In: Forest development in cold climates (Alden, J., Mastrantonio, J. L. and Odum, S., eds.): 511–527, New York

Liehl, E. (1958): Der Feldberg im Schwarzwald. Subalpine Insel im Mittelgebirge. Berichte zur Deutschen Landeskunde 22 (1): 1–28

Lindemann, R. (1972): Studien zur Geographie der Waldgrenzen im westlichen Norwegen. Diss. Philosophische Fakultät, Westfälische Wilhelms-Universität, Münster

Lindgren, D., Lindgren, K., Odin, H. and Eriksson, B. (1989): Temperaturen under vegetationsperioden i norra Sveriges skogsland 1933–1988. In: Skoglig Genetik och Växtfysiol., Sveriges Landbruksuniversitet, Arbetsrapport 30: 1–18

Lindsay, J. H. (1971): Annual cycle of leafwater potential in Picea engelmannii and Abies lasiocarpa at timberline in Wyoming. Arct. Res. 3 (2): 131–138

Linhart, Y. B. and Tomback, D. F. (1985): Seed dispersal by nutcrackers causes multi-trunk growth form in pines. Oecologia 67: 107–110

Litaor, M. I. (1987): The influence of eolian dust on alpine soils in the Front Range, Colorado. Soil Sci. Soc. Am. J. 51: 142–147

Little, E. L. (1979): Checklist of United States trees (native and naturalized). USDA For. Serv., Misc. Publ. 1146: 200 maps

Little, R. L. and Peterson, D. L. (1994): Regeneration of subalpine fir (Abies lasiocarpa) following fire: Effects of climate and other factors. Can. J. For. 24: 943–944

Lloyd, A. (2005): Ecological histories from Alaskan tree lines provide insight into future change. Ecology 86: 1687–1695

Lloyd, A. H. and Fastie, C. L. (2002): Spatial and temporal variability in the growth and climate response of treeline trees in Alaska. Clim. Change 52: 481–509

Lloyd, A. H. and Graumlich, L. J. (1997): Holocene dynamics of treeline forests in the Sierra Nevada. Ecology 78 (4): 1199–1210

Löffler, E. (1979): Ursprung und Verbreitung der Páramo-Grasländer in Ostneuguinea. Erdkunde 33: 226–236

Löffler, J. (2005): Reindeer grazing impact on arctic-alpine landscape in western Greenland and central and northern Norway. Geoöko 24: 1–18

Löffler, J. (2007): Reindeer grazing changes diversity patterns in arctic-alpine landscape of northern Norway. Die Erde 138 (3): 215–233

Löve, D. (1970): Subarctic and subalpine: Where and what? Arct. Alp. Res. 2: 63–73

Löve, D., McLellan, C. and Gamov, I. (1970): Coumarin and coumarin derivates in various growth types of Engelmann spruce. Svensk Botanisk Tidskrift 64: 284–296
Lowery, R. F. (1972): Ecology of subalpine tree clumps in the North Cascades Mountains of Washington. Unpubl. Ph.D. thesis, University of Washington, Seattle, WA
Luckman, B. H. and Kavanagh, T. A. (1998): Documenting the effects of recent climate change at treeline in the Canadian Rockies. In: The impact of climate variability on forests (Beniston, M. and Innes, J. L., eds.): 121–144, Heidelberg
Lüdi, W. (1961): Botanische Streifzüge durch die Rocky Mountains Nordamerikas. Berichte des Geobotanischen Instituts Rübel in Zürich 32: 217–236
Lundmark, T. and Hällgren, J. E. (1987): Effects of frost on shaded and exposed spruce and pine seedlings planted in the field. Can. J. For. Res. 17: 1197–1201
Luoto, M. and Seppälä, M. (2000): Summit peats ("peat cakes") on the fells of Finnish Lapland: Continental fragments of blanket mires ? Holocene 19 (2): 229–241
Luxmoore, R. J., O'Neill, E. G., Ellis, J. M. and Rogers, H. H. (1986): Nutrient uptake and growth response of Virginian pine to elevated atmospheric carbon dioxide. J. Environ. Qual. 15: 244–251
Lyr, H. and Hoffmann, G. (1967): Growth rates and growth periodicity of tree roots. In: International Review of Forestry Research 2 (Romberger, J. A. and Mikola, P., eds.): 181–236, New York/London
Lyr, H., Polster, H. and Fiedler, H.-J. (1967): Gehölzphysiologie. Jena
MacDonald, G. M. and Gajewski, K. (1992): The northern treeline of Canada. Geographical snapshots of North America, Commemoration the 27th Congress of the International Geographical Union and Assembly (MacDonald, G. and Janelle, D. G., eds.): 34–37
MacDonald, G. M., Case, R. A. and Szeicz, J. M. (1998a): A 538-year record of climate and treeline dynamics from the lower Lena river region of northern Siberia, Russia. Arct. Alp. Res. 30 (4): 334–339
MacDonald, G. M., Szeicz, J. M., Claircoates, J. and Dale, K. A. (1998b): Response of the central Canadian treeline to recent climate change. Ann. Ass. Amer. Geogr. 88 (2): 183–208
MacDonald, G. M., Velichko, A. A., Kremenetski, C. V., Borisova, O. K., Goleva, A. A., Andreev, A. A., Cwynar, L. C., Riding, R. T., Forman, S. L., Edwards, W. D., Aravena, R., Hammarlund, D., Szeicz, J. M. and Gattaulin, V. N. (2000): Holocene treeline history and climate change across northern Eurasia. Quatern. Res. 53: 302–311
MacMahon, J. A. and Andersen, D. C. (1982): Subalpine forests: A world perspective with emphasis on western North America. Prog. Phys. Geogr. 6: 368–425
Magee, T. K. and Antos, J. A. (1992): Tree invasion into a mountain-top meadow in the Oregon Coast Range, USA. J. Veg. Sci. 3: 485–494
Magnússon, S. H. and Magnússon, B. (2001): Effect of enhancement of willow (Salix spp.) on establishment of birch (Betula pubescens) on eroded soils in Iceland. In: Nordic mountain birch ecosystems (Wielgolaski, F. E., ed.), Man and Biosphere Series 27, New York/London: 317–329
Malanson, G. P., Butler, D. R., Fagre, D. B., Walsh, S. J., Tomback, D. F., Daniels, L. D., Resler, L. M., Smith, W. K., Weiss, D. J., Peterson, D. L., Bunn, A. G., Hiemstra, C. A., Liptzin, D., Bourgeron, P. S., Shen, Z. and Millar, C. I. (2007): Alpine treeline of western North America: Linking organism-to-landscape dynamics. Phys. Geogr. 28: 378–396
Malyshev, L. (1993): Levels of upper forest boundary in northern Asia. Vegetatio 109: 175–186

Maraun, M., Illig, J., Sandman, D., Krashevska, V., Norton, R. A. and Scheu, S. (2008): Soil fauna. In: Gradients in a tropical mountain ecosystem of Ecuador (Becke, E., Bendix, J., Kottke, I., Makeshin, F. and Mowandl, R., eds.), Ecol. Stud. 198: 181–192

Marchand, P. J. (1980): Causes of coniferous timberline in the northern Appalachian mountains. In: Mountain environments and subalpine tree growth (Benecke, U. and Davis, M. R., eds.), New Zeal. For. Serv. Tech. Paper 70: 231–246

Marchand, P. J. (1987): Life in the cold. An Introduction to winter ecology. University Press of new England, Hannover/London

Marchand, P. J. and Chabot, B. F. (1978): Winter water relations of the tree-line plant species, New Hampshire. Arct. Alp. Res. 10 (1): 105–116

Marek, R. (1910): Waldgrenzstudien in den österreichischen Alpen. Petermanns Geographische Mitteilungen, Ergänzungsheft 168

Mark, A. and Sanderson, F. R. (1962): The altitudinal gradient in forest composition, structure and regeneration in the Hollford valley, Fiordland. Proc. N. Z. Ecol. Soc. 9: 17–26

Mark, A., Dickinson, K. J. M. and Hofstede, R. G. M. (2000): Alpine vegetation, plant distribution, life forms, and environment in a perhumid New Zealand region. Oceanic and tropical high mountains affinities. Arct. Antarct. Alp. Res. 32 (3): 240–254

Marr, J. W. (1948): The forest-tundra ecotone in the east coast of Hudson Bay. Ecol. Monogr. 18: 117–144

Marr, J. W. (1967): Ecosystems of the east slope of the Front Range in Colorado. Univ. Colorado Studies, Series in Biology 8

Marr, J. W. (1977): The development and movement of tree islands near the upper limit of tree growth in the southern Rocky Mountains. Ecology 58: 1159–1164

Marr, J. W. and Marr, R. E. (1973): Environment and phenology in the forest-tundra ecotone, Front Range, Colorado (Summary). Arct. Alp. Res. 5: 65–66

Martin, E. V. and Clements, F. E. (1953): Studies of the effect of artificial wind on growth in Helianthus annuus. Plant Physiol. 10: 613–636

Martinelli, M. (1975): Water-yield improvement from alpine areas: The status of our knowledge. USDA For. Serv. Res. Paper RM-138, Rocky Mountain Forest and Range Experiment Station, Fort Collins, CO

Martini, J. A. and Palencia, J. A. (1975): Soils derived from volcanic ash in Central America: 1. Andepts. Soil Sci. 120 (4): 278–287

Maruta, E. (1994): Seedling establishment of Polygonum cuspidatum and Polygonum weyrichii var. alpinum at high altitudes of Mt. Fuji. Ecol. Res. 9: 205–213

Maruta, E. (1998): Winter water relations to timberline larch (Larix leptolepis Gord.) on Mt. Fuji. Trees 11: 119–126

Maruta, E., Nakano, T., Ishida, A., Iida, H. and Masuzawa, T. (1996): Water relations of Pinus pumila in the snow melting season at the alpine region of Mt. Tateyama. Proc. NIPR Symp. Polar Biol. 9: 335–342

Marx, D. H. (1973): Mycorrhizae and feeder root diseases. In: Ectomycorrhizae: Their ecology and physiology (Marks, G. C. and Kozlowski, T. T., eds.): 351–382, Academic Press, New York and London

Masuzawa, T. (1985): Ecological studies on the timberline of Mt. Fuji. I. Structure of plant community and soil development on the timberline. Bot. Mag. Tokyo 98: 15–28

Masuzawa, T. and Kimura, M. (1987): Role of stem and leaf in nitrogen and phosphorus economy of three deciduous trees at the timberline. In: Human impacts and management of mountain forests (Fujimori, T. and Kimura, M., eds.): 155–165, Forestry and Forest Products Research Institute, Ibaraki, Japan

Mattes, H. (1978): Der Tannenhäher im Engadin Studien zu seiner Ökologie und Funktion im Arvenwald. Münstersche Geographische Arbeiten 2

Mattes, H. (1982): Die Lebensgemeinschaft von Tannenhäher und Arve. Eidgenössische Anstalt für das forstliche Versuchswesen, Berichte 241

Mattes, H. (1985): The role of animals in cembran pine forest regeneration. In: Establishment and tending of subalpine forest: Research and management. Proceedings of the 3rd IUFRO workshop P 1.07-00 1984 (Turner, H. and Tranquillini, W., eds.), Eidgenössische Anstalt für das forstliche Versuchswesen, Berichte 270: 195–205

Mattheck, C. (1991): Trees. The mechanical design. Berlin/Heidelberg/New York/London/ Paris/Tokyo/Hong Kong/Barcelona/Budapest

Mattson, D. J. and Reinhart, D. P. (1994): Bear use of whitebark pine seeds in North America. In: Proc. int. workshop on subalpine stone pines and their environment: The status of our knowledge, St. Moritz, Switzerland, September 5–11, 1992 (Schmidt, W. C. and Holtmeier, F.-K., compilers): 212–220, USDA For. Serv., Intermtn. Res. Stat., Gen. Tech. Rep. INT-GTR 309

Mattson, D. J. and Reinhart, D. P. (1997): Excavation of red squirrel middens by grizzly bears in the whitebark pine zone. J. Appl. Ecol. 34: 926–940

Mattson, D. J., Blanchard, B. M. and Knight, R. R. (1991): Food habits of Yellowstone grizzly bears, 1977–1987. Can. J. Zool. 69: 1619–1629

Mattson, D. J., Kendall, K. C. and Reinhart, D. P. (2001): Whitebark pine, grizzly bears, and red squirrels. In: Whitebark pine communities. Ecology and restoration (Tomback, D. F., Arno, S. F. And Keane, R. E., eds.): 121–136, Washington, DC/Covelo, CA/London

Mattson, J. (1995): Human impact on the timberline in the far north of Europe. Prace Geograficzne 98: 41–55

Maurer, R. (1915): Einige Ergebnisse unserer höchsten Niederschlagssammler im Firngebiet. Meteorologische Zeitschrift

Mayer, H. (1965): Zur Waldgeschichte des Steinernen Meeres. Jahrbuch des Vereins zum Schutze der Alpenpflanzen und -Tiere 30: 100 120

Mayer, H. (1966): Waldgeschichte des Berchtesgadener Landes, Salzburger Kalkalpen. Forstwissenschaftliche Forschung, Beiheft zum Forstwissenschaftlichen Centralblatt 22: 1–42

Mayer, H. (1970): Waldgrenzstudien in den Berchtesgadener Kalkalpen. Mitteilungen der ostalpin-dinarischen Gesellschaft für Vegetationskunde 11: 109–120

Mayer, H. (1974): Wälder der Ostalpen. Stuttgart

Mayer, H. (1976): Gebirgswaldbau – Schutzwaldpflege. Stuttgart

Mayer, H. (1979): Ökosystem Lawinenschutzwald. Proc. IUFRO Seminar Mountains Forests Avalanches, Davos 1978: 281–299

Mayr, S., Gruber, A., Schwienbacher, F. and Dämon, B. (2003): Winter-embolism in a 'krummholz'-shrub (Pinus mugo) growing at the alpine timberline. Centralblatt für das gesamte Forstwesen 120 (1): 29–38

Mazepa, V. (2005): Stand density in the last millennium at the upper tree-line ecotone in the polar Ural Mountains. Can. J. For. Res. 35: 2082–2091

McCaughey, W. W. (1994): The regeneration process of whitebark pine. In: Proc. int. workshop on subalpine stone pines and their environment: The status of our knowledge, St. Moritz, Switzerland, September 5–11, 1992 (Schmid, W. C. and Holtmeier, F.-K., compilers): 179–187, USDA For. Serv., Intermtn. Res. Stat., Gen. Tech. Rep. INT-GTR 309

McCaughey, W. W., Schmidt, W. C. and Shearer, R. (1986): Seed-dispersal characteristics of conifers in the Inland Mountains West. Proc. conifer tree seed in the inland mountains west symposium, Bozeman, MT, USDA For. Serv. INT-203: 50–62

McDonald, G. I., Richardson, B. A., Zambino, P. J., Klopfenstein, N. B. and Kim, M.-S. (2006): Pedicularis and Castilleja are natural hosts of Cronartium ribicola in North America: A first report. For. Pathol. 36: 73–82

McQueen, D. R. (1976): The ecology of Nothofagus and associated vegetation in South America. Tuatara 22 (1): 38–68
McQueen, D. R. (1977): The ecology of Nothofagus and associated vegetation in South America. Tuatara 22 (3): 233–244
McVean, D. N. (1968): A year of weather records at 3480 m on Mt. Wilhelm, New Guinea. Weather 23: 377–381
McVean, D. N. (1973): Mountain climates of the southeastern Pacific. In: Altitudinal zonation in Malesia (Flenley, J. R., ed.): 47–57, Hull/Aberdeen
Meentemeyer, V. and Box, E. O. (1987): In: Landscape heterogeneity and disturbance (Turner, M. G., ed.), Ecol. Stud. 64: 15–34
Meinzer, F. C. and Goldstein, G. (1986): Adaptions of water and thermal balance in Andean giant rosette plants. In: On the economy of plant form and function (Givnish, T. J., ed.): 381–411, New York
Mellmann-Brown, S. (2002): The regeneration of whitebark pine in the timberline ecotone on Beartooth Plateau, Montana and Wyoming. Diss. Fachbereich Geowissenschaften, Mathematisch-Naturwissenschaftliche Fakultät der Westfälischen Wilhelms-Universität, Münster
Mellmann-Brown, S. (2005): Regeneration of whitebark pine in the timberline ecotone of the Beartooth Plateau, U.S.A.; Spatial distribution and responsible agents. In: Mountain ecosystems. Studies in treeline ecology (Broll, G. and Keplin, B., eds.): 97–115, Berlin/Heidelberg/New York
Meshinev, T., Apostolova, I. and Koleva, E. (2000): Influences of warming on timberline rising: A case study on Pinus peuce Griseb. in Bulgaria. Phytocoenologia 30 (3–4): 431–438
Messerli, B. and Hofer, T. (1992): Die Umweltkrise im Himalaya. Fiktion und Fakten. Geograpische Rundschau 44 (8): 435–445
Messerli, B. and Ives, J. D. (1984): Gonaga Shan (7556 m) and Yulongxue Shan (5596 m). Geographical observations in the Hengduan Mountains of Southwestern China. Erdwissenschaftliche Forschung XVIII: 55–77
Messerli, P. and Winiger, M. (1994): Climate, environmental change, and resources of the African mountains from the Mediterranean to the equator. Mt. Res. Develop. 12 (4): 315–336
Meurk, C. D. (1978): Alpine phytomass and primary productivity in Central Otago. New Zeal. J. Ecol. 1: 27–50
Mezhennyi, A. A. (1964): Biology of the nutcracker Nucifraga caryocatactes macrorhynchos on South Yakutia. Zool. Žurnal 43: 1679–1687
Michaelis, P. (1934a): Ökologische Studien an der alpinen Baumgrenze. II. Die Schichtung der Windgeschwindigkeit, Lufttemperatur und Evaporation über einer Schneefläche. Beiheft zum Botanischen Zentralblatt 52 B: 310–332
Michaelis, P. (1934b): Ökologische Studien an der alpinen Baumgrenze. III. Über die winterlichen Temperaturen der pflanzlichen Organe, insbesondere der Fichte. Beiheft zum Botanischen Zentralblatt 52 (B): 333–377
Michaelis, P. (1934c): Ökologische Studien an der alpinen Baumgrenze. IV. Zur Kenntnis des winterlichen Wasserhaushaltes. Jahrbuch für wissenschaftliche Botanik 80: 169–247
Michaelis, P. (1934d): Ökologische Studien an der alpinen Baumgrenze. V. Osmotischer Wert und Wassergehalt während des Winters in den verschiedenen Höhenlagen. Jahrbuch für wissenschaftliche Botanik 80: 337–362
Michiels, H. G. (1993): Die Stellung einiger Baum- und Straucharten in der Struktur und Dynamik der Vegetation im Bereich der hochmontanen und subalpinen Waldstufe der Bayrischen Kalkalpen. Forstliche Forschungsberichte München 135
Miehe, G. (1982): Vegetationsgeographische Untersuchungen im Dhaulagiri und Annapurna-Himalaya. Dissertationes Botanicae 66 (1/2), Vaduz

Miehe, G. (1984): Vegetationsgrenzen im extremen und multizonalen Hochgebirge. Erdkunde 38: 268–277

Miehe, G. (1989): Vegetation patterns of Mt. Everest as influenced by monsoon and föhn. Vegetatio 79: 21–32

Miehe, G. (1991): Der Himalaya, eine multizonale Gebirgsregion. In: Ökologie der Erde Bd. 4, Gemäßigte und arktische Zonen außerhalb Euro-Nordasiens (Walter, H. and Breckle, S., eds.): 181–230, Stuttgart

Miehe, G. and Miehe, S. (1994): Zur Waldgrenze in tropischen Gebirgen. Phytocoenologia 24: 53–110

Miehe, G. and Miehe, S. (1996): Die obere Waldgrenze in tropischen Gebirgen. Geographische Rundschau 48 (11): 670–676

Miehe, G. and Miehe, S. (2000): Comparative high mountain research on the treeline ecotone under human impact. Erdkunde 54: 34–50

Miehe, G., Miehe, S., Jiang, H. and Tsewang, O. (1998): Forschungsdefizite und –perspektiven zur Frage der potentiellen natürlichen Bewaldung in Tibet. Petermanns Geographische Mitteilungen 142 (3–4): 115–164

Miehe, S., Cramer, T., Jacobsen, J.-P. and Winiger, M. (1996): Humidity conditions in the western Karakoram as indicated by climatic data and corresponding distribution pattern of the montane and alpine vegetation. Erdkunde 50: 190–204

Miehe, S., Miehe, G., Jiang, H., Tsewang, O., Tsersen, T. and Yanli, T. (2000): Sacred forests of Southern-Central Xizang and their importance for the resaturation of forest resources (Contributions to ecology, phytogeography and environmental history of High Asia 2). Marburger Geographische Schriften 135: 228–249

Miehe, G., Miehe, S., Koch, K. and Will, M. (2003): Sacred forests in Tibet. Using geographical information systems for forest rehabilitation. Mt. Res. Develop. 23 (4): 324–328

Mikola, P. (1952): The effect of recent climatic variations on forest-growth in Finland. Fennia 75: 69–76

Mikola, P. (1971): Reflexion of climatic fluctuation in the forestry practises of Northern Finland. Rep. Kevo. Subarct. Res. Stat. 8: 116–121

Millar, C. I., Westfall, R. D., Delany, D. L., King, J. C. and Graumlich, L. J. (2004): Response of subalpine conifers in the Sierra Nevada, California, U.S.A., to 20th-century warming and decadal variability. Arct. Antarct. Alp. Res. 36 (2): 181–200

Miller, E. A. and Halpern, C. B. (1998): Effects of environment and grazing disturbance on tree establishment in meadows of the central Cascade Range, Oregon, USA. J. Veg. Sci. 9: 265–282

Minnich, R. A. (1984): Snow drifting and timberline dynamics on Mount San Gorgono, California, U.S.A. Arct. Alp. Res. 16 (4): 395–412

Mitchell, W. W. (1968): On the ecology of Sitka alder in the subalpine zone of south-central Alaska. In: Biology of alder. Proc. symp., Pacific Northwest Forest and Range Experiment Station, Portland, OR (Trappe, J. M. et al., eds.), Annual Meeting: Northwest Scientific Ass. 14: 154–167

Mitton, J. B. (1985): So grows the tree. Nat. Hist. 1: 58–104

Miyawaki, A. (1979): Vegetation und Vegetationskarten auf den Japanischen Inseln. In: Vegetation und Landschaft Japans (Miyawaki, A. and Okuda, S., eds.): 49–70, Yokohama

Moen, J., Aune, K., Edenius, L. and Angerbjörn, A. (2004): Potential effects of climate change on treeline position in the Swedish mountains. Ecol. Soc. 9 (1): 16 [online] URL: http://www.ecologyandsociety.org/vol9/iss1/art16

Moir, W. H., Rochelle, S. G. and Schoettle, A. W. (1999): Microscale pattern of tree establishment near upper treeline, Snowy Range, Wyoming, U. S. A. Arct. Alp. Res. 31 (4): 379–388

Moiseev, P. A. and Shiyatov, S. G. (2003): Vegetation dynamics at the treeline ecotone in the Ural highlands, Russia. Ecol. Stud. 167: 423–435

Moiseev. P. A., Van der Meer, M., Rigling, A. and Shevchenko, I. G. (2004): Effect of climate changes on the formation of Siberian spruce regenerations in subglotsy tree stands of the southern Urals. Russ. J. Ecol. 35 (3): 135–143

Monteith, J. L. (1975): Vegetation and the atmosphere 2, Case Studies. Academic Press, San Diego, CA

Mook, R. and Vorren, K.-D. (1996): The temperature climate at the altitudinal vegetation limits in Skibotn, northern Norway. Paläoklimaforschung 20: 61–74

Moore, R. M. and Williams, J. D. (1976): A study of subalpine wood-grassland boundary. Aust. J. Ecol. 1: 145–153

Mork, E. (1968a): Ökologiske undersökelser i fjellskogen be Kijölen forsöksomrade. Meddelser fra Norsk Institutt for Skogforskning 25: 463–614

Mork, E. (1968b): Ökologiske undersökelser i Nord-Norge. In: Skogbruk i Nord-Norge (Sveli, A., ed.), Nord-Norges Skogmannsforbund, Mosjön: 24–30

Moser, G., Röderstein, M., Soethe, N., Hertel, D. and Leuschner, C. (2008): Altitudinal changes in stand structure and biomass allocation of tropical mountain forests in relation to microclimate and soil chemistry. In: Gradients in a tropical mountain ecosystem of Ecuador (Becke, E., Bendix, J., Kottke, I., Makeshin, F. and Mowandl, R., eds.), Ecol. Stud. 198: 229–242

Moser, K. A. and MacDonald, G. M. (1990): Holocene vegetation change at treeline north of Yellowknife, Northwest Territory, Canada. Quatern. Res. 34: 227–239

Moser, M. (1964): Die Mykorrhizafrage bei der Anzucht von Forstpflanzen für das Hochgebirge. In: Forstsamengewinnung und Pflanzenanzucht für das Hochgebirge: 225–231, BLV Verlagegesellschaft München, Basel, Wien

Moser, M. (1967): Die ektotrophe Ernährungsweise an der Waldgrenze. Mitteilungen der forstlichen Bundes-Versuchsanstalt Wien 75: 357–380

Moser, M. (1982): Mycoflora of the transitional zone from subalpine forests to alpine tundra. Arct. Alp. Mycol. 1: 371–384

Müller, G. (1994): Bodenökologische Untersuchungen an der oberen Waldgrenze in Finnisch-Lappland (unter besonderer Berücksichtigung des Standortklimas). Diploma thesis Institut für Geographie, Westfälische Wilhelms-Universität, Münster

Müller, M. (1983): Bodenbildung auf Silikatunterlage in der alpinen Stufe des Oberengadins. Diss. ETH 7352, Zürich

Müller-Dombois, D. and Ellenberg, H. (1974): Aims and methods of vegetation ecology. Wiley, New York

Müller-Stoll, W. R. (1954): Beiträge zur Ökologie der Waldgrenze am Feldberg im Schwarzwald. Angewandte Pflanzensoziologie 2 (Festschrift E. Aichinger): 824–847

Murray, M. B., Smith, R. I., Leith, I. D., Fowler, D., Lee, H. S. J., Friend, A. D. and Jarvis, P. G. (1994): Effects of elevated CO_2, nutrition and climate warming on bud phenology in Sitka spruce (Picea sitchensis) and their impact on the risk of frost damage. Tree Physiol. 14: 691–706

Müterthies, A. (2002): Struktur und Dynamik der overen Grenze des Lärchen-Arvenwaldes im Bereich aufgelassener Alpweiden im Oberengadin. Arbeiten aus dem Institut für Landschaftsökologie, Westfälische Wilhelms-Universität, Münster 11

Müterthies, A. (2003): The potential timberline: Determination with dendrochronological methods. In: TRACE tree rings in archeology, climatology and ecology 1, Proc. Dendrosymposium 2002, April 11–13, Bonn (Schleser, G., Winiger, M., Bräuning, A.,

Gärtner, H., Helle, G., Jensma, E., Neuwirth, B. and Treydte, K., eds.), Schriftenreihe des Forschungszentrums Jülich, Reihe Umwelt, Band 33: 94–100

Nägeli, W. (1969): Waldgrenze und Kampfzone in den Alpen. Hespa-Mitteilungen 19 (1)

Nägeli, W. (1971): Der Wind als Standortsfaktor bei Aufforstungen in der subalpinen Stufe. Schweizerische Anstalt für das forstliche Versuchswesen, Berichte 47 (2): 33–147

Nakashinden, I. (1994): Japanese stone pine cone production estimated from cone scars, Mount Kisokomagatake, Central Japanese Alps. In: Proc. int. workshop on subalpine stone pines and their environment: The status of our knowledge (Schmidt, W. C. and Holtmeier, F.-K., compilers): 188–192, USDA For. Serv., Intermtn. Res. Stat., Gen. Tech. Rep. INT-GTR-309

Nara, K. and Hogetsu, T. (2004): Ectomycorrhizal fungi on established shrubs facilitate subsequent seedling establishment of successional plant species. Ecology 85 (6): 1700–1707

Neilson, R. P. and Chaney, J. (1997): Potential changes in the vegetation distribution in the United States. USDA Forest Service Global Change Research Program Highlights: 1991–95, NE-237: 66–79

Neuner, G. (2007): Frost resistance at the upper timberline. In: Trees at their upper limit. Treelife limitation at the alpine timberline (Wieser, G. and Tausz, M., eds.), Plant Ecophysiol. 5: 171–180

Neuvonen, S. and Wielgolaski, F. E. (2005): Herbivory in northern birch forests. In: Plant ecology, herbivory, and human impact in Nordic mountain birch forests (Wielgolaski, F. E., ed.): 183–189, Berlin/Heildeberg

Neuvonen, S., Virtanen, T., Nikula, A. and Vaarama, M. (1996): Climate change and the risk of forest insect outbreaks. In: The Finnish research programme on climate change, Final Report (Roos, J., ed.), Publications Acad. Finl. 4: 275–280

Neuvonen, S., Virtanen, T. and Nikula, A. (1998): Timberline, insect pests, and minimum winter temperatures: Applications of GIS. In: Research and management of the northern timberline region. Proc. Gustaf Sirén Symposium in Wilderness Centre Inari, September 4–5, 1997, The Finnish For. Res. Inst., Res. Papers 677: 17–21

Neuvonen, S., Niemelä, P. and Virtanen, T. (1999): Climate change and insect outbreaks in boreal forest: The role of winter temperatures. Ecol. Bull. 47: 63–67

Neuvonen, S., Ruohomäki, K., Bylund, H. and Kaitaniemi, P. (2001): Insect herbivory and herbivory effects on mountain birch dynamics. In: Nordic mountain birch ecosystems (Wielgolaski, F. E., ed.), Man and Biosphere Series 27: 207–222

Neuwinger, I. (1967): Zum Nährstoffgehalt in Vegetationseinheiten der subalpinen Entwaldungszone. Mitteilungen der forstlichen Bundes-Versuchsanstalt Wien 75: 269–304

Neuwinger, I. (1970): Böden der subalpinen und alpinen Stufe in den Tiroler Alpen. Mitteilungen der ostalpin-dinarischen Gesellschaft für Vegetationskunde 11: 135–150

Neuwinger, I. (1972): Standortsuntersuchungen am Sonnberg im Sellrainer Obertal, Tirol. Mitteilungen der forstlichen Bundes-Versuchsanstalt Wien 96: 177–207

Neuwinger, I. (1978): Vergleichende Untersuchungen von Almböden bei Badgastein. In: Ökologische Analysen von Almflächen im Gasteiner Tal, Veröffentlichungen des Österreichischen MaB-Hochgebirgsprogramms Hohe Tauern 2: 241–249

Neuwinger, I. (1980): Erwärmung, Wasserrückhalt und Erosionsbereitschaft subalpiner Böden. Mitteilungen der forstlichen Bundes-Versuchsanstalt Wien 129: 113–144

Neuwinger, I. (1984): Einfache Methoden zur ökologischen Bodenkennzeichnung. In: Recent investigations in the zone of aeration. Proc. Int. Symp. Munich, West-Germany, October 1984: 791–799

Neuwinger, I. (1986): Water relation and erodibilty in natural and disturbed mountain forest soils. In: The Temperate Forest Ecosystem, ITE Symposium No. 20, 5.-11. July 1986: 124–130

Neuwinger, I. and Czell, A. (1959): Standortsuntersuchungen in subalpinen Aufforstungsgebieten. I. Teil: Böden in den Tiroler Zentralalpen. Forstwissenschaftliches Centralblatt 78 (11/12): 327–372

Newesely, C., Tasser, E., Spadinger, P. and Cernusca, A. (2000): Effects of land-use changes on snow gliding processes in alpine ecosystems. Basic Appl. Ecol. 1: 61–67

Nichols, H. (1974): Arctic North America paleoecology: The recent history of vegetation and climate deduced from pollen analysis. In: Arctic and alpine environment (Ives, J. D. and Barry, R. G., eds.): 637–667, London

Nichols, H. (1975a): Palynological and paleoclimatic study of the late quaternary displacement of the boreal forest-tundra ecotone in Keewatin and Mackenzie N.W.T., Canada. Inst. Arctic and Alpine Res. Univ. Colo., Occ. Paper 15

Nichols, H. (1975b): The time perspective in northern ecology: Palynology and the history of Canadian boreal forests. Proc. Circumpolar Conf. on Northern Ecology, September 15–18, Ottawa: 157–164

Nichols, H. (1976): Historical aspects of the northern Canadian treeline. Arctic 29 (1): 18–47

Nicolussi, K., Bortenschlager, S. and Körner, C. (1995): Increase in tree-ring width in subalpine Pinus cembra from the central Alps that may be CO_2-related. Trees 9: 181–189

Nierhaus-Wunderwald, D. (1996): Pilzkrankheiten in Hochlagen, Biologie und Befallsmerkmale. Wald und Holz 10: 18–24

Nievergelt, B. (1966): Der Alpensteinbock (Capra ibex) in seinem Lebensraum. Mammalia depicta, Hamburg, Berlin

Nikolovski, T. (1970): Waldgesellschaften und Waldbäume an der oberen Grenze der Verbreitung in verschiedenen Gebirgssystemen der SR-Mazedonien. Mitteilungen der ostalpin-dinarischen Gesellschaft für Vegetationskunde 11: 151–160

Nikula, A. (1992): Animals as forest pests in Finnish-Lapland. Finnish Forest Institute Research Papers 451: 22–29

Nilsson, M. C. (1994): Separation of allelopathy and resource competition by the boreal dwarf shrub Empetrum hermaphroditum Hagerup. Oecologia 98: 1–7

Nilsson, M. C., Höberg, P., Zackrisson, O. and Wang, F. Y. (1993): Allelopathic effects by Empetrum hermaphroditum on development and nitrogen uptake by roots and mycorrhiza of Pinus silvestris. Can. J. Bot. 71: 620–628

Nimlos, T. J. and McConnell, R. C. (1965): Alpine soils in Montana. Soil Sci. 99 (5): 310–321

Ninot, J. M., Batllori, E., Carillo, E., Carreras, J., Ferré, A. and Gutiérrez, E. (2008): Timberline structure and limited tree recruitment in the Catalan Pyrenees. Plant Ecol. Divers. 1 (1): 47–57

Noble, D. L. and Alexander, R. R. (1977): Environmental factors affecting natural regeneration of Engelmann spruce in the central Rocky Mountains. For. Sci. 23 (4): 420–429

Noble, I. R. (1993): A model of response of ecotones to climate change. Ecol. Appl. 3: 396–403

Nola, P. (1994): A dendroecological study of larch at timberline in the Central Italian Alps. Dendrochronologia 12: 77–91

Nordell, K. O. and Karlssson, P. S. (1995): Resorption of nitrogen and dry matter prior to leaf abscisson – variation among individuals, sites and years in the mountain birch.. Funct. Ecol. 9: 326–333

Norton, D. A. and Schönenberger, W. (1984): The growth forms and ecology of Nothofagus solandri at the alpine timberline, Craigieburn Range, New Zealand. Arct. Alp. Res. 16 (3): 361–370

Nuorteva, P. (1963): The influence of Oporinia autumnata (Bkh.) (Lep. Geometridae) on the timberline in subarctic conditions. Ann. Ent. Fenn. 29: 270–277

Nüsser, M. (1998): Nanga Parbat (NW-Himalaya): Naturräumliche Ressourcenausstattung und humanökologisches Gefügemuster der Landnutzung. Bonner Geographische Abhandlungen 97

Oberdorfer, E. (1960): Pflanzensoziologische Studien in Chile. Ein Vergleich mit Europa. Weinheim

Oberdorfer, E. (1982): Die hochmontanen Wälder und subalpinen Gebüsche. In: Der Feldberg im Schwarzwald. Subalpine Insel im Mittelgebirge (Landesanstalt für Umweltschutz Baden-Württemberg, Institut für Ökologie, ed.), Die Natur- und Landschaftsschutzgebiete Baden-Württembergs 12: 317–364

Odland, A. (1996): Differences in the vertical distribution pattern of Betula pubescens in Norway and its ecological significance. Paläoklimaforschung 20: 42–59

Ohsawa, M. (1984): Differentiation of vegetation zones and species strategies in the subalpine region of Mt. Fuji. Vegetatio 57: 15–22

Ohsawa, M. (1990): An interpretation of latitudinal patterns of forest limits in south and east Asian mountains. J. Ecol. 78: 326–339

Oke, T. R. (1995): Boundary layer climates (2nd edition). Metheuen, London

Okitsu, S. (1997): Distribution, growth and regeneration of Larix gmelinii (Rupr.) Rupr. along the timberline ecotone of Mt. Dal'Nyaya Ploskaya, Central Kamchatka. Proc. NIPR Symp. Polar Biol. 10: 179–187

Okitsu, S. and Ito, K. (1984): Vegetation dynamics of the Siberian dwarf pine (Pinus pumila Regel) in the Teisetsu mountain range, Hokkaido, Japan. Vegetatio 58: 105–113

Okitsu, S. and Ito, K. (1989): Conditions for tree development of the Pinus pumila zone of Hokkaido, northern Japan. Vegetatio 84: 127–132

Oksanen, L. Moen, J. and Helle, T. (1995): Timberline patterns in northernmost Fennoscandia. Acta Bot. Fennica 153: 93–105

Oliver, E. G. H. (1988): Studies on the Ericoideae (Ericaceae). VI. The generic relationship between Erica und Philippia in southern Africa. Bothalia 18 (1): 1–10

Oosting, H. J. (1956): The study of plant communities. W.H. Freeman, San Francisco, CA

Oosting, H. J. and Reed, J. F. (1952): Virgin spruce-fir of the Medicine Bow Mountains, Wyoming. Ecol. Monogr. 22: 69–82

Örlander, G. (1993): Shading reduces both visible and invisible frost damage to Norway spruce seedlings in the field. Forestry 66: 27–36

Österreichische Akademie der Wissenschaften (1980): Untersuchungen an alpinen Böden in den Hohen Tauern 1974–1978, Stoffdynamik und Wasserhaushalt (Franz, H., managing ed.). Veröffentlichungen des Österreichischen MaB-Hochgebirgsprogramms Hohe Tauern 3

Österreichische Akademie der Wissenschaften (1985): Beiträge zu den Wechselbeziehungen zwischen den Hochgebirgsökosystemen und dem Menschen (Franz, H., managing ed.). Veröffentlichungen des Österreichischen MaB-Programms 9

Oswald, H. (1956): Beobachtungen über die Samenverbreitung bei der Zirbe (Pinus cembra). Allgemeine Forstzeitung 67: 200–202

Oswald, H. (1963): Verteilung und Zuwachs der Zirbe (Pinus cembra L.) in der subalpinen Stufe an einem zentralalpinen Standort. Mitteilungen der forstlichen Bundes-Versuchsanstalt Mariabrunn 60: 437–499

Oswald, H. (1969): Conditions forestières et potentialité de l'épicéa en haute Ardeche. Ann. Sci. For. 26: 183–224

Ott, E. (1978): Über die Abhängigkeit des Radialzuwachses und der Oberhöhen bei Fichte und Lärche von der Meereshöhe und Exposition im Lötschental. Schweizerische Zeitschrift für Forstwesen 129 (3): 169–193

Ott, E., Frehner, M., Frey, H.-U. and Lüscher, P. (1997): Gebirgsnadelwälder. Ein praxisorientierter Leitfaden für eine standortsgerechte Waldbehandlung. Bern/Stuttgart/Wien

Otto, H.-J. (1994): Waldökologie. Stuttgart
Ovaska, J. (1993): Compensatory photosynthesis as a response to foliage damage in birch (Betula L.). Rep. Dept. Biol. Univ. Turku 35 : 1–22
Ozenda, P. (1988): Die Vegetation der Alpen im europäischen Gebirgsraum. Stuttgart/New York
Ozenda, P. and Borel, J.-L. (1991): Les conséquences possibles des changements climatiques dans l'arc alpine. Rapport FUTURALP 1
Packee, E. C. (1995): Larch in North America. The Finnish Forest Res. Inst. Res. Papers 567: 161–172
Paijmans, K. and Löffler, E. (1972): High altitude forests and grasslands on Mt. Edward, New Guinea. J. Trop. Geogr. 34: 58–64
Pallmann, H. and Haffter, P. (1933): Pflanzensoziologische und bodenkundliche Untersuchungen im Oberengadin unter besonderer Berücksichtigung der Zwergstrauchgesellschaften der Ordnung Rhodoreta-Vaccinietalia. Berichte der Schweizerischen Botanischen Gesellschaft 42: 357–466
Palm, T. (1959): Följdverkningar av fjällbjörksmätarens härjning i Abiskodalen 1954–1956. En koleopterologisk undersökning somrarna 1958 och 1959. Entomol. Tidsskr. 80: 120–136
Parker, J. (1959): Seasonal variations in sugars of conifers with some observations on cold resistance. For. Sci. 5: 56–63
Parolini, J. D. (1995): Zur Geschichte der Waldnutzung im Gebiet des Schweizerischen Nationalparks. Doctoral thesis ETH No. 1187
Paton, D. M. (1988): Genesis of an inverted treeline associated with a frost hollow in south-eastern Australia. Aust. J. Bot. 36: 655–663
Patten, D. T. (1963a): Light and temperature influence on Engelmann spruce seed germination and subalpine forest advance. Ecology 44: 817–818
Patten, D. T. (1963b): Vegetational pattern in relation to environments in the Madison Range, Montana. Ecol. Monogr. 33: 375–406
Pattern, R. S. and Knight, D. H. (1994): Snow avalanches and vegetation pattern in Cascade Canyon, Grand Teton National Park, Wyoming, U.S.A. Arct. Alp. Res. 26: 35–41
Paulsen, J., Weber, U. M. and Körner, C. (2000): Tree growth near treeline: Abrupt or gradual reduction with altitude. Arct. Antarct. Alp. Res. 32 (1): 14–20
Payette, S. (1976): Succession écologique des forêts d'épignette blanche et fluctuations climatiques, Poste-de-la-Baleine, Nouveau Québec. Can. J. Bot. 54: 1394–1402
Payette, S. (1983): The forest tundra and present tree-line of the northern Québec-Labrador-Peninsula. Collect. Nordicana 47: 3–23
Payette, S. and Delwaide, A. (1994): Growth of black spruce at its northern range limit in Arctic Québec, Canada. Arct. Alp. Res. 26 (2): 174–179
Payette, S. and Gagnon, R. (1979): Tree-line dynamics in Ungava Peninsula, northern Québec. Holarct. Ecology 2: 239–248
Payette, S. and Lavoie, C. (1992): Black spruce growth forms as a record of a changing winter environment at treeline, Quebec, Canada. Arct. Alp. Res. 24 (1): 40–49
Payette, S. and Morenau, C. (1993): Holocene relict woodland at the eastern Canadian treeline. Quatern. Res. 39: 84–89
Payette, S., Deshaye, J. M. and Gilbert, H. (1982): Tree seed populations at the treeline in Rivière aux feuilles area, Northern Québec, Canada. Arct. Alp. Res. 14 (3): 215–221
Payette, S., Filion, L., Delwaide, A. and Begin, C. (1989): Reconstruction of the treeline vegetation response to long-term climate change. Nature 341: 429–432
Payette, S., Delwaide, A., Mourneau, C. and Lavoie, C. (1994): Stem analysis of a long-lived black spruce close at timberline. Arct. Alp. Res. 26 (1): 56–59

Pears, N. V. (1968): The natural altitudinal limit of forest in the Scottish Grampians. Oikos 19: 71–80
Pedersen, V. A., Linnell, J. D. C., Andersen, R., Andrén, H., Lindén, M. and Segerström, P. (1999): Winter lynx predation on semi-domestic reindeer Rangifer tarandus in northern Sweden. Wildlife Biol. 5 (4): 203–211
Peet, R. K. (1981): Forest vegetation of the Colorado Front Range. Vegetatio 45: 3–75
Pepin, N. (2000): Twentieth-century change in the climate record for the Front Range, Colorado, U.S.A. Arct. Antarct. Alp. Res. 32 (2): 135–146
Perkins, T. D., Adams, G. T. and Klein, R. M. (1991): Desiccation or freezing? Mechanisms of winter injury to red spruce foliage. Am. J. Bot. 78 (9): 1207–1217
Perry, Jr., J. P., Graham, A. and Richardson, D. M. (1998): The history of pines in Mexico and Central America. In: Ecology and Biogeography of Pinus (Richardson, D. M., ed.): 137–149, Cambridge University Press, Cambridge
Peschl, A. (1983): CO_2-Gaswechsel und Wasserhaushalt junger Pflanzen von Pseudotsuga menziesii und Abies grandis bei Frosttrocknis. Flora 173: 33–47
Peters, R. L. (1990): Effects of global warming on forests. For. Ecol. Manag. 35: 13–33
Peterson, D. L. (1998): Climate, limiting factors and environmental change in high-altitude forests of Western North America. In: The impact of climate variability in forests (Beniston, M. and Innes, J. L., eds.): 191–208, Heidelberg
Peterson, D. W. and Peterson, D. L. (1995): Climatic influences on radial growth of subalpine larch in the North Cascades. In: Proc. int. symp. ecology and management of larix forest: A look ahead, Whitefish, MT, October 5–9, 1992, USDA For. Serv., Intermtn. Res. Stn., Gen. Tech. Rep. GRT-INT 319: 268–271
Petrak, F. (1955): Über Phacidium infestans KARST., einen gefährlichen Parasiten der Zirbelkiefer und einige andere in seiner Gesellschaft wachsende Pilze. Sydowia Annales Mycologici 2 (9): 518–526
Pezzi, G., Bitelli, G., Ferrari, C., Girelli, V. A., Gusella, L., Masi, S. and Mognol, A. (2007): Pattern temporale del limite altitudinale die boshi di faggio nell'Appennino settentrionale. Un'analisi die dati fotogrammetrici. Forest@ 4 (1): 79–87. [online] URL: http://www.sisef.it/
Pfister, F., Walther, H., Erni, V. and Candrian, M. (1987): Walderhaltung und Schutzaufgaben im Berggebiet. Eidgenössische Anstalt für das forstliche Versuchswesen, Berichte 294
Pfister, R. D., Kovalchik, B. L., Arno, S. F. and Presby, R. C. (1977): Forest habitat types of Montana. USDA For. Serv., Gen. Tech. Rep. INT-34
Phares, R. E. and Crosby, J. S. (1962): Basal sprouting of fire-injured shortleaf pine trees. J. For. 60: 204–205
Pielke, R. A. and Vidale, P. L. (1995): The boreal forest and the polar front. J. Geophys. Res. 100 (25): 755–758
Piper, F. I., Cavieres, L. A., Reyes-Díaz, M. and Corcuera, L. J. (2006): Carbon sink limitation and frost tolerance control performance of the tree Kageneckia angustifolia D. Don (Rosaceae) at the treeline in central Chile. Plant Ecol. 185: 29–39
Pisek, A. (1952): Zur Kenntnis der Frosthärte alpiner Pflanzen. Die Naturwissenschaften 39: 73–78
Pisek, A. and Cartellieri, E. (1933): Zur Kenntnis des Wasserhaushaltes der Pflanzen. III. Alpine Zwergsträucher. Jahrbuch für wissenschaftliche Botanik 79: 131–190
Pisek, A. and Cartellieri, E. (1939): Zur Kenntnis des Wasserhaushaltes der Pflanzen. IV. Bäume und Sträucher. Jahrbuch für wissenschaftliche Botanik 88: 22–68

Pisek, A. and Schiessl, R. (1946): Die Temperaturbeinflußbarkeit der Frosthärte von Nadelhölzern und Zwergsträuchern an der alpinen Waldgrenze. Berichte des Naturwissenschaftlich-medizinischen Vereins in Innsbruck 47: 33–52

Pitschmann, H., Reisigl, H., Schiechtl, H. M. and Stern, R. (1970): Karte der aktuellen Vegetation von Tirol 1/100 000. I. Teil, Blatt 6, Innsbruck-Stubaier Alpen. Documents pour la carte de la végétation des Alpes VIII: 7–34

Pitterle, A. (1985): Performance of high altitude afforestations in the Alps since 1865: Analyses of damages and chances of success. In: Establishment and tending of subalpine forest: Research and management, Proc. 3rd IUFRO workshop P. 1.07-00, 1984 (Turner, H. and Tranquillini, W., eds.), Eidgenössische Anstalt für das forstliche Versuchswesen, Berichte 270: 241–250

Piussi, P. and Schneider, A. (1985): Die obere Wald- und Baumgrenze im Pfitschtal (Südtirol). Centralblatt für das gesamte Forstwesen 102 (4): 234–246

Platter, W. (1976): Wasserhaushalt, cutikuläres Transpirationsvermögen und Dicke der Cutinschichten einiger Nadelholzarten in verschiedenen Höhenlagen und nach experimenteller Verkürzung der Vegetationsperiode. Diss. Univ. Innsbruck

Plesnik, P. (1959): Die obere Waldgrenze in den Westkarpaten. Wissenschaftliche Zeitschrift der Martin-Luther-Universität, Halle-Wittenberg, Mathematik Naturwissenschaften 8 (2)

Plesnik, P. (1971): Horna Hranica leso vo vysokych av Belanskych Tatrách (German summary: Die obere Waldgrenze in der Hohen und Belauer Tatra: 203–238) Bratislava

Plesnik, P. (1972): Obere Waldgrenze in den Gebirgen Europas von den Pyrenäen bis zum Kaukasus. Erdwissenschaftliche Forschung IV: 73–92

Plesnik, P. (1973): La limite supérieure de la forêt dans les Hautes Tatras. Arct. Alp. Res. 5 (3, Pt. 2): 37–44

Plesnik, P. (1980): Vertikalna diferenciacia vegetaticie a horna hranica lesa na Klimandzare. Summary: The vertical differentiation of vegetation and the upper timberline on Kilimanjaro. Acta Facultatis Rerum Naturalium Universitas Comenianae Bratislava, Geographica 21: 29–50

Plesnik, P. (1991): System of timberlines on the earth. Geografický Casopis 43 (2): 134–149

Pohl, R. (1993): Die natürliche Wiederbewaldung aufgelassener Alpweiden im Oberengadin. Diploma thesis, Institut für Geographie, Westfälische Wilhelms-Universität, Münster

Pollmann, W. and Hildebrand, R. (2005): Structure and the composition of species in timberline ecotones of the southern Andes. In: Mountain ecosystems. Studies in treeline ecology (Broll, G. and Keplin, B., eds.): 117–151, Berlin/Heidelberg

Pontecorvo, G. and Bokhari, M. (1975): Hedge-like habit of Juniperus excelsa at high altitude on the Zagros Mountains in Iran. Proc. Roy. Soc. Lond. B 188: 507–508

Pott, R. (1992): Die Pflanzengesellschaften Deutschlands. Stuttgart

Pott, R. (1993): Farbatlas Waldlandschaften. Stuttgart

Pourtahmasi, K., Parsapajouh, D., Bräuning, A., Esper, J. and Schweingruber, F. H. (2007): Climatic analysis of pointer years in tree-ring chronologies from northern Iran and neighboring high mountain areas. Geoöko 28: 27–42

Pregitzer, K. S., Reed, D. D., Bornhorst, T. J., Foster, D. R., Mroz, G. D., McLachlans, J. S., Laks, P. E., Stokke, D. D., Martin, P. E. and Brown, S. E. (2000): A buried spruce forest provides evidence at the stand and landscape scale for the effects of environment on vegetation at the Pleistocene/Holocene boundary. J. Ecol. 88: 45–53

Price, L. W. (1978): Mountains of the Pacific Northwest, U.S.A.: A study in contrasts. Arct. Alp. Res. 10 (2): 465–478

Price, L. W. (1981): Mountains and Man. A study of process and environment. Berkeley, CA/Los Angeles, CA/London

Prutzer, E. (1967): Die Niederschlagsverhältnisse an der Waldgrenze. Mitteilungen der forstlichen Bundes-Versuchsanstalt Wien 75: 173–205

Puncer, I. and Zupančič, M. (1970): Vergleich der Vegetationsgrenzen beziehungsweise der Vegetationsprofile in verschiedenen Gebirgssystemen auf Karbonat- und Silikatunterlagen in Slovenien. Mitteilungen der ostalpin-dinarischen Gesellschaft für Vegetationskunde 11: 187–196

Qin, D., Majewski, P. A., Wake, C. P., Kang, S., Ren, J., Hou, S., Yao, T., Yang, Q., Jin, Z. and Mi, D. (2000): Evidence for recent climate change from ice cores in the central Himalaya. Ann. Glaciol. 31: 153–158

Rada, F. (1983): Mecanismos de resistencia a temperaturas congelantes en Espeletia spicata y Polylepis sericea. Unpubl. Masters Diss. Universidad de los Andes, Mérida, Venezuela (cited in Goldstein et al., 1994)

Rada, F., Goldstein, G., Azócar, A. and Meinzer, F. (1985): Daily and seasonal osmotic changes in a tropical treeline species. J. Exp. Bot. 36: 989–1000

Rada, F., Azóocar, A., Briceno, B., Gonzalez, J. and García-Núñnez, C. (1996): Carbon and water balance in Polylepis sericea, a tropical treeline species. Trees 10: 218–222

Rada, F., García-Núñez, C., Boero, C., Gallardo, M., Hilal, M., Gonzáles, J., Prado, F., Liberman-Cruz, M. and Azócar (2001): Low-temperature resistance in Polylepis tarapacana, a tree growing at the highest altitudes in the world. Plant Cell Environ. 24: 377–381

Rao, S. J., Iason, G. R., Hulbert, I. A. R., Elston, D. A. and Racey, P. A. (2003): The effect of sapling density, heather height and season on browsing by mountain hares on birch. J. Appl. Ecol. 40: 626–638

Raspopov, O. M. (2002): Global warming and regional tree-ring response at Kola Peninsula and Finnish Lapland. In: Abstracts prepared for Arctic Council, Northern Timberline Forests Workshop, Rovaniemi, Finland, May 23–25, 2002 (Arctic Centre, Univ. of Lapland, Finnish Forests and Park Service, Finnish Forest Research Institute, University of Joensuu, organizers): 14

Rathjens, C. (1972): Fragen der horizontalen und vertikalen Landschaftsgliederung im Hochgebirgssystem des Hindukush. Erdwissenschaftliche Forschung IV: 205–220

Rathjens, C. (1982): Geographie der Hochgebirge 1. Der Naturraum. Teubner Studienbücher, Stuttgart

Rauh, W. (1952): Vegetationstudien im Hohen Atlas und dessen Vorland. Sitzungsberichte der Heidelberger Akademie der Wissenschaften, Mathematisch-Naturwissenschaftliche Klasse 1

Rauh, W. (1988): Tropische Hochgebirgspflanzen. Wuchs- und Lebensformen. Heidelberg

Read, D. J. (1998): The mycorrhizal status of Pinus. In: Ecology and biogeography of Pinus (Richardson, D. M., ed.): 324–340, Cambridge University Press, Cambridge

Reardon, B. A., Pederson, G. T., Caruso, C. J. and Fagre, D. B. (2008): Spatial reconstructions and comparisons of historic snow avalanche frequency and extent using tree rings in Glacier National Park, Montana, U.S.A. Arct. Antarct. Alp. Res. 40 (1): 148–160

Rees, G., Brown, I., Mikkola, K., Virtanen, T. and Werkman, B. (2002): How can the dynamics of the tundra-taiga boundary be remotely monitored? Ambio Sp. Rep. 12: 56–62

Regel, C. (1938): Über die Depression der Waldgrenze in Griechenland. Repetitorium specierum novarum regni vegetabilis, Beiheft 100: 28–39

Regel, C. (1949): Klimaänderung und Vegetationsentwicklung im eurasiatischen Norden. Österreichische Botanische Zeitschrift 96: 369–398

Regel, C. (1950): Dynamik von Klima und Pflanzendecke in Nordeuropa. Berichte des Geobotanischen Forschungs-Institutes Rübel in Zürich: 11–23

Rehder, H., Beck, E. and Kokawo, J. O. (1988): The afroalpine plant communities of Mt. Kenya (Kenya). Phytocoenologia 16 (4): 433–463

Reichel, E. (1931): Die Niederschlagsverteilung in den Alpen. Jahrbuch des Deutsch-Österreichischen Alpenvereins 21–88

Reichstein, M., Bednorz, F., Broll, G. and Kätterer, T. (2000): Temperature dependence of carbon mineralisation: Concludings from a long-term incubation of subalpine soil samples. Soil Biol. Biochem. 32: 947–958

Reijmers, F. F. (1959): Birds of the cedar pine forest of South-Central Sibiria and their role in the life of cedar pine. Trudy Biol. Inst. Sibirsk. Otdel. Akad. Nauk. SSSR 5: 121–166

Reimoser, F. (1999): Schalenwild und Wintersport. Laufener Seminarberichte 6: 39–45

Reimoser, F. (2003): Steering the impacts of ungulates on temperate forests. J. Nat. Conserv. 10: 243–252

Reimoser, F., Mayer, H., Holzinger, A. and Zandl, J. (1987): Einfluss von Sommer- und Wintertourismus auf Waldschäden durch Schalenwild im Angertal (Bad Gastein). Centralblatt für das gesamte Forstwesen 104: 95–118

Renvall, A. (1912): Die periodischen Erscheinungen der Reproduktion der Kiefer an der polaren Waldgrenze. Acta Forestalia Fennica 1

Repo, T., Nilsson, J.-E., Rikala, R., Ryyppö, A. and Sutinen, M.-L. (2001): Cold hardiness of Scots pine (Pinus sylvestris L.) In: Conifer cold hardiness. Tree Physiology (Bigras, F. J. and Colombo, S. J., eds.): 1: 463–493, Kluwer, Dordrecht/Boston, MA/London

Resler, L. M. (2006): Geomorphic controls of spatial pattern and process at alpine treeline. Profess. Geogr. 58: 124–138

Resler, L. M. and Tomback, D. F. (2008): Blister rust prevalence in krummholz whitebark pine: Implications for treeline dynamics, Northern Rocky Mountains, Montana, U.S.A. Arct. Antarct. Alp. Res. 40 (1): 161–170

Resler, L. M., Butler, D. R. and Malanson, G. P. (2005): Topographic shelter and conifers establishment and mortality in an alpine environment, Glacier National Park, Montana. Phys. Geogr. 16: 112–125

Retzer, J. L. (1956): Alpine soils of the Rocky Mountains. J. Soil Sci. 7: 22–32

Retzer, J. L. (1974): Alpine soils. In: Arctic and alpine environment (Ives, J. D. and Barry, R. G., eds.): 771–802, London

Richardson, D. M. and Rundel, P. W. (1998): Ecology and biogeography of Pinus: An introduction. In: Ecology and biogeography of Pinus (Richardson, D. M., ed.): 3–46, Cambridge University Press, Cambridge

Richter, M. (2001): Vegetationszonen der Erde. Gotha/Stuttgart

Richter, M., Diertl, K.-H., Peters, T. and Bussmann, R. W. (2008): Vegetation structures and ecological features of the upper timberline ecotone. In: Gradients in a tropical mountain ecosystem of Ecuador (Beck, E., Bendix, J., Kottke, I., Makeshin, F. and Mosandl, R., eds.), Ecol. Stud. 198: 123–136

Rien, M., Spengler, T. and Richter, M. (1998): Klimaökologische Aspekte in Gebirgen der südwestlichen USA unter besonderer Berücksichtigung der White Mountains. Mitteilungen der Fränkischen Geographischen Gesellschaft 45: 301–333

Rikli, E. (1909): Die Arve in der Schweiz. Neue Denkschrif der Schweizerischen Naturforschenden Gesellschaft 44

Rikli, M. (1946): Das Pflanzenkleid der Mittelmeerländer. Bern

Robbins, R. B. (1970): Vegetation of the Goroka-Mount Hagen Area. In: Lands of the Goroka-Mt. Hagen Area, Territory of Papua and New Guinea (Haantjens, H. A., Compiler), Land Res. Series 27: 104–117

Robertson, A. (1986): Estimating windflow in hilly terrain from tamarack (Larix laricina (Du Roi) K. Koch) deformation. Int. J. Biomet. 30 (4): 33–349

Rochefort, R. M. and Peterson, D. L. (1991): Tree establishment in subalpine meadows of Mount Rainier National Park. Northwest Environ. J. 7 (2): 354–355

Rochefort, R. M. and Peterson, D. L. (1996): Temporal and spatial distribution of trees in subalpine meadows of Mount Rainier National Park, Washington, U.S.A. Arct. Alp. Res. 28 (1): 52–59

Rochefort, R. M., Little, R. L., Woodward, A. and Peterson, D. L. (1994): Changes in subalpine tree distribution in western North America: A review of climatic and other causal factors. Holocene 4 (1): 89–100

Roder, K. (1895): Die polare Waldgrenze. Diss. Philosophische Fakultät der Universität Leipzig, Dresden

Rohmeder, E. (1941): Die Zirbelkiefer (Pinus cembra) als Hochgebirgsbaum. Jahrbuch zum Schutze der Alpenpflanzen und -Tiere 13: 27–39

Roll-Hansen, F. (1989): Phacidium infestans. A literature review. Eur. J. For. Path. 19: 237–250

Roos, J. (ed.) (1996): The Finnish research programme on climate change: Final report. Helsinki

Ross, R. (1955): Some aspects of the vegetation of Ruwenzori. Webbia 11: 451–457

Rössler, O. and Löffler, J. (2007): Uncertainties of treeline alterations due to climatic change during the past century in the central Norwegian Scandes. Geoöko 28: 104–114

Roush, W., Munroe, J. S. and Fagre, D. B. (2007): Development of a spatial analysis method using ground-based repeat photography to detect changes in the alpine treeline ecotone, Glacier National Park, Montana, U.S.A. Arct. Antarct. Alp. Res. 39 (2): 297–308

Rübel, E. (1912): Pflanzengeographische Monographie des Berninagebietes. Botanisches Jahrbuch 47

Rubner, K. (1953): Die pflanzengeographischen Grundlagen des Waldbaus. Radebeul/Berlin

Rubner, K. and Reinhold, F. (1953): Das natürliche Waldbild Europas als Grundlage für einen europäischen Waldbau. Hamburg/Berlin

Rudberg, St. (1968): Wind erosion – preparation of maps showing the direction of eroding winds. Biul. Peryglacjalny 17: 181–193

Rundel, P. W. (1994): Tropical alpine climates. In: Tropical alpine environments. Plant form and function (Rundel, P. W., Smith, A. P. and Meinzer, F. C., eds.): 21–44, Cambridge University Press, Cambridge

Rundel, P. W., Meinzer, F. C. and Smith, A. P. (1994): Tropical alpine ecology: Progress and priorities. In: Tropical alpine environments. Plant form and function (Rundel, P. W., Smith, A. P. and Meinzer, F. C., eds.): 355–363, Cambridge University Press, Cambridge

Ruohomäki, K., Virtanen, T., Kaitaniemi, P. and Tammaru, T. (1997): Old mountain birches at high altitude are prone to outbreaks of Epirrita autumnata (Lepidoptera, Geometridae). Environ. Entomol. 26 (5): 1096–1104

Rupp, T. S., Chapin III, F. S. and Starfield, A. M. (2001): Modeling the influence of topographic barriers on treeline advance at the forest-tundra ecotone in northwestern Alaska. Climatic Change 48: 399–416

Ruthsatz, B. (1983): Der Einfluß des Menschen auf die Vegetation semiarider bis arider tropischer Hochgebirge am Beispiel der Hochanden. Berichte der Deutschen Botanischen Gesellschaft 96: 535–576

Ruwisch, K. (1983): Der Einfluß von Relief und Bodenbedeckung auf das bodennahe Windfeld im oberen Fourth of July Valley (Colorado Front Range). Diss. Mathematisch-Naturwissenschaftliche Fakultät, Westfälische Wilhelms-Universität, Münster

Rybnickova, E. and Rybnicek, K. (1993): Late quaternary forest line oscillations in the West Carpathians. Paläoklimaforschung 9: 187–194

Rychetnik, J. (1987): Projekt Stillberg, Schnee und Lawinen auf der Versuchsfläche. Eidgenössisches Institut für Schnee- und Lawinenforschung, Winterbericht 50: 184–194

Ryyppö, A., Repo, T. and Vapaavuori, E. (1998): Development of freezing tolerance in roots and shoots of Scots pine seedlings at non-freezing temperatures. Can. J. For. 28: 557V–565V

Saiko, H. and Masuzawa, T. (1987): Ecological studies on the timberline of Mt. Fuji, II. Primary production of Alnus maximowiczii dwarf forest. Bot. Mag. Tokyo: 349–363

Saito, T. (1971): Vegetation of the upper limit on half way circuit and its adjascent on Mt. Fuji. In: Report of the Scientific Survey of Mt. Fuji (Tsuya, H. et al., eds.): 639–656, Fujikuyu, Tokyo

Sakai, A. (1970): Mechanism of desiccation damage of conifers wintering in soil-frozen areas. Ecology 51: 657–664

Sakai, A. (1979): Freezing avoidance mechanism of primordial shoots of conifer buds. Plant Cell Physiol. 14: 1–9

Sakai, A. (1983): Comparative study on freezing resistance of conifers with special reference to cold adaptation and its evolutive aspects. Can. J. Bot. 61: 2323–2332

Sakai, A. and Larcher, W. (1987): Frost survival of plants. Springer, Berlin

Sakai, A. and Weiser, C. J. (1973): Freezing resistance of trees in North America with reference to tree regions. Ecology 54: 118–126

Sakai, A., Paton, D. M. and Wardle, P. (1979): Freezing resistance of temperate and sub-arctic conifers native to the southern hemisphere. Low Temp. Sci. Ser. B 37: 107–111

Salgado-Labouriau, M. L. (1984): Late-quaternary palynological studies in the Venezuelan Andes. Erdwissenschaftliche Forschung 18: 279–293

Salomons, J. B. (1986): Paleoecology of volcanic soils in the Columbian central cordillera (Parque Nacional Natural de los Nevados). Dissertationes Botanicae 95: 1–212

Salt, G. (1951): The Shira plateau of Kilimanjaro. Geogr. J. 117: 150–164

Sarmiento, G. (1986): Ecological features of climate in high tropical mountains. In: High altitude tropical biogeography (Vuilleumier, F. and Monasterio, M., eds.): 11–45, Oxford University Press, New York

Schäfer, E. (1938): Ornithologische Ergebnisse zweier Forschungsreisen nach Tibet. J. für Ornithology 86, Sonderheft

Scharfetter, R. (1938): Das Pflanzenleben der Ostalpen. Wien

Schickhoff, U. (1993): Das Kaghan-Tal im Westhimalaya (Pakistan). Bonner Geographische Abhandlungen 87

Schickhoff, U. (1995a): Himalayan forest-cover changes in historical perspective: A case study in the Kaghan valley, northern Pakistan. Mt. Res. Develop. 15 (1): 3–18

Schickhoff, U. (1995b): Verbreitung, Nutzung und Zerstörung der Höhenwälder im Karakorum und in angrenzenden Hochgebirgsräumen Nordpakistans. Petermanns Geographische Mitteilungen 1939 (2): 67–85

Schickhoff, U. (1996): Die Wälder der Nanga-Parbat-Region. Standortbedingungen, Nutzung, Degradation. Beiträge und Materialien zur Regionalen Geographie 8: 177–189

Schickhoff, U. (2005): The upper timberline in the Himalayas, Hindu Kush and Karakorum: A review of geographical and ecological aspects. In: Mountain ecosystems. Studies in treeline ecology (Broll, G. and Keplin, B., eds.): 275–354, Springer, Berlin/Heidelberg/New York

Schiechtl, H. M. (1967): Die Physiognomie der potentiellen natürlichen Waldgrenze und Folgerungen für die Praxis der Aufforstung in der subalpinen Stufe. Mitteilungen der forstlichen Bundes-Versuchsanstalt Wien 75: 5–55

Schinner, F. (1978): Die subalpine Waldgrenze und Bedeutung der Mykotrophie im Gasteiner Tal. In: Ökologische Analysen von Almflächen im Gasteiner Tal (Cernusca, A., ed.), Veröffentlichungen des Österreichischen MaB-Hochgebirgsprogramms Hohe Tauern: 311–314, Innsbruck

Schlagintweit, A. and Schlagintweit, H. (1854): Neue Untersuchungen über die physikalische Geographie und die Geologie der Alpen. Leipzig
Schlatter, A. J. (1935): Die Aufforstungen und Verbauungen des Oberengadins in den Jahren 1875–1934. Schweizerische Zeitschrift für Forstwesen 86: 309–328
Schmid, J. M., Mata, S. A. and Lynch, A. M. (1991): Red belt in lodgepole pine in the Front Range of Colorado. Research Note RM-503. USDA Forest Service, Rocky Mountain Forest and Range Experiment Station
Schmidt, E. (1936): Baumgrenzstudien am Feldberg im Schwarzwald. Tharandter Forstliches Jahrbuch 87: 143
Schmidt, G. (1969): Vegetationsgeographie auf ökologisch-soziologischer Grundlage. Leipzig
Schmidt, K. (1993): Winter ecology of nonmigratory Alpine red deer. Oecologia 95: 226–233
Schmidt, R. A. (1970): Locating snow fences in mountainous terrain. In: Snow removal and ice control research a model. Highway Res. Bord Special Report 115: 220–224
Schmidt, R. A. (1972): Sublimation of wind-transported snow a model. USDA For. Serv. Res. Paper RM-90, Rocky Mountain Forest and Range Experiment Station, Fort Collins, CO
Schmidt, W. C. (1995): Around the world with Larix: An introduction. In: Proc. int. symp. ecology and management of Larix forests. A look ahead. Whitefish, MT, October 5–9, 1992 (Schmidt, W. C. and McDonald, K. J., eds.): 6–18, USDA For. Serv., Intermtn. Res. Stn., Gen. Tech. Rep. GTR-INT-319
Schmidt, W. C. and McCaughey, W. W. (1990): Whitebark pine – a subalpine species needing silvicultural attention. In: Proc. Symp. whitebark pine ecosystems: Ecology and management of a high-mountain resource, USDA For. Serv., Intermtn. Res. Stat., Gen. Tech. Rep. INT-270: 373–374
Schmidt, W. C. and Shearer, R. C. (1995): Larix occidentalis: A pioneer of the North American West. In: Proc. Int. Symp. ecology and management of Larix forests: A look ahead. Whitefish, MT, October 5–9, 1995 (Schmidt, W. C. and McDonald, K. J., eds.): 33–37, USDA For. Serv., Intermtn. Res. Stat., Gen. Tech. Rep. GTR-INT-319
Schmidt-Vogt, D. (1990a): Die Höhenwälder im Jugal-Himal (Nepal): Waldtypen und Waldentwicklung unter dem Einfluß menschlicher Nutzung. 47. Deutscher Geographentag Saarbrücken, Tagungsbericht und wissenschaftliche Abhandlungen: 297–301
Schmidt-Vogt, D. (1990b): Fire in high altitude forests of the Nepal Himalaya. In: Fire in ecosystem dynamics, Proc. third int. symp. fire ecology, Freiburg, FRG, May 1989 (Goldammer, J. G. and Jenkins, M. J., eds.): 191–199
Schmithüsen, J. (1968): Allgemeine Vegetationsgeographie (3rd edition). Lehrbuch der Allgemeinen Geographie 4, Berlin
Schmitt, J. (1991): The vegetation of Aberdare National Park, Kenya. In: Hochgebirgsforschung 8 (Arbeitsgemeinschaft für Hochgebirgsforschung, München, ed.): 1–250
Schneider, T. R. (1962): Snow drifts and winter ice on roads. Tech. Transl. 1038, National Res. Council of Canada
Schöeneberger, P. J., Wysocki, D. A., Benham, E. C. and Broderson, W. D. (2002): Field book for describing and sampling soils. Version 2.0 Natural Resources Conservation Service, National Survey Center, Lincoln, NE
Schoettle, A, (1993): Comparison of shoot, needle, and physiological characteristics of erect and dwarf Bristlecone Pine (Pinus aristata) trees at treeline. Ann. Meeting Ecol. Soc. America 1993: Ecology and global sustainability, Madison, Wisconsin, U.S.A., July 31–August 4/93, Bull. Ecol. Soc. Amer. 74 (2nd supplement): 429
Schönenberger, W. (1975): Standortseinflüsse auf Versuchsaufforstungen an der alpinen Waldgrenze (Stillberg, Davos). Eidgenössische Anstalt für das forstliche Versuchswesen, Mitteilungen 51 (4): 359–428

Schönenberger, W. (1978): Ökologie der natürlichen Verjüngung von Fichte und Bergföhre in Lawinenzügen der nördlichen Voralpen. Eidgenössische Anstalt für das forstliche Versuchswesen, Mitteilungen 54 (3): 215–361
Schönenberger, W. (1981): Die Wuchsformen der Bäume an der alpinen Waldgrenze. Schweizerische Zeitschrift für Forstwesen 13 (3): 149–162
Schönenberger, W. (1984): Above-ground biomass of mountain beech (Nothofagus solandri (Hook, F.) Oerst. Var. Cliffortioides (Hook, F.)). Poole in different stand types near timberline in New Zealand. Forestry 57 (1): 59–73
Schönenberger, W. (1985): Performance of high altitude afforestation under various site conditions. In: Establishment and tending of subalpine forest: Research and management, Proc. 3rd IUFRO workshop 0.107-00, 1984 (Turner, H. and Tranquillini, W., eds.), Eidgenössische Anstalt für das Forstliche Versuchswesen, Berichte 270: 233–240
Schönenberger, W. (1986): Rottenaufforstung im Gebirge. Schweizerische Zeitschrift für Forstwesen 137: 501–509
Schönenberger, W. (2000): Silvicutural problems in subalpine forests in the Alps. In: Forests and sustainable mountain development: A state of knowledge report for 2000 (Proce, M. F. and Butt, N., eds.), IUFRO res. Series 5: 197–203, Wallinford/New York
Schönenberger, W. and Frey, W. (1988): Die wichtigsten Schäden in der Versuchsaufforstung Stillberg. In: Untersuchungen zur Ökologie und Technik der Hochlagenaufforstung, Schweizerische Zeitschrift für Forstwesen 139 (9): 805–811
Schönenberger, W. and Wasem, U. (1999): Der Beginn der Wiederbewaldung von Sturmwurfflächen im Gebirge. Ein Zwischenbericht. Forstliche Forschungsberichte München 176: 102–110
Schönenberger, W., Frey, W. and Leuenberger, F. (1990b): Ökologie und Technik der Aufforstung im Gebirge. Anregungen für die Praxis. Eidgenössische Anstalt für das forstliche Versuchswesen, Berichte 325
Schönenberger, W., Schweingruber, F. and Mattheck, C. (1994): Baumformen an der oberen Waldgrenze. Bündnerwald 1: 35–42
Schreiber, H. (1998): Waldgrenznahe Buchenwälder und Grasländer des Falakron und Pangäon in Nordostgriechenland. Arbeiten aus dem Institut für Landschaftsökololgie, Westfälische Wilhelms-Universität, Münster 4
Schröter, C. (1898): Über die Vielgestaltigkeit der Fichte (Picea abies LINK). Vierteljahresschrift der Naturforschenden Gesellschaft Zürich: 125–240
Schröter, C. (1926): Das Pflanzenleben der Alpen. Zürich
Schulte, M. (1994): Bodenökologische Untersuchungen an der oberen Waldgrenze in Finnisch-Lappland unter besonderer Berücksichtigung der Stickstoffnachlieferung. Diploma thesis, Institut für Landschaftsökologie., Westfälische Wilhelms-Univiversität, Münster
Schulze, E. D. (1980): Photosynthetic production and water relations in different plant forms as related to altitudinal gradients. In: Mountain environments and subalpine tree growth (Benecke, U. and Davis, M. R., eds.), New Zeal. For. Serv. Techn. Paper 70: 213–229
Schulze, E. D., Mooney, H. A. and Dunn, E. L. (1967): Wintertime photosynthesis of bristlecone pine (Pinus aristata) in the White Mountains of California. Ecol. 48: 1044–1047
Schulze, E. D., Chapin III, F. S. and Gebauer, G. (1994): Nitrogen nutrition and isotope differences among life forms at the northern treeline of Alaska. Oecologia 100: 406–412
Schuster, W. S. F. and Mitton, J. B. (1991): Relatedness within clusters of a bird-dispersed pine and the potential for kin interactions. Heredity 67: 41–49
Schütz, H.-U. (1998): Untersuchungen zur Ökologie von "ribbon forests" der Colorado Front Range (Rocky Mountains, U.S.A.) unter besonderer Berücksichtigung von Thomomys talpoides (Geomyidae). Diss. Mathematisch-Naturwissenschaftliche Fakultät, Westfälische Wilhelms-Universität, Münster

Schütz, H.-U. (2005): Pocket gopher actor under the stage. Studies on Niwot Ridge, Colorado Front Range, U.S.A. In: Mountain ecosystems. Studies in treeline ecology (Broll, G. and Keplin, B., eds.): 153–180, Heidelberg/New York
Schwarz, R. (1983): Simulationsstudien zur Theorie der oberen Waldgrenze. Erdkunde 37: 1–11
Schwarz, W. (1970): Der Einfluß der Photoperiode auf das Austreiben, die Frosthärte und die Hitzeresistenz von Zirben und Alpenrosen. Flora 159: 258–285
Schweinfurth, U. (1957): Die horizontale und vertikale Verbreitung der Vegetation im Himalaya. Bonner Geographische Abhandlungen 20
Schweinfurth, U. (1962): Mt. Egmont Taranaki. Erdkunde XVI (1): 34–48
Schweinfurth, U. (1966): Neuseeland Beobachtungen und Studien zur Pflanzengeographie und Ökologie der antipodischen Inselgruppe. Bonner Geographische Abhandlungen 36
Schweinfurth, U. (1968): Vegetation of the Himalayas. Mountains and rivers of India (21. Int. Geogr. Congr.), Calcutta
Schweinfurth, U. (1978): Geoökologische Beziehungen zwischen der temperierten Zone der Südhalbkugel und den Tropengebirgen im australasiatischen Sektor. Erdwissenschaftliche Forschung 11: 29–48
Schweinfurth, U. (1980): Die neuseeländischen Inseln in dreidimensionaler Sicht. In: Höhengrenzen in Hochgebirgen (Jentsch, Ch. and Liedtke, H., eds.) Arbeiten aus dem Geographischen Institut des Saarlandes 29: 251–261
Schweinfurth, U. (1988): Pinus in Southeast Asia. Beiträge zur Biologie der Pflanzen 63: 253–269
Schweingruber, F.-H. (1980): Dichteschwankungen in Jahrringen von Nadelhölzern in Beziehung zu klimatisch-ökologischen Faktoren, oder das Problem der falschen Jahrringe. Eidgenössische Anstalt für das forstliche Versuchswesen, Berichte 213
Schweingruber, F.-H., Albrecht, H., Beck, M., Hessel, J., Joos, K., Keller, D., Kontic, R., Lange, K., Niederer, M., Nippel, C., Spang, S., Spinnler, A., Steiner, B. and Winkler-Seifert, A. (1986): Abrupte Zuwachsschwankungen in Jahrringabfolgen als ökologische Indikatoren. Dendrochronologia 4: 125–183
Scott, P. A., Bentley, C. V., Fayle, D. C. F. and Hansell, R. I. C. (1987a): Crown forms and shoot elongations of white spruce at the treeline, Churchill, Manitoba, Canada. Arct. Alp. Res. 19: 175–186
Scott, P. A., Hansell, R. I. C. and Fayle, D. C. F. (1987b): Establishment of white spruce populations and responses to climatic change at the treeline, Churchill, Manitoba, Canada. Arct. Alp. Res. 19 (1): 45–51
Scott, P. A., Hansell, R. I. C. and Erickson, W. R. (1993): Influences of wind and snow on northern tree-line environments at Churchill, Manitoba, Canada. Arctic 46 (4): 313–323
Scott, P. A., Lavoie, C., MacDonald, G. M., Sveinbjörnsson, B. and Wein, R. W. (1997): Climate change and the future position of arctic treeline. In: Global change and arctic terrestrial ecosystems (Oechel, W. C., Callaghan, T., Gilmanov, T., Holten, J. I., Maxwell, B., Molau, U. and Sveinbjörnsson, B., eds.): 245–265, New York
Scuderi, L. A. (1987): Late holocene upper timberline variation in the southern Sierra Nevada. Nature 325: 242–244
Sedlag, P. (1995): Tiergeographie. Urania Tierreich, Leipzig/Jena/Berlin
Seibert, P. (1983): Human impact on landscape and vegetation in the Central High Andes. In: Man's impact on vegetation (Holzner, W. et al., eds.): 261–276, The Hague
Seibert, P. and Menhofer, X. (1991): Die Vegetation des Wohngebietes der Kallawaya und des Hochlandes von Ulla-Ulla in den bolivianischen Anden. Phytocoenologia 20 (2): 145–276

Sendtner, O. (1854): Die Vegetationsverhältnisse Südbayerns. München
Senn, J. (1999): Tree mortality caused by Gremmeniella abietina in a subalpine afforestation in the central Alps and its relationships with duration of snow cover. Eur. J. Pathol. 29: 65–74
Senn, J. and Schönenberger, W. (2001): Zwanzig Jahre Versuchsaufforstung Stillberg: Überleben und Wachstum einer subalpinen Aufforstung in Abhängigkeit vom Standort. Schweizerische Zeitschrift für Forstwesen 152 (6): 226–246
Seppälä, M. and Rastas, J. (1980): Vegetation map of northernmost Finland with special reference to subarctic forest limit and natural habitats. Fennia 158 (1): 41–61
Shi, P., Körner, C. and Hoch, G. (2006): End of season carbon supply status of woody species near the treeline in western China. Basic Appl. Ecol. 7: 370–377
Shigo, A. L. (1985): Compartmentalization of decay in trees. Sci. Am. 252 (4): 96–103
Shimin, W. and Shengxian, Z. (1995): Ecological and geographical distribution of Larix and cultivation of its major species in southwestern China. In: Ecology and management of Larix forests: A look ahead, Proc. int. symp., Whitefish, MT, October 5–9, 1992, USDA For. Serv., Intermtn. Res. Stn., Gen. Tech. Rep. GTR-INT 319: 38–40
Shiyatov, S. G. (1966): Time of dispersion of Siberian larch seeds in north-western part of its areal space and role of that factor in mutual relation between forest and tundra (cited in Mazepa, 2005; Original text in Russian). Publication of the Sverdlovsk Branch of the All-Union Botanical Society 4: 109–113
Shiyatov, S. G. (1993): The upper timberline dynamics during the last 1350 years in the polar Ural Mountains. Paläoklimaforschung 9: 195–203
Shiyatov, S. G. (2000): Climate dependent dynamics of the upper timberline and forest-tundra ecosystems during the last 1350 years in the Polar Ural mountains, Russia. int. conf. dendrochron. for the third millennium, April 2–7, 2000, Mendoza, Argentina: 241
Shoji, S., Dahlgren, R. and Nanzyo, M. (1993a): Terminology, concepts and geographic distribution of volcanic ash soils. In: Volvanic ash soils. Genesis, properties and utilization (Shoji, S., Nanzyo, M. and Dahlgren, R. A., eds.): 1–5, Amsterdam
Shoji, S., Dahlgren, R. and Nanzyo, M. (1993b): Genesis of volcanic ash soils. In: Volcanic ash soils. Genesis, properties and utilization (Shoji, S., Nanzyo, M. and Dahlgren, R. A., eds.): 37–71, Amsterdam
Simms, H. R. (1967): On the ecology of Herpotrichia nigra. Mycologia 59: 902–909
Simpson, N. B. (1979): A revision of the genus Polylepis (Rosaceae; Sanguisorbeae). Smithsonian Contributions to Botany 43 (cited in Goldstein et al., 1994)
Singh, J. S. and Singh, S. P. (1987): Forest vegetation of the Himalaya. Bot. Rev. 53: 80–192
Siren, G. (1958): Jajapoimintoja Lapin metsien tulevaisuudesta. Terra 70 (2) (cited in Hustich, 1958)
Siren, G. (1961): Skogsgränstallen som indikator för klimafluktuationerna in norra Fennoskandien under historik tid. Comm. Ins. For. Fenn. 54 (2)
Siren, G. (1963): Tree rings and climate forecast. New Scient. 4: 20
Sirois, L. and Payette, S. (1991): Reduced postfire tree regeneration along a boreal forest-forest-tundra transect in northern Québec. Ecology 72 (2): 619–627
Skottsberg, C. (1910): Übersicht über die wichtigsten Pflanzenformationen Südamerikas S von 41 °S, ihre geographische Verbreitung und Beziehungen zum Klima. Botanische Ergebnisse der schwedischen Expedition nach Patagonien und dem Feuerlande, 1907–1909, I. Kungliga Svenska Vetensksakpsakademiens Handlingar 46 (3): 1–28
Skottsberg, C. (1916): Die Vegetationsverhältnisse längs der Cordillera de los Andes südlich von 41 °S Br. Ein Beitrag zur Kenntnis der Vegetation in Chile, Westpatagonien, dem Andinen Patagonien und Feuerland. Ergebnisse der schwedischen Expedition nach Patagonien und dem Feuerlande, 1907–1907. V. Kungliga Svenska Vetenskapsakademiens Handlingar 56 (5): 1–366

Skottsberg, C. (1960): Remarks on the plant geography of the southern cold temperate zone. Proc. Roy. Soc. Lond. B 152: 447–457
Skre, O. (1972): High temperature demands for growth and development in Norway Spruce (Picea abies (L.) Karst.) in Scandinavia. Meldinger fra Norges Landbrukshögskole 51 (7): 1–29
Skre, O. (1988): Frost resistance in forest trees: A literature survey. Meddelser fra Norsk Institutt for Skogforskning 40 (9)
Skre, O. (1993): Growth of mountain birch (Betula pubescens ERH.) in response to changing temperature. In: Forest development in cold climate (Alden, K., Mastrantonio, J. L. and Ødum, S., eds.), NATO ASI Series A 244, Plenum Press, New York
Skre, O., Baxter, R., Crawford, R. M. M., Callaghan, T. V. and Fedorkov, A. (2002): How will the taiga-tundra interface respond to climate change. Ambio Sp. Rep. 12: 37–46
Skre, O., Taulavuori, K., Taulavuori, E., Nilsen, J., Igeland, B. and Laine, K. (2003): Critical levels of winter temperature for survival and growth in mountain birch populations. In: Human interactions with the mountain birch ecosystem: Implications for sustainable development (HIBECO) (Laine, K., Skre, O. and Wielgolaski, F.-E., eds.), Final report for the project EU FP5, QLK5-CT-1999-01515: 89, Oulu
Skre, O., Taulavuori, K., Taulavuori, E., Nilsen, J., Igeland, B. and Laine, K. (2008): The importance of hardening and winter temperature for growth in mountain birch populations. Environ. Exp. Bot. 62: 254–266
Slankis, V. (1973): Hormonal relationships in mycorrhizal development. In: Ectomycorrhizae: Their ecology and physiology (Marks, G. C. and Kozlowski, T. T., eds.): 231–298, New York
Slatyer, R. O. (1976): Water deficits in timberline trees in the Snowy Mountains of south-eastern Australia. Oecologia (Berl.) 24: 357–366
Slatyer, R. O. (1988): Alpine and valley bottom treelines. In: The scientific significance of the Australian Alps, Cap. 10 (Good, R., ed.), Nat. Parks Liason Committee and the Austr. Acad. Sci., Canberra
Slatyer, R. O. and Noble, I. R. (1992): Dynamics of montane treelines. In: Landscape boundaries: Consequences for biotic diversity and ecological flows (Hansen, A. and Di Castri, F., eds.), Ecol. Stud. 92: 346–359
Slot, M., Wirth, C., Schumacher, J., Mohren, G. M. J., Shibistova, O., Lloyd, J. and Ensminger, J. (2005): Regeneration pattern in boreal Scots pine glades linked to cold-induced photoinhibition. Tree Physiol. 25: 1139–1150
Smith, A. P. (1974): Bud temperature in relation to nyctinastic leaf movement in an Andean giant rosette plant. Biotropica 6: 263–265
Smith, C. M., Wilson, B., Rasheed, S., Walker, R. C., Carolin, T. and Shepherd, B. (2008): Whitebark pine and white pine blister rust in the Rocky Mountains of Canada and northern Motana. Can. J. For. Res. 38: 982–995
Smith, D. M., Larson, B. C., Kelty, M. J. and Ashton, P. M. S. (1997): The practice of silviculture. Applied forest ecology (9th edition). New York
Smith, J. M. B. (1975): Mountain grasslands of New Guinea. J. Biogeogr. 2 (1): 27–44
Smith, J. M. B. (1977a): Vegetation and microclimate of east- and westfacing slopes in the grasslands of Mt. Wilhelm, Papua New Guinea. J. Ecol. 65: 39–53
Smith, J. M. B. (1977b): An ecological comparison of two tropical high mountains. J. Trop. Geogr. 44: 71–80
Smith, J. M. B. (1980): Ecology of the high mountains of New Guinea. In: The afroalpine flora of New Guinea (Van Royen, O., ed.): 11–132, Vaduz
Smith, W. K. and Knapp, A. K. (1990): Ecophysiology of high elevation forests. In: Plant biology of the Basin and Range (Osmond, C. B., Pitelka, L. F. and Hidy, G. M., eds.), Ecol. Stud. 80: 87–142

Smith, W. K., Germino, M. J., Hancock, T. E. and Johnson, D. M. (2003): Another perspective on altitudinal limits of alpine timberlines. Tree Physiol. 23: 1101–1112

Smit-Spinks, B., Swanson, B. T. and Markhart, A. H., III. (1985): The effect of photoperiod and thermoperiod on cold acclimation and growth of Pinus sylvestris. Can. J. For. Res. 15: 453–460

Soil Survey Staff (1990): Keys to soil taxonomy, 4th Ed. Soil Manag. Supp. Serv. Tech. Monogr. 6. Virginia State Univ., Blacksburg, Virginia

Solomon, A. M. (1989): Measures for the protection of forests in temperate zones. Enquête-kommission, Vorsorge zum Schutze der Erdatmosphäre, Hearing on "Schutz der Wälder in mittleren und nördlichen Breiten", 8. Juni 1989: 141–147

Solomon, S., Qin, D., Manning, M., Alley, R. B., Bernstsen, T., Bindoff, N. L., Chen, Z., Chidthaisong, A., Gregory, J. M., Hegerl, G. C., Heimann, M., Hewitson, B., Hoskins, B. J., Joos, F., Jouzel, J., Kattsov, V., Lohmann, U., Matsuno, T., Molina, M., Nicholls, N., Overpeck, J., Raga, G., Ramaswamy, V., Ren, J., Rusticucci, M., Somerville, R., Stocker, T. F., Whetton, P., Wood, R. A. and Wratt, D. (2007): Technical summary. In: Climate Change 2007. The physical science basis. Contribution of working group I to the fourth assessment report of the Intergovernmental Panel on Climate Change (Solomon, S., Qin, D., Mannin, M., Chen, Z., Marquis, M., Averyt, K. B., Tignor, M. and Miller, H. L., eds.): 19–91, Cambridge University Press, Cambridge

Sowell, J. B., Koutnik, D. L. and Lansing, A. J. (1982): Cuticular transpiration of whitebark pine (Pinus albicaulis) within a Sierran Nevadan timberline ecotone, U.S.A. Arct. Alp. Res. 14 (2): 97–103

Söyrinki, N. (1939): Studien über die generative und vegetative Vermehrung der Samenpflanzen in der alpinen Vegetation Petsamo-Lapplands. Annales Botanici Societas Zoologicae Botanaicae Fennocae "Vanamo" 11 (1); 14 (1)

Spatz, G., Weis, G. B. and Dolar, D. M. (1978): Der Einfluß von Bewirtschaftungsveränderungen auf die Vegetation von Almen im Gasteiner Tal. In: Ökologische Analysen von Almflächen im Gasteiner Tal (Cernusca, A., ed.), Veröffentlichungen des Österreichischen MAB-Hochgebirgsprogramms Hohe Tauern 2: 163–180

Spomer, G. G. and Salisbury, F. B. (1968): Eco-physiology of Geum turbinatum and implications concerning alpine environments. Bot. Gazette 129: 33–49

Squeo, F. A., Rada, F., Acozar, A. and Goldstein, G. (1991): Freezing tolerance and avoidance in high tropical Andean plants: Is it equally represented in species with different plant height? Oecologia 86: 378–382

Šrutek, M. and Lepš, J. (1994): Variation in structure of Larix olgensis stands along the altitudinal gradient on Paerktusan, Chanbaishan, North Korea. Arct. Alp. Res. 26 (2): 166–173

Šrutek, M., Doležal, J. and Hara, T. (2002): Spatial structure and associations in a Pinus canariensis population at the treeline, Pico del Teide, Tenerife, Canary Islands. Arct. Antarct. Alp. Res. 34 (2): 201–210

Stadtmüller, T. (1987): Cloud forests in the humid tropics. Turrialba, Costarica

Stahelin, R. (1947): Factors influencing the natural restocking of high altitude burns in coniferous trees in the central Rocky Mountains. Ecology 24: 19–30

Stålfelt, M. G. (1944): The water consumption of the spruce. Kungl. Lantbruksakademiens Tidskrift 83: 1–83

Steiner, M. (1935): Winterliches Bioklima und Wasserhaushalt an der alpinen Waldgrenze. Bioklimatische Beiblätter 2: 57–65

Stern, R. (1966): Der Waldrückgang im Wipptal. Mitteilungen der forstlichen Bundersversuchsanstalt Mariabrunn 70

Stevens, G. C. and Fox, J. F. (1991): The causes of treeline. Ann. Rev. Ecol. Syst. 22: 177–191
Stimm, B. (1985): Untersuchungen und sproßbürtige Bewurzelung der Fichte (Picea abies (L.) KARST.). Diss. Forstwissenschaftliche Fakultät Universität München
Stimm, B. (1987): Morphologisch-anatomische Untersuchungen zur Ablegerbildung und sproßbürtigen Bewurzelung der Fichte (Picea abies (L.) KARST.). Flora 179: 421–443
Stockton, C. W. (1984): An alternative hypothesis to direct CO_2 fertilization as a cause of increased tree growth during 1850–1980 in Central Nevada. Preliminary Report, Laboratory of Tree Ring Research, University of Arizona, Tucson
Stoffel, J. R. (1938): Das Hochtal Avers. Die höchstgelegene Gemeinde Europas. Zofingen
Strain, B. R. and Cure, J. D.(1985): Direct effects of increasing carbon dioxide on vegetation. DOE/ER-0238. U. S. Dept. Energy, Carbon Dioxide Res., Div. Washington, DC
Strasburger, E., Noll, F., Schenck, H. and Schimper, A. F. W. (1991): Lehrbuch der Botanik für Hochschulen (33rd edition, revised by Sitte, P., Ziegler, H., Ehrendorfer, F. and Bresinsky, A.) Stuttgart/Jena/New York
Streule, A. and Häsler, R. (2006): Windschutz für junge Bäume in subalpinen Aufforstungen an stark windexponierten Standorten, "Pru dal Vent" (Alp Grüm, GR) Eidgenössische Anstalt für Wald, Schnee und Landschaft WSL, Birmensdorf
Stugren, B. and Popovici, J. (1991): Ökologie in Siebenbürgen. In: Naturwissenschaftliche Forschungen über Siebenbürgen IV (Heltmann, H., ed.): 227–244, Cologne/Weimar/Vienna
Sturm, H. (1998): The ecology of the Páramo region in tropical high mountains. Hildesheim/Berlin
Stützer, A. (2000): Die Wald- und Baumgrenze der Saualpe: Ein Vergleich alter und neuer Bilder. Forstwissenschaftliches Centralblatt 119: 20–31
Stützer, A. (2002): Zwischen subalpinem Wald und alpiner Tundra. Eine Studie zur Struktur und Dynamik des Fichten-Waldes auf der Saualpe (Kärnten). Wulfenia 9: 89–104
Stützer, A. (2004): Fotografien als Zeugen der Vegetationsgeschichte. Eine Waldgrenzstudie aus Kärten. Wulfenia 11: 63–68
Stützer, A. (2005): Bildsequenzen als Zeugen der Vegetationsdynamik in der subalpinen-alpinen Höhenstufe der Koralpe (Kärnten/Österreich). Wulfenia 12: 127–138
Succow, M. (1989): Die mittelasiatischen Hochgebirge. In: Hochgebirge der Erde und ihre Pflanzen und Tierwelt (Klotz, R., ed.): 187–204, Leipzig/Jena/Berlin
Suffling, R. (1989): Climate change, forest fires and the boreal forest. Landscape ecological impact of climate change on boreal/(sub)arctic regions, with emphasis on Fennoscandia (Koster, E. A. and Boer, M. M., compilers), Discussion report prepared for the European Conference on Landscape Ecological Impact of Climatic Change, to be held in Lunteren, The Netherlands, December 3–7, 1989: 112–124
Sumoninen, O. and Olofsson, J. (2000): Impacts of semi-domestic reindeer on structure of tundra and forest communities in Fennoscandia: A review. Annales Zoologici Fennici 37: 233–249
Sundqvist, M. K., Björk, R. G. and Molau, U. (2008): Establishment of boreal forest species in alpine dwarf-shrub heath in subarctic Sweden. Plant Ecol. Divers. 1 (1): 67–75
Supan, A. (1994): Grundzüge der physischen Erdkunde. Leipzig
Sutinen, M.-L., Ritari, A., Holappa, T. and Kujala, K. (1997): Seasonal changes in soil temperature and in the frost hardiness of Scots pine roots under subarctic conditions. In: Int. symp. on physics, chemistry, and ecology of seasonally frozen soils, June 10–12, 1997, Fairbanks, AK (Islander, I. K., Wright, E. A., Radke, J. K., Shamatt, B. S., Groenvelt, O. H. and Hinzman, L. B., eds.), Cold Regions Research and Engineering Laboratory Special Report 97-10: 513–517, Hanover, New Hampshire

Sutinen, M.-L., Arora, R., Wisniewski, M., Ashworth, E., Strimbeck, R. and Palta, J. (2001): Mechanisms of frost survival and freeze-damage in nature. In: Conifer cold hardiness. Tree physiology (Bigras, F. J. and Colombo, S. J., eds.): 1: 89–119, Kluwer, Dordrecht/Boston, MA/London

Sutter, E. and Amann, F. (1953): Wie weit fliegen vorratssammelnde Tannenhäher? Ornithologischer Beobachter 50: 89–90

Sveinbjörnsson, B. (1993): Climate and growth of mountain birch near the treeline in Northern Sweden and Iceland. Paläoklimaforschung 9: 57–67

Sveinbjörnsson, B. (2000): North American and European treelines: External forces and internal processes controlling position. Ambio 29 (7): 388–395

Sveinbjörnsson, B., Kauhanen, H. and Nordell, O. (1996): Treeline ecology of mountain birch in the Torneträsk area. Ecol. Bull. 45: 65–70

Swedberg, K. C. (1961): The coniferous ecotone of the east slope of the northern Oregon Cascades. Unpubl. Ph.D. thesis, Oregon State University, Corvallis, OR

Szeicz, J. M. and MacDonald, G. M. (1995): Recent white spruce dynamics at the subarctic alpine treeline of northwestern Canada. J. Ecol. 63: 873–885

Tabler, R. D. (1975): Estimating the transport and evaporation of blowing snow. In: Snow management on Great Plains, Symposium (Bismarck, North Dakota, July 1975). Proc. Great Plains Agric. Counc. Publ. 73: 84–104

Tabler, R. D. and Schmidt, R. A. (1973): Weather conditions that determine snow transport distance at a site in Wyoming. International Symposia on the role of snow distribution and ice in hydrology. Symposium on properties and processes, Banff, Alberta, Canada, September 1972

Takahashi, K. (1944): Die Baum- und Waldgrenze im Hida-Gebirge (Japanische Nordalpen). Ein Beitrag zur Baum- und Waldgrenze Ostasiens. Jpn. J. Ecol. 12 (3): 269–343

Takahashi, K. (2003): Effects of climatic conditions on shoot elongation of alpine warf pine (Pinus pumila) at its upper and lower altitudinal limits in Central Japan. Arct. Antarct. Alp. Res. 35 (1): 1–7

Takei, F. (1995): Distribution and management of Japanese larch. In: Ecology and management of Larix forests: A look ahead. Proc. int. symp., Whitefish, MT, October 5–9, 1992 (Schmidt, W. C. and McDonald, K. J., compilers): 58–65, USDA For. Serv., Intermtn. Res. Stn., Gen. Tech. Rep. GTR-INT 319

Tammaru, U., Kaitaniemi, P. and Ruohomäki, K. (1995): Oviposition choices of Epirrita autumnata (Lepidoptera: Geometridae) in relation to its eruptive dynamics. Oikos 74: 296–304

Tasanen, T., Norokorpi, Y., Sepponen, P. and Juntunen, V. (1998): Monitoring timberline dynamics in northern Lapland. In: Research and management of the northern timberline region, Proc. Gustaf Siren Symp. Wilderness Center Inari, September 4–5, 1997 (Tasanen, T., ed.), The Finnish For. Res. Inst., Res. Papers 677: 75–86

Taschler, D. and Neuner, G. (2004): Summer frost resistance and freezing patterns measured in situ in leaves of major plant growth forms in relation to their upper distribution boundary. Plant Cell Environ. 27 (6): 737–746

Taschler, D., Beikircher, B. and Neuner, G. (2004): Frost resistance and ice nucleation in leaves of five woody timberline species measured in situ during shoot expansion. Tree Physiol. 24: 331–337

Tausz, M. (2007): Photo-oxidative stress at the timberline. In: Trees at their upper limit. Treelife limitation at the alpine timberline (Wieser, G. and Tausz, M., eds.), Plant Ecol. 5: 181–195

Teeri, J. A. and Barret, P. E. (1975): Detritus transport by wind in a high arctic terrestrial ecosystem. Arct. Alp. Res. 7 (4): 387–391

Ten Houte de Lange, S. M. (1978): Zur Futterwahl des Steinbocks (Capra ibex L.). Eine Untersuchung an der Steinbockkolonie am Piz Albris bei Pontresina. Zeitschrift für Jagdwissenschaft 24: 113–138

Tenow, O. (1996): Hazards to a mountain birch forest. Abisko in perspective. Ecol. Bull. 45: 104–114

Tenow, O., Bylund, H. and Strong, D. R. (1995): Multiple mechanisms in cycles of the autumnal moth on mountain birch. In: Long-term interactions between the autumnal moth and mountain birch: The roles of resource, competitors, natural enemies, and weather (Bylund, H., ed.). Diss. Sveriges Lantbruksuniversitet Uppsala

Tenow, O., Bylund, H., Nilssen, A. C. and Karlsson, P. S. (2005): Long-term influences of herbivores on northern birch forests. In: Plant ecology, herbivory, and human impact in Nordic mountain birch forests (Wielgolaksi, F. E., Neuvonen, S. and Thannheiser, D., eds.), Ecol. Stud. 180: 165–181

Tessier, L., de Beaulieu, J.-L., Couteaux, M., Edouard, J.-L., Ponel, P., Rolando, C., Thinon, M., Thomas, A. and Tobolski, K. (1993): Holocene paleoenvironments at the timberline in the French Alps – a multidisciplinary approach. Boreas 22: 244–254

Thannheiser, D., Meier, K.-D. and Wehberg, J. (2003): Klimabedingte und anthropogene Veränderungen im Fjellbirkenwald in Nordnorwegen. Norden 15: 213–220

Theurillat, J.- P. and Guisan, A. (2001): Potential impact of climate change on vegetation in the European Alps: A review. Climate Change 50: 77–109

Theurillat, J.- P., Felber, F., Geissler, P., Gobat, J.-M., Fierz, M., Fischlin, A., Küpfer, P., Schussel, A., Velutti, C. and Zhao, G.-F. (1998): Sensitivity of plant and ecosystems of the Alps to climate change. In: Views from the Alps: Regional perspectives on climate change (Cebon, P., Dahinden, U., Davies, H. C., Imboden, D. and Jaeger, C. C., eds.): 225–308, MIT Press, Cambridge, MA

Thinon, M. and Talon, B. (1998): Ampleur de l'anthropisation des étages supérieurs dans les Alpes du sud: Donnés pédoanthracologiques. Écologie 29: 323–328

Thompson, W. F. (1990): Climate related landscapes in world mountains: Criteria and map. Zeitschrift für Geomorphologie, Supplement Band 78

Thorn, C. E. (1978): The geomorphic role of snow. Ann. Assoc. Am. Geogr. 68 (3): 414–425

Thórsson, A. T., Salmela, E. and Anamthawat-Jónsson, K. (2001): Morphological cytogenetic, and molecular evidence from introgressive hybridization in birch. J. Hered. 92: 404–408

Thouret, J. C. (1989): Suelos de la Cordillera Central, transecto Parque Los Nevados. In: Studies on Tropical Andean Ecosystems 3 (Van Der Hammen, T., Diaz, S. and Alvarez, V. J., eds.): 293–441, Vaduz

Tikhomirov, B. A. (1962): The treelessness of the tundra. Polar Rec. 11: 24–30

Tilman, D. (1988): Plant strategies and the dynamics and structure of plant communities. Princeton University Press, New Jersey, NJ

Timonen, M. (2002): Is climate warming at the pine timberline? In: Abstracts prepared for Arctic Council, Northern Timberline Forests Workshop, Rovaniemi, Finland, May 23–25, 2002 (Arctic Centre, Univ. of Lapland, Finnish Forest and Park Service, Finnish Forest Res. Institute, Univ. of Joensuu, organisers): 22

Timoney, K. P., La Roi, G. H., Zoltai, S. C. and Robinson, A. L. (1992): The high subarctic forest of northwestern Canada: Position, width, and vegetation gradients in relation to climate. Arctic 45 (1): 1–9

Tinner, W. and Theurillat, J.-P. (2003): Uppermost limit, extent, and fluctuations of the timberlines and treeline ecocline in the Swiss Central Alps during the past 11500 years. Arct. Antarct. Alp. Res. 35: 158–169

Tinner, W. and Kaltenrieder, P. (2005): Rapid response of high-mountain vegetation to early Holocene environmental changes in the Swiss Alps. J. Ecol. 93: 936–947

Tinner, W., Ammann, B. and Germann, P. (1996): Treeline fluctuations recorded for 12500 years by soil profiles, pollen, and plant macrofossils in the Central Swiss Alps. Arct. Alp. Res. 28: 131–147

Tollner, H. (1949): Der Einfluß großer Massenerhebung auf die Lufttemperatur und die Ursachen der Hebung der Vegetationsgrenzen in den inneren Ostalpen. Archiv für Meteorologie, Geophysik und Bioklimatologie, Serie B 1: 347–372

Tolmachev, A. I. (1970): Die Erforschung einer entfernten "Waldinsel" in der Großlandtundra. Colloquium Geographicum 12: 98–103

Tomback, D. F. (1977). The behavioral ecology of Clark's Nutcracker (Nucifraga columbiana) in the eastern Sierra Nevada. Ph.D. thesis, University of California, Santa Barbara, CA

Tomback, D. F. (1978): Foraging strategies of Clark's Nutcracker. The Living Bird 16: 123–161

Tomback, D. F. (1982): Dispersal of whitebark pine seeds by Clark's Nutcracker: A mutualism hypothesis. J. Anim. Ecol. 51: 451–467

Tomback, D. F. (1989): The broken circle: Fire, birds and whitebark pine. In: Wilderness and wildfire, Wilderness Institute, Montane Forestry and Range Experiment Station, University of Montana, Misc. Publ.: 14–17

Tomback, D. F. and Linhart, Y. B. (1990): The evolution of bird-dispersed pines. Evol. Ecol. 4: 185–219

Tomback, D. F. and Resler, L. M. (2008): Invasive pathogens at alpine treeline: Consequences for treeline dynamics. Phys. Geogr. 28: 397–418

Tomback, D. F. and Schuster, W. F. (1994): Genetic population structure and growth form distribution in bird-dispersed pines. In: Proc. int. workshop on subalpine stone pines and their environment: The status of our knowledge, USDA For. Serv., Intermtn. Res. Stat., Gen. Tech. Rep. INT-GTR-309: 43–50

Tomback, D. F., Sund, S. K. and Hoffmann, L. (1993): Post-fire regeneration of Pinus albicaulis: Height-age relationships, age structure, and microsite characteristics. Can. J. For. Res. 23: 113–119

Tomback, D. F., Holtmeier, F.-K., Mattes, H., Carsey, K. S. and Powell, M. L. (1994): Tree clusters and growth form distribution in Pinus cembra, a bird-dispersed pine. Arct. Alp. Res. 25 (4): 374–381

Tømmervik, H., Wielgolaski, F. E., Neuvonen, S., Solberg, B. and Høgda, K. A. (2005): Biomass and production on a landscape level in the mountain birch forests. In: Plant ecology, herbivory and human impact in northern mountain birch forests (Wielgolaski, F. E., ed.), Ecol. Stud. 180: 53–70

Tranquillini, W. (1957): Standortsklima, Wasserbilanz und CO_2-Gaswechsel junger Zirben (Pinus cembra L.) an der alpinen Waldgrenze. Planta 49: 612–661

Tranquillini, W. (1958): Die Frosthärte der Zirbe unter besonderer Berücksichtigung autochthoner und aus Forstgärten stammender Jungpflanzen. Forstwissenschaftliches Centralblatt 77: 89–105

Tranquillini, W. (1959): Die Stoffproduktion der Zirbe (Pinus cembra) an der Waldgrenze während eines Jahres. Planta 54: 107–151

Tranquillini, W. (1965): Über den Zusammenhang zwischen Entwicklungszustand und Dürreresistenz junger Zirben (Pinus cembra L.) im Pflanzgarten. Mitteilungen der forstlichen Bundes-Versuchsanstalt 66: 2411–271

Tranquillini, W. (1967): Über die physiologischen Ursachen der Wald- und Baumgrenze. Mitteilungen der forstlichen Bundes-Versuchsanstalt Mariabrunn 75: 457–487

Tranquillini, W. (1974): Der Einfluß von Seehöhe und Länge der Vegetationszeit auf das cutikuläre Transpirationsvermögen von Fichtensämlingen im Winter. Berichte der Deutschen Botanischen Gesellschaft 87: 175–184

Tranquillini, W. (1976): Water relations and alpine timberline. In: Water and plant life (Lange, O. L., Kappen, L. and Schultze, E. D., eds.), Ecol. Stud. 19: 473–491
Tranquillini, W. (1979a): Physiological ecology of the alpine timberline – tree existence at high altitudes with special reference to the European Alps. Ecol. Stud. 31
Tranquillini, W. (1979b): Über die Frostgefährdung von Fichten in verschiedener Höhenlage. Wissenschaftliche Mitteilungen Meteorologisches Institut, Universität München (Festschrift A. Baumgartner) 35: 51–57
Tranquillini, W. (1980): Winter desiccation as the cause for alpine timberline. In: Mountain environments and subalpine tree growth (Benecke, U. and Davis, M. R., eds.), New Zeal. For. Serv. Techn. Paper 70: 263–267
Tranquillini, W. (1982): Frost drought and its ecological significance. In: Encyclopedia of plant physiology (new series) 12 B (Water relations and carbon assimilation) (Lange, O. L., Noble, P. S., Osmond, C. B. and Ziegler, H., eds.): 379–400, Berlin
Tranquillini, W. (2001): Alpine Waldgrenze. In: Die subalpine und alpine Vegetation in Tirol. Exkursionsführer, 50. Jahrestagung der Floristisch-soziologischen Arbeitsgemeinschaft, 7.-10. Juli in Innsbruck (Erschbamer, B. and Grabherr, S., eds.): 42–45, Innbruck
Tranquillini, W. and Holzer, K. (1958): Über das Gefrieren und Auftauen von Coniferennadeln. Berichte der Deutschen Botanischen Gesellschaft 71: 143–156
Tranquillini, W. and Plank, A. (1989): Ökophysiologische Untersuchungen an Rotbuchen (Fagus sylvatica L.) in verschiedenen Höhenlagen Nord- und Südtirols. Centralblatt für das gesamte Forstwesen 161 (3): 225–246
Tranquillini, W. and Turner, H. (1961): Untersuchungen über die Pflanzentemperaturen in der subalpinen Stufe mit besonderer Berücksichtigung der Nadeltemperaturen der Zirbe. Mitteilungen der forstlichen Bundes-Versuchsanstalt Mariabrunn 59: 127–151
Tranquillini, W. and Unterholzner, R. (1968): Dürreresistenz und Anpflanzungserfolg von Junglärchen verschiedenen Entwicklungszustandes. Centralblatt für das gesamte Forstwesen 85 (2): 97–110
Tranquillini, W., Lechner, F., Oberarzbacher, P., Unterholzner, L. and Holzer, K. (1980): Über das Höhenwachstum von Fichtenklonen in verschiedener Seehöhe. Mitteilungen der forstlichen Bundes-Versuchsanstalt Wien 129: 7–25
Trappe, J. M. and Fogel, R. D. (1977): Ecosystematic functions of mycorrhizae. In: The belowground ecosystem: A synthesis of plant-associated processes. Range Sci. Dept. Sci. Ser. 26: 295–214, Colorado State University, Fort Collins, CO
Treter, U. (1996): Gebirgswaldsteppe in der Mongolei. Geographische Rundschau 48 (11): 655–659
Treter, U., Block, J. and Kastner, R. (2002): Ergebnisse dendrochronologischer und dendroökologischer Analysen der Lärchenwälder in der Gebirgswaldsteppe der nordwestlichen Mongolei. In: Zum Stand der Anwendung der Dendrochronologie in den Geowissen-schaften (Bräuning, A., ed.), Stuttgarter Geographische Studien 133: 83–98
Troll, C. (1939): Das Pflanzenkleid des Nanga Parbat. Wissenschaftliche Veröffentlichungen des Deutschen Museums für Länderkunde Leipzig, Neue Folge 7
Troll, C. (1941): Studien zur vergleichenden Geographie der Hochgebirge der Erde. Bericht über die 23. Hauptversammlung der Gesellschaft von Freunden und Förderern der Rheinischen Friedrich-Wilhelms-Universität, Bonn: 49–69
Troll, C. (1948a): Der subnivale und periglaziale Zyklus der Vegetation. Erdkunde 2: 1–22
Troll, C. (1948b): Der asymmetrische Aufbau der Vegetationszonen und Vegetationsstufen auf der Nord- und Südhalbkugel. Jahresbericht des Geobotanischen Instituts Rübel in Zürich für 1947: 46–83
Troll, C. (1955a): Der jahreszeitliche Ablauf des Naturgeschehens in den verschiedenen Klimagürteln der Erde. Studium Generale 8 (12): 713–733
Troll, C. (1955b): Der Mt. Rainier und das Mittlere Cascadengebirge. Erdkunde IX (4): 264–286

Troll, C. (1955c): Forschungen in Zentralmexico 1954. Die Stellung des Landes im dreidimensionalen Landschaftsbau der Erde. Deutscher Geographentag in Hamburg, Tagungsbericht und wissenschaftliche Abhandlungen: 191–213
Troll, C. (1958): Tropical mountain vegetation. Proc. 9th Pac. Sci. Congr. 1957 (20): 37–46
Troll, C. (1959): Die tropischen Gebirge. Ihre dreidimensionale klimatische und pflanzengeographische Zonierung. Bonner Geographische Abhandlungen 25
Troll, C. (1961): Klima und Pflanzenkleid der Erde in dreidimensionaler Sicht. Die Naturwissenschaften 48: 332–348
Troll, C. (1964): Die klimatische und vegetationsgeographische Gliederung des Himalaya-Systems. In: Khumbu Himal. Ergebnisse des Forschungsunternehmens Nepal Himalaya (Hellmich, W., ed.): 353–388, Berlin/Göttingen/Heidelberg
Troll, C. (1968): The cordilleras of the tropical Americas. Aspects of climatic, phytogeographical and agrarian ecology. – Colloquium Geographicum 9: 15–56
Troll, C. (1972): The upper limit of aridity and the arid core of High Asia. Erdwissenschaftliche Forschung IV: 237–243, Wiesbaden
Troll, C. (1973): The upper timberlines in different climatic zones. Arct. Alp. Res. 5 (3, Pt. 2): 3–18
Troll, C. (1978): Polylepis – Hagenia – Leucosidea, eine merkwürdige Konvergenz von Gehölztypen in der tropischen und subtropischen Gebirgsvegetation Südamerikas und Afrikas. In: Geoecological relations between the southern temperate zone and the tropical mountains (Lauer, W., ed.), Erdwissenschaftliche Forschung XI: 561–563
Tschermak, L. (1950): Waldbau auf pflanzengeographisch-ökologischer Grundlage. Wien
Tuhkanen, S. (1980): Climatic parameters and indices in plant geography. Acta Phytogeographica Suecica 67: 1–110
Tuhkanen, S. (1992): The climate of Tierra del Fuego from a vegetation geographical point of view and its ecoclimatic counterparts elsewhere. Acta Bot. Fenn. 145: 1–64
Tuhkanen, S. (1993): Treeline in relation to climate, with special reference to oceanic areas. In: Forest development in cold climates. Proc. NATO advanced workshop, 1991, June 18–23, 1993, Langarvatn, Iceland (Alden, J. et al., eds.): 115–134
Tuhkanen, S. (1999): The northern timberline in relation to climate. In: Sustainable development in northern timberline forests. Proceedings of the timberline workshop, May 10–11, 1998 in Whitehorse, Canada (Kankaanpää, S., Tasanen, T. and Sutinen, M.-L., eds.). The Finnish Forest Res. Inst., Res. Papers 734: 29–61
Turner, H. (1958): Maximaltemperaturen oberflächennaher Bodenschichten an der alpinen Waldgrenze. Wetter und Leben 10 (1/2): 1–12
Turner, H. (1961): Die Niederschlags- und Schneeverhältnisse. Mitteilungen der forstlichen Bundes-Versuchsanstalt Mariabrunn 59: 265–315
Turner, H. (1968): Über "Schneeschliff" in den Alpen. Wetter und Leben 20 (9/10): 192–200
Turner, H. (1970): Grundzüge der Hochgebirgsklimatologie. In: Die Welt der Alpen: 170–182, Innsbruck, Frankfurt
Turner, H. (1971): Mikroklimatographie und ihre Anwendung in der Ökologie der subalpinen Stufe. Annalen der Meteorologie, Neue Folge 5: 275–281
Turner, H. (1980): Types of microclimate at high elevations. In: Mountain environments and subalpine tree growth (Benecke, U. and Davis, M. R., eds.), New Zealand For. Serv., For. Res. Inst. Techn, Paper 70: 21–26
Turner, H. (1985): Das Gebirgsprogramm "Wiederherstellung der oberen Waldgrenze" – Anmerkungen zu einem langfristigen interdisziplinären Forschungsvorhaben. – In: 100 Jahre Eidgenössische Anstalt für das forstliche Versuchswesen 1885–1985, Teil 2, Die EAFV heute und morgen, Meinungsäußerungen zur Gestaltung forstlicher Forschung (Bossard, W., ed.). Mitteilungen der Eidgenössischen Anstalt für das forstliche Versuchswesen 61 (2): 757–765

Turner, H. and Blaser, P. (1977): Mikroklima, Boden und Pflanzen an der oberen Waldgrenze. Eidgenössische Anstalt für das forstliche Versuchswesen, Berichte 173
Turner, H. and Streule, A. (1983): Wurzelwachstum und Sprossentwicklung junger Koniferen im Klimastress der alpinen Waldgrenze, mit Berücksichtigung von Mikroklima, Photosynthese und Stoffproduktion. In: Wurzelökologie und ihre Nutzanwendung, Int. Symp. Gumpenstein 1982: 617–635, Bundesanstalt Gumpenstein, Irdning
Turner, H., Rochat, P. and Streule, A. (1975): Thermische Charakteristik von Hangstandorts-typen im Bereich der oberen Waldgrenze (Stillberg, Dischmatal bei Davos). Mitteilungen der Eidgenössischen Anstalt für das forstliche Versuchswesen 51 (1): 95–119
Turner, H., Häsler, R. and Schönenberger, W. (1982): Contrasting microenvironments and their effects on carbon uptake and allocation by young conifers near alpine treeline in Switzerland. In: Proc. IUFRO Workshop P 1.07-00 Ecology of Subalpine Zones, August 2–3, 1982, Corvallis, OR (Waring, R. H., ed.): 22–30
Turner, M. G. (1989): Landscape ecology: The effect of pattern on process. Ann. Rev. Ecol. Syst. 10: 171–197
Tyron, P. R. and Chapin III, F. S. (1983): Temperature control over root growth and root biomass in taiga forest trees. Can. J. For. Res. 13: 827–833
Ukkola, R. (1995): Trampling tolerance of plants and ground cover in Finnish Lapland, with example from the Pyhätunturi National Park. In: Environmental aspects of the timberline in Finland and in the Polish Carpathians (Heikkinen, O., Obrebska-Starkel, B. and Tuhkanen, S., eds.): 91–110, Krakow
Ulmer, W. (1937): Über den Jahresgang der Frosthärte einiger immergrüner Arten der alpinen Stufe sowie der Zirbe und Fichte. Jahrbuch für Wissenschaftliche Botanik 84: 553–592
Ungerson, J. and Scherdin, G. (1968): Jahresgang von Photosynthese und Atmung unter natürlichen Bedingungen von Pinus silvestris L. an ihrer Nordgrenze in der Subarktis. Flora, Abteilung B, 157: 391–434
Vaarama, A. and Valanne, T. (1973): On the taxonomy, biology and origin of Betula tortuosa Ledeb. Rep. Kevo Subarct. Res. Stat. 10: 70–84
Vale, T. R. (1981): Tree invasion in montane meadows in Oregon. The Amer. Midl. Nat. 105: 61–69
Van der Meer, M., Hagedorn, F., Schweingruber, F. H., Rigling, A. and Moiseev, P. (2004): Dynamik der alpinen Waldgrenze im südlichen Ural (Russland). Die Erde 135 (2): 151–174
Van Gardeningen, P. R., Grace, J. and Jeffree, C. E. (1991): Abrasive damage by wind to the needle surface of Picea sitchensis (Bong.) Carr. and Pinus sylvestris L. Plant Cell Environ. 14: 185–193
Van Miegroet, H., Hysell, M. T. and Johnson, A. D. (2000): Soil microclimate and chemistry of spruce-fir tree islands in Northern Utah. Soil Sci. Soc. Am. J. 64: 1515–1525
Van Steenis, C. C. G. J. (1953): Papuan Nothofagus. J. Arnold Arboretum 34, add. note 35
Van Steenis, C. C. G. J. (1972): The mountain flora of Java. Leiden
Vander Wall, S. B. and Balda, R. P. (1977): Co-adaptations of the Clark's Nutcracker and the Pinon Pine for efficient seed harvest and dispersal. Ecol. Monogr. 47: 89–11
Vander Wall, S. B. and Balda, R. P. (1981): Ecology and evolution of food storage behaviour in conifer-seed caching corvids. Zeitschrift für Tierphysiololgie 56: 217–242
Väre, H. (2001): Mountain birch taxonomy and floristics of mountain birch woodlands. In: Nordic mountain birch ecosystems (Wielgolaski, F., ed.), Man and Biosphere Series 27: 35–46
Väre, H., Ohtonen, R. and Mikkola, K. (1996): The effect and extent of heavy grazing by reindeer in oligotrophoc pine heaths in northeastern Fennoscandia. In: Relationships between spatial heterogeneity in vegetation communities and in the belowground eco-systems (Väre, H., ed.), Acta Ouluensis A 275: 1–14
Veblen, T. T., Ashton, D. H., Schlegel, F. M. and Veblen, A. T. (1977): Plant succession in a timberline depressed by volcanism in south-central Chile. J. Biogeogr. 4: 275–294

Veblen, T. T., Donosco, C., Schlegel, F. M. and Escobar, B. (1981): Forest dynamics in south central Chile. J. Biogeogr. 8: 211–247

Veijola, P. (1998): The northern timberlines and timberline forests in Fennoscandia. The Finnish For. Res. Inst., Res. Papers 672

Veit, H. (1993): Holocene solifluction in the Austrian and southern Tyrolean Alps: Dating and climatic implications. In: Solifluction and climatic variation in the Holocene (Frenzel, B., ed.), Paläoklimaforschung 11: 23–32

Velázquez, A., Toledo, V. M. and Luna, I. (2000): Mexican temperature vegetation. In: North American terrestrial vegetation (Barbour, M. G. and Billings, W. D., eds.): 573–592, Cambridge University Press, New York

Venn, K. (1970): Studies on particular dieback of terminal shoots of Pinus sylvestris L. Meddelser fra det Norske Sokogsforsöksvesen 27

Venn, K. (1979): Winter vigour in Picea abies (L.) Karst. VI. Frost injury to needles of Norway spruce seedlings exposed to experimental freezing during midwinter. Rep. Norw. For. Res. Inst. 35: 1–21

Venn, K. (1993): Red belts in boreal forests. In: Forest pathological research in northern forests with special reference to abiotic stress factors (Jalkanen, R., Aakto, T. and Lahti, M.-L., eds.). Extended SNS meeting in forest pathology in Lapland, Finland, 3–7 August, 1992, Finnish For. Res. Inst., Res. Papers 451: 50–54

Verweij, P. A. (1995): The páramo of Los Nevados National Park. In: Spatial and temporal modelling of vegetation patterns. Burning and grazing in the páramo of Los Nevados. In: Studies on Tropical Andean Ecosystems 3 (Van Der Hammen, T., Diaz, S. and Alvarez, V. J., eds.): 293–442, Vaduz

Viereck, L. A. (1973): Ecological effects of river flooding and forest fires on permafrost in taiga of Alaska. In: North American contribution, Permafrost International Conference (Yakutsk, USSR, 13–28 July 1973) (Nat. Acad. Sci., Washington, DC, ed.): 60–67

Viereck, L. A. (1979): Characteristics of tree-line communities in Alaska. Holarct. Ecol. 2: 228–238

Viereck, L. A. and Van Cleve, K. (1984): Some aspects of vegetation and temperature relationships in the Alaskan taiga. In: The potential effects of carbondioxide induced climatic changes in Alaska (McBeath, J. H., ed.), Univ. Alaska-Fairbanks Misc. Publ. 83 (1): 129–143

Villalba, R., Boninsegna, J. A., Veblen, T. T., Schmelter, A. and Rubulis, S. (1997): Recent trends in tree-ring records from high elevation sites in the Andes of northern Patagonia. In: Climatic change at high elevation sites (Diaz, H. F., Beniston, M. and Bradley, R. S., eds.): 193–222, Dordrecht

Virtanen, T., Neuvonen, S. and Nikula, A. (1998): Modelling topoclimatic patterns of egg mortality of Epirrita autumnata (Lepidoptera: Geometridae) with Geographical Information System: Predictions for current and warmer climatic scenarios. J. Appl. Ecol. 32: 1–12

Vincent, G. (1938): Topographie der Wälder der tschechoslowakischen Republik, I. Hohe Tatra. Recueil des Travaux des Intsituts pour la Recherche Agronomomique de la Républ. Tschécoslov. 113, Praha

Vogel, A. (1996): Beobachtungen zur Regeneration der Vegetation nach Ascheeruption am Hudson-Vulkan im südlichen Chile. In: Beiträge aus den Arbeitsgebieten am Institut für Landschaftsökologie (Holtmeier, F.-K., ed.), Arbeiten aus dem Institut für Landschaftsökologie der Westfälischen Wilhelms-Universität, Münster 1: 107–114

Vogel, A. (2000): Beobachtungen zur Regeneration der Vegetation nach Ascheeruptionen am. Hudson-Vulkan im südlichen Chile. Ergebnisse weltweiter ökologischer Forschung.

Beiträge des 1. Symposiums der A. W. Schimper-Stiftung von H. und E. Walter (Breckle, S.-W., Schweizer, B. and Arndt, U., eds.): 271–284
Von Wissmann, H. (1960): Stufen und Gürtel der Vegetation und des Klimas in Hochasien. Erdkunde 14 (4): 249–272
Von Wissmann, H. (1961): Stufen und Gürtel der Vegetation und des Klimas in Hochasien und seinen Randgebieten. Erdkunde 15: 19–41
Von Wissmann, H. (1972): Die Juniperus-Gebirgswälder in Arabien. Ihre Stellung zwischen dem borealen und tropisch-afrikanischen Florenreich. Erdwissenschaftliche Forschung 4: 157–176
Wace, N. M. (1961): The vegetation of Gough Island. Ecol. Monographs 31: 337–367
Wace, N. M. and Holdgate, M. W. (1958): The vegetation of Tristan da Cunha. J. Ecol. 46: 593–620
Wade, J. E. and Hewson, E. W. (1979): Trees as local climatic indicator. J. Appl. Met. 18 (9): 1182–1187
Wade, J. E. and Hewson, E. W. (1980): A guide to biological wind prospecting. Dept. Atmospheric Sciences, Oregon State University, Corvallis, OR
Walder, U. (1983): Ausaperung und Vegetationsverteilung im Dischmatal. Mitteilungen der Eidgenössische Anstalt für das forstliche Versuchswesen, Mitteilungen 59 (2): 80–212
Walker, D. (1968): A reconaissance of the non-arboreal vegetation of the Pindaunde catchment, Mount Wilhelm, New Guinea. J. Ecol. 56: 445–466
Walker, D. and Flenley, J. R. (1979): Late Quaternary vegetational history of the Enga District of Upland Papua New Guinea. Phil. Trans. Roy. Soc. B 286: 265–344
Walker, D. A., Billings, W. D. and De Molenar, J. G. (2001): Snow-vegetation interactions in tundra environments. In: Snow ecology (Jones, H. G., Pomeroy, J. W., Walker, D. A. and Hoham, R. W., eds.): 266–324, Cambridge University Press, Cambridge
Walker, L. R. (1999): Patterns and processes in primary succession. In: Ecosystems and disturbed ground (Walker, L. R., ed.): 16: 585–610, Elsevier, Amsterdam/Lausanne/New York/Oxford/Shannon/Singapore/Tokyo
Walsh, S. J., Malanson, G. P. and Butler, D. R. (1992): Alpine treeline in Glacier National Park, Montana. Geogr. Snapshots of North America. Commemorating the 24th Congress of the Internat. Geogr. Union and Assembly (Janelle, D., ed.): 167–171
Walsh, S. J., Butler, D. R., Allen, T. R. and Malanson, G. P. (1994): Influence of snow pattern and snow avalanches on the alpine treeline ecotone. J. Veg. Sci. 5: 657–672
Walter, H. (1964): Die Vegetation der Erde in öko-physiologischer Betrachtung. Bd. I, Die tropischen und subtropischen Zonen. Stuttgart
Walter, H. (1968): Die Vegetation der Erde in öko-physiologischer Betrachtung. Bd. II, Die gemäßigten und arktischen Zonen. Stuttgart
Walter, H. (1970): Vegetationszonen und Klima. Stuttgart
Walter, H. (1971): Ökologische Verhältnisse und Vegetationstypen in der intermontanen Region des westlichen Nordamerikas. Zoologisch-Botanische Verhandlungen: 111–123
Walter, H. (1973): Die Vegetation der Erde in öko-physiologischer Betrachtung. Bd I, Die tropischen und subtropischen Zonen. Stuttgart
Walter, H. and Breckle, S. (1984): Ökologie der Erde 2, Spezielle Ökologie der tropischen und subtropischen Zonen Stuttgart
Walter, H. and Breckle, S. (1986): Ökologie der Erde 3, Spezielle Ökologie der gemäßigten und arktischen Zonen Euro-Nordasiens, Zonobiom VI–IX. Stuttgart
Walter, H. and Breckle, S. (1991): Ökologie der Erde 4, Spezielle Ökologie der gemäßigten und arktischen Zonen außerhalb Euro-Nordasiens. Stuttgart

Walter, H. and Medina, E. (1969): Die Bodentemperaturen als ausschlaggebender Faktor für die Gliederung der subalpinen und alpinen Stufe in den Anden Venezuelas (Vorläufige Mitteilung). Berichte der Deutschen Botanischen Gesellschaft 82: 275–281
Walters, S. M. (1964): Betula. In: Flora Europaea 1 (Tutin, T. G. et al., eds): 57–58, Cambridge University Press, Cambrige
Wang, T., Zhang, Q.-B. and Ma. K. (2006): Treeline dynamics in relation to climatic variability in the central Tianshan Mountains, northwestern China. Glob. Ecol. Biogeogr. 15: 406–415
Wardle, J. A. (1970): Ecology of Nothofagus solandri. New Zeal. J. Bot. 8: 494–649
Wardle, P. (1962): Subalpine forest and scrub in the Tararua Range. Trans. Roy. Soc. New Zeal. Bot. 1 (6): 77–89
Wardle, P. (1963): Growth habits of New Zealand subalpine shrubs and trees. New Zeal. J. Bot. 1: 18–47
Wardle, P. (1964): Facets of the distribution of the forest vegetation in New Zealand. New Zeal. J. Bot. 2: 352–366
Wardle, P. (1965a): Significance of xeromorphic features in humid subalpine environments in New Zealand. New Zeal. J. Bot. 3: 342–343
Wardle, P. (1965b): A comparison on alpine timberlines in New Zealand and North America. New Zeal. J. Bot. 3 (2): 113–135
Wardle, P. (1968): Engelmann spruce (Pice engelmannii Engel.) at its upper limits on the Front Range, Colorado. Ecology 49: 483–495
Wardle, P. (1971): An explanation for alpine timberline. New Zeal. J. Bot. 9 (3): 371–402
Wardle, P. (1973): New Zealand timberlines. Arct. Alp. Res. 5 (3, Pt. 2): 127–135
Wardle, P. (1974): Alpine timberlines. In: Arctic and Alpine Environment (Ives, J. D. and Barry, R. G., eds.): 371–402, Metheuen, London
Wardle, P. (1977): Japanese timberlines and some geographic comparisons. Arct. Alp. Res. 9 (3): 249–258
Wardle, P. (1978): Ecological and geographical significance of some New Zealand growth forms. Erdwissenschaftliche Forschung 11: 531–536
Wardle, P. (1980): Tree species and growth forms at timberline in different parts of the world. In: Mountain environments and subalpine tree growth (Benecke, U. and Davis, M. R., eds.), New Zeal. For. Serv. Techn. Paper 70: 175–177
Wardle, P. (1981a): Winter desiccation of conifer needles simulated by artificial freezing. Arct. Alp. Res. 13: 419–423
Wardle, P. (1981b): Is the alpine timberline set by physiological tolerance, reproductive capacity, or biological interactions? Proc. Ecol. Soc. Aust. 11: 53–66
Wardle, P. (1984): The New Zealand beeches: Ecology, utilisation and management. New Zealand Forest Service, Wellington
Wardle, P. (1985a): New Zealand timberlines. 1. Growth and survival of native and introduced tree species in the Craigieburn Range, Canterbury. New Zeal. J. Bot. 23: 219–234
Wardle, P. (1985b): New Zealand timberlines. 2. A study of forest limits in the Crow Valley near Arthur's pass, Canterbury. New Zeal. J. Bot. 23: 235–261
Wardle, P. (1985c): New Zealand timberlines. 3. A synthesis. New Zeal. J. Bot. 23: 263–271
Wardle, P. (1986): Alpine vegetation of New Zealand: A review. In: Flora and fauna of Australasia, ages and origins (Barlow, B., ed.): 45–61, CSIRO Australia, Leiden
Wardle, P. (1991): Vegetation of New Zealand. Cambridge University Press, Cambridge
Wardle, P. (1993): Causes of alpine timberline: A review of the hypotheses. In: Forest development in cold climates, Proc. NATO advanced workshop, 1991, June 18–23, Langarvatn, Iceland (Alden, J. N., Odum, S. and Mastrantonio J. L., eds.), New York
Wardle, P. (1998): Comparison of alpine timberlines in New Zealand and the Southern Andes. Royal Soc. New Zeal. Misc. Publ. 48: 69–90

Wardle, P. (2007): New Zealand forest to alpine transitions in global context. Arct. Antarct. Alp. Res. 40 (1): 240–249
Wardle, P. and Coleman, M. C. (1992): Evidence for rising upper limits of four native New Zealand forest trees. New Zeal. J. Bot. 30: 303–314
Wargo, P. M. and Shaw, C. G. (1985): Armillaria root rot: The puzzle is being solved. Plant Dis. 69: 826–822
Watson, A. (1991): Increase of people on Cairn Gorm plateau following easier access. Scott. Geogr. Mag. 107: 99–105
Watt, A. S. and Jones, E. W. (1948): The ecology of the Cairngorms. Part 1. The environment and the altitudinal zonation. J. Ecol. 36 (2): 283–304
Wegener, A. (1923): Das Wesen der Baumgrenze. Meteorologische Zeitschrift: 371–372
Weih, M. (1998): The nitrogen economy of Mountain birch. Comprehensive Summaries of Uppsala Dissertations from the Faculty of Science and Technology 351, Uppsala
Weih, M. and Karlsson, St. (1999): The nitrogen economy of mountain birch seedlings: Implications for winter survival. J. Ecol. 87: 211–219
Weinert, E. (1985): Die aktuelle Gebirgsvegetation am Horn von Afrika. Flora 176: 231–243
Weis, G. B., Spatz, G. and Dunz, K. (1982): Zur Wiederbewaldung aufgelassener Almen. Natur und Landschaft 57 (7/8): 256–260
Weisberg, P. J. and Baker, W. L. (1995): Spatial variation in tree seedling and krummholz growth in the forest-tundra ecotone of Rocky Mountain National Park, Colorado, U.S.A. Arct. Alp. Res. 27 (2): 116–129
Weischet, W. (1955): Geländeklimate der Niederrheinischen Bucht und ihrer Rahmenlandschaften. Eine geographische Analyse subregionaler Klimadifferenzierung. Münchener Geographische Hefte 8
Weiser, C. J. (1970): Cold resistance and injury in woody plants. Science 169: 1269–1278
Weiwen, H. (1994): The prehistoric human occupation of the Qinghai-Xizang Plateau. Göttinger Geographische Abhandlungen 95: 201–219
Wesche, K. (2002): Structure and dynamics of Erica forest at tropical-African treelines. Berichte der Reinhold-Tüxen-Gesellschaft 14: 145–160
Wesche, K., Miehe, G. and Kaeppeli, M. (2000): The significance of fire for Afroalpine ericaceous vegetation. Mt. Res. Develop. 20 (4): 340–347
Wesche, K., Cierjacks, A., Assefa, Y., Wagner, S., Fetene, M. and Hensen, I. (2008): Recruitment of trees at tropical alpine treelines: Erica in Africa versus Polylepis in South America. Plant Ecol. Divers. 1 (1): 35–46
Wettstein, W. von (1946): Alpenlärchenrassen. Der Züchter 17: 40–44
Whitmore, T. C. (1975): Tropical rain forest of the Far East. Clarendon, Oxford
Widmann, P. (1991): Zur Synökologie von Wild und Waldvegetation. Allgemeine Forstzeitung 8: 382–384
Wielgolaski, F. E. (ed.) (2001): Nordic mountain birch ecosystems. Man and Biosphere Series 27: 241–249
Wielgolaski, F. E., Thannheiser, D., Neuvonen, S. and Karlsson, P. S. (2003): Sustainable use of mountain birch forests. Scenarios and recommendations. In: Human interactions with the mountain birch ecosystems: Implications for sustainable development (HIBECO) (Laine, K., Skre, O. and Wielgolaski, F. E., eds.). Final Report for the project EU FP5, QLK5-CT-11999-01515: 118–121
Wielgolaski, F. E., Larlsson, P. S., Neuvonen, S. and Thannheiser, D. (eds.) (2005): Plant ecology, herbivory and human impact in Nordic mountain birch forests. Ecol. Stud. 180: 343–356
Wieser, G. (1997): Carbon dioxide gas exchange of cembran pine (Pinus cembra) at the alpine timberline during winter. Tree Physiol. 17: 473–477

Wieser, G. (2007): Limitation by an insufficient carbon assimilation and allocation. In: Trees at their upper limit. Treelife limitation at the alpine timberline (Wieser, G. and Tausz, M., eds.), Plant Ecophysiology 5, Springer, Dordrecht

Wieser, G. and Tausz, M. (2007): Trees at their upper limit. Treelife limitation at the alpine timberline. Plant Ecophysiology 5, Springer, Dordrecht

Wieser, G., Gigele, T. and Pausch, H. (2005): The carbon budget of an adult Pinus cembra tree at the alpine timberline in the Central Austrian Alps. Eur. J. For. Res. 124: 1–8

Williams, N. W., Losleben, M., Caine, N. and Greenland, D. (1996): Change in climate and hydrochemical responses in a high-elevation catchment in the Rocky Mountains, USA. Limnol. Oceanogr. 41 (5): 939–946

Wilmanns, O. (1973): Ökologische Pflanzengeographie (4th edition). Quelle & Meyer, Heidelberg

Wilmanns, O. (1998): Ökologische Pflanzensoziologie (6th edition). Quelle & Meyer, Heidelberg/Wiesbaden

Wilmanns, O. and Ebert, J. (1974): Aktuelle und potentielle Grenzen des Latschengürtels im Quellgebiet des Lech (Vorarlberg). Bericht über das Internationale Symposium der Internationalen Vereinigung für Vegetationskunde, 8–11. April 1968: 207–218

Wilmanns, O. and Müller, K. (1977): Zum Einfluß der Schaf- und Ziegenbeweidung auf die Vegetation im Schwarzwald. Bericht über das Internationale Symposium der Internationalen Vereinigung für Vegetationskunde, Vegetation und Fauna (Rinteln 12. 15. 4. 1976): 465–479, Vaduz

Wilmanns, O., Bogenrieder, A. and Nakamura, Y. (1985): Vergleichende Studien des Pinus-Krummholzes in den japanischen und europäischen Alpen. Tuexenia, Neue Serie 5: 335–358

Wilson, C., Grace, J., Allen, S. and Slack, F. (1987): Temperature and stature: A study of temperatures in montane vegetation. Funct. Ecol. 1: 405–413

Wilson, J. W. (1958): Dirt on snow patches. - J. Ecol. 46: 191–198

Wilson, J. W. (1959): Notes on the wind and its effects in arctic-alpine vegetation. J. Ecol. 47: 415–427

Winiger, M. (1979): Bodentemperaturen und Niederschlag als Indikatoren einer klimatisch-ökologischen Gliederung tropischer Gebirgsräume. Methodische Aspekte und Anwendbarkeit diskutiert am Beispiel des Mt. Kenya (Ostafrika). Geomethodica 4: 121–150

Winiger, M. (1980): Wärmemangel- und Trockengrenzen der Vegetation. 42. Deutscher Geographentag 1979, Göttingen: 301–303

Winiger, M. (1981): Zur thermisch-hygrischen Gliederung des Mount Kenya. Erdkunde 35: 248–263

Winiger, M. and Menz, G. (1993): Klima und Vegetation in Kenya. Erfassung von Ressourcen mit Hilfe von digitaler Bildverarbeitung und geographischen Informationssystemen. Trierer Geographische Studien 9: 333–350

Winkler, D. (1994): Die Waldvegetation in der Ostabdachung des Tibetanischen Hochlandes. Dreidimensionale Vegetationszonierung mit Bodenuntersuchungen am Beispiel des Jiuzhai-Tales in NNW-Sichuan (NO-Kham). Berliner Geographische Abhandlungen, Beiheft 2

Winkler, D. (1997): Waldvegetation in der Ostabdachung des Tibetanischen Hochlandes und die historische und gegenwärtige Entwaldung. Erdkunde 51: 143–163

Withington, C. L. and Sanford, R. L. (2007): Decomposition rates of buried substrates increase with altitude in the forest-alpine tundra ecotone. Soil and Biochemistry 39: 68–75

Witmer, U., Filliger, P., Kunz, S. and Küng, P. (1986): Erfassung, Bearbeitung und Kartieren von Schneedaten in der Schweiz. Geographica Bernensia G25
Wodzicki, K. (1963): Introduced mammals in New Zealand forests. Acta Biologica Cracoviensia Ser. Zool. 6: 11–134
Wohlgemut, T., Bugmann, H., Lischke, H. und Tinner, W. (2006): Wie rasch ändert sich die Waldvegetation als Folge von raschen Klimaveränderungen? Forum für Wissen 2006: 716
Wolf, A., Callaghan, T. V. and Larson, K. (2008): Future changes in vegetation and ecosystem function of the Barents region. Climate Change 87 (1-2): 51–73
Wolfsegger, M. and Posch, A. (1980): Der Wasserhaushalt von Böden am Südhang des Hochtores (Hohe Tauern). In: Untersuchungen an alpinen Böden in den Hohen Tauern 1974–1978, Stoffdynamik und Wasserhaushalt (Franz, H., managing ed.), Veröffentlichungen des Österreichischen MaB-Hochgebirgsprogramms Hohe Tauern 3: 223–242, Innsbruck
Woodward, A. (1998): Relationship among environmental variables and distribution of the species at high elevation in the Olympic Mountains. Northwest Sci. 72 (1): 10–22
Woodward, A., Schreiner, E. G. and Silsbee, D. G. (1995): Climate, geography, and tree establishment in subalpine meadows of the Olympic Mountains, Washington, U.S.A. Arct. Alp. Res. 27 (3): 217–225
Wooldridge, G. L. (1989): Wind patterns and snow depths in high altitude complex terrain as determined from coniferous tree deformation. USDA For. Serv., Rocky Mountain Experiment Station (Fort Collins), Final Report Order 43-82FT-8-159
Wooldridge, G. L., Musselman, R. C., Sommerfeld, R. A., Fox, D. G. and Connel, B. H. (1996): Mean wind patterns and snow depths in an alpine-subalpine ecosystem as measured by damage to conifers. J. Appl. Ecol. 33: 100–108
Wraber, M. (1970): Die obere Wald- und Baumgrenze in den slowenischen Hochgebirgen in ökologischer Betrachtung. Mitteilungen der ostalpin-dinarischen Gesellschaft für Vegetationskunde 11: 135–248
Wright, R. D. and Mooney, H. A. (1965): Substrate-oriented distribution of bristlecone pine in the White Mountains of California. Am. Midl. Nat. 73 (2): 257–284
Yanagimachi, O. and Ohmori, H. (1991): Ecological status of Pinus pumila scrub and the lower boundary of the Japanese alpine zone. Arct. Alp. Res. 23 (4): 424–435
Yoshimura, M. (1971): Die Windverbreitung im Gebiet des Mt. Fuji. Erdkunde XXV (3): 195–198
Yoshino, M. (1966): Wind-shaped trees as indicators of micro- and local climate wind situation. Biometeorol. 2: 1007–1005
Yoshino, M. (1973): Wind-shaped trees in the subalpine zone in Japan. Arct. Alp. Res. 5 (3, Pt. 2): 115–126
Young, K. R. (1993): Tropical timberlines: Changes in forest structure and regeneration between the Peruvian timberline margins. Arct. Alp. Res. 25 (3): 167–174
Zeuner, F. E. (1949): Frost soils on Mt. Kenya and the relation of frost soils to aeolian deposits. J. Soil Sci. 1: 20–32
Zimina, R. P. (1973): Upper forest boundary and the subalpine belt in the mountains of the southern USSR and adjacent countries. Arct. Alp. Res. 5 (3, Pt. 2): 29–32
Zimmermann, M. H. and Brown, C. L. (1971): Trees. Structure and function. Berlin/Heidelberg/New York
Zimmermann, N. E., Bolliger, J., Gehrig-Fasel, J., Guisan, A., Kienast, F., Lischke, H., Rickebusch, S. and Wohlgemut, T. (2006): Wo wachsen die Bäume in 100 Jahren? Forum für Wissen 2006: 63–71
Zinserling, G. D. (1934): Geografia rastitelnogo pokrova Severospada jevropeiskoi casti SSSR (Ref.: Geographie der Vegetationsdecke des Nordwestens des europäischen Teils der U.d.S.S.R.). Akad. Nauk SSSR., Geomorf. Inst., Ser. Fys.-Geogr. 4

Zoller, H. (1967a): Postglaziale Klimaschwankungen und ihr Einfluß auf die Waldentwicklung Mitteleuropas einschließlich der Alpen. Berichte der Deutschen Botanischen Gesellschaft 80 (10): 690–696

Zoller, H. (1967b): Holocene fluctuations of cold climate in the Swiss Alps. Rev. Palaeobot, Palynol. 2: 267–269

Zoltai, S. C. (1975): Structure of subarctic forests on hummocky permafrost terrain in northwestern Canada. Can. J. For. Res. 5 (1): 1–9

Zotov, V. D. (1938): Some correlations between vegetation and climate in New Zealand. New Zeal. J. Sci. Technol. 19: 474–487

Zschokke, H. (1804): Die Alpenwälder. Gotta-sche Buchhandlung, Tübingen

Zschokke, H. (1805): Beobachtungen im Hochgebirge auf einer Alpenreise im Sommer 1803. Isis, Wochenschrift von deutschen und schweizerischen Gelehrten Bd. I/II, Zürich

Zschokke, H. (1806): Der Gebürgsförster. Bd. I

Zuber, R. (1995): Von urwüchsigen Gestalten und Überlebensstrategien. Bündnerwald 6: 35–40

INDEX

GPSR Compliance

The European Union's (EU) General Product Safety Regulation (GPSR) is a set of rules that requires consumer products to be safe and our obligations to ensure this.

If you have any concerns about our products, you can contact us on ProductSafety@springernature.com

In case Publisher is established outside the EU, the EU authorized representative is:

Springer Nature Customer Service Center GmbH
Europaplatz 3
69115 Heidelberg, Germany

Batch number: 10607204

Printed by Printforce, the Netherlands